Molecular Reaction Dynamics

Molecular Reaction Dynamics is a brand new version of the text by Levine and Bernstein. The book delivers an updated treatment of this fundamental topic. An appreciation of how chemical reactions occur and their control is essential to chemists and to those in interdisciplinary fields such as materials and nanoscience, drug design, and astrochemistry. The first half of the book describes experimental techniques for initiating and probing reaction dynamics and the essential insights gained. The second part explores key areas including photoselective chemistry, stereochemistry, chemical reactions in real time, and chemical reaction dynamics in solutions and interfaces. Typical of the new challenges are molecular machines, enzyme action, and molecular control. With problem sets included, this book is aimed at advanced undergraduate and graduate students studying chemical reaction dynamics, as well as physical chemistry, biophysics, and materials science.

RAPHAEL LEVINE is Max Born Professor of Natural Philosophy at the Hebrew University of Jerusalem and Distinguished Professor of Chemistry at the University of California, Los Angeles. He is active in the area of chemical reaction dynamics and his published scientific work has earnt the recognition of the Israel Prize and the Wolf Prize. He is a member of the Israel National Academy of Sciences and a foreign member of the National Academy of Sciences of the United States and of Academiae Europaeae.

Molecular Reaction Dynamics

Raphael D. Levine

CAMBRIDGE
UNIVERSITY PRESS

CAMBRIDGE
UNIVERSITY PRESS

University Printing House, Cambridge CB2 8BS, United Kingdom

One Liberty Plaza, 20th Floor, New York, NY 10006, USA

477 Williamstown Road, Port Melbourne, VIC 3207, Australia

4843/24, 2nd Floor, Ansari Road, Daryaganj, Delhi - 110002, India

79 Anson Road, #06-04/06, Singapore 079906

Cambridge University Press is part of the University of Cambridge.

It furthers the University's mission by disseminating knowledge in the pursuit of
education, learning and research at the highest international levels of excellence.

www.cambridge.org
Information on this title: www.cambridge.org/9780521140713

© R. D. Levine 2005

First published 2005
Paperback edition 2009

A catalogue record for this publication is available from the British Library

Library of Congress Cataloging in Publication data
Levine, Raphael D.
Molecular reaction dynamics / Raphael D. Levine.
    p.   cm.
Includes bibliographical references and index.
ISBN 0 521 84276 X
1. Molecular dynamics.   2. Chemical reaction, Conditions and laws of.   I. Title.
QD461.L66   2004
541′.394 – dc22   2004045662

ISBN 978-0-521-84276-1 Hardback
ISBN 978-0-521-14071-3 Paperback

Additional resources for this publication at www.cambridge.org/Levine

I am grateful to the many people who, over the years, joined me
to watch the molecules dance to the tune of time.
This book is dedicated to Mira who is able to make us join the dance.

The cover illustration is by the late Israeli physical chemist and artist Jacob Wilf. Jacob was my friend and we had many scientific discussions. One result is that he has drawn several paintings depicting themes from Molecular Reaction Dynamics such as harpoon reactions, stereodynamics and cluster impact. The painting shown on the cover is titled 'surprisal analysis' and was dedicated to me by the artist. The topic of surprisal analysis is discussed in Section 6.4.2. The works of Wilf on a variety of scientific themes were exhibited at the Israel National Academy of Sciences and other Institutions. Many paintings by the late Jacob Wilf can be seen at http://jacob.wilf.org/

Contents

Preface

Molecular reaction dynamics unfolds the history of change on the molecular level. It asks what happens on the atomic length and time scales as the chemical change occurs. This book is an introduction to the field.

Molecular reaction dynamics has become an integral part of modern chemistry and is set to become a cornerstone for much of the natural sciences. This is because we need a common meeting ground extending from nanoscale solid state devices through material and interface chemistry and energy sciences to astrochemistry, drug design, and protein mechanics. For some time now the quantitative understanding on the molecular level has provided this common ground. At first, the scaffolding was the concept of the molecular structure. Once we understood the spatial organization we felt that we had an entry to real understanding. The required input was provided by the different experimental methods for structure determination and, from the theory side, by quantum chemistry and by equilibrium statistical mechanics. But now we want more: not just the static structure, we also ask how this structure can evolve in time and what we can do to control this evolution. We want to write the history of the change or, better yet, to be a conductor and orchestrate the motion. This is what this book is about.

In going from statics to dynamics we need new experimental tools and also theoretical machinery that allows for the dependence on time. This means that the stationary states that are usually the subject of an introductory quantum mechanics course have to be extended to non-stationary ones. Fairly often, classical dynamics is sufficient to describe the time evolution but there are a number of interesting exceptions. Non-equilibrium statistical mechanics is necessary to describe systems with many degrees of freedom and their far-from-equilibrium pattern formation.

Molecular reaction dynamics is not yet able to do all that has to be done. There are places where we lack understanding of the principle and not only of the details of a particular family of processes. Indeed, as we move into more complex systems the gaps in our understanding are wider than the passes. As just two examples, we do not have a complete understanding of the atmospheric chemistry of the outer planets nor can we describe how an enzyme mobilizes chemical energy to its active site. But we do have enough of the basics in place that it is a good time to stop and survey where we are, where we need more work on the foundations, and where there are whole areas that call for applications, where different subjects need to be better connected, and what new families of processes are there to be deciphered. This book is a primer for what we already know.

As was the original (1974) intention, this book seeks to describe why a particular experiment was carried out, what we have learned, what concepts are necessary to describe and understand the experiment, and how we move forward. The problems that follow each chapter provide additional applications and illustrations. A concept that is much more prominent in the present version of the book is *coherence*, and we bring it in as soon as possible. Much recent progress has come through the outstanding development of computational means. These include not only the ability to compute the forces between atoms at realistic accuracy, but also the computation of the (classical or quantal) motion subject to these forces and the ability to visualize the resulting dynamics. Our debt to these developments will be clear throughout the book, but what we will be concerned with is what we have learned rather than how to implement a computation. The need for visualization arose not only because of the increasing concern with more complex systems, but also because of the technological ability to achieve a time resolution sufficient to probe intramolecular motions. Instead of just imagining how the reaction unfolds in time, we can directly image the transformation experimentally. In a different dimension, the experimental ability to image the distribution of the products of the reaction in space has a major impact. We are almost ready to be able to image in both space and time. Another key initiative is the bold forays into dynamics in the condensed phase and interfaces. The integration of our understanding of gas phase, isolated collision dynamics and of dynamical proceedings dressed by their environment is already making promising progress. Because the chemical change is localized in space and time we can often think of a change in a complex system as a reaction center "solvated" by the rest of the system. Therefore, issues similar to dynamics in the condensed phase arise in the need for rational drug design or the understanding and development of molecular machines and other applications where the molecules are large.

This book is based on my class notes at the Hebrew University of Jerusalem and at the University of California, Los Angeles. The level is that of senior undergraduate or graduate students. The prerequisite is a class in chemical kinetics. Some familiarity with spectroscopy and with statistical mechanics is beneficial but not essential, and introductory material is provided where necessary. The scope of the book is more than can be covered in a lecture course of one semester. The first six chapters develop the tools and illustrate their applications. The examples are usually simple ones that can be used to make the point. The development in these chapters is linear, there are sections that can be skipped, but the order of topics is sequential. There are people who will want to get as quickly as possible to Chapter 5. This is understandable, but I recommend first to go at least through Sections 2.1, 2.2, and 2.3. In the following six chapters the text is arranged around applications where each chapter has a common theme. This part of the book offers a choice of material because the different chapters are almost, but not quite, independent of one another. Starred sections take you away from the main line of development.* There are endnotes that provide more details and also cite original sources for the results quoted. References to review-type articles are provided to enable further reading. (A complete bibliography, with titles, is at the very end of the book.) Revision problems with hints follow each chapter. Some of these problems are easy but others are not.

---

* Both in class and in writing I use too many footnotes. I hope that it does not distract you too much.

Acknowledgments

The text is a completely rewritten version of Levine and Bernstein, *Molecular Reaction Dynamics* (1974). In this task I have received indispensable advice and encouragement from R. N. Zare (Stanford) and J. L. Kinsey (Rice). I am very grateful to them and at the same time I wish to clearly state that all remaining shortcomings in presentation and coverage are entirely my responsibility. Chapter 10, on stereodynamics, plainly shows my indebtedness to Richard Zare. As I was writing, Tamar Raz was preparing an abbreviated version of the text, in Hebrew, for distance learning by senior undergraduates of the Open University of Israel. The feedback from Tamar has been essential. I also acknowledge the critical help of Micha Asscher, Michal Ben-Nun, Richard Bersohn, Eleanor Campbell, Mervin Hanson, Robert Gordon, Mark Marshall, Izhack Oref, Eliyahu Pollak, Françoise Remacle, Sanford Ruhman, Benjamin Schwartz, Tamar Seideman, and Yehuda Zeiri. They have read and commented on one or more chapters and did their very best to help me make the text clearer and more accurate. Here, too, I bear complete responsibility for the final version. I have profited from detailed information communicated to me by Stephen Bradforth, David Chandler, Wilson Ho, Kendal Houk, Todd Martinez, Gilbert Nathanson, Gabor Somorjai, and Steven Stolte. Graduate students Hadas Amiezer, Ayelet Gross, and Dan Steinitz have each helped in essential ways. Many colleagues and students have provided additional insights and advice. Mrs E. Guez has followed the evolution of this book from its addressograph plates typed in 1972.

I have tried to make sure that proper reference is provided for specific results cited in the text. I recognize that I have probably failed fully to do so, and I apologize beforehand. I would be grateful to anyone putting me right on this, as well as on any other aspect. I welcome comments (rafi@fh.huji.ac.il).

This book would not have been possible without the many years of fruitful collaboration with the late Richard Bernstein.

I thank the US Air Force Office of Scientific Research, the Volkswagen Foundation, the United States–Israel Binational Science Foundation, the German–Israeli Binational Science Foundation, the James Franck Program, and the Humboldt Foundation for the support of my work on molecular reaction dynamics. Directly and indirectly this support was critical to my being able to write this book.

I am most grateful to Toshi Suzuki (鈴木 俊法), Professor of chemistry at the Department of Chemistry of Kyoto University and Kiyohiko Someda (染田 清彦), Professor at the Graduate School of Arts and Sciences of the University of Tokyo, the translators of the book to Japanese, for kindly providing a detailed set of typos and of suggestions for improved presentations.

Chapter 1
Understanding chemical reactions at the molecular level

"**chem·i·stry** (kem′ i strë), *n., pl.* **-tries**. The science that deals with or investigates the composition, properties, and transformations of substances and various elementary forms of matter." The dictionary definition emphasizes chemical transformation as a central theme of chemistry.

By the end of the nineteenth century, the young science of physical chemistry had characterized the dependence of the rate of the chemical transformation on the concentrations of the reactants. This provided the concept of a chemical reaction rate constant k and by 1889 Arrhenius showed[1] that the temperature dependence of the rate constant often took on the simple form $k = A \exp(-E_a/RT)$, where A is referred to as the pre-exponential factor and E_a as the activation energy. Arrhenius introduced the interpretation of E_a as the energetic barrier to the chemical rearrangement. Only later did we understand that reactions also have steric requirements and that the Arrhenius A factor is the carrier of this information.

It was next realized that the net transformation often proceeds by a series of elementary steps. A key progress was the identification of the reaction mechanism, which is a collection of elementary processes (also called elementary steps or elementary reactions) that leads to the observed stoichiometry and explains how the overall reaction proceeds. A mechanism is a proposal from which you can work out a rate law that agrees with how the observed rate of the reaction depends on the concentrations. The fact that a mechanism "explains" the experimental results, however, is not a proof that the mechanism is correct. Bulk kinetic studies are carried out at a controlled temperature, that is, under conditions of thermal equilibrium. The measured thermal rate constants refer to an average over all accessible reactant states weighted by the populations of those states at that temperature. As might be expected, such averages hide detailed information about what factors really cause the reaction to proceed. What we need is the ability to examine the individual processes and preferably to do so with selection of the energetic (and orientation) states of the reactants.

One of the greatest challenges in chemistry is to devise experiments that can reveal how chemical transformations occur that are otherwise hidden behind thermal averages and multi-step mechanisms **and** to develop the theoretical framework for describing and understanding these chemical changes. With this book you are invited to a dance of molecules. With an appreciation of the dance steps comes the power to understand and predict chemical behavior, if not become a molecular choreographer.

1.1   What is molecular reaction dynamics?

Reaction dynamics is the study of the molecular level mechanism of elementary chemical and physical processes. It seeks to understand what actually takes place at that level when a change, chemical or physical, occurs. As an example, when molecular chlorine gas is introduced into a vessel containing bromine vapor, a chemical reaction does take place, and it can be monitored in time by a change in the color. The net chemical change in the vessel is $Cl_2 + Br_2 \rightarrow 2BrCl$. The reaction rate is observed to be of second order, that is, the rate of disappearance of Cl_2 or Br_2 is first order in the concentration of each reactant. Yet on the molecular level the elementary reaction

$$Cl_2 + Br_2 \rightarrow 2BrCl$$

does not take place. In other words, when a single chlorine molecule strikes a single bromine molecule, the two molecules bounce off each other without exchanging atomic partners, and this fact has been demonstrated experimentally.[2]

Molecular reaction dynamics is the study of elementary processes and the means of probing them, understanding them, and controlling them. We will also apply molecular reaction dynamics to reactions in solution and to reactions on surfaces, exploring the elementary steps in catalysis. As a bridge between the gas and condensed phase we discuss clusters of molecules. Molecular reaction dynamics is not limited to neutral reagents and products but also includes positively and negatively charged species (cations and anions), either in their bare state or solvated in solution. Biochemical reactions provide important examples of processes where electrostatic effects are central. Current rational drug design includes the consideration of the approach of the intended drug to its receptor and how both are modified as a result of their interaction. Molecular reaction dynamics has applications in all branches of chemistry because chemists are not content just to prepare a desired product. Nor is it sufficient to optimize conditions such as temperature or solvent or catalyst so as to get high reaction rate and purity. Chemists nowadays require a molecular-level **understanding** of reactivity.

Molecular reaction dynamics is becoming relevant well outside the traditional boundaries of chemistry and increasingly addresses technological issues. The reason is that from modern genetics to size-reducing nanoscience the molecular point of view provides a unified framework. First we needed to understand the structural aspects. But these are now well in hand and we are increasingly trying to unravel the time-history of the event. The need to understand change on the molecular scale is now common throughout the natural sciences.

The study of dynamics allows us to raise additional questions, questions that do not quite make sense when we study the net change, at or near equilibrium, in the bulk. For example, we can wonder if exciting vibrational motion in either or both reactant diatomic molecules will make the $Cl_2 + Br_2$ four-center reaction proceed more rapidly. In bulk chemical kinetics, when the reactants are intentionally arranged to be in thermal equilibrium, a particular mode cannot be energized preferentially. To learn about a selective role of internal energy in promoting a reaction it is necessary to work under non-equilibrium conditions.[3]

Section 1.2 provides a case study of the kind of new questions raised by dynamics. In doing so, it also points to where we are going to go in the following chapters. We will, for example, discuss how lasers can act so as to prepare the reactants or, better yet, to access directly the transition region from reactants to products.[4] When the laser is intense enough it can even alter completely the dynamical course. As an example, an intense laser field can be tailored to alter the ratio of products in two dissociation pathways of acetophenone:[5]

Different options for control using lasers feature throughout our road ahead.

Spectroscopy provides essential information about the structure of molecules through radiation–matter interactions. The application of spectroscopy, technically made possible using lasers, to molecular dynamics has allowed us to extend the asking of structural questions into the time domain, even for times comparable to or shorter than the periods of molecular vibrations. We will seek to understand how the reactants evolve over time into products. In so doing we must recognize that during a chemical transformation molecules must become less rigid and more floppy. It is the electrons, which are fast moving compared to the slower nuclei, that set up the energy landscape for the motion of the nuclei. Sometimes the electrons may not move quite as fast as we assume.

The technical details of the experimental and theoretical methods of molecular dynamics can be intricate but the concepts are simple. An understanding of these concepts – the ability to read the language – is all that is necessary to be able to view the very process of chemical change. This book is a primer of the language for expressing chemical transformations as dynamical events, proceeding in space and time.

1.1.1 Much of chemistry is local: from the elementary act to complex systems

We begin our journey with elementary events. Our first example will be a simple chemical transformation, a hydrogen atom transfer between two atoms, in the gas phase. There are numerous systems that are chemically more exciting; say, the mechanism of C—H bond activation[6] by metal complexes, in solution, or how does an enzyme transfer chemical energy liberated at a localized site to a functionally relevant receptor site?[7] We adopt a bottom-up approach that almost all of chemistry is local in character; even a complex process is a sequence of elementary steps, each involving only a few atoms. Just as organic chemists break a complete synthesis into its essential steps (Corey, 1991), so we want to chart what are the possibly few atom events that, played in rapid succession, make up a complex reaction.

Chemistry is local because chemical forces are short range. An atom sees only its immediate surroundings. It is therefore possible to break the evolution from reactants to products into simpler steps. Our first task is to examine and understand the elementary dynamical events; only then can we build up to more complex processes.

A key factor in our ability to understand complex systems is the coming of age of modern computational chemistry.[8] It is the fast motion of the electrons that determines the forces that act on the nuclei. Quantum chemistry provides the methods for analyzing electronic structure and thereby allows the determination of the equilibrium configuration for the nuclei and the energy of the electrons at that point.[*] In the same computation we can also determine the forces at that point and not only the potential. This allows the computation of the frequencies for the vibrations about the stable equilibrium. Next, methods have been introduced that enable us to follow the line of steepest descent from reactants to products and, in particular, to determine the stationary points along that route, and the forces at those points.[9] Our ability to do so provides us with the means for quantitative understanding of the dynamics.

1.2 An example: energy disposal in an exoergic chemical reaction

Typical of the kind of information that is available from the experimental techniques of molecular dynamics is the determination of the *energy disposal* in exoergic atom–molecule exchange reactions. One example of such a system is the H-atom transfer reaction

$$Cl + HI \rightarrow ClH + I$$

In the course of this reaction the relatively weak HI bond is broken and replaced by the stronger HCl bond. The reaction liberates chemical energy, as shown

---

[*] The electronic energy is the potential energy for the nuclei. The change in the electronic energy when the nuclei are displaced is the force. See also Sections 5.0.1 and 7.0.2.

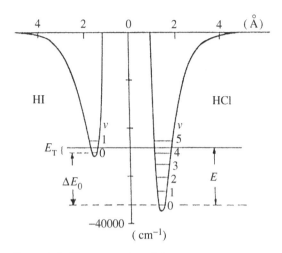

Figure 1.1 The energetics of the reaction $Cl + HI \rightarrow ClH + I$. The plot is drawn so as to have a zero of energy common to the reactants and products. This is achieved by taking the zero of energy when all three atoms are at rest and far apart from each other. The reactants or products, where two atoms are bound, are then below the zero. The *exoergicity*, ΔE_0, of the reaction is the difference in the bond dissociation energies. Here ΔE_0 is negative because the new bond is stronger. The figure further shows the vibrational levels for the old, HI, or the new, HCl, bonds. For this purpose the potential energies of the HI and HCl bonds are plotted as a function of the bond stretch coordinate, $1 \text{Å} = 10^{-10} \text{ m} = 0.1 \text{ nm}$. If the molecular reactant is cold, i.e., an HI molecule in its ground state, then the energy of the reactants is just the translational energy E_T of the relative motion of Cl and HI. As shown in the figure, this energy is sufficient to form the products up to and including the fourth vibrational state of HCl.

graphically in Figure 1.1. The *exoergicity* is about 134 kJ mol^{-1} or, equivalently, see Appendix, 32 kcal mol^{-1}, or about 1.39 eV. This amount of energy is large by chemical standards when we recall that the H—H bond energy is about 435 kJ mol^{-1}. Do not confuse the exoergicity of the reaction with the *exothermicity* (or heat) of a **bulk** chemical process. In the bulk there are subsequent collisions where the nascent products collide with other molecules. We focus attention on the elementary chemical event. We are dealing with the nascent products of an **isolated** triatomic system, ClHI. We ask: when this system of three atoms evolves into the products HCl + I, where is the liberated energy to be found?

In a short while we discuss how to obtain an experimental answer to our question. What is important is the idea that we center attention on the isolated system and ask to probe the products prior to their engaging in any further action. The question is: immediately after the reactive collision of Cl + HI is over, how is the energy distributed among the reaction products? Even if both products are formed in their electronic ground states, we need to determine the partitioning of the excess energy of the chemical reaction into the three remaining modes of *energy disposal*. They are: vibration of HCl; rotation of HCl; and relative

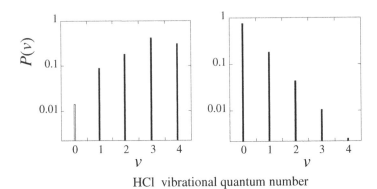

HCl vibrational quantum number

Figure 1.2 Two distributions, $P(v)$, of vibrational states of HCl, drawn on a log scale vs. the vibrational quantum number v. Left: the distribution measured for the nascent HCl product of the Cl + HI reaction (adapted from the observed chemiluminescence (in the infrared) of the vibrationally excited HCl(v) by D. H. Maylotte, J. C. Polanyi, and K. B. Woodall, *J. Chem. Phys.* **57**, 1547 (1972). See also Polanyi (1987)). The observed distribution of HCl(v) immediately after the reaction is **qualitatively** different from the distribution at thermal equilibrium, which is shown on the right. The distribution at thermal equilibrium is exponentially decreasing with the excess vibrational energy of HCl. Not so for the distribution on the left, which is loosely described as showing a *population inversion*.

translation of I and HCl recoiling from one another.[10] But what is the distribution of energy among these three modes?

When a reaction is studied in the "bulk" gas phase, the nascent products soon collide with other molecules, energy is transferred upon collision (thus becoming effectively partitioned among all molecules), and the overall reaction exoergicity is finally liberated in its most degraded form, i.e., heat. In macroscopic terms, the reaction is exothermic, i.e., $\Delta H^0 < 0$. The microscopic approach of molecular dynamics, however, is concerned with the outcome of the **individual** reactive collisions. The experimental challenge, as discussed in Section 1.2.5, is to arrest the *collisional relaxation* of the nascent reaction products and to probe them as they exit from the reactive collision. In this sense, it is customary to speak about the nascent or newborn reaction products.

1.2.1   Distribution of products' energy states

Figure 1.2 shows on the left a typical experimental result, illustrating the distribution of energy among the vibrational states of nascent HCl. A vibrational quantum of HCl is about 35.5 kJ mol^{-1}, so that a large fraction of the available energy (134 kJ mol^{-1} + the thermal energy of the reactants) goes into vibrational excitation of HCl, and thus, by difference, only a small fraction into translational recoil of HCl and I or into the rotation of HCl.

The vibrational distribution on the left of Figure 1.2 can be compared with that on the right, which is expected when a reaction is run under "bulk" conditions and the system has run to equilibrium. Then a Boltzmann equilibrium distribution would be produced: the most probable state is $v = 0$ and the relative populations decline exponentially with the vibrational quantum number. Of course, the bulk population does not arise from a single elementary process but rather from a succession of energy-degrading collisions of the vibrationally energy-rich HCl molecules with various other molecules.

Molecular dynamics in its "purist" approach tries to seek out (and understand) the truly elementary events. Thus it is more interested in the left than in the right panels of Figure 1.2. It is, however, concerned not only with the primary reactive collision process but also with the subsequent non-reactive, inelastic energy-transfer steps that take the system from the nascent distribution of products to the fully relaxed one. The Cl + HI system is not exceptional. Many exoergic reactions release a substantial part of their energy into internal modes of product excitation.[11] A key problem facing us is to understand this observation in terms of the forces that act during the collision. In this introductory case study we use a model.

1.2.2 Simple view of products' energy disposal: the spectator

We need a model, oversimplified of course, which will at least provide a simple interpretation of the observed energy disposal in the Cl + HI → HCl + I reaction. Let us try to take advantage of any "handle" that may help us approximate the dynamics of the problem.

One aspect of the collision is that it involves the transfer of a light atom (H) between two much heavier atoms. Recall that the I atom is about three times heavier than the Cl atom, which in turn is more than 30 times heavier than the H atom. The ClHI triatomic system is similar in that respect to the H_2^+ molecule-ion, where it is the light electron that "mediates" between the two heavy protons. When a molecule undergoes an electronic transition we obtain insight into the distribution of the final vibrational states from the Franck–Condon principle, discussed in detail in Chapter 7. The principle says that during the very fast electronic transition the heavy nuclei do not change their momentum, the nuclei merely act as spectators during the rapid electronic rearrangement. A *spectator* is someone who is not involved, i.e., does not feel any impulse. From Newton's second law we expect that a spectator is likely to be "heavy," because its momentum is resilient to change, the mass being the measure of the inertia to change. It thus follows that a spectator has a constant momentum.

To apply similar ideas to the present problem we must assume that the H transfer reaction is over in a time short compared to the time required for the heavy nuclei to move substantially. The model is then that the heavy iodine atom acts as a spectator during the (rapid) transfer of the light hydrogen atom to the heavy chlorine atom. This means that the final momentum of the I atom after the collision $(\mathbf{p}'_I)$ is essentially the same as the initial momentum of the I atom:

$$\mathbf{p}'_I = \mathbf{p}_I \tag{1.1}$$

It is easy to realize that this spectator model can account for the observation that very little of the reaction exoergicity is released as translational energy of the products. The Cl atom approaches the HI molecule with a particular momentum and "captures" the H. But the H atom is so light that the momentum of the I atom is left nearly unchanged, and so too is that of the Cl atom, which is part of the HCl product. But if energy is to be conserved without altering the translational motion, it follows that the exoergicity of the reaction must be "deposited" in the internal motion of the HCl. The quantitative version of our conclusion is the subject of Problem B. Here we proceed to look for additional experimental evidence that can lend support to the model.

1.2.3   Products' angular distribution

The spectator model makes a statement about vectors, namely that not only the magnitude but also the **direction** of the momentum of the iodine atom is unchanged in the collision (Newton's second law requires a force to change the direction of the momentum). Hence, the product iodine atom should appear in the same direction as that of the incident HI while the product HCl will appear in the direction of the incident chlorine atom. Leaving the details for later, it is sufficient to say that this description, which we colorfully call the *spectator stripping* picture, is qualitatively the behavior found experimentally. The product HCl appears mainly in the "forward" direction (i.e., in the direction of the incident chlorine atom). Note again that such a statement is only possible because we are focusing our attention on the isolated collision.[12] In the bulk, the products would soon collide with other molecules and very rapidly lose all memory of their nascent direction of motion.

The observation that the angular distribution of products is rather anisotropic implies that no long-lived ClHI intermediate "complex" is formed. If the reaction duration were long compared to the period of rotation of such an intermediate, almost all memory of the initial directions of the reactants would be erased and the products' angular distribution would not distinguish between the forward and backward directions. An insertion reaction for which this is the case is shown later in Figure 1.4. If there were a long-lived intermediate we would also not expect a very specific energy disposal because then there would be time for the energy to become approximately "equipartitioned" among the different modes of this intermediate.

1.2.4 From specific energy disposal to the mode-selective control of chemical reactions

Consider the endoergic reaction

$$I + HCl \rightarrow IH + Cl$$

The reverse exoergic $Cl + HI$ reaction is observed specifically to populate the final vibrational states of HCl. Using our model and any other means, can we predict the *energy requirements* of this endoergic reaction and, in particular, can we enhance the reaction rate by a selective preparation of the reactants?

Since we are dealing with an isolated collision, the reaction endoergicity has to be supplied by the initial energy of the reactants, I and HCl. This energy can be provided by the relative translational energy of the colliding pair and/or by the internal energy of HCl. When the energy of the I and HCl reactants just exceeds the endoergicity, not enough energy is available to form a vibrationally excited HI product and the final momentum of the HI product is also small. If, Eq. (1.1), the momentum of the I atom is to be nearly unchanged during the collision, the necessary energy for the reaction cannot be provided by the relative translational energy of the reactants, I + HCl, for this would require a high initial momentum of the I reactant relative to the center of mass. The reaction endoergicity, at least just above the energy threshold for reaction, must therefore be provided by the initial internal energy of the HCl.

The conclusion of a selective energy requirement, which is based on the model, can be much strengthened by consideration of the principle of *microscopic reversibility*.[13] Recall that the experiments on $Cl + HI$ showed that at low energies vibrationally **cold** HI leads mainly to the formation of vibrationally **hot** HCl, with only a smaller fraction of the energy released as translation. Because vibrationally cold HCl is formed with a very low probability in the forward reaction, it follows that for the reverse reaction involving the collision of vibrationally cold HCl with an I atom at high translational energies, most collisions are nonreactive. By this we mean that reaction will occur only rarely on such collisions. In contrast, collisions of vibrationally hot HCl molecules with I atoms will be fruitful even at low translational energies.

Selective preparation* of reactant energy states as a means for controlling not only the rate but also the chemical nature of the products has now been

---

* How can reactions take place starting with bulk thermal reactants for which the proportion of molecules in the higher vibrational states is exponentially small? It is a requirement of chemical kinetics that reaction rates be measured for reactants that are maintained in thermal equilibrium. If necessary, a buffer gas is added whose role is to insure that thermal equilibrium is maintained, by collisions. In the bulk the very few vibrationally hot, i.e., excited, HCl molecules react with I atoms produced by thermal dissociation of I_2. This displaces the remaining HCl molecules from their thermal equilibrium because the mean vibrational energy is now lower. Collisions with the buffer gas restore the thermal equilibrium or, on a molecular level, collisions repopulate the higher vibrational states of HCl and also dissociate I_2 molecules. Next, the vibrationally hot HCl molecules are preferentially removed by reaction with I atoms. Equilibrium needs to be restored, and so on. All this is hidden when we just focus attention on the thermal reaction rate constant.

well demonstrated. As an example, the rate of the H + HOH $\rightarrow$ H$_2$ + OH reaction is enhanced when the H—O stretch motion is excited.[14] Say now that instead of H$_2$O one considers HOD where D is the heavy isotope of H. The O—H and O—D vibrations have sufficiently different frequencies that the two modes represent nearly independent vibrations of HOD. The reactions of H atoms with H—OD excited with four vibrational quanta produce primarily H$_2$ + OD, whereas reactions of H atoms with HO—D excited with five vibrational quanta produce primarily HO + DH:

$$H + \begin{matrix} H \\ \diagdown \\ \diagup \\ D \end{matrix} O \rightarrow \begin{cases} H_2 + OD & \text{H—OD excited by four quanta} \\ HO + DH & \text{HO—D excited by five quanta} \end{cases}$$

The model introduced in Problem G concludes that the bond that is unaffected is a spectator in such reactions. Here we follow a more chemical argument: OH is isoelectronic with the F atom. The F + H$_2$ exoergic reaction selectively populates the vibrational states of the HF product. The H + H—(OH) reaction is expected to have similar forces to the H + HF reaction and the masses are also similar. Therefore, by microscopic reversibility, H—(OH) vibrational excitation should promote the endoergic H + H—(OH) reaction.

1.2.5   The experiment

The entire discussion in this section is based on the experimental determination of the vibrational energy partitioning in the nascent reaction products. The original experiment by John Polanyi and his coworkers was a *tour de force*. The background and the early results are described in his Nobel Prize lecture (Polanyi, 1987); for an early account, see Polanyi (1972). Nowadays we could use a *pump–probe technique*. What we want is to be able to probe the nascent products before they undergo any relaxation by interaction with the surroundings. This condition can be achieved if we can slow time down so as to catch the products as soon as they emerge from the reactive collision. The technique of *pump and probe*, which uses two fast laser pulses delayed in time with respect to one another, can achieve this. We need a precursor that, upon photolysis, promptly dissociates to yield the reactive atom or radical. The photolysis pulse, known as the pump, is "on" for only a brief time interval. How brief we will see shortly, but the necessary short time pulses are nowadays routinely available. A short time after the pump a second short laser pulse, known as the probe, interrogates the nascent products. The experiment is in the bulk so that the time interval between the pump and probe pulses needs to be of the order of the time between two successive collisions that a molecule undergoes in the bulk. In this way we insure that the products arise from only those reactive atoms that have undergone one, reactive, collision. Prior to this one collision the reactive atoms have not been deflected or slowed down by collisions with other molecules. The same short delay between the pulses also

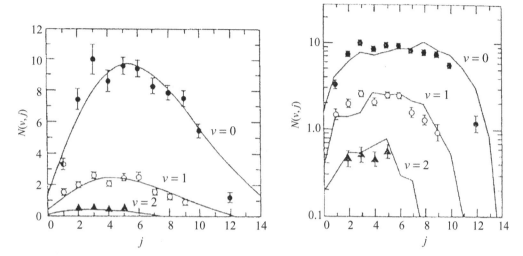

Figure 1.3 HD rotational and vibrational state distributions measured for the $H + D_2$ reaction at a collision energy of 1.3 eV. The energy is determined by the recoil energy of the H atom in the photodissociation of HI at a wavelength where it dissociates primarily to ground state I atoms. The experimental results shown [adapted from D. P. Gerrity and J. J. Valentini, *J. Chem. Phys.* **81**, 1298, (1984) and Valentini and Phillips (1989)] used CARS spectroscopy to determine the state of HD. E. E. Marinero, C. T. Rettner, and R. N. Zare, *J. Chem. Phys.* **80**, 4142 (1984) used resonance enhanced multiphoton ionization, REMPI, for this purpose.[15] The figure also shows curves. Those on the left are the so-called linear surprisal representation, see Section 6.4. The plot on the right shows the same experimental data on a logarithmic scale. The curves [adapted from N. C. Blais and D. G. Truhlar, *J. Chem. Phys.* **83**, 2201 (1985)] are a dynamical computation by the method of classical trajectories, Section 5.2.

insures that the products have not had the time to undergo subsequent collisions that relax the energy distribution. Probing detects truly nascent products.

The experimental desideratum is that the probe laser follows the pump laser within one time interval between two successive collisions. Either laser pulse needs therefore to be individually shorter than that time interval. The frequency of collisions of course depends on the pressure in the bulk sample. The lower the pressure the less frequent are the collisions and the less demanding are the conditions on the duration of the laser pulses. On the other hand, the lower the pressure the fewer the molecules in the sample and the lower is their measured response to the probe laser. Chapter 2 will demonstrate that laser pulses below about 50 ns are sufficiently short.

Figure 1.3 shows a complete vibrational and rotational state population distribution for the nascent HD product of the $H + D_2 \rightarrow DH + D$ reaction studied by the pump and probe technique.

An experimental arrangement that uses many of the techniques that we have mentioned is shown schematically in Figure 1.4.

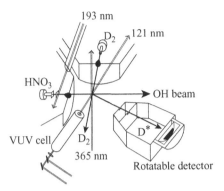

Figure 1.4 A crossed molecular beams arrangement for determining both the occupancy of the internal states and the angular distribution in space of the products of the $D_2 + OH \rightarrow D + DOH$ reaction. Within a vacuum chamber, not shown, OH radicals are produced by photolysis of HNO_3 using pulsed light at a wavelength of 193 nm. The OH radicals are collimated by a skimmer and cross a beam of D_2 molecules at a right angle. The relative collision energy is 6.6 kcal mol^{-1}. The D atoms are detected by electronically exciting them from the 1s to the 2p state using UV light, at a wavelength of 121 nm (VUV). Another pulsed laser, at 365 nm, excites the short-lived 2p state to a state of high principal quantum number, shown as D*. These excited D atoms are longer living[16] and are counted by a detector whose angle with respect to the molecular beams can be changed. This determines the products' angular distribution. The internal energy of HOD is determined by conservation of energy from the kinetic energy of the D atoms that is measured by the time they take to fly from the pulse-initiated reaction to the detector that is 29 cm away. [Adapted from H. Floyd Davis; see Strazisar *et al.* (2000). Lin *et al.* (2003) discuss the $F + CD_4$ reaction in a crossed beam arrangement.]

The reaction exoergic $D_2 + OH \rightarrow D + DOH$ is observed to form DOH primarily in the $v = 2$ state of the D—O stretch with only a small amount of bending excitation and essentially no energy in the OH vibration. This is as expected from the discussion of the reversed reaction in Section 1.2.4. The DOH product is backwards scattered as is DF from the $D_2 + F$ reaction, which is consistent with a mechanism where reaction occurs when the two reactants run head-on into one another.

1.2.6   Launching the system in the transition state region: the first steps toward control

The *transition state region* separates the reactants from the products (Polanyi and Zewail, 1995). For the atom–diatom reactions that we discussed, it is the region where the system is most like a triatomic molecule. It is an unstable triatomic because it proceeds to evolve into the separated products. Quite often it has only a fleeting existence, as indicated by the anisotropic angular distribution of the products showing that there is a memory of how the system was formed. We cannot therefore keep the system in the transition state region for long but we can **launch** it there and seek to see how it evolves. As an example, we earlier drew an analogy between a light H atom exchange between two heavy atoms and the light electron that is exchanged between the two protons in H_2^+. Is ClHI at all similar to H_2^+? One can have a look, for example, in the following way.[17] The negative ion ClHI$^-$ is a stable precursor and can be prepared in the gas phase. It is known that in this negative ion the light H atom is sandwiched by the two heavy atoms. Therefore, if we could only form the neutral species at the same atomic arrangement as the ion, we would be in the transition state region of the neutral triatomic system.

The extra electron of ClHI$^-$ can be rapidly detached by a short laser pulse. We then form a neutral triatomic species that finds itself with the three atoms close by. We know the energy of the neutral ClHI because the kinetic energy of the outgoing electron can be measured and the laser wavelength is known. We also have a good idea about its geometry because the electron departs so rapidly that the nuclei are where they were in the ground vibrational state of the cold ClHI$^-$ molecule that we started with. Launching systems into non-equilibrium geometries so as to probe the subsequent dynamics is a theme that will recur in this book all the way to reactions in condensed phases.

1.2.7 The steric requirements of chemical reactions

Not only energy is required to drive a reaction successfully. Almost always there is also a preferred direction of attack. The steric requirements of the reaction is a theme that takes us all the way to the docking of a drug at an active site of an enzyme. The experimental study that directly demonstrates such requirements started with colliding beam experiments that use oriented methyl iodide molecules. It is found that the reaction probability for the "favorable" configuration, Rb + ICH$_3$, meaning that the Rb atom approaches from the I end, is significantly greater than that for the "unfavorable" orientation, Rb + H$_3$CI. A graphical summary of the experimental results[18] is provided in Figure 1.5.

An iodine atom is bulky compared to the CH$_3$ group so that the methyl can shield the iodine from the attack only when it is directly in the way. For our earlier example, the Cl + HI reaction, the steric hindrance will be more effective and the cone of acceptance for reaction is expected to be much narrower because the Cl atom needs to reach the small H atom, which we expect to be effectively shielded by the bulky I atom. The computational evidence is that in less than a third of all Cl + HI collisions can the Cl atom come within the cone of acceptance for reaction, a cone spanned by the H atom.

There is much more to stereodynamics, the topic of the entire Chapter 10. Much of the recent work in the gas phase uses lasers to prepare oriented reactants

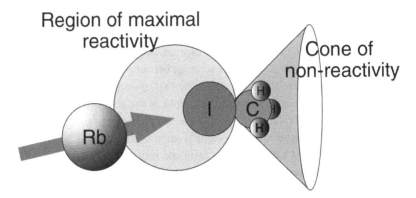

Figure 1.5 The cone of approach of Rb to CH_3I that does not lead to reaction to yield RbI, determined by experiments where Rb atoms approach oriented CH_3I molecules. The C, H, and I atoms are drawn according to their conventional size. The angle between the relative velocity of Rb and CH_3I and the axis of CH_3I is γ, the angle of attack. The probability of reaction to produce RbI is highest for Rb coming towards the I end and decreases when Rb comes in sideways. Reaction is vanishingly small for an approach in the region labeled as the cone of non-reactivity [adapted from Parker and Bernstein (1989)].

or probe the products.[19] Here we take a look at another steric aspect: is the collinear approach always the favored one?

1.2.7.1 Abstraction vs. insertion

The reactions we have discussed so far are all what a chemist will call abstractions. These reactions are characteristic of atoms from the first or seventh columns. Abstraction reactions typically are specific in their energy and angular disposal. As an example of a different behavior we show the angular distribution of the products where an (electronically excited) $O(^1D)$ atom inserts into the H—H bond. This results in the formation of a water (H—O—H) molecule and we expect the H—O—H bending motion to be energy-rich because in an insertion the O atom needs to attack in a direction perpendicular to the H—H bond and much energy is made available due to a replacement of one H—H bond by two O—H bonds. Unlike the molecules we usually encounter, here we deal with a very energy-rich water molecule. It can and does fall apart,[20] but it will stay bound for a while. Why? Imagine the gyrations of this energy-rich species. Because of the insertion mode, the energy is initially made available to the H^OH bending motion. To form the products we need an H atom to separate from OH. In other words we need energy in an O—H stretch mode. It takes a while before the energy made available by the formation of H^OH is channeled also into the stretch modes. When an O—H bond finally breaks it hardly remembers if the O atom initially came from the right and the H_2 from the left, or vice versa. The products' angular distribution will therefore exhibit a forward–backward symmetry[21] as shown in Figure 1.6.

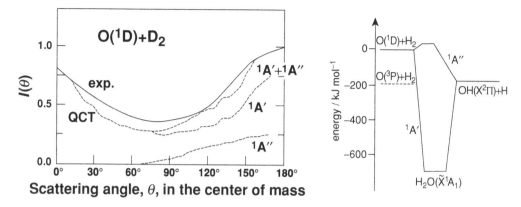

Figure 1.6 Left: observed angular distribution (solid line) of the OD product from the O(^{1}D) + D$_2$ insertion reaction [adapted from Casavecchia *et al.* (1998)]. The scattering angle θ is defined as the angle that the velocity of the departing OD product makes with the velocity of the incident O atom in the center of mass system, Section 2.2.7. Shown on the right are the energies of the different species. The low-energy path to products proceeds via a D$_2$O intermediate. In the collision we form an energy-rich D$_2$O molecule that has enough energy either to dissociate back to the reactants or to proceed to the OD + D products. The observed angular distribution of OD shows a slight preference for backwards scattering. This is because OD can also be formed by abstraction when the approach of O(^{1}D) to D$_2$ is collinear and this route does not go through D$_2$O as an intermediate but proceeds via the electronically excited ^{1}A'' state [for more on the abstraction channel see Y.-T. Hsu, J.-H. Wang, and K. Liu, *J. Chem. Phys.* **107**, 2351 (1997)]. We expect that when the O atom runs head-on into D$_2$, OD will scatter backwards. Dynamical computations (dashed line, QCT) by the method of classical trajectories, Chapter 5, verify that the abstraction reaction contributes primarily in the backwards direction and that the OD formed by the insertion reaction indeed shows no forwards–backwards preference with respect to the incident O atom. This is further discussed in Chapters 4 and 10.

1.2.8   The time scales of the chemical change

The discussion so far has emphasized the selection of the reactants and the inter-rogation of products. From this information we infer the molecular-level details of what must have happened as the reactants evolved to the products. Can we view the motion during the very chemical act? If this approach were techni-cally feasible we could probe the reaction as it unfolded. The entire Chapter 8 is devoted to such, technically demanding, experiments. An understanding requires examining the implications of the Heisenberg time–energy uncertainty principle. This principle implies that by probing the system over a short time interval we lose the ability to precisely know the energy of the system. However, for the time scales of interest to us, Table 1.1, the resulting uncertainty in energy is not only tolerable but also sufficient to allow us to localize the system in space. The localization comes about because an uncertainty in energy means an uncertainty in momentum. For the simple case of a free particle of mass m, $E = p^2/2m$ so that

Table 1.1   *Time scales for fast and ultrafast motions*

Time (s)	Intramolecular time scales		Chemical time scales
10^{-16}	period of electronic motion in low n orbitals, increases as n^3		
femto 10^{-15}			
10^{-14}	vibrational motion; fast for stretch motions, particularly so in hydrides, slower for heavy atoms and/or shallow wells	duration of reaction; fast for direct reactions or dissociation on a repulsive potential but slow to much slower for sticky collisions that proceed via an intermediate. Can be comparable to time between collisions	time between collisions in the liquid phase (pressure and viscosity dependent)
10^{-13}			
pico 10^{-12}			intramolecular energy redistribution; faster at higher internal energies
10^{-11}	rotational motion slower for larger molecules with large moments of inertia		
10^{-10}			time between collisions in the gas phase (decreases with increasing pressure)
nano 10^{-9}	radiative decay from electronically excited states		

On the left are periods of intramolecular motions that are relevant to chemistry. On the right are the different perturbations that can result in a physical or chemical change. The duration of a chemical reaction spans a wide range, from the very fast direct reactions and direct photochemical bond breaking to the much longer times when the energy-rich species [e.g., D_2O formed by a collision of $O(^1D) + D_2$ as shown in Figure 1.6, or by excitation of a stable molecule] live for a while before breaking apart. When we discuss biological function or molecular motors we will encounter even longer time scales.

$\delta E = (p/m)\delta p$. The uncertainty in energy is $\delta E = \hbar/\Delta t$, where Δt is the required time resolution. Hence we can localize the particle to within $\delta x = \hbar/\delta p = v\Delta t$, where $v = p/m$ is the velocity. Here then is what we need from our laser pulse. Its duration, Δt, needs to be short enough so that we can localize the motion within the range of distances that we want to probe. In other words, what we need are pulses short compared to the intramolecular time scale of the motion we want to follow. Two such characteristic times suggest themselves. One is the duration of the reactive event; the time it takes the reactants to rearrange into products. This time is often in the range of a few hundreds of femtoseconds (fs) or less (100 fs is the time needed to cover a distance of 0.3 Å when moving at a thermal velocity of $3 \cdot 10^4$ cm s^{-1}). If our time resolution is better than 100 fs, we can already watch the receding products. Achieving even shorter times, lesser than a vibrational period, will allow us to watch bound intramolecular motions. Some relevant time scales are shown in Table 1.1. As shown, periods of motions characteristic of molecules can span quite a wide range. For a protein, the stretch motions, say the C—H modes, can be rather fast, of the order of 10 fs, while the skeletal deformations are far slower. Even for electronic motion the range is wide. Electrons of low principal quantum number n move much faster than the motion of atoms. This is not necessarily true for electronically excited states. States of high n, known as Rydberg states, have a particularly slow-moving electron.

Quantum mechanically, a system that is localized in space and time is not in a stationary state. A non-stationary wave function, known as a *wave-packet*, changes with time and is the solution of the Schrödinger time-*dependent* equation of motion, rather than the more familiar time-*independent* equation, as discussed further in Chapters 7 and 8. Here we just note that the superposition of states in quantum mechanics allows us to write a non-stationary state as a linear combination of stationary states. For example, the uncertainty in energy that we noted means that states of different energy (and momentum) contribute to such a linear combination. We have δE as the range of energies of the stationary states that make significant contributions to the linear combination that is the non-stationary state.

The required technology is currently available[22] and is making much headway in providing real-time dynamics, even in biochemical systems. The development of ever-shorter pulses keeps making progress but the time–energy uncertainty principle tells us to examine what we want because when we probe a system over a very short time, say a time sufficient to probe electronic motion, we have a corresponding serious loss in the energy resolution. So, in principle, we are forced to make a choice: how short a time resolution do we really require? How serious is this choice? Let us discuss electronic excitation of a diatomic molecule as in Figure 1.7. The separation in energy between two vibrational states of the upper well is $h\nu$, where ν is the vibrational frequency. The inverse of the vibrational frequency is the period of the vibration. If our time resolution is better than a vibrational period, our energy resolution is inherently poorer than the spacing between two adjacent vibrational states of the upper state. Several vibrational states will then contribute to the linear combination that is the non-stationary state prepared by the short time pulse, as shown in Figure 1.7. The advantage is that such a state is localized within the potential well, unlike the more familiar stationary vibrational states, states that are delocalized over the entire allowed range (see Problem H).

Using ultrafast pump excitation we can launch such a localized wave-packet in the transition state region of a chemical reaction and then probe the temporal evolution[23] toward the products (Bernstein and Zewail, 1988; Zewail, 1996, 2000).

1.2.9 Reaction dynamics in solution and on surfaces

Finally, we will examine our understanding of the dynamics in the condensed phase, where much of real chemistry takes place. We discuss both reactions in solution and reactions on surfaces, with special attention paid to the new features that are not present in the gas phase.

Chemists traditionally use the environment in which the reaction takes place to control reaction rates and the branching between different possible products. We

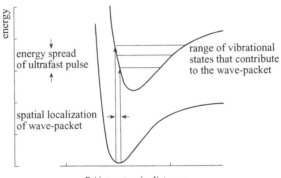

Figure 1.7 Preparing a localized vibrational state by an electronic excitation of a diatomic molecule using an ultrashort laser pulse. Shown are the interatomic potentials vs. distance for the ground and electronically excited states. By the uncertainty principle the light pulse spans a range of frequencies. Those frequencies that contribute significantly are within the range delineated by the two vertical arrows that originate from the vibrational ground state. The energy width of the laser pulse shown is small compared to the electronic excitation energy but is larger than a vibrational spacing. Therefore the pulse prepares the molecule in a definite excited electronic state but in a range of final vibrational states, as shown. Note that in accordance with the Franck–Condon principle, discussed in detail in Chapter 7, we do not allow the nuclei to change their relative separation R during the fast electronic excitation. The localized vibrational wave-packet $\psi(R, t)$ formed in the excited electronic state can be written as a linear combination of stationary (and delocalized) vibrational states, $\psi_v(R)$, $\psi(R, t). = \Sigma_v\, A_v\, \exp(-iE_v t/\hbar)\psi_v(R)$, where v is the vibrational quantum number and the energies E_v are shown. The contribution of each vibrational state is specified by the amplitudes A_v. Problem H shows that the amplitudes can be chosen such that the initial wave-packet is localized and that for motion on a harmonic potential it remains localized as it evolves over time.

want to discuss the different ways for thinking about the role of the environment. At the same time we need to alert you to the fact that what can be measured for reactions in solution and how to implement these measurements is different from in the gas phase and needs its own discussion (Fleming, 1986; Cong and Simon, 1994).

In solution the first key notion is that of *solvation*, the incessant interaction with the solvent. The essential qualitative aspect is the cage that the solvent builds around the reactants or products, as shown in Figure 1.8.

If the chemical reaction is fast, the observed rate of change is determined not by the crossing of the chemical barrier by the caged reactants but by the rate of the reactants diffusing towards one another and getting into the cage.

Solvation also has an energetic aspect. This allows us to understand the rate at which reactions of ions in solution, such as electron transfer, take place: the rate is governed by the need of the solvent to reorganize. The quantitative expression of this idea is the Marcus theory[24] that we will discuss in Chapter 11.

Gas phase: photodissociation

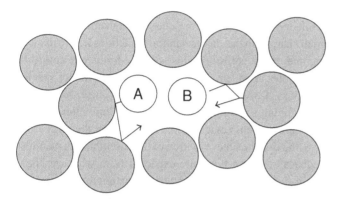

Solution phase: photodissociation and germinate recombination

Figure 1.8 Contrasting direct photodissociation in the gas phase and in a solvent [adapted from Schwartz *et al.* (1994)]. As we shall discuss, the coming back together of the two fragments owing to the "fence" presented by the solvent may initially be coherent in that the wave-packet describing the relative motion has not yet dephased,[25] see Problem H. On a longer (>picosecond) time scale the recombination will be diffusive. When the fragments are polyatomic a diffusive recombination means that the fragments will lose their relative orientation. They can even recombine to a different isomer of the parent.

The discussion of solvation emphasizes the motion of the solvent with respect to the solute. But for an activated chemical reaction to take place we need to cross a chemical barrier. How does the solvent affect this crossing? We sketch a unified point of view, where both the solvent and the solute are allowed to evolve during the chemical change.

Like a liquid, a surface can act as an energy sink or source. A surface often forms strong, directed, chemical bonds with the absorbed molecules. This modifies the reactants but also the structure of the surface. There are therefore several linked stages, with different time scales, in any overall surface-assisted reaction. In this book we emphasize the unraveling of the dynamics of some elementary processes,[26] but recognize that we have not quite reached the stage where our fundamental understanding is sufficient for the very many technological applications of surface processes, partly so because the morphology of the surfaces that are involved is more complex than the ideal surfaces that are used in careful laboratory studies. Progress in probing local surface structure promises that much further progress is possible, including the ability not only to probe but also to control the position of individual molecules on the surface (Ho, 1998; Hla and Rieder, 2003).

*1.2.9.1 Chaos and spatiotemporal pattern formation

What we want to do is first understand and second control the chemical process. In closing this introduction we have to take note that nature will cooperate with us, but not all the way. In this book the term "chaotic" will be used in several places. It will arise in two main contexts. First, even in the description of an isolated, individual, collision it can be that the classical dynamics is chaotic, meaning that rather small changes in the classical initial conditions lead to marked differences in the outcome of the collision. This severely limits our ability to control because real molecules are quantum mechanical and we cannot specify initial conditions as tightly as classical mechanics allows. Of course, we will put this to advantage by developing statistical theories. But the limitation must be borne in mind. Next, when we discuss macroscopic systems that are away from equilibrium we will also find that the time evolution of the ensemble can manifest nonlinearities and feedbacks that are not quite expected from what we know about systems near equilibrium.[27] We close by pattern formation for reactions on surfaces (Imbihl and Ertl, 1995) as an example where the role of dynamics has been elucidated. The surface structure is not static and it responds to the chemical reaction that is taking place. This active role of the environment brings us closest to how the feedback mechanisms regulate biochemical processes. Unraveling the molecular-level dynamics of such complex systems is already an active subject. The unprecedented structural information that is becoming available means that the understanding of the dynamics can only gather momentum and that control is forthcoming.

1.2.10   The road ahead

Our ability to develop simple models and to predict trends is, of course, based on familiarity with a large body of experimental results and more elaborate theoretical developments. The aim of the present book is to provide an introduction to the necessary background to be able to understand such results in the field of molecular dynamics. Much of the chemistry that is interesting to us takes place in solution or on the surface of a catalyst or in close proximity to a protein, etc. Yet we began with an isolated bimolecular collision in the gas phase. This is because our first task is to marshal the evidence for our claim that chemistry is local, meaning that the configurational change that we call a chemical reaction occurs over a short range. We will need to set the distance scale for both physical and chemical changes. Then we will examine the different processes that can take place and their time scales. We will pause to develop the tools for describing such processes. Next we shall try to exercise control.[28] As our understanding grows we discuss ever more complex systems.

Appendix: Units

Table A.1   *Useful physical constants (rounded)*

Designation of quantity	Symbol	Value	Units SI[a]	cgs
Avogadro's number	N_A	6.0221	10^{26} kmol^{-1}	10^{23} mol^{-1}
Atomic mass unit (^{12}C-12)	amu	1.6606	10^{-27} kg	10^{-24} g
Electron charge/mass ratio	e/m_e	1.7588	10^{12} C kg^{-2}	10^7 emu g^{-3}
Electron charge	e	1.6022	10^{-19} C	10^{-20} cmu
Electron mass	m_e	9.1095	10^{-31} kg	10^{-28} g
Bohr radius (a.u.)	a_0	5.2918	10^{-11} m	10^{-9} cm
Rydberg constant	R_x	1.09737	10^7 m^{-1}	10^5 cm^{-1}
Speed of light in a vacuum	c	2.99792	10^8 m s^{-1}	10^{-3} cm^{-1}
Planck's constant	h	6.6262	10^{-34} J s	10^{-27} erg s
Dirac's $\hbar$ ($h/2\pi$)	$\hbar$	1.05459	10^{-34} J s	10^{-27} erg s
Hartree (a.u.)	au	4.3598	10^{-18} J	10^{-11} erg
Gas constant	R	8.3144	10^3 J kmol^{-1} K^{-1}	10^3 erg mol^{-1} K^{-1}
Boltzmann's constant (R/N_A)	k	1.38066	10^{-23} J K^{-1}	10^{-16} erg K^{-1}

Adapted from *Pure Appl. Chem.* **51**, 1 (1979).
[a] SI: Système International d'Unités (International System of Units), adopted in 1960. Special symbols for units: C, coulomb; J, Joule; K, K.

Table A.2   *Useful conversion factors*[a]

Length
    1 ångström (Å) $= 10^{-10}$ m $= 10^{-1}$ nm;
    1 micron (μm) $= 10^{-6}$ m
Force
    1 newton (N) = 1 kg ms^{-2} [$= 10^{-5}$ dyn $= 10^{-5}$ g cm s^{-2}]
Pressure
    1 pascal (Pa) = 1 N m^{-2} $= 10^{-5}$ bar [$= 10$ dyn cm^{-2}];
    1.01325×10^5 Pa [= 1 atm $= 1.01325 \times 10^6$ dyn cm^{-2} = 760 torr]
Energy
    1 joule (J) = 1 kg m^2 s^{-2} $= 10^2$ erg [= 0.239006 cal]

[a] The familiar designations (units) enclosed to brackets are not part of the International System of Units (SI)

Table A.3   *Energy conversion factors (approximate)*[a]

	erg	J	cal	eV	au	cm⁻¹	Hz	K	kJ mol⁻¹	kcal mol⁻¹
1 erg =	1	1.0000(−7)	2.390(−8)	6.241(+11)	2.294(+10)	5.034(+15)	1.5092(+26)	7.243(+15)	6.022(+13)	1.4393(+13)
1 joule (J) =	1.0000(+7)	1	2.390(−1)	6.241(+18)	2.294(+17)	5.034(+22)	1.5092(+33)	7.243(+22)	6.022(+20)	1.4393(+20)
1 cal =	4.1840(+7)	4.1840	1	2.611(+19)	9.597(+17)	2.106(+23)	6.315(+33)	3.031(+23)	2.520(+21)	6.022(+20)
1 eV =	1.6022(−12)	1.6022(−19)	3.829(−20)	1	3.675(−2)	8.065(+3)	2.418(+14)	1.1605(+4)	9.648(+1)	2.306(+1)
1 hartree (au) =	4.360(−11)	4.360(−18)	1.0420(−18)	2.721(+1)	1	2.195(+5)	6.580(+15)	3.158(+5)	2.626(+3)	6.275(+2)
1 cm⁻¹ =	1.9865(−16)	1.9865(−23)	4.748(−24)	1.2399(−4)	4.556(−6)	1	2.998(+10)	1.4388	1.1963(−2)	2.859(−3)
1 Hz =	6.626(−27)	6.626(−34)	1.5837(−34)	4.136(−15)	1.5198(−16)	3.336(−11)	1	4.799(−11)	3.990(−13)	9.537(−14)
1 °K (K) =	1.3807(−16)	1.3807(−23)	3.300(−24)	8.617(−5)	3.167(−6)	6.950(−1)	2.084(+10)	1	8.314(−3)	1.9871(−3)
1 kJ mol⁻¹ =	1.6606(−14)	1.6606(−21)	3.969(−22)	1.0364(−2)	3.809(−4)	8.359(+1)	2.506(+12)	1.2027(+2)	1	2.390(−1)
1 kcal mol⁻¹ =	6.948(−14)	6.948(−21)	1.6606(−21)	4.337(−2)	1.5936(−3)	3.498(+2)	1.0486(+13)	5.032(+2)	4.184	1

[a] Numbers in parentheses denote powers of 10 by which the entry is to be multiplied.

Problems

A. The rare gas ion Ar^+ reacts with H_2 to form ArH^+. The reaction is exoergic. Provide one physical model to illustrate why the nascent product will be vibrationally excited. To continue with this reaction go to Problem B and then to C. The $F + H_2$ reaction also leads to nascent HF molecules that are vibrationally excited. What chemical argument can be used to make this observation further support the model?

B. Spectator stripping: a quantitative version of Section 1.2.2. It is easy to control and to measure the velocity of an ion. Say we study the final (relative) kinetic energy of $ArH^+ + H$ as a function of the initial (relative) kinetic energy of Ar^+ and H_2. Argue that spectator behavior, Eq. (1.1), implies that roughly half of the initial kinetic energy appears as products' translation. In the general $A + BC$ case you should get that with the A atom as the spectator the final (primed) kinetic energy is related to the initial one as

$$E'_T = [m_A m_C/(m_A + m_B)(m_B + m_C)]E_T$$

C. As in Problems A and B but now let us give the incident Ar^+ ion even higher initial kinetic energies and detect the products. Beyond a certain energy, not too high, there is a steep drop in the formation of ArH^+ [K. M. Ervin and P. B. Armentrout, *J. Chem. Phys.* **83**, 166 (1985)]. Why? What is happening to the production? The bond energy of H_2 is about $435\,kJ\,mol^{-1}$. That of ArH^+ is about $370\,kJ\,mol^{-1}$. (a) What is the exo- (or endo-) ergicity of the reaction of Ar^+ and H_2? (b) Estimate the energy at which formation of bound ArH^+ drops off.

D. Chemical kinetics of four-center reactions. In the family of "four-center" reactions $H_2 + X_2 \rightarrow 2HX$, X = halogen, there are significant variations in the bond strengths of X_2. The X = I case has the simplest rate law and the others proceed by a chain mechanism (Steinfeld *et al.*, 1999; Houston, 2001). For $H_2 + I_2$ the proposed mechanism [J. H. Sullivan, *J. Chem. Phys.* **46**, 73 (1967); G. Hammes and B. Widom, *J. Chem. Phys.* **96**, 7621 (1974)] is a rapid dissociation equilibrium $I_2 \rightleftharpoons 2I$ followed by I forming a weakly bound complex with H_2, $I + H_2 \rightleftharpoons IH_2$, which then reacts with another I atom, $I + IH_2 \rightarrow 2HI$. Show that this mechanism accounts for the overall kinetics being of the second order, first order in H_2 and first order in I_2. Propose experimental tests for this mechanism.

E. A dynamical study of $X_2 + Y_2$ four-center reactions, X, Y halogens. As in Problem D, here too there are variations among the different possible reactants, differences that reflect differences in bond energies. For $F_2 + I_2$, molecular beam scattering [C. C. Kahler and Y. T. Lee, *J. Chem. Phys.* **73**, 5122 (1980)] has shown that for collisions at energies above $17\,kJ\,mol^{-1}$ the primary process is $F_2 + I_2 \rightarrow FI_2 + F$. At a collision energy above about $30\,kJ\,mol^{-1}$ the FI_2 product dissociates to I + IF. In the bulk it is considered that the relatively weak F_2 bond allows for there being thermally generated F atoms that lead to IF formation via

$$F + I_2 \rightarrow I + IF\ ^*(B) \xrightarrow{h\nu} IF(X)$$

where X denotes the ground state and B is an electronically excited state of IF. The B state that is at an energy of about $225 \, \text{kJ mol}^{-1}$ above the X state decays by emission of light. This explains the chemiluminescence observed in the bulk gas-phase reaction of $F_2 + I_2$. (a) In what spectral range do we expect to detect the chemiluminescence due to the B to X transition? (b) The dissociation energies of F_2 and I_2 are 1.59 and 1.54 eV, respectively (it will be necessary to convert units). What is the maximal possible internal energy of FI_2 formed at the threshold for the reaction $F_2 + I_2 \rightarrow FI_2 + F$? (c) Provide an estimate for the energy of dissociation of FI_2 to $FI + I$?

*F. This problem is starred not because it is difficult but because you will need to be careful about transformation from the laboratory coordinates to the coordinates suitable to describe the relative motion. You may prefer to return to it after Section 2.2.7. As in Section 1.2.5 we initiate a reaction with a barrier by creating fast-moving atoms by photolysis of a precursor. For the $D + CH_4 \rightarrow DH + CH_3$ reaction the threshold energy has been determined to be 0.65 eV, while ND_3 has a bond dissociation energy to $ND_2 + D$ of $110 \, \text{kcal mol}^{-1}$. (a) Assuming that ND_3 is used as the photolytic hot atom source and that the thermal energy distribution of both ND_3 and CH_4 can be neglected, show that the longest photolysis wavelength that can produce the HD product is 220 nm. Along the way show that the D atom takes most of the energy available when ND_3 is photolyzed. (b) Experimentally, detection of HD requires using a shorter wavelength, say about 200 nm. What does this imply about the dynamics of photodissociation of ND_3?

*G. The unaffected bond is a spectator. In the $A + BCD \rightarrow AB + CD$ reaction it is often the case that the vibrational energy of CD is hardly changed during the reaction. Why? Make a structural model as follows: (a) Introduce coordinates that allow you to write the kinetic energy as a sum of uncoupled terms. For the reactants these can be the C–D distance, the distance of B to the center of mass of CD, and the distance of A to the center of mass of ABC. Express the kinetic energy in these coordinates. See D. W. Jepsen and J. O. Hirschfelder, *Proc. Natl. Acad. Sci. USA* **45**, 249 (1959). (b) For the products the coordinates can be the C—D and A–B distances and the distance from the center of mass of AB to the center of mass of CD. Express the kinetic energy in these coordinates. (c) Next, examine the kinetic energy of C—D motion and show that it is uncoupled to the other motions and is unaffected by the rearrangement. Where is the approximation? It is that kinetic energy is not necessarily conserved if there are forces acting. In Chapter 10 we will call the above "a *kinematic model*," see Problem B therein.

*H. A localized vibrational wave-packet as a linear combination of delocalized vibrational states. Figure 1.7 discussed the preparation of a localized vibrational state, a state that vibrates in the potential well in a manner similar to a classical particle. If the well is harmonic the wave function will remain localized indefinitely. Realistic molecular potentials are anharmonic so that after a few oscillations the state will delocalize. Even in the harmonic case, external perturbations such as collisions with other molecules will lead to delocalization. (a) By plotting the non-stationary wave function at different times, or analytically, show that in the harmonic case the initially localized state remains localized. To do so it is easiest if the amplitude of the vth stationary state is $A_v = \alpha^v/\sqrt{v!}$ where α is a complex number. (b) Show that the width in frequency is $v|\alpha|$ where v is the harmonic frequency of the potential well and that $|\alpha|$ determines the most probable vibrational state in the wave-packet. (c) In the graphical approach show that other choices for the amplitudes do not necessarily lead to a localized state. (d) In the analytical approach compute the mean position $\langle \psi(R,t)|R|\psi(R,t)\rangle$ of the wave-packet and show that it behaves like a classical harmonic oscillation. (e) In the graphical approach use anharmonic vibrational energy levels, with an anharmonicity parameter x_e, $E_e/hcv = (v + 1/2) - x_e (v + 1/2)^2$, to show that after some time the linear combination will start to *dephase*, meaning that it becomes less localized. Explore that behavior as a function of x_e recalling that $x_e \ll 1$ since $1/x_e$ is the (finite but large) number of bound vibrational states of the anharmonic potential.

Notes

1 For more on the development of the Arrhenius equation see Logan (1982), Laidler (1984), Further discussion of the temperature dependence of the reaction rate constant is in Section 3.1.

2 D. L. King, D. A. Dixon, and D. R. Herschbach, *J. Am. Chem. Soc.* **96**, 3328 (1974). This diatom–diatom elementary exchange reaction belongs to a family known as four-center reactions. Woodward and Hoffmann (1970) have explained why considerations of orbital symmetry suggest that such reactions will have high energy barriers. We will return to this point in Chapters 5 and 8. The early understanding of four-center reactions is reviewed by Bauer (1978, 1979). One can observe facile four-center reactions if the reactants are ionic, D. L. King and D. R. Herschbach, *Faraday Disc. Chem. Soc.* **55**, 331 (1973), or if the reaction is catalysed. For $H_2 + D_2 \rightarrow 2HD$ on transition metals see M. F. Bertino and J. P. Toennies, *J. Chem. Phys.* **110**, 9186 (1999), Table I in particular.

3 Chemical kinetics studies the rate of reactions in systems that are in thermal but not in chemical equilibrium. One can thereby examine the dependence of the reaction rate on the concentration of the reactants. Dynamics allows for extending such studies to systems that are neither in thermal nor in chemical equilibrium. One can thereby examine the reaction rate constant for reactants in particular internal energy states.

4 See Zewail and Bernstein (1988), Zare (1998).

5 In this reaction the strong laser field controls not only the breaking but also the forming of bonds because the toluene product cannot be formed without a rearrangement. The experiment is by Levis *et al.* (2001). Hurley and Castleman (2001) provide a commentary. A review of the experimental capabilities is Brixner and Gerber (2003).

6 See Bergman (1984), Bromberg *et al.* (1997). C—H or C—C bond activation by electron-rich species is an example of an insertion reaction; see Section 1.2.7.1 and R. Z. Hinrichs, J. J. Schroden, and H. Floyd Davis, *JACS* **125**, 860 (2003). It is also an example of a reaction in which electrons are transferred, i.e., an oxidation–reduction reaction. We will encounter many examples of such reactions.

7 It can be argued that there are processes, particularly reactions in solution, or on a surface of a catalyst or, from the biological side, reactions at the active site of an enzyme, that proceed very much like a well-orchestrated dance. Different players must each do their part at the right place in the sequence. Taking as an example reactions in solution, a topic that we discuss in more detail in Chapter 11, it is necessary for the reactants to come together, become caged at the foothills of the barrier to the chemical reaction, accept a fluctuation of the solvent to scale the barrier, and slide into a cage on the other side, etc. Even then we will try to distinguish stages in this teamwork. It should also be noted that our view is biased because we only look at the successful events, those where reaction did take place. There are many more failures – the reactants failed to get caged, or they did get caged and repeatedly failed to scale the barrier – but we only pay attention to the rare and fruitful events. The environment (enzyme, solvent, catalyst) needs to be quite fluxional during the overall process, but often this synchronous action is a series of local steps. See, e.g., Armstrong *et al.* (2003). There are collective processes, see Section 12.3 in particular, but we will seek to uncover the underlying molecular mechanism.

8 General references include Schatz and Ratner (1993), Head-Gordon (1996), Szabo and Ostlund (1996), Schleyer (1998), Levine (2000), Simons (2001), Cramer (2002). Special references to organic chemistry: Bernardi *et al.* (1996), Robb Garavelli *et al.* (2000). Surface chemistry has also received much attention: Garrison and Srivastava (1995), Musaev and Morokuma (1996), Radeke and Carter (1997), Greeley *et al.* (2002). For density functional theory see Kohn *et al.* (1996), Tse (2002).

9 For an introduction see Pulay (1995), Pulay and Baker (2001).

10 It is essential to note that the liberated energy cannot be found in the motion of I and HCl individually but only in their relative motion. The reason is that, in the world view that we discuss, the triatomic system is isolated. Therefore the only forces that act are the forces that the three atoms exercise on one another. The absolute position of the atoms does not matter, only their relative distances come into consideration. Technically this means that no force acts on the center of mass (c.m.) of the triatomic system and therefore the kinetic energy of the center of mass is conserved. The energy of the center of mass is the same for the reactants and products and it will be convenient to set it equal to zero, so that the c.m. is at rest.

11 One practical implication of this observation is that such reactions can be used as the pumping mechanism for *chemical lasers*. Skipping many important points, a *laser* is an amplifier. It provides a big output signal from a small input but the energy required for doing so must be supplied. For most lasers, the energy needs to be brought in rather quickly. It is not a trivial matter to transmit electrical energy in large doses. The faster the current flows, the greater are the losses by heating. To achieve a high energy flux, chemical energy, which can be made to flow in tubes, is a very attractive option. Chemical lasers, that is, lasers pumped by chemical energy, are therefore interesting. The other observation about lasers is that lasing requires *population inversion*: there need to be more molecules populating the emitting quantum state than there are molecules in the final quantum state of the radiative transition. Reactions leading to a specific energy disposal can provide such (extreme) non-equilibrium conditions since there can be more molecules in the upper, emitting state.

12 Molecular beam scattering (Bernstein, 1982; Scoles, 1986; Herschbach, 1987) is the method that is often used for this purpose. As shown in Figure 1.4, it can not only determine the overall product angular distribution but also resolve it for the separate

contributions from each internal state of the products (Lee, 1987; Casavecchia *et al.*, 1999; Casavecchia, 2000; Liu, 2001). High-resolution laser spectroscopic methods, as applied in Chapter 7, determine the absorption of a particular product's state and so can, by invoking the Doppler shift, determine also the direction of the final velocity (Houston, 1989, 1995).

13 Loosely phrased, this principle states that if a transition from A to B is possible, then so is the reverse transition from B to A. By implication, if A does not go to B, B does not go to A. The reason behind this statement is that in both classical and quantum mechanics time flows equally well in either direction so that any process can always be stopped and reversed. The microscopic dynamics is fully reversible so that any process and its time inverse satisfy the basic equation of motion. This result is clearly not applicable in the macro world and so the principle seems counterintuitive. In a strictly mechanical world view, reversibility is the rule and it is our everyday experience that time has a direction that is in need of an explanation.

14 Several overviews seek to place this work in context: Zare (1998), Crim (1999), Schatz (2000), Strazisar *et al.* (2000), Valentini (2001). Schatz discusses in particular the experimental work of Strazisar *et al.* (2000), as shown in Figure 1.4, and the quantum mechanical computations of Zhang *et al.* (2000) and of Pogrebnya *et al.* (2000). Castillo (2002) is a review of theoretical work and comparison with earlier experiments. Brouard *et al.* (2002) compare the effect of H atom translation vs. H_2O stretch excitation. The exit channel is reviewed by Loomis and Lester (1997). The experiments we discuss are those of A. Sinha, M. C. Hsiao, and F. F. Crim, *J. Chem. Phys.* **94**, 4928 (1991) and M. J. Bronikowski, W. R. Simpson, and R. N. Zare, *J. Phys. Chem.* **97**, 2204 (1993).

15 More recent such determinations include K. D. Rinnen, D. A. V. Kliner, and R. N. Zare, *J. Chem. Phys.* **91**, 7514 (1989) and B. D. Bean, F. Fernandez-Alonso, and R. N. Zare, *J. Phys. Chem. A* **105**, 2228 (2001). L. Schnieder *et al.*, *J. Chem. Phys.* **107**, 6175 (1997) measured the H atom recoil energy by a technique called Rydberg tagging, as discussed in Chapter 7. Such spectroscopic techniques are able nowadays to determine not only the internal state but also the direction of motion of HD, see for example Althorpe *et al.* (2002) and Chapters 7 and 10. For resonances in this reaction see also Manolopoulos (2002).

16 This is called Rydberg tagging and is further discussed in Chapter 7. See Schnieder *et al.* (1991, 1995). The VUV light is generated in the cell by a nonlinear process in Kr vapor where two laser beams interact.

17 The actual experiment for $ClHI^-$ and other XHI^- systems is by S. E. Bradforth *et al.*, *J. Chem. Phys.* **92**, 7205 (1990). It verifies that in the transition state region the light H atom hops several times between the two heavy atoms. For a general overview of this approach see Neumark (1993, 2001). For an application to the $H + H_2O$ reaction that we discussed in Section 1.2.4, using the H_3O^- anion, see Zhang *et al.* (2002). For organic anions see Wenthold and Lineberger (1999).

18 The experiment selects the angle γ between the initial relative velocity v of the reactants and the direction of the CH_3I bond. Then it detects only those reactive collisions where the direction of v was preferentially along the approach coordinate of Rb and CH_3I. How do we manage such a detection? By looking only at those RbI molecules that were backscattered. These rebounding molecules must come from a collision where the Rb atom ran head-on towards CH_3I.

19 This is possible because the plane of the electric field of the laser light can be defined by a polarizer. The interaction between a molecule and the light depends on the angle between

the molecular transition dipole μ and the electric field $\mathbf{E}$ of the light as $\mu \cdot \mathbf{E}$ where the dot denotes the scalar product of the two vectors (Greene and Zare, 1982; Orr-Ewing and Zare, 1994; Alexander and Zare, 1998).

20 Some of the energy-rich water molecules go back to the reactants but most of them dissociate to OH + H. Figure 1.4 that shows the angular distribution of the products also shows the energy levels. The dissociation to OH + H is more exoergic and a chemist will say "it is reasonable that the bond that breaks is the weakest." This is a good rule of thumb, and we will discuss why it is usually so but at the same time we will point out that there can be exceptions, when entropy and not only energy matters. For more on the $O(^1D) + H_2$ reaction see Simons (1997) and Alexander *et al.* (1998).

21 The angle of scattering is defined as the angle between the initial relative velocity $\mathbf{v}$ and the final relative velocity $\mathbf{v}'$. These two vectors define the plane of the collision and in that plane, a long-living, energy-rich ABC molecule dissociates with the products scattered equally in all directions. Under usual conditions the collision has a cylindrical symmetry about $\mathbf{v}$. When we rotate the vector $\mathbf{v}'$ about $\mathbf{v}$ as an axis, nearly forward and backward scatterings have the tips of the $\mathbf{v}'$ vectors concentrated in a narrow cone. For sideways scattering the vectors span out a wide ring. Consequently, the angular distributions fan out at the equator and bunch together near the poles, as shown in Figure 1.4. See Chapter 4.

22 The response time of available electronic devices is longer than the timescales shown in Table 1.1. So how can we measure such short times? The point is that, as in Section 1.2.5, we perform a pump–probe experiment. The first "pump" pulse prepares the system and the second pulse, delayed with respect to the pump pulse, probes the system. We monitor the response to the probe pulse as a function of the delay. It is the time delay between the two pulses that needs to be shorter than the timescale that we want to study. Each pulse on its own needs therefore to be even shorter. The commercially available ultrashort-pulse Ti–sapphire laser has played a key role in enabling the technology to be widely applied.

23 Like short-pulse experiments, scattering theory too can be implemented in the time domain. As a first step we can think of a classical trajectory, as discussed in Chapter 5. The trajectory specifies the position of each atom, as a function of time, as the system evolves from reactants to products. Is such a trajectory not the analog of a full-time resolved experiment? Not quite. Just as for the experiment, the implications of the time–energy uncertainty principle need to be faced. At each time instance along the trajectory we specify both the position of the atom and its velocity. In classical mechanics we must know the velocity in order to compute where the atom will be at the next instance of time. But the precise specification of position and momentum is, in principle, unacceptable in quantum mechanics. As discussed in Chapter 5, we can mimic this quantum uncertainty by running not one trajectory but a swarm of trajectories with a distribution of initial conditions. Alternatively, we can do the real thing, that is, time-propagate a wave function whose energy is imprecisely specified. Such a wave-packet is a linear combination of stationary states where each component has a definite energy. But the wave-packet has only a mean energy with a distribution of energy about the mean that is determined by the range of energies of the stationary states that contribute to it, as in Figure 1.7. See also Problem H.

24 For more on electron transfer, see Schatz and Ratner (1993), Barbara *et al.* (1996), Billing and Mikkelsen (1996), Marcus (1997), Caldin (2001).

25 A bound classical trajectory moving back and forth in an anharmonic potential has a period that is a shade longer the higher is its energy. A localized swarm of classical trajectories of somewhat different energies will therefore slowly deform and delocalize as the trajectories with shorter periods surge ahead. A wave-packet being the analog of such a swarm undergoes a similar delocalization known as dephasing. This dephasing which is due to anharmonicity of the potential is often as important as dephasing due to external perturbations.

26 For further reading on surface chemistry see Somorjai (1981, 1994), Greeley *et al.* (2002), Kolasinski (2002), Darling *et al.* (2001), Tully (2000), Gellman (2000), Thomas and Thomas (1997), Weinberg (1996), Rettner *et al.* (1996), Ertl (1982).

27 For oscillations and pattern formation in bulk systems, see Scott (1994), Hunt *et al.* (1990), Gray and Scott (1994), Noyes (1989), Epstein and Pojman (1998), Scott (2001).

28 We will have much to say on control using lasers. For an introductory account that emphasizes the quantal aspects see Brumer and Shapiro (1995). A detailed coverage is provided by Rice and Zhao (2000) and Shapiro and Brumer (2003).

Chapter 2
Molecular collisions

The collision of particles without internal structure is the simplest model for interacting molecules. For the bulk, the model can account for the deviations from ideal gas behavior all the way to the formation of clusters. For systems in thermal disequilibrium the model describes the relaxation back to equilibrium. What is missing from the model is chemistry, that is the internal atomic configuration of the molecules. We will not forget this key point but we need to develop a language for thinking about how reactions take place. In order to undergo a reaction, the two reactants need to get close to one another and it is this approach motion, unhindered by any environment, that is discussed in this chapter. The angular distribution of the particles as they exit from the collision serves as a probe for the forces that acted between them when they were close by. Knowledge of these interactions is also needed for the prediction of the properties of liquids and solids and for understanding the conformation of large molecules.

In this chapter we use a two-body, A + B, point of view. But we have to leave the familiar vibration and rotation of a bound diatomic AB and go to the unbound or continuum motion. It is the vibrational displacement that is unbound. The rotation of the bound diatomic remains a rotation of the A–B axis but we will have to recognize that during the collision the A–B center-to-center distance varies over a wide range. So we cannot use the rigid rotor approximation that for strongly bound diatomics simplifies the description by allowing a separation of the vibrational and rotational motions. Here the two motions are more closely entwined.

This chapter is a prerequisite for the discussion of collisions that result in a chemical reaction, that comes immediately in Chapter 3. Where essential, we supplement the use of classical mechanics with quantum mechanical considerations. Modern computers can easily keep track of motions of planets or of satellites and equally can handle the motion of particles during a collision, whether classical or quantum mechanical. However, this chapter is not about the numerical simulation of a collision. It is about understanding the relevant factors and developing a language for thinking about chemical reactions in the gas phase. Building on this background we will, in later chapters, discuss reactions in solution and on surfaces.

2.1 Molecules have a finite size

We now discuss how to characterize the finite size of a molecule. The probe for the size of a molecule will be another molecule. The probing is by the action of the mutual force between the two molecules. It is only through this force that molecules "see" one another. In the absence of a force each molecule moves in a straight line (Newton's first law). Such a force-free model is sufficient to derive the properties of an ideal gas at equilibrium. It cannot however account for the properties of real gases nor offer a mechanism for the establishment of equilibrium. The interaction between two molecules leads to a collision – the two molecules deviate from their straight-line trajectories. As a molecule moves through the gas, every so often it undergoes a collision with some other molecule. The *mean free path* measures how far, on average, a molecule moves before it undergoes another collision.

Our immediate purpose is to measure the mean free path by detecting the deviation of a molecule from a straight-line trajectory. Next we shall argue that the mean free path must surely depend on the density of the other molecules. If there are few molecules available to collide with, the mean free path will necessarily be long and vice versa. The *collision cross-section* results when we factor out the role of the density of target molecules in determining the mean free path. What remains is the cross-section, a molecular measure of size, with units of area. The final point of this section is that this size does depend on the energy with which the two molecules collide and that this dependence reflects the influence of the intermolecular potential on the distance that the molecules can approach one another.

2.1.1 Direct determination of the mean free path by a scattering experiment

The essential ingredients of an experiment for measuring the mean free path are shown schematically in Figure 2.1. Molecules (of type A) exit from the source and collimator with a well-defined velocity and so move as a beam.[1] This molecular beam passes through a scattering chamber that contains a gas (of type B) known as the target molecules. For simplicity we ignore the thermal motion of the target molecules so the target molecules are static. The intensity of the beam is reduced as some of the A molecules are deflected out of the beam owing to collisions with the B molecules.

Newton's first law is directly invoked in the analysis of the scattering experiment: an A molecule is said to have collided with a target molecule if it changed its direction of velocity and thereby leaves the well-collimated beam. In other words, the experiment demonstrates that a force acted between A and B molecules and it measures the resulting attenuation of the beam of A molecules. Here is where we define when a **collision** occurs: it is when a force acted between the two molecules.

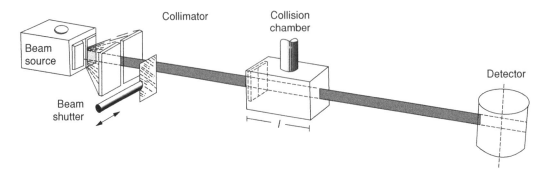

Figure 2.1 Measurement of the mean free path (schematic). A collimated beam[1] of molecules of type A passes through a collision chamber of length l, containing a scattering gas of type B molecules. The flux of transmitted beam molecules is measured (at the detector) as a function of the density n_B of the B molecules. From the attenuation of the A beam we can calculate the mean free path via Eq. (2.4) and the collision cross-section via Eq. (2.7).

2.1.2   Quantitative analysis of the scattering experiment

The direction of the beam of A molecules is taken as the x axis. The *flux* is defined as the number of beam molecules crossing a unit area (perpendicular to the direction of the beam) per unit time. The flux, $I(x)$, at distance x along the beam can be expressed in terms of the velocity v of the beam molecules as

$$I(x) = vn_A(x) \tag{2.1}$$

where $n_A(x)$ is the number density of A molecules (number of molecules per unit volume) in the beam, at position x. Equation (2.1) is the key to all discussions of transport phenomena in gases. The velocity v is the distance covered by a beam molecule in a unit time interval. Therefore the molecules that will cross a unit area in the next time interval are those that, at the beginning of the time interval, are up to a distance v from that unit area. Those molecules are confined to a box with a base, oriented perpendicular to the x axis, that has a unit area and a side of numerical value v. There are n_A beam molecules per unit volume and hence vn_A molecules in the box:

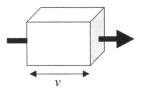

$$v$$

Owing to collisions with molecules of the target gas, A molecules are deflected from the beam and so the beam flux decreases down the length of the scattering cell. The fractional loss in beam intensity when an A molecule traverses a short distance Δx is determined by the likelihood of a collision with a B molecule in the short distance Δx and so must be proportional to Δx. We write the (dimensionless) probability of a collision occurring between x and $x + \Delta x$ as $\Delta x / \lambda$ and note that, by its definition, λ must have the dimension of a distance so that $\Delta x / \lambda$ is a dimensionless quantity. The fractional decrease in the beam flux, for sufficiently small interval Δx, can therefore be written as

$$\frac{I(x) - I(x + \Delta x)}{I(x)} = -\frac{\Delta I}{I} = \frac{\Delta x}{\lambda} \tag{2.2}$$

The minus sign is needed because the flux, $I(x)$, is decreasing as the distance x increases. We shall shortly show that λ is the *mean* free path, namely the average distance that a beam molecule travels along the x axis before it collides with another molecule.

Dividing both sides of Eq. (2.2) by Δx and taking the limit $\Delta x \to 0$ yields

$$\frac{-dI/dx}{I} = -\frac{d \ln I}{dx} = \frac{1}{\lambda} \tag{2.3}$$

Integration of Eq. (2.3) gives the expected result

$$I(x) = I(0) \exp(-x/\lambda) \tag{2.4}$$

where $I(0)$ is the flux of the beam as it enters the cell at $x = 0$. The beam flux is thus an exponentially decreasing function of the length of the scattering path, a result similar to that for a beam of light attenuated by absorption (there it is known as the Beer–Lambert law).

We now argue that, for a well-designed experiment, the length l of the scattering cell must be comparable to the mean free path, which implies that a beam molecule undergoes on average about one collision as it transverses the scattering cell: at the detector the flux of A molecules is $I(l) = I(0) \exp(-l/\lambda)$. To determine the mean free path, this flux is compared to the flux measured without B molecules in the cell, which is the same as the entering flux $I(0)$. The ratio, $I(l)/I(0) = \exp(-l/\lambda)$, determines the mean free path. But, like any other measuring device, the detector has a finite dynamic range. If it needs to measure both $I(l)$ and $I(0)$ in a reliable manner they cannot differ by more than one or two orders of magnitude. It follows that the length l of the cell must be comparable to the mean free path λ (if the length is much longer almost no exiting flux can be detected). As we discuss in Section 2.1.4, this condition places an operational limitation on the density of B molecules in the cell. Section 2.1.3 provides an explicit proof that the mean free path λ equals the average distance that an A molecule travels before it collides with a B molecule.

*2.1.3 The mean free path and the probability of a collision

We compute the **mean** free path and show that it equals λ.

From Eq. (2.2), the probability that a molecule collides between x and $x +$ Δx is $\Delta x / \lambda$. The probability of no collision between x and $x + \Delta x$ is then

$1 - \Delta x/\lambda$. We denote the probability of no collision in the entire interval 0 to x by $p(x)$. Therefore we can write the probability of no collision in the interval 0 to $x + \Delta x$ as

$$p(x + \Delta x) = p(x)(1 - \Delta x/\lambda) \qquad (2.5a)$$

On the other hand, when Δx is small enough

$$p(x + \Delta x) = p(x) + (\mathrm{d}p/\mathrm{d}x)\,\Delta x \qquad (2.5b)$$

This gives that $\mathrm{d}p/\mathrm{d}x = -p(x)/\lambda$ or, as is consistent with Eq. (2.4), $p(x) = \exp(-x/\lambda)$.

The probability $q(x)\mathrm{d}x$ that the first collision of an A molecule is in the interval x to $x + \mathrm{d}x$ is the probability, $p(x)$, that no previous collision occurred up to x, times the probability that a collision does take place between x and $x + \mathrm{d}x$, namely $\mathrm{d}x/\lambda$. Note that this probability, $q(x) = p(x)/\lambda$, is properly normalized, $\int_0^\infty q(x)\mathrm{d}x = 1$, that is, summing over all intervals $\mathrm{d}x$ yields unity as the probability that an A molecule will collide somewhere along its entire flight path. The mean distance that an A molecule covers before its first collision is then the sum, over all intervals, of the distance x weighted by the probability $q(x)$ that an A molecule collides for the first time in the interval x to $x + \Delta x$:

$$\langle \text{free path} \rangle = \int_0^\infty x q(x)\mathrm{d}x = \int_0^\infty x p(x)(\mathrm{d}x/\lambda)$$

$$= \int_0^\infty x \exp(-x/\lambda)(\mathrm{d}x/\lambda) = \lambda \qquad (2.6)$$

2.1.4 The collision cross-section

In this section we go from the mean free path to the collision cross-section. We take this route because we expect the probability $\Delta x/\lambda$ of a collision in the short interval Δx to be proportional to the number density n_B of the scattering gas in the cell. We want to factor out this dependence. First, the experimental evidence. Figure 2.2 shows experimental results obtained for a beam of CsCl molecules scattered by two target gases, (a) Ar and (b) CH_2F_2, in a cell of length l. Plotted is $I(l)/I(0)$ on a logarithmic scale vs. the scattering gas pressure, showing that the mean free path is an inverse function of the target gas pressure, $\lambda \propto P_B^{-1}$ ($n_B = P_B/kT$). This behavior is just what we should expect if the beam flux is attenuated by scattering collisions with target gas molecules.

To factor out the dependence of the mean free path on the density of the other molecules we define the *collision cross-section* σ (with dimensions of area) by

$$\lambda = (\sigma n_B)^{-1} \qquad (2.7)$$

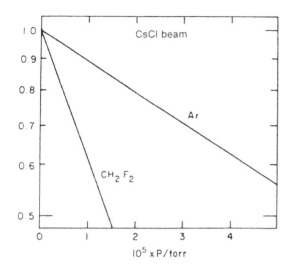

Figure 2.2 Attenuation data, $I(l)/I(0)$, on a logarithmic scale, vs. pressure, for the scattering of a thermal ($T \approx 1100 \, \mathrm{K}$) beam of CsCl by Ar atoms and by the polar CH_2F_2 molecules in a 44 mm cell. The log of the transmission is observed to decrease linearly with the pressure of the target gas [adapted from H. Schumacher, R. B. Bernstein, and E. W. Rothe, *J. Chem. Phys.* **33**, 584 (1960)].

This definition can be used to rewrite Eq. (2.4) as $I(x) = I(0) \exp(-x\sigma n_B)$. Shortly we estimate a cross-section and thereby show that at ordinary densities in the gas phase the mean free path is significantly longer than the range of the force between molecules. This is what we intuitively expect for the motion of molecules in a gas.

The collision cross-section is a measure of the "size" of the two colliding molecules. On the macro scale this means that the larger the cross-section, the smaller the mean free path so the more likely it is for a molecule in the beam of molecules to experience a collision. In Eq. (2.8) below, we rephrase this relation to show that when the cross-section is larger, the molecular collisions are more frequent.

In this chapter we are equally concerned with the micro-scale interpretation of the cross-section as a measure of the size of molecules. For example, Figure 2.2 shows that the mean free path of a CsCl molecule is significantly shorter for the CH_2F_2 target gas than for Ar, so the cross-section for the collision of CsCl with CH_2F_2 is appreciably greater than that for collisions with Ar. This is intuitively to be expected: the operational definition of a collision is that one molecule collides with another when a force acts between them. We need to make it quantitative, but, loosely speaking, the longer the range of the force, the more likely is a collision. The very polar molecule CsCl has a longer range of attraction to another polar molecule than to a spherical atom.

The implication that the cross-section depends on the range and strength of the intermolecular force carries with it an obvious corollary: the magnitude of the collision cross-section is a property of the two molecules that are colliding. Strictly speaking, a molecule does not have a "size" whose value is independent of how we probe it. We do not have a standard against which the sizes of all molecules are measured. Of course, an Ar atom will generally be smaller than a polar molecule but that is because Ar is a rare gas atom with closed shells of electrons and as such is compact and, typically, weakly interacting.

To estimate the mean free path under standard conditions, let us set d as the range of the intermolecular force so that πd^2 is a reasonable guess for the magnitude of the collision cross-section. Using Eq. (2.7), the mean free path, in units of d, is given by $\lambda/d = 1/n\pi d^3$ where n is the number of B molecules per unit volume. At s.t.p., there are Avogadro's number of molecules per 22.4 l, $1\,l = 10^3$ cm^3. So $1/n \approx 22.4$ (l mol^{-1})$\cdot 10^3$ (cm^3 l^{-1})$\cdot(10^8$ (Å cm^{-1}))$^3/6 \cdot 10^{23}$ (molecules mol^{-1}) $\cong 3.7 \cdot 10^4$ (Å^3 molecule^{-1}) $= 37$ (nm^3 molecule^{-1}). Taking $d \approx 5$ Å $= 0.5$ nm leads to $\sigma \cong 400$ Å$^2 = 4$ nm^2 and to $\lambda \cong 470$ Å $\approx 100d$ at a pressure of one atmosphere. This mean free path is significantly longer than the range, d, of the force between molecules, as is consistent with our physical picture of a dilute real gas: most of the time a molecule is unperturbed by other molecules. But the mean free path at atmospheric pressure is rather short by comparison to any realistic macroscopic width l of the scattering cell. The pressure of the scattering gas in the scattering cell shown in Figure 2.2 needs to be quite low compared to ordinary conditions so that the mean free path is correspondingly longer.

2.1.5 The rate of molecular collisions

In the macro world, the physical interpretation of the cross-section σ is that of an effective area whose size determines the number of collisions that a particular molecule undergoes per unit time as it moves in a gas. The image is as follows. In our mind we place a circular area σ drawn around the center of the molecule A (we pick a circle because all possible geometrical positions of the molecule are possible). The area is in a plane perpendicular to the direction of motion. As the molecule moves through the gas, the area σ sweeps, per unit time, a cylinder whose volume is $v\sigma$. There are $n_B v\sigma$ B molecules in this cylinder.

The number, ω, of collisions per unit time of a particular A molecule in a gas of number density n_B is used to operationally define the cross-section:

$$\omega = v n_B \sigma \tag{2.8}$$

ω is known as the *collision frequency*. From Eq. (2.7) $\omega = v/\lambda$, which is a result that is consistent for a molecule of speed v that travels an average distance λ between collisions. Note that in Eq. (2.8) v is the relative velocity of the molecules A and B, because we assumed that the thermal motion of the B molecules is negligible. If this is not so we need to perform an average over all possible values of the relative velocity as in Eq. (2.11).

Another quantity of interest is Z, the number of all bimolecular collisions per unit volume and unit time. If there are n_A A molecules per unit volume, the number of collisions per unit volume and unit time is $Z = n_A \omega = n_A n_B v \sigma$. We can factor out the density dependence of Z by defining

$$k = v\sigma \qquad (2.9)$$

k has the same dimensions as the rate constant of a reaction of second order, except that here we are counting all collisions and not only the reactive ones. Problem E proves that the flux along the beam decreases as

$$-\frac{dI}{dx} = k n_A n_B \qquad (2.10)$$

and that this relation recovers the result $k = v\sigma$.

The collision rate constant $k = k(v)$ is not quite a conventional bimolecular rate constant. It refers to collisions for which the relative velocity (here the beam velocity) is well defined both in magnitude and direction. The usual "thermal" rate coefficient is useful when we have a gas at thermal equilibrium so that there is a distribution of the velocities of the molecules. We show in Chapter 3 that this requires averaging of $k(v)$ over the relative velocity distribution for the gases at thermal equilibrium:

$$k(T) = \langle k(v) \rangle = \langle v\sigma \rangle \qquad (2.11)$$

Here the brackets denote an average over the velocity distribution in a gas at the temperature T. It is tempting to replace the average in Eq. (2.11) by $\langle v \rangle \sigma$. Textbooks of physical chemistry sometimes do not resist the urge to do so. This estimate is fine when we need to make rough calculations or if we invoke the hard-sphere approximation, as is discussed next. Otherwise, there are reasons to expect the cross-section to be dependent on the relative velocity of the colliding molecules. Experiments definitely concur. The result[2] $k(T) = \langle v \rangle \sigma$ must therefore be regarded either as an approximation or as a definition of an effective cross-section as $k(T)/\langle v \rangle$, see Problem E.

2.1.6 Molecules as hard spheres

The approximation of molecules as *hard spheres* can be quite realistic on the one hand and very misleading on the other, particularly so in the thermal energy range. What is always true is that it is a useful approximation because the model is

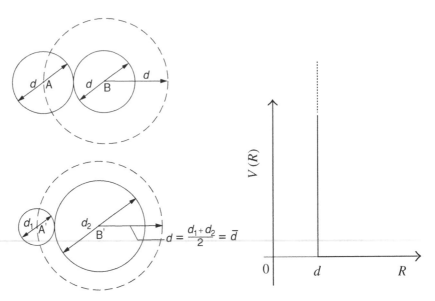

Figure 2.3 Left panel: collision of hard spheres. Upper: hard spheres of equal radius $d/2$ centered at A and B. The excluded volume (dashed), drawn centered at B, is of radius d. Lower: hard spheres of unequal diameters, d_1 and d_2. The "equivalent" exclusion sphere is of radius $d = (d_1 + d_2)/2$. The hard-sphere potential as a function of the center-to-center separation is shown in the right panel.

such a simple, appealing one. Two molecules interact as hard spheres if no force acts until the separation of the two centers decreases to some definite value, say d. At that point the molecules cannot approach any closer. They are impenetrable or, technically, they repel one another with an infinite force. The distance d can be interpreted as the sum of the radii of the two hard (or rigid) spheres. In other words, collision occurs whenever the center of one molecule approaches to within an "excluded volume" sphere of radius d about the second molecule. Figure 2.3 emphasizes that the radius of the excluded volume sphere is the sum of the radii of the two molecules.

Armed with the hard-sphere model we can make the definition of a cross-section more transparent. Imagine that an exclusion sphere is centered around each beam molecule. Corresponding to this sphere is a circle of radius d in the plane perpendicular to the beam velocity. Thus the beam molecule sweeps out a cylinder of volume $\pi d^2 \Delta x$ as it moves a distance Δx through the target gas. If the center of a target molecule lies within that volume a collision will occur and the beam molecule will be deflected off the x axis and therefore lost to the detector. From the volume of the cylinder swept we see that for the hard-sphere model

$$\sigma = \pi d^2 \tag{2.12}$$

so the collision cross-section is literally the area of a beam molecule as seen by a target molecule (and vice versa).

The hard-sphere interaction has a clearly defined and energy-independent range, namely d, and so the collision cross-section has the exact value πd^2. Even when the interaction is not that of hard spheres, it is the case that the cross-section can be interpreted in such a form, where now d is an effective range of the interaction. To do so we need to know more about realistic forces between molecules.

2.1.7 Realistic short-range repulsion

Bulk matter is compressible. So one expects that molecules are not strictly hard spheres. But in the condensed phase, where adjacent molecules are quite near to one another, matter is not very compressible. The repulsion between molecules is perhaps not infinitely steep as for hard spheres, but the quite limited compressibility suggests that it is quite steep. So the model of hard spheres seems to need some fine-tuning in that the range d of the repulsion can decrease with increasing energy of the approaching molecules. The direct experimental evidence[3] is that the collision cross-section for beam scattering at a well-defined relative velocity decreases slowly with increasing energy, as is shown in Figure 2.4.

It follows that the short-range repulsive force $F(R)$ between real molecules is not "infinitely hard," i.e., the potential $V(R)$ is not a vertical "wall" at some critical value of the center-to-center separation R of the colliding particles (e.g., $R = d$). Rather, there is a steeply increasing repulsion that is often approximated by an exponential form with a range parameter[4] ρ and strength A:

$$V_{\text{repulsive}}(R) = A \exp(-R/\rho) \qquad \text{short-range repulsion} \qquad (2.13)$$

Given the potential we can compute the force as the negative gradient of the potential,[*] $F(R) = -\mathrm{d}V(R)/\mathrm{d}R$:

$$F(R) = \frac{A}{\rho} \exp(-R/\rho) \qquad \text{short-range repulsion} \qquad (2.14)$$

This verifies that the short-range potential is repulsive because the force acts so as to increase the separation R.

2.1.8 Toward realistic interatomic potentials

The steeply rising repulsion at short range cannot be the full story. As the temperature is lowered, bulk gaseous matter shows increasing evidence for

---

[*] This result is very central in all that follows. Given the potential what we want to understand is the motion: how does the distance change with time? For this we need the force. In one dimension this is straightforward, $F(R) = -\mathrm{d}V(R)/\mathrm{d}R$. By Chapter 5 we will have more than one coordinate. There, the force is a vector, the gradient of the potential, and the gradient is in the direction of the greatest change.

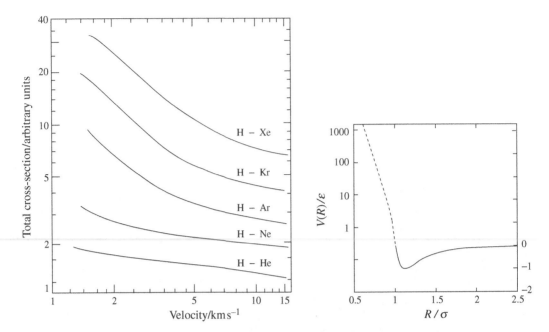

Figure 2.4 Left: typical decrease of the scattering cross-section with increasing collision velocity. The time spent sampling any distance interval depends inversely on the speed, so that at the higher collision energies the measurements are sensitive mainly to the repulsive part of the potential (because the collision partners spend more time in the region where the potential energy is comparable to the total energy). This log–log plot of $\sigma(v)$ vs. v refers to the elastic scattering of H by the different rare gases showing the increase in size of the atom as one goes down a column of the periodic table. The right panel is a plot of the potential vs. the separation. Note that the potential is plotted such that the repulsive part, $V(R) > 0$, shown as a dotted line, is on a logarithmic scale. There is also an attractive part of the potential, shown as a solid line, and it is plotted on a linear scale. The plot of the potential uses scaled variables that allow the same R-dependence to be used for all H–rare gas cases by a suitable choice of the range and strength of the interaction [adapted from R. W. Bickes *et al.*, *Faraday Disc. Chem. Soc.* **55**, 167 (1973)].

attraction between molecules. Scattering experiments at low (thermal-range) energies characterize the attractive part of the potential. The energy dependence of the cross-section demonstrates that a realistic intermolecular potential, Figure 2.5, has an attractive, possibly weak but long-range contribution typically of the form

$$V_{\text{attractive}}(R) \sim -C/R^6 \qquad \text{long-range attraction} \qquad (2.15)$$

A potential well is present even in the absence of chemical binding forces. It results from the competition between the physical long-range attractive forces and

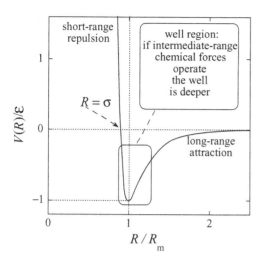

Figure 2.5 Schematic drawing of a realistic intermolecular potential (heavy curve). The minimum of the attractive well is located at R_m. We set the energy scale such that the interatomic potential vanishes when the two particles are very far apart. Therefore the potential is negative in the region of the well and its depth is ε, $V(R_m) = -\varepsilon$. The functional form plotted is that due to Lennard-Jones, $V(R) = 4\varepsilon$ $[(\sigma/R)^{12} - (\sigma/R)^6]$. Closer in than R_m the force is repulsive. For $R < \sigma$, where $V(R = \sigma) = 0$, the potential is nearly impenetrable and is almost hard-sphere like, that is, it requires a large increase in energy to access shorter interparticle distances.

the short-range repulsion. We call the long-range attraction "physical" because it operates in the region where the electronic wave functions of the individual atoms do not yet appreciably overlap. Therefore the physical attraction lacks chemical specificity but it always contributes.[5] If there are chemical forces, they are stronger but they only come into play when the atoms are closer in. A typical chemical well is deeper by an order of magnitude or more than a well due to physical forces. For example, the fit to the scattering data shown in Figure 2.4 leads to well depths, ε, ranging from about 0.2 meV for H—He to 6.8 meV for H—Xe. The chemical well of H—H is about 4.7 eV deep and it occurs at a significantly shorter atom–atom distance.

This section was called realistic "interatomic" potentials for good reason. We did improve upon the hard-sphere model by adding a qualitative correction, namely, the long-range attraction, and by having a more reasonable, somewhat softer repulsion. But the potentials that we have so far discussed still have a major limitation: they were taken to depend only on the distance R between the centers of the two interacting molecules. We were still discussing particles without internal chemical structure. The anisotropic shape of molecules means that a molecule does not look the same from all possible directions of approach.

Realistic intermolecular potentials depend not only on the separation of the two colliding molecules but also on their relative orientation. We will have much to say on the steric aspects of the interaction because molecular recognition and biological selectivity depend on it.

2.1.9 Simplistic approach to long-range interatomic and intermolecular forces

Even rather simple theory can be of help to us in providing useful information on at least the long-range part of the potential. Chemical binding forces ("exchange forces") begin to contribute significantly only at intermediate separations where the charge clouds of the approaching molecules begin to overlap.[6] The short-range repulsion is primarily a reflection of the fact that the energy increases if we bring too many electrons into a small volume. Next we turn to a simplified discussion of the long-range physical forces. We discuss two contributions. There can be classical *electrostatic interactions* and these depend on the charge distribution of the two molecules. Furthermore, there is always the additional attractive "London" or "van der Waals" force that is due to quantal electronic correlation between any two polarizable systems. For molecules without charge or permanent dipole moments these long-range quantal or *dispersion forces* dominate.

We begin with the classical electrostatic interactions. These are strong when one of the molecules (i) has a charge (is an ion) or (ii) has a permanent dipole moment. The polarizability of the second molecule can then interact with the electric field owing to the permanent charge (or the dipole) of the first molecule, giving rise to the *induction energy*. At a distance R, the field due to an ion of charge q is $E = q/R^2$ so the induction energy for an ion interacting with a molecule of polarizability α is

$$V_{\text{induction}}(R) = -\frac{1}{2}\alpha E^2 = -\frac{\alpha q^2}{2R^4} \quad \text{long-range attraction to an ion of charge } q \quad (2.16)$$

This attraction both has a longer range and is stronger so that the dispersion term, which is, in principle, always contributing, is neglected by comparison. This strong long-range attractive potential has a dominant effect on the dynamics of reactions of charged species and can significantly reduce the barrier to a chemical rearrangement when one reactant is an ion.

The electric field due to a permanent dipole μ at a large distance R away from it is $E = \mu/R^3$. The induction energy of this dipole with a spherical molecule of polarizability α is

$$V_{\text{induction}}(R) = -\frac{\alpha\mu^2}{2R^6} \equiv -\frac{C_{\text{ind}}}{R^6} \quad \text{long-range attraction to a permanent dipole} \quad (2.17)$$

Other expressions for these so-called asymptotic intermolecular potentials are available for dipole–dipole cases, etc., including explicit orientation-dependent terms.[7] A simple example is the interaction of an ion with a molecule that has a permanent dipole moment μ. In this case, the induction energy (2.17) has an extra term, $\mu \cdot \mathbf{E}$, where $\mathbf{E}$ is the field of the ion, which is in the direction of the relative position vector $\mathbf{R}$ and the dot represents the scalar product of two vectors. Using the angle γ between the permanent dipole and the relative separation:

$$V(R) = -\mu \cdot \mathbf{E} - \frac{1}{2}\alpha E^2 = -\frac{\mu q \cos \gamma}{R^2} - \frac{\alpha q^2}{2R^4}$$

(2.18)

Furthermore, since the ion is also polarizable, there is the need to add a third induction term like Eq. (2.17) that, however, is weaker.

When the atoms or molecules have no permanent charge or dipole moment, there can still be a quantum mechanical long-range attractive dispersion force. A simple picture is formulated when we recognize that even when there is no permanent dipole moment, there can be a fluctuating dipole that averages out to zero. We can associate such a fluctuation with the transition dipole moment of an allowed electronic transition of the atom or molecule. Consider two atoms, where, for simplicity, each atom has only one excited state with *transition dipole moment*[8] μ from the ground state. The field due to this dipole at a large distance R away from the first atom is $E_1 = \mu_1/R^3$. When we place the second atom, with polarizability α_2, at a distance R from the first atom we gain in energy $-\alpha_2 E_1^2/2$. A similar gain is achieved due to the polarizability of atom one being in the field of atom two. The total gain in energy is thus

$$V_{\text{dispersion}}(R) = -\frac{1}{2}\left(\alpha_2 E_1^2 + \alpha_1 E_2^2\right) = -\frac{\alpha_2 \mu_1^2 + \alpha_1 \mu_2^2}{2R^6}$$

$$\equiv -\frac{C_{\text{disp}}}{R^6} \qquad \text{long-range attraction} \qquad (2.19)$$

The energy gain due to the interaction of the polarizability of one system and the field of the transition dipoles of the other is the so-called dispersion energy. For atom–atom or atom–diatom interactions typical values of the dispersion constant C_{disp} are in the range[9] $10^{-57}-10^{-59}$ erg cm^6. It is a correlation effect because the electronic polarizability of one molecule is responding to the instantaneously fluctuating transition dipole on another molecule. It does not average out because it is the square of the transition dipole that enters Eq. (2.19). Despite the weakness of the dispersion interaction it is chemically relevant. This is particularly so when there are many atoms in proximity, as when two biological molecules approach one another. The exquisite specificity of enzymes for particular substrates is guided by such forces, as is the tertiary structure of biopolymers.

Note that two factors enter into the simple estimate (Eq. (2.19)) for the strength of the dispersion force. One is the polarizability. As a rough guide, the polarizability is a measure of the volume of the system. Larger systems (and this includes electronically excited states of smaller systems) are more polarizable and hence have stronger long-range attraction. The other factor is the transition dipole μ. It is larger when there is an allowed (= strong) electronic transition from the ground state. For example, molecules that are deeply colored have large transition dipole moments in the visible region of the spectrum.

The knowledge of the long-range part of the potential often suffices to account for most of the dynamical behavior of suitable systems provided that the main observable effect comes from glancing collisions that sample only large inter-molecular separations. We shall provide several such applications.

2.1.10   Sources of interaction potentials

Accurate potential functions for chemically stable diatomic molecules are well known thanks to extensive studies in the field of spectroscopy. The potential in the vicinity of the equilibrium position of the (usually deep) well has been determined for both ground and excited molecular states. Such studies have also yielded information on repulsive parts of potential curves, especially when light absorption or emission leads to dissociation, as in Chapter 7. Other important sources have been the measured temperature dependence of the deviation from ideal gas behavior and of transport coefficients, and from lattice energies and compressibility of solids.

The situation is not as satisfactory for inter*molecular* potentials, that is potentials that depend on orientation and on internal degrees of freedom and for the potential of interaction between weakly interacting pairs (the "van der Waals potential"). The scattering method is providing accurate van der Waals potentials and also yields direct information on repulsive interactions over extended ranges of interparticle separations. Simultaneously, *ab initio* quantum theory computations of interaction energies are feasible. In what follows we look at both deviations from ideal gas behavior that were an early source on the realistic shape of the potentials for chemically unbound pairs and on the newer results from scattering studies.

*2.1.10.1   Deviations from ideal gas behavior and intermolecular forces

A useful measure of the interaction between molecules in the bulk gas is the *second virial coefficient* $B(T)$ (units of volume per mole). It is defined by

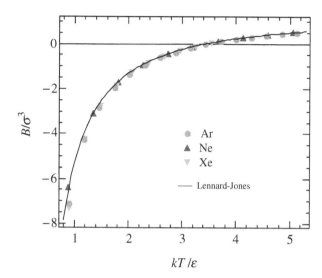

Figure 2.6 Second virial coefficient, $B(T)$, in units of σ^3, for the three rare gases as identified in the insert. By using reduced variables as in Eq. (2.22), the experimental results for different gases, points, can be plotted on a common scale. For hard spheres the second virial coefficient, $B(T)$, is the excluded volume, but for realistic systems note the change of sign: B becomes negative at lower temperatures and this is a signature of the attractive part of the potential.[11] The curve is a theoretical computation for a Lennard-Jones potential, Eq. (2.24) [adapted from Ben-Amotz *et al.* (2003)]. For a thorough overview of the route from the potential to deviations from ideal gas behavior, see Hirschfelder *et al.* (1954) and Maitland *et al.* (1987).

the *virial expansion*, which is a series expansion, in powers of 1/volume, of the deviations from ideal gas behavior. For one mole we write

$$\frac{PV}{RT} = 1 + \frac{B(T)}{V} + \frac{C(T)}{V^2} + \cdots \tag{2.20}$$

To interpret the first correction[10] term we start, in the spirit of the hard-sphere model, by subtracting from the volume V of the container an excluded volume, V_e, that is not available for the molecules so that $P(V - V_e) = RT$. Then, by expanding $1/(1 - x)$ as $1 + x + \cdots$ we derive $PV/RT = 1 + (V_e/V) + \cdots$ This is a virial expansion where V_e, the excluded volume per mole, is the second virial coefficient. For hard spheres, consideration of the geometrically determined, excluded volume or the use of Eq. (2.12) yields $B(T) = (2/3)N_A \pi d^3 > 0$, where N_A is Avogadro's number. Experimentally $B(T)$ does indeed tend to a positive, T-independent, asymptote at higher temperatures and this reflects the short-range repulsion. But at low temperatures $B(T)$ is often negative, Figure 2.6. We interpret the negative value of $B(T)$ and statistical mechanics provides the quantitative validation, Eq. (2.21), as indicating the presence of a net, overall weak attraction between molecules. As a result of this attraction, molecules prefer to be in the vicinity of one another. It is only thermal agitation that prevents them from condensing.

Other "macro" properties, such as transport processes, are also sensitive to the intermolecular forces and have been used to help establish the potentials. The problem is the "leverage" that the macro measurements provide. It is not as good as we would like because equilibrium values are measured as a function of temperature. It is by changing the temperature that different regions of the potential are given different weights and so their importance can be gauged. The second virial coefficient is a good example. It can be computed from the potential as

$$B(T) = -2\pi N_A \int_0^\infty [\exp(-V(R)/kT) - 1]R^2 \, dR \qquad (2.21)$$

But it is not realistic to invert this equation to express the potential as a function of $B(T)$ because we can only measure over a rather limited range of temperatures.[12] Instead, the procedure is to assume a flexible functional form for the potential, compute $B(T)$, and then vary the shape of the potential until a satisfactory reproduction of the data is achieved. Such a fit is shown in Figure 2.6.

*2.1.10.2 Potential curves from beam scattering

Measurements of the velocity dependence of the collision cross-section at higher energies determine the purely repulsive part of the potential. The results obtained, e.g. for rare gases and relatively inert molecules, are that the potentials were essentially of the exponential form, Eq. (2.14), and the "slopes" were in accord with expectations based upon crystal energy and compressibility data, etc., as well as the best available *ab initio* computations. At lower energies, the results are used to determine the long-range "tail" and the attractive "well" of the potential. The best probe of the potential is, however, the angular distribution as discussed below.

Figure 2.7 shows the resulting potentials for the homonuclear rare-gas pairs. Note the increasing well depth and range of the potential with increasing polarizability of the atoms (the polarizability is a measure of the deformability as well as the "volume" of the electron cloud of the atom), recall the discussion in Section 2.1.9.

It has been found that within a family of similar diatoms (e.g., rare-gas pairs, alkali systems, etc.) the resulting potentials can be scaled (approximately) to a common "reduced" functional form. We can define a reduced separation by $z \equiv R/R_m$ where R_m is the position of the minimum in $V(R)$, and a reduced potential energy by $V^* \equiv V/\varepsilon$ where ε is the well depth. Then it is found that $V^*(z)$, related to $V(R)$ by

$$V(R) = \varepsilon V^*(R/R_m) \qquad (2.22)$$

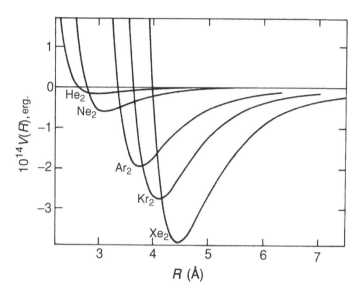

Figure 2.7 Potential wells for the homonuclear rare-gas pairs derived from angular distribution results [adapted from J. M. Farrar, T. P. Schafer, and Y. T. Lee, In *Transport Phenomena*, J. Kestin (ed.), AIP Conference Proceedings **11** (1973)]. In the region of the well these potentials can be scaled to a common functional form V^* (R/R_m) as in Eq. (2.22).

has a nearly universal shape, at least for the family of related atoms or molecules.

This explains the success of the well-known "rule of corresponding states" in the macro world: deviations from ideal gas behavior depend on two parameters that characterize the particular system, but otherwise have a universal shape. For example, Eq. (2.21) for the second virial coefficient can be expressed in reduced units as

$$B(T) = -2\pi R_m^3 N_A \int_0^\infty [\exp(-V^*(z)/T^*) - 1]\,z^2\,\mathrm{d}z \qquad (2.23)$$

This shows that for an interatomic potential function that satisfies Eq. (2.22) the virial coefficient, in units of $(2\pi/3)R_m^3$, is a universal function of the reduced temperature $T^* = kT/\varepsilon$ as shown for the rare gases in Figure 2.6 where the theoretical curve is for a Lennard-Jones potential,[13] which is the function plotted in Figures 2.4 and 2.5:

$$V(R) = 4\varepsilon\left(\left(\frac{\sigma}{R}\right)^{12} - \left(\frac{\sigma}{R}\right)^{6}\right) = \varepsilon\left(\left(\frac{R_m}{R}\right)^{12} - 2\left(\frac{R_m}{R}\right)^{6}\right) \qquad (2.24)$$

Parameters of the Lennard-Jones 6–12 potential for rare gases are shown in Table 2.1. These values give the fit shown in Figure 2.6. As discussed above,

Table 2.1 *Lennard-Jones 6–12 potential*
parameters[a]

	σ (nm)	ε/k_B (K)
Ne—Ne	0.28	34.8
Ar—Ar	0.34	117
Xe—Xe	0.39	225.6

[a] Obtained by the fit shown in Figure 2.6.

note how both the depth of the well and the range of the interaction increase with increasing size of the atoms.

To further discuss the role of the "size" of the atom in determining the interatomic potential we can think of electronically excited atoms or molecules.[14] Excited states have more weakly bound electrons and so are more diffuse and therefore have longer-range forces compared to other molecules.

2.1.11 On to collision dynamics

Before we turn to discussion of intermolecular potentials that depend also on the orientation of the colliding molecules and their internal deformations, we need to examine the dynamics of the collision of structureless particles. For this simple case the only outcome of the collision is that the particles deflect one another from their original trajectories. The collision is said to be elastic.

At low energies, collisions between rare-gas atoms are elastic, but for molecules (which possess internal structure) the collisions may also induce changes in internal energy or even bring about chemical rearrangement. For the present we confine our attention to simple elastic scattering, the key features of which are essential to our understanding of the more complicated classes of collision phenomena.

2.2 The approach motion of molecules

We discuss the motion of two molecules as they approach one another. Our ultimate goal is to understand chemical reactions. Toward this goal we need to know how close the two molecules will approach and how much energy is available to their relative motion. A chemical reaction requires the reactants to approach one another and the ideas of this section are applied to reactivity in Chapter 3. The same tools will allow us to examine the entire course of the collision of structureless particles, the theme of Chapter 4. By Chapter 5 we will fully recognize the internal structure of the colliding molecules.

This section has two parts. We first discuss the approach motion using the conservation of energy as our tool. Our purpose is to argue that in every collision

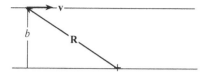

Figure 2.8 The relative position vector **R** and the velocity **v** a long time before the collision. The impact parameter b is indicated. In this figure the cross represents the origin. In the absence of a force between the two particles, **v** remains in the same direction, shown as a light line so the tip of the vector **R** lies on that line. We take $t = 0$ at the point of closest approach where the relative separation R equals the impact parameter b.

there is a centrifugal force that acts to prevent the molecules from getting too close in. This force is controlled by two variables, the miss-distance of the colliding particles, to which we give the technical name of the *impact parameter*, and the kinetic energy of the molecules before the collision when they are far apart. By the end of Section 2.2.5 we have set up all that is required for building models of chemical reactivity in Chapter 3. We next define the center-of-mass system and discuss the kinematics of elastic collisions.

In this section we center attention on the relative distance R of the colliding A + B molecules. The motion of the coordinate R looks like that of the motion of a single particle with mass $\mu = m_A m_B / (m_A + m_B)$, usually called the *reduced mass*. This is discussed in textbooks of classical mechanics and in Section *2.2.7 below.

2.2.1 The classical trajectory and the impact parameter

We want to describe the relative motion of two particles, A and B. Connecting the two is a vector, **R**, pointing from A to B. For our purpose, what is interesting is how the length of this vector, the relative distance R, varies with time. The form of $R(t)$ is what we mean by a classical trajectory. Before the collision R is quite large, and the relative velocity **v** is unperturbed because there is no force. Figure 2.8 shows the two vectors, **R** and **v**, at a time long before the collision. They are drawn such that both vectors are in the plane of the paper. There is always a plane defined by two vectors, so it is not a restriction. What we prove below is that, for structureless particles, the two vectors **R** and **v** remain confined to this particular plane throughout the collision.

The striking distance or *impact parameter** b is also defined in Figure 2.8. Conceptually b is a miss-distance, that is b is defined so that it is how near to one

---

* To connect the impact parameter with more familiar notions, start with the *angular momentum* vector **L** defined as $\mathbf{L} = \mathbf{R} \times \mu\mathbf{v}$. Here $\times$ denotes the vector product. We show below that for a potential that depends only on the distance R, **L** is conserved. **L** is a vector and so both its direction and its magnitude are constant. The direction of **L** is perpendicular to both **R** and **v**. The constancy of the direction of **L** confines the motion to a plane. The magnitude L of the angular momentum is

another the particles A and B approach, when there is no force acting between them. Technically, b is the component of the vector $\mathbf{R}$ that is perpendicular to $\mathbf{v}$ before the collision. If $b = 0$ the two particles run into one another head-on. Otherwise, the larger is b the less is the initial velocity directed along $\mathbf{R}$ and the more will the two particles miss one another because the approach motion is more off-center.

We begin by turning off the force between the two molecules. Newton's first law tells us that in the absence of a force the velocity $\mathbf{v}$ will continue to be along the same direction. At any time t we can therefore use the Pythagorean theorem to write the distance R as a function of time:

$$R^2 = v^2 t^2 + b^2 \qquad \text{a classical trajectory in the absence of a force} \qquad (2.25)$$

or, in a vector form, $\mathbf{R} = \mathbf{v}t + \mathbf{b}$ (where $\mathbf{v}$ and $\mathbf{b}$ are at right angles to each other and $\mathbf{b}$ is the component of $\mathbf{R}$ that initially is perpendicular to $\mathbf{v}$). Here we chose the origin of the time axis so that the collision lasts from $-\infty$ to ∞ and the two particles are closest at $t = 0$. When no force is acting, the distance of closest approach is b, which is therefore given the technical name of the impact parameter. If there is no force between the two particles, the vector $\mathbf{b}$ remains perpendicular to $\mathbf{v}$ for all times. If there is a force, we can still say that $\mathbf{b}$ is to be specified before the collision as the component of $\mathbf{R}$ that is perpendicular to $\mathbf{v}$.

Note that during the approach and receding motion of the two particles their relative separation vector changes its direction as the collision proceeds. We can see this from Figure 2.8. The tip of the vector $\mathbf{R}$ moves along the thin line. We can imagine this by letting two fists be the two particles. As the particles fly by one another, $\mathbf{R}$ undergoes a complete reversal of its direction. In Section 4.2 we show this from equations of motion.

Once we allow the two particles to interact, the trajectory will no longer be a straight line and the impact parameter will not be exactly equal to the distance of closest approach. But the impact parameter will remain well defined as the striking distance: it specifies, before the collision, how large is the component of $\mathbf{R}$ that is perpendicular to the initial velocity $\mathbf{v}$.

computed by taking the component of $\mathbf{R}$ perpendicular to $\mathbf{v}$, which equals b, times the magnitude of $\mathbf{v}$, times μ, $L = \mu v b$. In classical mechanics $L = \mu v b$ assumes a continuous range of values. In quantum mechanics $L = \hbar l$ [strictly $L = \hbar \sqrt{l(l+1)}$], where l is the dimensionless discrete angular momentum quantum number. This is the same quantum number l that is used to label the electronic orbitals of the hydrogen atom. (The internal dynamics of the non-relativistic hydrogen atom corresponds to the relative motion of two structureless particles.) We pursue this close analogy further in Section 4.3, where we introduce orbitals for the collision, and these allow for a quantal description of the approach motion. Here we just recall that s ($l = 0$) state electrons can get closer to the nucleus while states with higher ($p, d, f \ldots \Leftrightarrow l = 1, 2, 3 \ldots$) angular momentum are kept further away, increasingly so as l increases.

The colliding particles are, by assumption, structureless. So the magnitude of the velocity is the same before and after the collision. But due to the force the direction of the velocity need not be conserved. The final direction of the velocity provides such an important diagnostic for what has happened during the collision that we will devote the entire Section 4.2 to a quantitative analysis.

2.2.2 The centrifugal barrier and the effective potential

To examine the classical trajectory we use the conservation of energy. Before the collision, for structureless particles, the total energy E equals the kinetic energy $E_T = \mu v^2/2$. Here μ is the reduced mass of the collision pair. For the moment we remain with the assumption that the force has been turned off. The kinetic energy is the rate of change of the vector $\mathbf{R}$ with time, $K = \mu(\mathrm{d}\mathbf{R}/\mathrm{d}t)^2/2$. But, during the collision, as the particles approach, it is not only the magnitude, $R(t)$, that is changing. The direction of the vector $\mathbf{R}$ is also changing. The kinetic energy is the sum of these two contributions, even though we are only interested in the change of the magnitude $R(t)$ of the vector $\mathbf{R}$. There are two contributions to the kinetic energy, whether there is a force acting or not. We can identify the two terms from Eq. (2.25). It gives that $\mathrm{d}R^2/\mathrm{d}t = 2R(\mathrm{d}R/\mathrm{d}t) = 2v^2t$. Using $vt = (R^2 - b^2)^{1/2}$ the kinetic energy of the motion along the line of centers is found to be $(\mu/2)(\mathrm{d}R/\mathrm{d}t)^2 = E_T(1 - b^2/R^2)$. Without a force, the kinetic energy before the collision has to equal the kinetic energy during the collision. This conservation of energy condition implies that

$$E_T = (\mu/2)v^2 = (\mu/2)\left(\frac{\mathrm{d}R}{\mathrm{d}t}\right)^2 + \frac{E_T b^2}{R^2} \qquad \text{without a force} \qquad (2.26)$$

Equation (2.26) equates the kinetic energy before the collision, where $R \to -\infty$, to the kinetic energy at a finite value of R. Before the collision all the kinetic energy is due to the two molecules approaching one another. But as the two molecules get closer there is another contribution, the kinetic energy arising from the rotation of the interparticle distance $\mathbf{R}$. Equation (2.26) identifies this centrifugal energy term as $E_T b^2/R^2$. When the particles are far apart, meaning that $R \gg b$, the entire kinetic energy originates from the first term in Eq. (2.26), representing the particles approaching one another. However, as they get closer in, more and more of the energy is being spent on the rotation of the relative distance. By $R = b$ all the kinetic energy is in the rotation and $(\mathrm{d}R/\mathrm{d}t) = 0$, so that the particles cannot get any closer. We rederive these results later; see also Problem H. For now we just point out that Eq. (2.26) provides a quantitative statement of the result that is seen graphically in Figure 2.8: in the absence of force, the classical trajectory is restricted to $R \geq b$; the impact parameter is the distance of closest approach. One sees that from Eq. (2.26) because the kinetic energy of the approach motion $(\mu/2)(dR/dt) = E_T (1 - b^2/R^2)$ cannot be negative and so we recover the implication of Figure 2.8, $R^2 \geq b^2$. In quantum mechanics,[15] see Section 4.3, the motion is not restricted to a region where the kinetic energy of relative motion needs to be positive.

When there is a force acting between the particles, the conservation of energy reads:

$$\text{Energy} = K + V(R) = \frac{1}{2}\mu \left(\frac{dR}{dt}\right)^2 + E_T b^2/R^2 + V(R)$$
$$= E_T \qquad (2.27)$$

The first line is the total energy being equal to the sum of the kinetic and potential energies. The result is true throughout the collision. By our convention we take the zero of the potential energy $V(R)$ when the two particles are far apart. The second line is the value of the total energy before a collision of structureless particles. Conservation of energy requires that the two lines are equal to one another.

To focus attention on the trajectory $R(t)$ it is convenient to consider the centrifugal energy and the potential energy summed together, and call the sum an *effective potential*:

$$V_{\text{eff}}(R) = V(R) + E_T b^2/R^2 \qquad (2.28)$$

The energy is then the sum of the kinetic energy along the lines of centers and the effective potential

$$E_T = \frac{1}{2}\mu \dot{R}^2 + V_{\text{eff}}(R) \qquad (2.29)$$

and we have thereby the energy as a function only of the scalar quantity $R(t)$. Of course, this makes the effective potential a function of the energy E_T and not only of the position. But the kinetic energy along the lines of centers is anyway a function of E_T and E_T is a constant of the motion so it really makes no essential difference.

Equation (2.29) shows the role of the potential between the molecules in determining the *kinetic energy along the lines of centers R*:

$$\frac{1}{2}\mu \dot{R}^2 = E_T - V_{\text{eff}}(R) = E_T(1 - b^2/R^2) - V(R) \qquad (2.30)$$

In the range of R where the potential $V(R)$ is attractive, $V(R) < 0$, more kinetic energy is available than for a free motion. This is particularly significant for collisions of ions with molecules. The additional kinetic energy provided by the unusually strong electrostatic potential is sometimes sufficient to surmount the barrier to a chemical reaction so that a collision of even translationally quite cold reactants can lead to a reaction. In the range of R where the potential $V(R)$ is repulsive less kinetic energy is available than for a free motion. The region where $E_T(1 - b^2/R^2) - V(R) < 0$ is forbidden in classical (but not in quantal) mechanics.

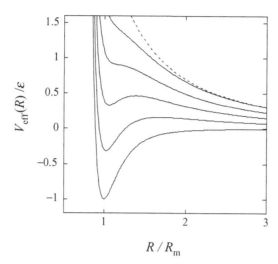

Figure 2.9 A plot of the effective potential for several (increasing) values of the impact parameter. At high b values the well in the potential $V(R)$ is filled in and the effective potential is purely repulsive. The dashed line is the centrifugal barrier alone for the last case.

The centrifugal energy acts as a repulsive contribution to V_{eff}, often known as the *centrifugal barrier*, as shown in Figure 2.9. It is a repulsive barrier because it acts against the approach of the colliding particles. The origin of this barrier is as follows. As the two particles approach each other, their centrifugal energy (due to the rotation of the relative separation), $E_T b^2 / R^2$, increases. Because the total energy is conserved, this increase is at the expense of the potential energy and the radial kinetic energy. Eventually we reach the smallest separation, R_0, see Figure 2.10, where the kinetic energy along the lines of centers of the two particles vanishes.

2.2.2.1 The distance of closest approach

The *turning point* or *distance of closest approach*, R_0, is that function of the impact parameter given as the solution of the implicit equation

$$E_T = V(R_0) + E_T b^2 \Big/ R_0^2 \tag{2.31}$$

In the absence of a potential $R_0 = b$, as mentioned earlier. For the hard-sphere model, Eq. (2.31) yields the expected result

$$R_0 = \begin{cases} b, & b > d \\ d, & b \leq d \end{cases} \qquad \text{hard spheres} \tag{2.32}$$

shown also in Figure 2.10(b).

For more realistic interatomic potentials R_0 depends upon both b and E_T, but for large enough values of the impact parameter it can be seen from Eq. (2.31) or Figure 2.10 that $R_0 \to b$, independent of the total energy. This result is often

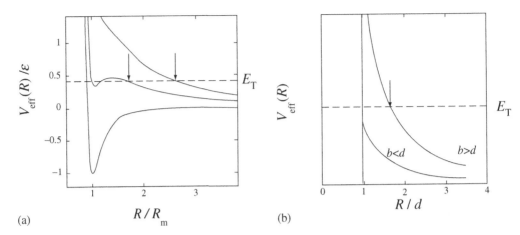

(a)   (b)

Figure 2.10 A graphical construction for determining the distance of closest approach, R_0, and the region allowed for the motion in classical mechanics. (a) For a realistic potential $V(R)$. The effective potential $V_{eff}(R)$ is shown vs. R, in reduced units for a head-on collision, $b = 0$, and for an intermediate and a high value of the impact parameter b. R_m is the equilibrium distance and ε is the well depth. For $b = 0$ the effective potential equals the potential $V(R)$. The motion in classical mechanics is confined to those values of R for which $E_T \geq V_{eff}(R)$, see Eq. (2.30). Equality is at the distance of closest approach $R = R_0$ and for higher values of b this value of R is identified by an arrow. A collision starts at the far right of the figure with $R(t)$ decreasing until R_0 when it starts to increase and move towards the left. Note that for the intermediate and high b cases the motion fails to reach the well region, $R_0 > R_m$, and samples only the long-range part of the potential. For the intermediate value of b it is seen that there is also an option for a bound classical motion inside the well, to the left of R_0. More on this in Section 4.3.5. (b) Same construction but for a hard-sphere potential where, for $R > d$, the effective potential is just the centrifugal barrier $E_T b^2 / R^2$. For this case no force due to the hard-sphere potential is applied if $b > d$ and $R_0 = b$, so the spheres do not "collide" but move as free particles. For any $b < d$, the motion reaches the impenetrable hard core of the potential and $R_0 = d$.

a useful approximation when dealing with large impact parameter collisions.* A graphical construction for solving (2.30) is shown in Figure 2.10. At very low bs the motion can cross the region of the well and get all the way to the inner repulsive part of the potential. At somewhat higher bs, and particularly so for lower energies, the motion can meet a turning point that is further out and it will not reach the region of the potential well. This is due to the repulsion from the centrifugal barrier. If there is a chemical attraction between the particles then Chapter 3 points out that such a motion fails to reach the region of the chemical

---

* At large bs $R_0 \to b$ because at larger Rs the centrifugal barrier, that decreases as R^{-2}, overwhelms any long-range attraction that falls as R^{-n}, $n > 2$.

forces. At higher bs there can be multiple roots when we solve (2.31). Classically, the approach motion of the reactants begins at a large separation that progressively gets smaller, and it is the largest root of (2.31) that is the turning point for the approach motion and the start of the receding motion. The other two roots are further discussed in Section 4.3.5 where, due to quantum mechanical tunneling, they are shown to allow for collisions of longer duration.[16]

We have completed our first task. The relative motion during the collision is fully specified in terms of one (scalar) function of time, $R(t)$. From here on we turn to examining the motion, keeping in mind that we need to specify the initial conditions not only by the velocity (or kinetic energy) but also by the impact parameter.

2.2.3 On the centrifugal force

The full implications of the concept of a centrifugal barrier will only be evident when we discuss reactive collisions. The point for the elastic case is that the long-range potential between molecules is attractive. If there is no impediment, the molecules will get as close as possible, namely up to their short-range, steep, repulsive interaction. The model of hard-sphere collisions will then be quite realistic. It is the centrifugal barrier that prevents the close approach for collisions with non-zero impact parameters (and we shall see that these are the overwhelming majority). This role is familiar as, for example, it prevents the Moon (or any satellite) from falling to Earth despite the very strong gravitational attraction.

What we called "the distance of closest approach" is what was already known to the ancients as the perihellion of the orbit of a planet around the Sun. Modern chemists know this concept in terms of how close the electrons get to the nucleus of the atom. For collisions under a realistic potential, if the impact parameter is quite low the molecules will get all the way in to the range of the repulsive forces. For high-impact parameters the molecules will only sample the long-range attraction and not penetrate much beyond $R = b$. But for any impact parameter the colliding particles will feel a mutual force. The only exceptions are potentials whose influence extends over only a finite range, such as a hard-sphere potential.

Rather than speak of the centrifugal barrier it is common to speak of the force, minus the gradient of the potential,

$$\mu \frac{d^2 R}{dt^2} = -\frac{dV_{eff}}{dR} = -\frac{dV(R)}{dR} + 2\frac{E_T b^2}{R^3} \qquad (2.33)$$

The *centrifugal force*, $2E_T b^2 / R^3$, is repulsive (that is, always positive in sign) and acts to force the two particles apart. Its role is more important as the initial energy or the impact parameter[17] increases.

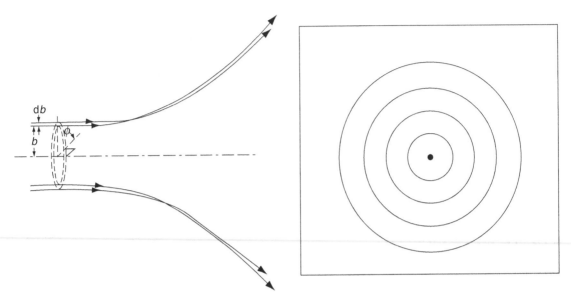

Figure 2.11 Left: the construction of the ring through which all collisions with an impact parameter in the range b to $b + db$ must cross. Under ordinary conditions all initial values of the azimuthal angle ϕ are equally probable. Right: the dartboard in the plane perpendicular to the initial velocity. The rings are drawn for collisions with all impact parameters in the range 0 to 1, 1 to 2, 2 to 3, etc. Note how in this dartboard image the areas of the ring increase as b increases.

2.2.4   The micro view of the cross-section

We are about to provide a microscopic definition of the collision cross-section. Two ingredients come in. One is the definition of a "collision." On the basis of Newton's laws of motion we took it that a collision occurs whenever two molecules exercise a force on one another. The outcome of the collision can be that a chemical reaction took place, or only that the two molecules deflected from their unperturbed straight-line motion, or anything in between. Whatever the outcome, when a force due to the potential acted, a collision is said to have taken place. Nor need this force be repulsive, and indeed the long-range part of the force is in general weakly attractive, Section 2.1.8. The other point is that the cross-section is that area, drawn in a plane perpendicular to the initial velocity, that the relative motion of the molecules needs to cross if a collision is to take place.

The impact parameter is defined in a plane that is perpendicular to the initial velocity **v**. In that plane we can imagine a circular target area, Figure 2.11. The center of the target is at $b = 0$. A collision with an impact parameter that is in the range b to $b + db$ has to cross the target through a ring of radius b and width db.[18] The area of such a ring is $2\pi b \, db$. This bullseye or dartboard picture is shown on the right of Figure 2.11. We have defined the cross-section as an area, in a plane perpendicular to the relative velocity $\mathbf{v}$, such that the relative separation vector $\mathbf{R}$ had to cross that area for a collision to take place. Hence, when the impact parameter is in the range b to $b + db$, the collision cross-section is the area of the ring on the target plane:

$$d\sigma = 2\pi b\, db \qquad \text{differential collision cross-section} \qquad (2.34)$$

When we specify only the (magnitude of the) initial velocity of the colliding particles, collisions can occur with all possible values of b, hence

$$\sigma = \int 2\pi b\, db \qquad \text{total collision cross-section} \qquad (2.35)$$

The integral is over the entire range of values of b that lead to a collision. For example, for the hard-sphere model, a deflection of the trajectory caused by a force can only take place for $b \leq d$ (Problem I). Hence, for rigid spheres we recover our earlier result

$$\sigma = \int_0^d 2\pi b\, db = \pi d^2 \qquad (2.36)$$

But what about more realistic potentials whose long-range attraction has an unbounded range? Then, whatever is the initial value of b, the trajectory will undergo some deflection. Therefore, a collision can be said to have taken place. The range of b values leading to a collision is then zero to infinity and the integral in (2.35) diverges! The collision cross-section is infinite. The answer is that here is a case where classical mechanics fails in a qualitative way.[19] It is possible to argue that a failure is to be expected; that we have specified that a collision took place even when the Heisenberg uncertainty principle does not allow us to make this observation. Where and why did we run into conflict with the uncertainty principle? It will be easier to answer that after we know a shade more about the deflection of the trajectory during the collision.

*2.2.4.1 On controlling the impact parameter

The impact parameter is a key player in determining the outcome of the collision. Yet our ability to control it is far more limited than we would like. In Section 2.2.5 and in detail in Section 4.2 we will do the next best thing and ask, a posteriori, what was the impact parameter of the collision. But can we control its value beforehand?

To control the impact parameter the collision must be started with a restricted range of angles between the relative velocity and the relative position. This kind of atomic resolution control of geometry is achievable, for example, by a variant of the experiment we discussed in Section 1.2.5. There we prepared a reactive atom by photolysis of a precursor in the bulk, and so the precursor was randomly

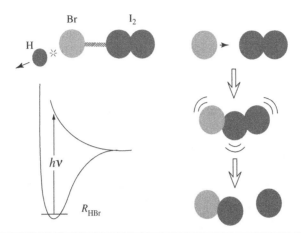

Figure 2.12 Controlling the impact parameter in the Br+I_2 reaction. Here the anisotropy of the van der Waals potential is such that the preferred configuration is H—Br·I—I. Upon photodissociation of HBr the H atom moves away rapidly while the Br atom approaches I_2 slowly (why slowly?). The BrI_2 species vibrates for a while, see Figure 8.5, before it dissociates to BrI and I. The bound and repulsive potentials involved in the initial photodissociation of the H—X bond are also shown [adapted from Wittig *et al.* (1988)].

oriented with respect to the reactant. Now we intend to restrict the geometry of approach. To do so recall that even in the absence of chemical binding there is a well in the potential between molecules. As we shall discuss, mixed bound dimers such as XH·OCO can be prepared where X is a halogen atom and the dot indicates that the dimer is a bound state of a weak long-range potential.* The XHO configuration is nearly but not exactly linear because the bending potential is shallow.

In the experiment,[20] a pulse of a UV laser breaks the HX bond. This bond scission sets the H atom moving in the opposite direction to X, that is toward the OCO molecule. The reaction H + OCO → HO + CO can then take place. The H atom is moving toward the OCO in the direction of the old HX bond and this defines the direction of the relative velocity. It is not necessarily exactly in the direction of the O—C bond, but it is quite a bit constrained.[21] The results can then be compared to photodissociating HX in the gas phase in the presence of CO_2. In that case the orientation of the H—X axis is random with respect to the O—C—O.

Dissociation of XH·OCO has the same purpose as the photodetachment of a stable anion as described in Section 1.2.6. Both experiments probe spatial aspects

---

* This potential is anisotropic and it also determines the preferential angle for the X—H—O configuration. Even if the equilibrium is for a collinear arrangement, the bending potential is quite shallow so that the zero-point motion of the bending spans a broad range of bent configurations.

of the dynamics by starting from *constrained geometries*. See Figure 2.12 for an illustration. What is being controlled in such experiments is the magnitude of the impact parameter. Photodissociation experiments on cold molecules, as discussed in Chapter 7, provide another way to experimentally limit the range of the angular momentum. Chapter 10 addresses the control of the **direction** of an angular momentum vector.

2.2.5 Qualitative examination of the deflection function

Owing to the force acting between them, two colliding molecules will deflect from their otherwise straight-line motion. This section provides a simple discussion of how this deflection depends on the impact parameter of the collision. Since we typically do not select the value of the impact parameter, this discussion allows us to do the next best thing, namely to infer what was the impact parameter from the observed deflection after the collision. A key concept in the discussion will be how the centrifugal barrier acts so as to keep the two molecules from getting too close. Hence this section is a good overview of the material of Section 2.2. Sections 4.1 and 4.2 will provide many quantitative details. In particular we show therein how the deflection characterizes the intermolecular force and provide a quantum view of the scattering. For now we do a poor person's view. We ask "what is of the essence" in both classical and quantal scattering.

For structureless particles, the magnitude of the relative velocity must be the same before and after the collision. This is because the total energy before or after is the kinetic energy of the relative motion and so the kinetic energy is conserved. But since a force did act during the collision, the direction of the relative velocity will change. The *deflection* is the angle between the initial and final relative velocity.

That the deflection must vary with the impact parameter follows from quite simple considerations: for a very low impact parameter the velocity is directed practically along the position vector so the molecules collide nearly head-on. The centrifugal barrier is rather low so they can reach the inner steep repulsion and so rebound from one another. Their velocity reverses its direction. As $b \to 0$, the deflection $\to \pi$. Low-impact-parameter, head-on, collisions result in backward scattering. At the other extreme, for a very high impact parameter the molecules can barely approach one another. They are kept far apart by the centrifugal force and can only sample the weakly attractive tail-end of their interaction. If b is indeed large, the molecules hardly deviate from a straight line, the resulting deflection is quite small, and the scattering is in the forward direction.

The correlation low $b \leftrightarrow$ backward scattering, high $b \leftrightarrow$ forward scattering is the essence. To be complete one more point must augment it: other things being equal, there will be more collisions with higher impact parameters.* It is straightforward to make this quantitative. The cross-section for collisions with an impact parameter in the range b to $b + db$ increases linearly with b, $d\sigma = 2\pi b\, db$. Therefore, many more collisions result in forward as opposed to backward scattering.

Can we interpolate between very high and very low bs? We do so in detail in Section 4.1, but the essence is clear. If we allow b to be large but not very large the molecules will sample more of their long-range attraction. The deflection will be bigger. To a zeroth-order approximation the distance of closest approach is b and so when b is comparable to R_m, the equilibrium distance of the potential, the molecules sample their maximal attraction and the deflection will be large. The scattering will be quite intense. As discussed below and in more detail in Section 3.2.5, this is known as *rainbow scattering* and it provides a signature of the well in the potential. When the impact parameter is even smaller than R_m, the molecules begin to feel the short-range repulsion, culminating in fully backward scattering as $b \to 0$.

The qualitative considerations are summarized in the plot shown in Figure 2.13. The result of such considerations is the angle of deflection $\chi(b)$ as a function of impact parameter. Chapter 4 shows that for not small impact parameters such a plot reflects the shape of the potential in the sense that $\chi(b) \propto V(b)/E$.

2.2.6 Rainbow scattering and the quantum mechanical interference of different trajectories

We need much of Chapter 4 to do full justice to the topic of the rainbow and other aspects of classical and semiclassical scattering. What follows is the essence.

The terminology *rainbow scattering* arises because the quantum mechanical wavelike nature of matter is necessary for the complete understanding of the effect. But we first deal with the purely classical aspects. When the impact parameter is in the narrow range $b, b + \delta b$ the final velocity is deflected by an angle in the range $\chi(b), \chi(b) + (d\chi(b)/db)\delta b$. At the rainbow, where $d\chi(b)/db = 0$, there is a small but finite range of impact parameters that give rise to scattering into the same deflection. This can also be seen from Figure 2.12 or, quantitatively, from the definition $d\chi(b)/db = 0$ at $b = b_r$. The approximation $\chi(b) \propto V(b)/E$ identifies the rainbow impact parameter as the equilibrium distance of the potential and also explains why the optical rainbow is observed at different scattering

---

* Surprising? Not to a chemist. The impact parameter is a measure of the angular momentum. The structure of the periodic table of the elements and much else depends on there being $2l + 1$ distinct quantum states when the angular momentum has the value $l\hbar$. The origin of this quantum degeneracy is physically the same. There are different possible directions in space for the angular momentum vector and these are distinguished by the value, m, of the projection of the vector, $-l \le m \le l$, m an integer. In classical mechanics m is a continuous variable so one has a ring, as shown in Figure 2.11, where the tip of the vector lies on the ring. In terms of the azimuthal angle ϕ, $m = \cos\phi$.

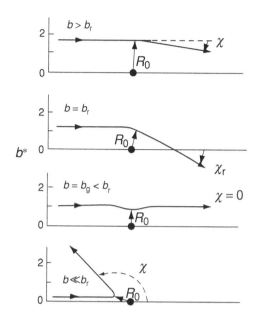

Figure 2.13 Trajectories for collisions at different impact parameters showing the deflection χ at different values of $b^* = b/\sigma$ where σ, see Eq. (2.24), is the distance where the potential switches from attractive to repulsive. For large initial b the trajectory is a shade pulled in by the long-range force. The deflection is maximal at the *rainbow*, which is at $b_r \approx R_m$. For closer-in approach the trajectory begins to sample the repulsive potential. At what is called the *glory* the net deflection is zero because the initial attraction is fully counterbalanced by the repulsion closer in. Below the glory, $b < b_g$, repulsion dominates and the scattering is backwards. R_0 is the distance of closest approach.

angles for light of different colors.* In classical mechanics the scattering intensity is infinite at the rainbow. Quantum mechanically the intensity is finitely large and oscillatory. Why is there a qualitative difference between the two results?

The presence of the well in the potential means that impact parameters somewhat below and somewhat above the rainbow impact parameter lead to the same angle of deflection. Approximating the potential about its minimum as a harmonic well, the approximation $\chi(b) \propto V(b)/E$ suggests that near the rainbow $\chi(b) = \chi_r + c(b - b_r)^2$ where c depends on the collision energy. There are then two different trajectories, with impact parameters just above and just below b_r, that scatter into the same angle.

In quantum mechanics, when two classically distinct histories give rise to the same observable outcome, they are said to *interfere*. The textbook example of interference is the two-slit experiment[22] where an electron hits a wall after going through one or another slit in a screen. The interference pattern observed in this experiment was instrumental in validating the wave nature of light. Here we do

---

* The color is determined by the wavelength, which is inverse to the frequency. The energy scales as the frequency.

not need the slits. The dynamics of the collision selects two different impact parameters, i.e., two rings in the circular target of Figure 2.11, that lead to the same deflection.

Interference means that we must treat the trajectories as light rays. That is, the scattering intensity is the square of a scattering amplitude, $I(\theta) = |f(\theta)|^2$. If there are two rays we must add up the amplitudes of two waves and take the absolute value squared of the sum: $|f_1(\theta) + f_2(\theta)|^2$. This is not the same as the classical result that requires adding the two intensities: $|f_1(\theta)|^2 + |f_2(\theta)|^2$. The difference is the interference term that comes about because the quantal amplitudes f are complex numbers, numbers that have both an absolute value and a phase: $f_n = |f_n|\exp(i\delta_n)$. Therefore

$$
\begin{aligned}
I(\theta) &= |f_1(\theta) + f_2(\theta)|^2 \\
&= |f_1(\theta)|^2 + |f_2(\theta)|^2 + f_1(\theta)f_2^*(\theta) + f_2(\theta)f_1^*(\theta) \\
&= |f_1(\theta)|^2 + |f_2(\theta)|^2 + 2|f_1(\theta)f_2(\theta)|\cos\delta_{1,2} \quad (2.37)
\end{aligned}
$$

The *interference term* depends on the difference in the phases of the two amplitudes. It can be constructive or destructive, depending on the sign of the cosine term.

For particles as heavy as atoms the phase, δ, of an amplitude is typically large compared to 2π and it varies rapidly with energy. The interference term is therefore a very oscillatory function of initial conditions and any imperfection in the energy resolution leads to its averaging out.

The rapid interference pattern means that under normal circumstances it is often justifiable to neglect interference effects. Near the rainbow the circumstances are not normal because the two interfering trajectories, both of which lead to the same deflection, are quite similar in their initial conditions and so the difference in their phase is small and leads to observable effects.

Another example where interference leads to observable effects is the finite value of the total cross-section. As discussed in Section 2.2.2, for a realistic potential with its long-range tail, the purely classical total cross-section is infinite. This is because, no matter how large the impact parameter, there is always a classical deflection so that one can discern that a collision took place. Quantum mechanics prevents us from distinguishing a trajectory that is barely deflected from one that has not sampled the potential at all. Such interference occurs when the paths of the two trajectories differ by a wavelength λ. For the trajectory with an impact parameter b, that did deflect by a small angle $\delta\theta$, the extra path length is $b\,\delta\theta$. The largest deflection that can be resolved is therefore when $b\,\delta\theta \approx \lambda$. Since the deflection $\delta\theta$ is a function of b, this provides a cutoff on the maximal impact parameter for which one can say that a collision did take place. See Section 4.3.4 for the quantitative details. There the value of the total cross-section is derived using the Heisenberg uncertainty principle. We therefore emphasize that interference and quantum mechanical uncertainty are two ways of stating the same idea: classical alternatives cannot always be told apart. They can interfere and when they do there is a resulting uncertainty that is quantified by the uncertainty principle.

At this point you are faced with a wide choice. We know enough already to start discussing reactive collisions. To do so, go to Chapter 3. To know more about the approach motion, carry on to the end of this section. To proceed directly to a discussion of the scattering, go to Chapter 4.

*2.2.7   The center-of-mass system

A complete description of the paths of two point particles undergoing a collision in three-dimensional space requires a specification of 2 (particles) × 3 (position coordinates of a particle) = 6 scalar coordinates as a function of time. If, as is often the case, there are no external forces operating on the system, the force between the point particles depends only upon the relative separation R between them. For the purpose of our understanding it is therefore sufficient to describe the collision in terms of this single scalar coordinate $R(t)$. To be able to do so we[23] work in the so-called center-of-mass (c.m.) system, i.e., a coordinate system in which the c.m. of the colliding particles is at rest. This choice of reference frame already means that we need only to keep track of a single vector, namely, $\mathbf{R}$, the relative coordinate of the two particles. The vector $\mathbf{R}$ has three components so we still need to keep track of three scalar functions of time. In Chapter 3 we show systematically how to reduce the problem to keeping track of only one scalar function, namely the magnitude R of the distance between the particles. For many purposes, the tools we already have are sufficient: first we argued that there is a privileged plane, defined by $\mathbf{R}$ and the initial velocity, to which the collision is confined and then we used the conservation of energy to obtain an equation of motion, Eq. (2.30), that specifies the scalar velocity with which the particles approach (or recede) in terms of $R(t)$, the collision energy, and the impact parameter.

*2.2.7.1   Kinematics in the center-of-mass system

Kinematics means the description of the motion. Here we consider the motion of two point particles, 1 and 2, in the laboratory and in the c.m. system. In the laboratory the origin of the coordinate system is fixed and the location of the two particles is described by two vectors $\mathbf{R}_1$ and $\mathbf{R}_2$. The (vector) velocity is, as usual, the rate of change of the position vector, for example, $\mathbf{v}_1 = d\mathbf{R}_1/dt$ and similarly for $\mathbf{v}_2$. The relative position vector of the two molecules is $\mathbf{R} = \mathbf{R}_1 - \mathbf{R}_2$. Therefore, the *relative velocity* $\mathbf{v} \equiv d\mathbf{R}/dt$ is the difference

$$\mathbf{v} = \mathbf{v}_1 - \mathbf{v}_2 \qquad (2.38)$$

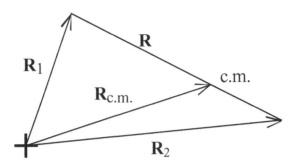

Figure 2.14 Position coordinates of the two colliding particles as well as the position of the center of mass $\mathbf{R}_{\text{c.m.}}$ and the relative distance $\mathbf{R}$.

The position of the center of mass is defined to be at $\mathbf{R}_{\text{c.m.}}$, $M\mathbf{R}_{\text{c.m.}} \equiv m_1\mathbf{R}_1 + m_2\mathbf{R}_2$, $M = (m_1 + m_2)$ where m_1 and m_2 are the masses of the two particles. If we want to, the definition of $\mathbf{R}_{\text{c.m.}}$ allows us to refer the position of each atom to the center of mass. For example, $\mathbf{R}_1 = \mathbf{R}_{\text{c.m.}} - (m_2/M)\mathbf{R}$ and similarly for $\mathbf{R}_2$, as shown in Figure 2.14. Note that the center of mass is nearer to the heavier of the two particles. This result, while trivial, serves to explain many interesting features in the dynamics of collisions.

The c.m. system is defined by the condition that the center of mass is at rest, $d\mathbf{R}_{\text{c.m.}}/dt = 0$. From the expression for the position vectors of the atoms we determine the velocities, $\mathbf{u}_i$, in the c.m. system in terms of the velocities in the laboratory, $\mathbf{v}_i = d\mathbf{R}_i/dt$ and the relative velocity $\mathbf{v} = d\mathbf{R}/dt$:

$$\mathbf{v}_1 = d(\mathbf{R}_{\text{c.m.}} - (m_2/M)\mathbf{R})/dt = (d\mathbf{R}_{\text{c.m.}}/dt) - (m_2/M)\mathbf{v} \equiv \mathbf{v}_{\text{c.m.}} + \mathbf{u}_1$$
$$\mathbf{v}_2 = (d\mathbf{R}_{\text{c.m.}}/dt) + (m_1/M)\,\mathbf{v} \equiv \mathbf{v}_{\text{c.m.}} + \mathbf{u}_2$$
$$\mathbf{v}_{\text{c.m.}} \equiv (d\mathbf{R}_{\text{c.m.}}/dt) = (m_1/M)\,\mathbf{v}_1 + (m_2/M)\,\mathbf{v}_2 \qquad (2.39)$$

Equation (2.39) specifies the velocity $\mathbf{u}_i$ of particle i with respect to the center of mass. The definition of the c.m. system, or the condition of conservation of linear momentum, tells us that $m_1\mathbf{u}_1 + m_2\mathbf{u}_2 = 0$. In other words, irrespective of the laboratory conditions, in the c.m. system the lighter particle is faster moving. Similarly note that, as one would expect, the relative velocity $\mathbf{v}$ of the two particles has the same value in the laboratory and in the center-of-mass system:

$$\mathbf{v} \equiv d\mathbf{R}/dt = \mathbf{v}_1 - \mathbf{v}_2 = \mathbf{u}_1 - \mathbf{u}_2 \qquad (2.40)$$

The kinetic energy of two particles in the laboratory system is

$$K_{\text{laboratory}} = (1/2)m_1\mathbf{v}_1^2 + (1/2)m_2\mathbf{v}_2^2 \qquad (2.41)$$

The kinetic energy in the c.m. system has the same form:

$$K = (1/2)m_1\mathbf{u}_1^2 + (1/2)m_2\mathbf{u}_2^2 \qquad \text{c.m. system} \qquad (2.42)$$

Finally, note that we can write the kinetic energy in the c.m. system as the kinetic energy of the relative motion

$$K = (1/2)\mu v^2 \qquad \text{c.m. system} \tag{2.43}$$

by introducing μ, the *reduced mass*, as the mass associated with the relative velocity $\mathbf{v}$:

$$\mu \equiv m_1 m_2/(m_1 + m_2) \tag{2.44}$$

We leave it as an exercise to show that the kinetic energy in the laboratory, $K_{\text{laboratory}}$, can be written as the kinetic energy for the motion of the center of mass plus the kinetic energy of the relative motion, Eq. (2.42) or (2.43).

*2.2.7.2   Kinematics in velocity space: the Newton diagram

The transformation of the velocities from the laboratory fixed system of coordinates to the center-of-mass system is often represented graphically in a manner that has come to be called a Newton diagram. This is an often used representation because it allows us to show not only the velocities before but also **after** the collision. Thereby one can read from the diagram the final velocities of the two particles in the laboratory fixed system of coordinates. An experiment measures velocities in the laboratory while we discuss the collision in the center-of-mass system. Therefore a Newton diagram, Figure 2.15, is a useful transcription.

The first stage of the construction of the diagram, Figure 2.15(a), specifies the center-of-mass system. It shows the velocity $\mathbf{v}_{\text{c.m.}}$ of the center of mass, the relative velocity $\mathbf{v}$, and also the velocities of the two particles, $\mathbf{u}_1 = (m_2/M)\mathbf{v}$, $\mathbf{u}_2 = (m_1/M)\mathbf{v}$. After the collision the relative velocity $\mathbf{v}'$ is in a different direction, specified by the deflection angle θ as shown. In the c.m. system the final velocities of the two particles are $\mathbf{u}_1' = -(m_2/M)\mathbf{v}'$, $\mathbf{u}_2' = +(m_1/M)\mathbf{v}'$, and these are also drawn. Finally, the diagram shows the construction of the laboratory scattering angle Θ, assuming that it is particle number 1 that is detected. What is required is the velocity $\mathbf{v}_1'$ of particle 1, after the collision, in the laboratory fixed system. Since the velocity of the center of mass is unchanged, what we need is the post-collision form of Eq. (2.39):

$$\begin{aligned}
\mathbf{v}_1' &= d\mathbf{R}_{\text{c.m.}}/dt - (m_2/M)\,\mathbf{v}' \equiv \mathbf{v}_{\text{c.m.}} + \mathbf{u}_1' \\
\mathbf{v}_2' &= d\mathbf{R}_{\text{c.m.}}/dt + (m_1/M)\,\mathbf{v}' \equiv \mathbf{v}_{\text{c.m.}} + \mathbf{u}_2'
\end{aligned} \tag{2.39'}$$

The Newton diagram shown in Figure 2.15(b) is for the simplest scattering problem, where the particles are structureless. Therefore the scattering is elastic – the final velocity $\mathbf{v}'$ can be in a different direction from the initial velocity but its magnitude is unchanged. There is a circle about the c.m., such that the final velocities $\mathbf{u}_i'$ are confined to lie on it. Our construction is valid for structureless particles. When the colliding particles have internal structure they can gain (or lose)

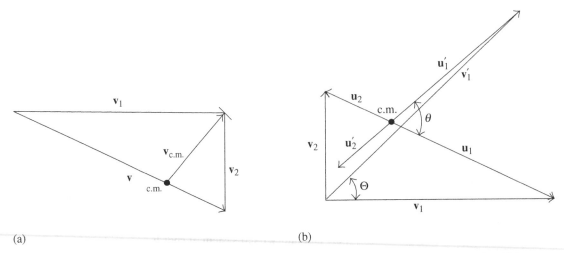

(a)   (b)

Figure 2.15 Two particles moving initially at right angles to one another. This corresponds to the most common experimental arrangement when two beams of molecules are arranged to intersect in a crossed beam experiment. One can make a similar diagram for other angles. (a) The initial construction, defining the center-of-mass system; $\mathbf{v}$ is the relative velocity vector, $\mathbf{v} = \mathbf{v}_1 - \mathbf{v}_2$, see Eq. (2.38). The position of the center of mass is at the dot. This dot partitions the relative velocity into two contributions, $\mathbf{v} = \mathbf{u}_1 - \mathbf{u}_2$, the two velocities in the c.m. system. (b) The Newton diagram. This diagram shows also the final velocities in a collision that resulted in scattering in the c.m. system, at the angle θ. This angle defines the direction of the final relative velocity with respect to the initial direction. Say that the detector is set for particle number 1. Then, given the final velocity $\mathbf{u}_1'$ of particle 1 in the c.m. system we construct the final velocity of particle 1 in the laboratory system as $\mathbf{v}_1' = \mathbf{v}_{c.m.} + \mathbf{u}_1'$. The detector is located in the laboratory and reports a scattering angle Θ as the laboratory angle of scattering (Θ is the angle between the initial and final laboratory velocity of particle 1). It is necessary to construct the $\theta \Leftrightarrow \Theta$ correspondence in order to convert scattering as measured in the laboratory to intensity in the center-of-mass system. See also the text below.

translational energy at the expense of their internal excitation. Equation (2.39′) is still valid but the final velocity $\mathbf{v}'$ can change in both magnitude and direction. There would then be different circles where the radius of the circle is determined, through conservation of energy, by the final internal state and the location on a given circle determines the scattering angle for that state. A second generalization is required when a chemical reaction is possible. Then we need to put primes also on the two masses m_i in Eq. (2.39′) because the masses of the emerging particles are not the same.

Imaging methods (Heck and Chandler, 1995; Houston, 1996; Suits and Continetti, 2001; Whitaker, 2003) are making it possible to literally view the Newton diagram, see Section 7.1.3.1. In this connection it should be noted that Eq. (2.39′) is a statement of conservation of momentum after the event. So the final products need not come from a full collision. They can also result from a "half collision", that is, from the photodissociation of a parent molecule into two fragments, where now $\mathbf{v}_{c.m.}$ is the initial velocity of the parent molecule in the laboratory.

Problems

A. The density and the mean free path. (a) Show, from the ideal gas laws, that the pressure P of a dilute gas in thermal equilibrium is related to the number density n by $P = nm\langle v^2 \rangle/3$ where m is the mass of the molecules and $\langle v^2 \rangle$ is the average squared velocity. (b) Under standard conditions ($0°C$, 1 atm) the density of a pure gas is $1.429 \cdot 10^{-3}$ g cm^{-3}. Suggest the possible chemical identity of the gas and compute the root mean squared velocity at room temperature. (c) Assuming a collision cross-section of 40 Å^2 (an interaction radius of 1.8 Å), compute the mean free path. (d) Discuss whether the computed mean free path is large or small by providing a suitable distance for comparison. With this result check if our assumption that the gas is dilute enough to be nearly ideal is reasonable. (e) Plot the mean free path against pressure. At what low pressure will a molecule be likely to survive one second before colliding with another molecule? (We gave you a mixed bag of units. That is how real life is.)

B. More on density. The vapor pressure of solid Ba at equilibrium at a temperature of 585 K is 10^{-8} torr and that of solid K at 300 K is $2 \cdot 10^{-8}$ torr. (a) Compute the number densities of the two vapors. (b) The measured density of Ba agrees with the computed result but it is about twice the computed value for K. What is a possible explanation?

C. Derive Eq. (2.1) and hence compute the flux of O_2 molecules incident on the wall of a room under ordinary conditions. Is the flux of N_2 molecules the same and if not, how different is it? In Chapter 12 we will conclude that this flux is rather high. Can you suggest what is the number against which we will compare the flux to decide when it is high?

D. Measuring the mean free path. Using the details provided with Figure 2.2 compute the number density of the scattering gas in the range of pressures shown and hence compute the collision cross-section for both Ar and CH_2F_2. Both are large but the cross-section for collisions with CH_2F_2 is about five times larger than that for Ar. Why? If we would instead use $CHF=CHF$ as a target gas, do we expect a difference in the cross-section for the cis and trans isomers? They are, after all, similar in size. The experimental answer is a cross-section of about 1400 Å^2 for trans $CHF=CHF$, a shade smaller than that for CH_2F_2, and 2800 Å^2 for cis. Conclude that electrostatic forces are long range and that the collision cross-section is a joint property of both partners. Non-polar collision partners have significantly smaller cross-sections.

E. The rate constant for bimolecular collisions. In the text we defined the number, Z, of collisions per unit volume and unit time as $Z = n_A \omega = n_A n_B v \sigma$, where n_A and n_B are the number densities of the two gases. By analogy to chemical kinetics, where the corresponding number of *reactive* collisions is written as $kn_A n_B$ and k is called the reaction rate constant, we can call $k = v\sigma$ the collision rate constant. (a) If we define k as the loss of flux in the experiment of Figure 2.1, Eq. (2.10), show that this recovers the result $k = v\sigma$. (b) The collision cross-section is energy dependent. The collision rate constant for thermal partners is therefore given by $k = \langle v\sigma \rangle$ where the averaging is over a thermal distribution of the relative velocity v.* It is however customary to define an effective collision cross-section by $k = \langle v\sigma \rangle \equiv \langle v \rangle \sigma_{eff} \cdot \langle v \rangle = (8k_B T/\pi\mu)^{1/2} \cdot k_B$ where $k_B = R/N_A$ is Boltzmann's constant, namely the gas constant per molecule and not per mole. Show that in practical units

$$k \text{ (cm}^3 \text{ mol}^{-1} \text{ s}^{-1}) \cong 8.76 \cdot 10^{11} (T/\mu \text{ (amu)})^{1/2} \sigma_{eff}(\text{Å}^2)$$
$$\omega \text{ (s}^{-1}) \cong P \text{ (atm)} (298/T) 4.11 \cdot 10^{-8} k \text{ (cm}^3 \text{ mol}^{-1} \text{ s}^{-1})$$

(c) Go to a source of chemical kinetics rate data and conclude that our k above is significantly larger than a typical bimolecular reaction rate constant. Why? Some bimolecular reactions are known to be fast. Can their rate constants be larger than the collision rate constant? (No, but why not?)

F. The Lennard-Jones (12,6) functional form, Eq. (2.24), is often used to approximate interatomic potentials. (a) What is the equilibrium distance for the potential in units of σ? (b) Show that near its minimum the potential can be expanded in a harmonic form $V(R) \cong -\varepsilon + \frac{1}{2}k(R - R_m)^2$. (c) Express the force constant k in terms of the parameters of the potential and show that in reduced units it has the value 72. (d) Compute the vibrational frequency of the harmonic motion at the bottom of the well as a function of the mass and express it in reduced units.

G. The Morse functional form, $V(R) = \varepsilon [1 - \exp(-(R - R_m)/2\rho)]^2 - \varepsilon$, is also often used. This has the short-range exponential form as in Eq. (2.13). (a) Show that the Morse potential satisfies the scaling law (2.22). (b) The Morse potential is more flexible because the range parameter ρ can be fitted independently of the position, R_m, of the minimum. In particular verify that this allows fitting the harmonic frequency, the anharmonicity, and the rigid-rotor rotational constant as three independent parameters. In this fashion, data from spectroscopy can be used directly to infer the potential (Herzberg, 1950).

H. The partitioning of the kinetic energy during a collision, Eq. (2.26). Plot as a function of time throughout the approach and receding motion, the kinetic energy along the line of centers of the two partners and the centrifugal kinetic energy. Assume that no forces are acting and make your plot simpler by choosing

* Prove that if A and B both have a thermal distribution of velocities at the same temperature T then their relative velocity also has a thermal distribution at the same temperature T. You will need to equate the expression for the kinetic energy in the laboratory to that in the center of mass system. See Section 2.2.7. Next prove that if A and B both have a thermal distribution of velocities but at two different temperatures, then their relative velocity also has a thermal distribution and determine the temperature T for the distribution of the relative velocity.

the origin of the time axis to be zero at $R = b$. It is possible to express time in a reduced (dimensionless) form. Suggest the physical meaning of the variable with dimension "time" that you need to use as a scale. For further discussion of this problem see Section 4.1.2.

I. Use the conservation of energy condition, Eq. (2.27), to show that for a collision of hard spheres the particles do not feel any force if the impact parameter is larger than the hard-sphere diameter, $b > d$. Hence derive Eq. (2.36).

J. The Heisenberg uncertainty principle for position and momentum in the same direction is $\delta x \cdot \delta p > h$ (recall that these uncertainties are inherent. We cannot reduce them by a better designed experiment). We cannot accept an uncertainty in position larger than b itself. Hence derive a bound for the uncertainty in the de Broglie wavelength of the relative motion.

K. Orbiting. At low collision energies there will be more than one solution for the distance R_0 of closest approach as determined from Eq. (2.31). The easy way to see it is to solve Eq. (2.31) by graphical means, i.e., to plot $V_{\text{eff}}(R) = V(R) + E_T b^2/R^2$ vs. R and see where $V_{\text{eff}}(R) = E_T$. The largest root is the turning point of the collision. The other two roots are the inner and outer turning points for a state that is bound in the inner hollow of the effective potential. Such a bound A–B molecule can dissociate by tunneling. As the collision energy is increased the turning point eventually coincides with the very top of the barrier in the effective potential. Above this energy there is only one turning point. For a Lennard-Jones (6–12) potential, determine this energy. The collision at this energy will just manage to crawl over the top of the barrier. It will do so very, very slowly and the collision partners will have lots of time to rotate about one another. Hence "orbiting."

L. Collision cross-section for hard spheres. The hard-sphere potential has no attractive part. We can retain much of the simplicity of the model but modify the potential, as shown in Problem D of Chapter 5, by adding to the repulsive part a square well of finite range. Evaluate the collision cross-section for this potential. As a first step, plot the effective potential.

Notes

1 Collimation of a molecular beam is a technical subject in its own right, discussed in detail in Scoles (1988) and Pauly (2000). Well-collimated beams with a flux of over 10^{15} particles s^{-1} cm^{-2} are available.

2 Textbooks sometimes write this result as $\omega = k(T)n_B = \langle v \rangle n_B \sigma = \langle v \rangle/\lambda$ and then conclude that $\lambda = 1/n_B \sigma = kT/\sqrt{2}P\sigma$. The factor of $\sqrt{2}$ is not needed for scattering of beam molecules because the velocity v of the beam is typically larger than the mean thermal velocity. The relative velocity in an A–B collision is therefore also about **v**. *If* we are interested in collisions in the bulk gas phase and if both A and B have a thermal velocity distribution, recall the Maxwell–Boltzmann equation, at the same temperature **and** A and B have the same mass, then the mean thermal relative velocity is $\sqrt{2}$ higher than the mean thermal velocity of A or B alone.

3 With the molecular beam technique it is possible to determine the velocity dependence of the collision cross-section. The distribution of the velocities of the beam molecules can be measured first without and then with a scattering gas, i.e., in the absence and presence of attenuation by scattering. One way to accomplish the velocity measurement is by the *time of flight* method. Here the beam is pulsed by a mechanical shutter near the scattering cell (the switching on of the beam also triggers the time-base for detection). The beam intensity at the detector at a time t later is due to molecules of speed $v = L/t$, where L is the path length from the beam shutter to detector. Comparing the transmitted intensity (for a given speed v) in the absence and presence of the target gas in the cell, one can determine the dependence of the attenuation and hence of the collision cross-section upon velocity.

4 Typically $0.1 \leq \rho \leq 0.4$ Å. The smaller is ρ the steeper is the repulsion. For R small but of the order of ρ, the repulsion is due mainly to the increased kinetic energy of the electrons when they are forced to be confined to a small volume. (A quantum mechanical phenomenon originating from the Pauli exclusion principle.) As R is decreased further the coulomb repulsion between the nuclei (otherwise "screened" by the electrons) begins to contribute significantly. The repulsion cannot be strictly exponential because an exponential has a finite value (i.e., A) at $R = 0$. But an exponential increase, see dotted curve Figure 2.4, realistically reflects the actual R-dependence for $R > \rho$.

5 Intermolecular forces determine directly the dynamics and, indirectly, the properties of bulk matter. Books, such as Maitland *et al.* (1987), are devoted to this important subject. The articles in Hirschfelder (1967) are still a valuable source. The more recent input from spectroscopy of weakly bound dimers is reviewed by Hutson (1990) and Nesbitt (1994).

6 All intermolecular forces are ultimately electrostatic in nature and are due to the coulombic interactions between the charges on the electrons and on the nuclei. To compute the potential energy $V(R)$ we need to clamp the partners at a distance R apart and average over the motion of the electrons. What we mean by a long-range physical force is the result of this averaging when R is so large that electrons belong to one species or the other. To compute the chemical force we need to be at a shorter distance, where the wave functions of the two species begin to significantly overlap. It is then necessary to properly antisymmetrize the electronic wave function of the whole system so as to allow exchange of electrons. This is why the chemical contribution is known as the exchange force.

7 A neutral molecule will, in general, have a charge distribution that is not spherical. In addition to a dipole moment it can have higher multipole moments, quadrupole, etc. Each such higher moment will generate its own electric field. These will decrease faster with distance and are therefore neglected in the first approximation.

8 The transition dipole moment characterizes the strength of the electronic transition (its square is, up to constants, the rate constant for the transition known as the Einstein B coefficient). In the semiclassical limit we can think of the transition dipole moment as a dipole oscillating at the frequency of the transition, rather like the antenna of a radio transmitter. The dispersion force is due to the fluctuating field around the atom (or molecule). While the dipole averages out to zero, its interaction energy with another molecule does not.

9 To estimate the dispersion constant, books relate the transition dipole to the polarizability, where, in atomic units, $\alpha = \mu^2/\Delta E$ and ΔE is the energy of the transition. So $C_{\text{disp}} \approx \alpha_1\alpha_2\Delta E$. To estimate the transition energy books often use the ionization potential of the molecule with $\Delta E \approx I_1 I_2/(I_1 + I_2)$. The polarizability is, as an order of magnitude, the

volume of the molecule, say 10 Å³. As a typical transition energy we take 3 eV ≈ $5 \cdot 10^{-12}$ erg, leading to $C_{disp} = 5 \cdot 10^{-58}$ erg cm⁶. Strictly speaking one needs to sum over all transitions of either atom, $C_{disp} = \sum_{i,j} |\mu_{1,i}|^2 |\mu_{2,j}|^2 / \Delta E_{ij}$, where ΔE_{ij} is the sum of the i and j transition energies.

10 The rigorous interpretation is in terms of the difference between the actual density of molecules (around some reference molecule) and the density corresponding to a uniform distribution.

11 That the potential must have an attractive part, where $V(R) < 0$, follows from Eq. (2.21) because only if $(\exp(-V(R)/kT) - 1) > 0$ can the second virial coefficient be negative.

12 At a too low T the gas will condense, and measuring above a few hundred degrees is problematic because of properties of materials.

13 This particular functional form is known as the 6–12 potential because other powers are also possible. We wrote the 6–12 potential using two common alternatives for the distance scale factor. One is the equilibrium distance R_m and the other is the point, σ, where there is the onset of repulsion, namely where the function $V(R)$ changes sign: $R_m = 2^{1/6}\sigma$. As seen in Figure 2.4, an R^{-n}, $n > 8$, term can mimic the exponential rise of the short-range repulsion.

14 There are fewer such studies because excited electronic states can interact according to more than one potential energy curve. In other words, several different electronic states of the diatom dissociate to the same electronic state of the separated atoms. This problem occurs also for open-shell, ground-state atoms and makes the analysis somewhat more difficult.

15 The lighter the mass, the more is quantum dynamics willing to allow sampling of regions with negative kinetic energies. This is known as *tunneling*. Electrons tunnel. The far heavier nuclei do tunnel but only over rather short distances. So it is only under special circumstances that tunneling can make a qualitative difference to the dynamics. The first instance is in Section 4.3.5.

16 This so-called *orbiting* happens when there can be more than one solution for the turning point of the trajectory and is taken up in Chapter 4. This is of importance at low collision energies, typically lower than the well depth. In classical mechanics the signature of orbiting is the deflection due to the top of the barrier in the effective potential, the point where the centrifugal repulsion is just balanced by the attraction due to the potential $V(R)$.

17 This is familiar to a driver entering a turn in the road. You know that the instinctive reaction is to brake before taking the turn. If you are a racing driver you also know that you will try to cut the corner so that your velocity is as great as possible along **R** (i.e., reduce the impact parameter b). Traffic engineers take a different approach. They bank the road so that the resultant of the force of gravity and the centrifugal force acts in the direction perpendicular to the road (Figure 2.16).

18 Unless we exercise a very special control, all initial orientations of **b** are equally probable. So there can be many equivalent trajectories that differ only in the orientation of the plane of the collision. Different collision planes correspond to different azimuthal angles in the plane of the target.

19 If the long-range attraction between the molecules decreases more slowly than R^{-2} (and a coulomb potential between two oppositely charged ions is the more common example of this rare situation), then the divergence of the cross-section is real. The same divergence is

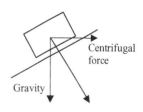

Figure 2.16

also known in statistical thermodynamics, where the partition function for a purely coulombic attraction, e.g., a hydrogen atom, is divergent.

20 See Wittig *et al.* (1988), Shin *et al.* (1991).

21 A situation where molecules are possibly even more tightly aligned is when they are adsorbed on a surface. When the precursor and the coadsorbed reactant have a definite location with respect to one another, both the orientation and the impact parameter are restricted. The precursor is photodissociated to initiate the reaction. See Figure 12.9.

22 A particularly good discussion of interference is in Chapters 1 and 2 of Volume III of the Feynman Lectures on Physics (Feynman *et al.*, 1966).

23 Computers easily keep track of the position of many atoms where each coordinate evolves in time. Therefore, for numerical purposes one often works in a laboratory fixed Cartesian system of coordinates where the description of a collision requires three more space coordinates, those of the center-of-mass. In particular, this simplifies the expressions for the forces acting on the internal motions. The conservation of, for example, the center-of-mass energy can then be used as a check on the numerical stability of the computation.

Chapter 3
Introduction to reactive molecular collisions

We examine in this chapter how the motion of the reactants as they approach each other governs chemical reactivity. This allows us to use a two-body point of view where the internal structure of the colliding species is not explicitly recognized. All that we can do therefore is lead the reactants up to a reaction. But we will not be able to describe the chemical rearrangement itself nor to address such questions as that of energy disposal in the products. Chapter 5 takes up these themes. On the other hand, without the approach of the reactants there cannot be a bimolecular reaction. The tools already at our disposal are sufficient to discuss this approach motion. As expected, the striking distance that we have called the impact parameter will be a key player. We do all of this in Section 3.2. What we will obtain is information about the dependence of the reaction cross-section on the collision energy.

In chemical kinetics one characterizes the role of energy in chemical reactivity by the temperature dependence of the reaction rate constant. In Section 3.1 we review the input from chemical kinetics – the Arrhenius representation of the rate constant – then go from the rate constant to the reaction cross-section. Next we go in the opposite direction, from the microscopic reaction cross-section to the macroscopic rate constant. What we obtain thereby is the Tolman interpretation of the activation energy as the (mean) excess energy of those collisions that lead to reaction.

3.1   The rate and cross-section of chemical reactions

This section is a review of the macroscopic notions we are familiar with from chemical kinetics.[1] Our final purpose is to build rate constants from the bottom up and therefore to describe the rate of chemical reactions in systems that are not in thermal equilibrium. Thermally equilibrated reactants are more typical of the laboratory than of the real world and, even in the laboratory, it takes care and attention to insure that the reactants are indeed thermally equilibrated. Outside of the laboratory, whether in the internal combustion engine (which fires many thousands of times per minute), in the atmosphere, or in outer regions of space, this is not the case.

The second theme that we begin to explore is the role of energy in promoting chemical reactions. Macroscopically this is characterized by the temperature dependence of the reaction rate as summarized by the activation energy. We can already guess that the underlying root is the collision energy dependence of the reaction cross-section. The new feature is that certain reactions have cross-sections that decrease with increasing energy. In fact, this is rather typical for reactions with large cross-sections.[2]

3.1.1   The thermal reaction rate constant

The rate of an elementary gas-phase bimolecular reaction, say

$$Cl + CH_4 \rightarrow HCl + CH_3$$
$$O + CS \rightarrow S + CO$$
$$F + HCl \rightarrow Cl + HF$$

is characterized by the thermal reaction rate constant $k(T)$, which is a function of the temperature only. This rate constant is "constant", meaning that it is not a function of time, and it is a measure of the rate of depletion of the reactants (that are kept in a thermal bath) or the rate of appearance of the products. Because the reaction is bimolecular, it is a second-order rate constant, e.g.

$$-\frac{d[F]}{dt} = \frac{d[Cl]}{dt} = k(T)[F][HCl] \qquad (3.1)$$

For our purpose it is essential to emphasize that the thermal reaction rate is defined only when the experiment does maintain a thermal equilibrium for the reactants. If necessary, the reaction needs to be slowed down, say by the addition of a buffer gas, so that non-reactive collisions rapidly restore the reactants to thermal equilibrium. If this is not possible, Appendix 3.A introduces the reaction rate constant under more general conditions. Under non-equilibrium conditions the rate constant defined through Eq. (3.1) may however depend on other variables such as the pressure and even on time.

The experimental temperature dependence of the thermal reaction rate constant is often represented in an Arrhenius form

$$k(T) = A \exp(-E_a/k_B T) \qquad (3.2)$$

We write the Boltzmann constant k with a subscript B to avoid confusion with the reaction rate constant. If the activation energy is specified per mole, the exponent needs to be written as E_a/RT where $R = N_A k_B$ is the gas constant and N_A is Avogadro's number.

If measurements are carried out over a wide range in $1/T$ there is no reason to expect that either the Arrhenius "A factor" or the activation energy E_A is independent of temperature. Therefore a strict definition of the activation energy is as the local (meaning possibly T-dependent) slope of the Arrhenius plot of $\ln k(T)$ vs. $1/T$

$$E_a \equiv -k_B \frac{\mathrm{d}\ln k(T)}{\mathrm{d}\frac{1}{T}} = k_B T^2 \frac{\mathrm{d}\ln k(T)}{\mathrm{d}T} \tag{3.3}$$

This is the definition that we shall use.

The chemical change as observed in a macroscopic experiment is the result of many molecular collisions. In Section 3.1.2 we define the reaction cross-section from the macroscopic observable rate of reactive collisions. This is sufficient to discuss the dependence of the cross-section on the collision energy. Then we discuss the temperature dependence of the reaction rate as arising from this energy dependence. To understand the origin of the energy needs of chemical reactions, we provide, in Section 3.2, a microscopic interpretation of the reaction cross-section. Finally, a caveat. In this chapter we are not taking account of the internal structure of the reactants. This key topic has to wait until Chapter 5. The only concession is that an appendix provides an extension of chemical kinetics to the important special case where we do resolve internal states.

3.1.2 The reaction cross-section – a macroscopic view

We define the reaction cross-section, σ_R, in a way suggested by the definition of the total collision cross-section (Section 2.1.5). For molecules colliding with a well-defined relative velocity v, the reaction cross-section is defined such that the chemical reaction rate constant $k(v)$ is given by

$$k(v) = v\sigma_R \tag{3.4}$$

Here, $k(v)$ is the reaction rate constant for a specific, well-defined, relative velocity. It is not the same quantity as the thermal rate constant. We can imagine measuring it by passing a beam of reactant A molecules through a scattering cell as in Figure 2.1. Then, the loss of flux due to *reactive* collisions is given by*

$$-(\mathrm{d}I/\mathrm{d}x)_R = k(v)n_A n_B = I n_B \sigma_R \tag{3.5}$$

Here, as in Eqs. (2.1)–(2.4), $I(x)$ is the flux of beam molecules A at position x and n_B is the number density of the target B molecules. However, not all collisions need to lead to reaction. While we can conclude that $\sigma_R \leq \sigma$, it is not enough to measure the attenuation of the parent beam, because an appreciable loss of intensity can result from *non*-reactive scattering. We must specifically determine the loss of flux due to reactive collisions. Experimentally, this is easy to do for reactions producing ions: the ions are simply collected by the application of an electric field. One interesting class of reactions of this sort is that of the endothermic

---

* Later on we shall generalize Eq. (3.5), for example looking at those reactive collisions where the products scatter in a particular direction.

collisional ionization type; here two neutral molecules collide to form ions, e.g.

$$K + Br_2 \rightarrow K^+ + Br_2^-$$

without atom exchange, or

$$N_2 + CO \rightarrow NO^+ + CN^-$$

Another type of reaction involving both reactant and product ions is the so-called ion–molecule class, e.g.

$$H_2^+ + He \rightarrow HeH^+ + H$$

Of course it is not enough just to collect the ionic or neutral products; it is also necessary to identify their chemical nature. This is often achieved by mass spectrometric methods. Such identification is essential when several different reaction paths are possible, e.g.

$$K + Br_2 \rightarrow \begin{cases} KBr + Br \\ K^+ + Br_2^- \\ K^+ + Br^- + Br \end{cases}$$

and one needs to determine the *branching ratio* or the relative contribution of each process to the total reaction cross-section.

3.1.2.1 The energy threshold of reaction

We embark on our study of the role of energy in chemical dynamics by examining the dependence of the reaction cross-section on the translational energy* of the colliding partners. Our first consideration is the operational concept of the *threshold energy*, E_0, as the minimum energy needed for the reaction to take place. The reaction cross-section vanishes for energies below this threshold value. For endothermic reactions, the conservation of energy implies that there is a minimal energy for reaction to take place. For example, for the ion–molecule reaction $H_2^+ + He \rightarrow HeH^+ + H$, the minimal energy expected on thermochemical grounds is the difference between the binding energies of the reactants and products: $E_0 = D_0(H_2^+) - D_0(HeH^+) = 2.65 - 1.84 = 0.81 \, eV$. The experimental results, shown in Figure 3.1, are that this minimal energy is indeed the threshold. It is further seen that the reaction cross-section increases rapidly as the translational energy increases above E_0. This behavior is typical for reactions with an energy threshold.

Reactions can have a finite energy threshold that is higher than the thermochemical threshold, meaning that σ_R is effectively zero below some threshold energy even though the reaction is thermodynamically "allowed." One then speaks of an activation barrier that needs to be surmounted for reaction to

---

* As before, only the *relative* translational energy is of importance. For a binary collision, the motion of the center of mass cannot affect the outcome of the collision.

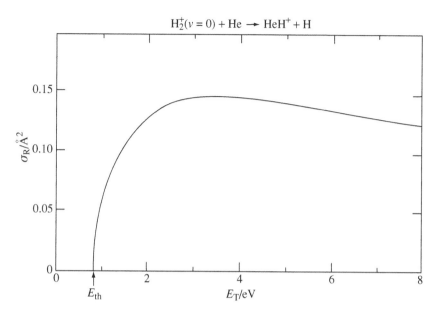

$$H_2^+(v=0) + He \rightarrow HeH^+ + H$$

Figure 3.1 Translational energy dependence of the reaction cross-section, $\sigma_R(E_T)$ for the $H_2^+(v=0) + He \rightarrow HeH^+ + H$ reaction [adapted from T. Turner, O. Dutuit, and Y. T. Lee, *J. Chem Phys.* **81**, 3475 (1984)]. For this ion–molecule reaction the observed threshold energy is equal to the minimal possible value, the endoergicity of the reaction. Exoergic ion–molecule reactions often have no threshold.[3] By exciting the vibrations of the H_2^+ reactant the cross-section for the reaction above can be considerably enhanced.

take place. Clear examples are thermoneutral exchange reactions with an energy threshold. The (actually, a shade endoergic, Problem A) reaction

$$H + D_2(v=0) \rightarrow D + HD$$

has a threshold energy of about $30 \, kJ \, mol^{-1}$ ($\approx 0.3 \, eV$). Thus, while *all* endoergic reactions necessarily have an energy threshold, many exoergic reactions also have an effective energy threshold. The reaction energy threshold E_0 can be no lower than the minimum energy ΔE_0 thermochemically required for the reaction, but may be higher or even significantly higher.

An important class of reactions, of particular interest in atmospheric chemistry (aeronomy) and, in general, for interstellar chemistry, is that of the exoergic ion–molecule reactions, e.g.

$$N^+ + O_2 \rightarrow \begin{cases} NO^+ + O \\ N + O_2^+ \end{cases}$$

Such reactions often show no threshold energy[3] and the reaction cross-sections are found to be a decreasing function of the translational energy, roughly as

$$\sigma_R(E_T) = A E_T^{-1/2} \tag{3.6}$$

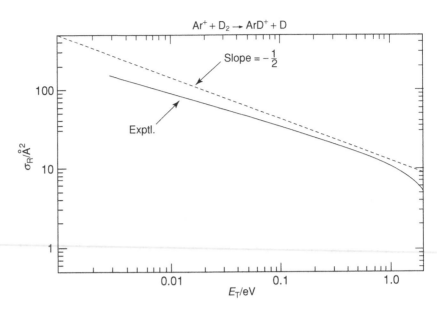

Figure 3.2 Log–log plot of the translational energy dependence of the reaction cross-section when there is no apparent threshold energy. The solid curve is an experimental result for the system $Ar^+ + D_2 \rightarrow ArD^+ + D$; the dashed curve has a slope of $-1/2$, cf. Eq. (3.28), and the potential shown in Figure 3.7 [adapted from K. M. Ervin and P. B. Armentrout, *J. Chem. Phys.* **83**, 166 (1985)]. For the drop in the reaction cross-section at higher energies see Problem J.

as shown in Figure 3.2. It is the preference for low collision energies that makes ion–molecule reactions so important for the synthesis of molecules in the interstellar medium.

3.1.2.2 *Translational energy requirements of chemical reactions*
On the basis of the translational energy requirements of chemical reactions we can thus make the following rough correlation.

(1) Reactions that have an energy threshold (this necessarily includes all endoergic reactions) have a reaction cross-section that is an increasing function of the translational energy in the post-threshold region. This is the case chemists are more familiar with. It gives rise to a positive Arrhenius activation energy, as discussed next.

(2) Reactions that proceed without any apparent energy threshold (and this includes some, but not all, exoergic reactions) often have a reaction cross-section that is a decreasing function of the translational energy. However, as the translational energy is increased, other, previously endoergic, reaction paths become allowed. These have a threshold and their cross-section will increase with energy, at the expense of the previously allowed reaction.

To rationalize these correlations we turn in Section 3.2 to the microscopic interpretation of the reaction cross-section and the concept of the reaction probability. Before that we reiterate that the energy requirements of chemical reactions appear, in the macro world, as the temperature dependence of the reaction rate constant.

3.1.2.3 The temperature dependence of the reaction rate constant

The translational energy dependence of the reaction cross-section translates into the temperature dependence of the reaction rate constant. The procedure is clear: take $k(v) = v\sigma_R$, Eq. (3.4), and average it over a thermal distribution of velocities, $k(T) = \langle v\sigma_R(v) \rangle$. We wrote $\sigma_R(v)$ as a reminder that the reaction cross-section can depend on the collision velocity.

Sometimes the thermal averaging $\langle v\sigma_R(v) \rangle$ required to compute $k(T)$ is easy to implement. For example, for ion–molecule reactions for which, cf. Eq. (3.6), $v\sigma_R \propto$ constant, $k(T)$ is **independent** of temperature.* At other times, the averaging needs to be carried out. Explicitly, it means evaluating an integral over a Maxwell–Boltzmann velocity distribution $f(v)$ of the (collision-energy-dependent) reaction cross-section

$$k(T) = \int v\sigma_R f(v)\,dv = (\mu/2\pi k_B T)^{3/2} \int v\sigma_R \exp(-\mu v^2/2k_B T)4\pi v^2 dv \quad (3.7)$$

It is often convenient to change the variable of integration to the collision energy $E_T = \mu v^2/2$. Then

$$k(T) = (8k_B T/\pi\mu)^{1/2} \int (E_T/k_B T)\sigma_R \exp(-E_T/k_B T)\,d(E_T/k_B T) \quad (3.8)$$

where the factor in front of the integral is the thermal (relative) velocity and the integral itself has the dimension of a cross-section.**

3.1.2.4 The Tolman interpretation of the activation energy: the reactive reactants

The temperature dependence of the reaction rate is determined by the integrals (3.7) or (3.8). The integrand is a product of a rapidly decreasing function, the Boltzmann factor, $\exp(-E_T/k_B T)$, and the energy-dependent cross-section. For reactions with an energy threshold, in the post-threshold regime the cross-section will be a rapidly increasing function of the energy, Figure 3.3.

It is then very clear that increasing the temperature will allow the integrand, and hence the reaction rate, to have a higher value. Problem D uses a saddle point (approximate) integration to show that for an integrand that is a product of an increasing and a decreasing function, the integral can be approximated by an Arrhenius form, Eq. (3.2). Here we proceed to show, following Tolman, that there is always an exact interpretation of the activation energy, Eq. (3.3), as the excess energy of the collisions that do lead to reaction:

$$E_a \equiv \langle E_T \rangle_{\text{reactive reactants}} - \langle E_T \rangle_{\text{all reactants}} \quad (3.9)$$

---

* How thoughtful nature was to make ion–molecule reactions so that there can be chemistry in the cold regions of space where many molecules are ionized by cosmic and UV radiation.

** Note however that the integral is not a thermally averaged cross-section. Problem C shows that the integral in (3.8) is an average of the cross-section over the thermal *flux density* of molecules.

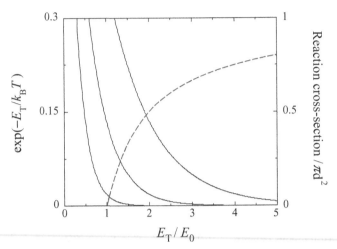

Figure 3.3 The origin of the Arrhenius temperature dependence as reflecting the rise of the reaction cross-section past the energy threshold (Menzinger and Wolfgang, 1969). The two factors in the integrand in Eq. (3.8) are plotted against energy. The Boltzmann factor, $\exp(-E_T/k_B T)$, is shown for an intermediate, high, and quite high temperature ($k_B T = E_0/4$, $E_0/2$, and E_0, respectively). The energy dependence of the reaction cross-section above the threshold energy, shown as a dashed line, is $\sigma_R \propto (1 - E_0/E_T)$, a result derived in Section 3.2.7 below. As a function of the collision energy, the exponential decline of the Boltzmann factor readily overwhelms the increase of the reaction cross-section. Note how at a higher temperature the integrand is exponentially larger because of the shift in the Boltzmann curve. The figure also illustrates why thermal measurements can provide only a limited leverage for determining the energy dependence of the reaction cross-section. It is only the tail-end of the Boltzmann distribution and the threshold behavior of the cross-section that contribute to the thermal reaction rate.

Here *reactive reactants* means those reactants that do react and the average is over a thermal distribution. The proof is directly from the definition of the activation energy. The proof begins by examining Eq. (3.7) and concluding that the entire T-dependence of $k(T)$ is due to the Boltzmann factor. After some mathematical manipulations* the result is

$$
\begin{aligned}
\langle E_T \rangle_{\text{reactive reactants}} &\equiv \langle E_T \rangle_{\text{thermal reactants weighted by reaction rate}} \\
&= \int E_T \frac{(E_T \sigma_R) \exp(-E_T/k_B T)\, dE_T}{\int (E_T \sigma_R) \exp(-E_T/k_B T)\, dE_T} \\
&\equiv \int E_T P_{\text{rr}}(E_T)\, dE_T
\end{aligned}
\tag{3.10}
$$

---

* To differentiate $k(T)$ with respect to T, what is needed is to differentiate the thermal Boltzmann factor. Care is needed at one place. The Boltzmann factor carries with it a normalization (the fore-factor in the last term in (3.7)) and this normalization is T-dependent. So it too needs to be differentiated and this is the origin of the second factor in Eq. (3.9).

Here $P_{\mathrm{rr}}(E_{\mathrm{T}})$, defined by Eq. (3.10), is the (normalized) distribution of the reactive reactants. This is the distribution of energy for those pairs of reactants that will react. The average energy of all colliding pairs is not Eq. (3.10) but Eq. (3.11)

$$\langle E_{\mathrm{T}}\rangle_{\text{all reactants}} \equiv \langle E_{\mathrm{T}}\rangle_{\text{thermal reactants}}$$
$$= \int E_{\mathrm{T}} \frac{E_{\mathrm{T}}^{1/2}\exp(-E_{\mathrm{T}}/k_{\mathrm{B}}T)\,\mathrm{d}E_{\mathrm{T}}}{\int E_{\mathrm{T}}^{1/2}\exp(-E_{\mathrm{T}}/k_{\mathrm{B}}T)\,\mathrm{d}E_{\mathrm{T}}} \tag{3.11}$$

When we recognize that, strictly speaking, the activation energy is really a thermal average many puzzles go away. The activation energy is not, for example, the minimal energy for reaction. It is closely related to it because the reactive reactants need to have at least the minimal energy, but they can have more than the minimal amount and indeed the reaction cross-section typically rises steeply in the post-threshold regime, Figure 3.1. Nor does the activation energy have to be independent of temperature.* Atypically the activation energy can even be negative if low-energy collisions preferentially lead to reaction.[4]

3.A Appendix: Reaction rate under non-equilibrium conditions

On a microscopic, molecular level, we can write for a state-selected and state-resolved reaction

$$\mathrm{F} + \mathrm{HCl}(i) \rightarrow \mathrm{Cl} + \mathrm{HF}(j)$$

where i and j label the different internal states. We assume here that the kinetic energy has a thermal distribution. The observed bulk reaction rate at complete thermal equilibrium is then a sum over the rates of reaction of the HCl molecules, in all possible states, with F atoms:

$$-\frac{\mathrm{d}[\mathrm{F}]}{\mathrm{d}t} = \sum_i k_i(T)[\mathrm{F}][\mathrm{HCl}(i)] \tag{A.3.1}$$

Here $k_i(T)$ is the reaction rate constant for a selected state of the reactants. Whether the system is in thermal equilibrium over the internal states of HCl or not, we can rewrite Eq. (A.3.1) as

$$-\frac{\mathrm{d}[\mathrm{F}]}{\mathrm{d}t} = \sum_i k_i(T)[\mathrm{F}]\frac{[\mathrm{HCl}(i)]}{[\mathrm{HCl}]}[\mathrm{HCl}] = k(T)[\mathrm{F}][\mathrm{HCl}] \tag{A.3.2}$$

where [HCl] is the total concentration of HCl. By comparison with Eq. (3.1) this defines a reaction rate constant, $k(T)$, as

$$k(T) = \sum_i k_i(T)\frac{[\mathrm{HCl}(i)]}{[\mathrm{HCl}]} \equiv \sum_i p_i k_i(T) \tag{A.3.3}$$

---

* Show, Problem E, that this dependence tells us about the entropic requirements of the reaction. We provide a microscopic view of such requirements in Chapter 5.

Here p_i, as defined by Eq. (A.3.3), is the relative population ($=$ mol fraction) of HCl molecules in the state i

$$p_i = \frac{[HCl(i)]}{[HCl]} \tag{A.3.4}$$

In general p_i can depend on time and so Eq. (A.3.3) does not necessarily define a rate *constant*. If however the system is in thermal equilibrium at the temperature T then p_i is the Boltzmann factor for the state i and the averaging in Eq. (A.3.3) defines the familiar thermal reaction rate constant. The point is that Eq. (A.3.3) is valid whether the internal states of the HCl molecules are or are not in thermal equilibrium.

Already in Section 1.2.4 we have seen that the reaction rate does depend on the *internal excitation* of the reactants. See also Figure A3.1. Yet the measurement of $k(T)$ for reactions in thermal equilibrium can give no indication of such an effect (for a gas in thermal equilibrium, we are unable to vary the p_is independently). In other words, at thermal equilibrium we are unable to state-select the reactants. It is only by imposing non-equilibrium reactant distributions that we can characterize the role of reactant excitation. Otherwise, when we vary T we vary **both** the occupations p_i of the different internal states of HCl and the kinetic energy of the collision (which is why the state-selected rates $k_i(T)$ are T-dependent). The measurement of $k(T)$ only cannot tell the two apart, without making assumptions.

So far we have shown that the observed reaction rate constant is an average over the rate constants for the selected state reactants. If we do state-resolve the products then

$$\frac{d[HF(j)]}{dt} = -\sum_i k_{ij}(T)[F][HCl(i)] \tag{A.3.5}$$

and proceeding as before

$$k(T) = \sum_{i,j} p_i(T)k_{ij}(T) = \sum_i p_i(T)\left(\sum_j k_{ij}(T)\right) = \sum_i p_i(T)k_i(T) \tag{A.3.6}$$

In summary: the rule is **asymmetric**. To define the overall rate constant we are to *sum* over the states of the products but to *average* over the internal states of the reactants.

The experimental evidence, as reiterated in Figure A3.1, is that the detailed reaction rate constants $k_{ij}(T)$ do in general depend upon both the initial and final states (i and j). These state-to-state reaction rate constants also depend on the collision energy and hence on the translational temperature T. Our purpose is to reduce this extensive averaging. Among the features that we would want to explore would be: (a) the "energy requirements" of chemical reaction. In particular, the threshold energy or the minimum energy required for the reaction to occur[5] and the variation of the reactivity with the reactants' translational (and internal)

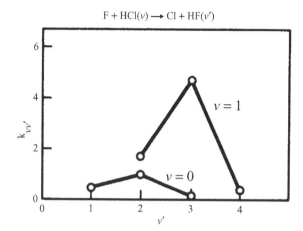

Figure A3.1 Influence of the vibrational state of the reactant molecule upon the distribution of vibrational states of the product molecule, for the reaction $F + HCl(v) \rightarrow Cl + HF(v')$. Plotted is $k_{vv'}$ vs. v' for $v = 0$ and $v = 1$. Note the strong effect of reagent vibration upon the overall reaction rate also [the reaction is about five times more efficient from the $v = 1$ state; adapted from J. L. Kirsch and J. C. Polanyi, *J. Chem. Phys.* **57**, 4498 (1972)].

energy. (b) The *steric effect* or the variation of the reactivity with the relative orientation of the reactants. (c) The energy disposal into the products. (d) The angular distribution of the products after they have separated from the region of interaction. We have made a start on (c) and (d) already in Chapter 1, but before we continue on this road we need to know how to characterize quantitatively the reaction rate under non-equilibrium conditions.

The unraveling of the averaging that goes into the definition of a thermal reaction rate constant shows that its temperature dependence is not quite a simple matter. The state-to-state reaction rate constants vary with temperature because the state-to-state cross-section depends on the collision energy. In addition, for a system in thermal equilibrium, the populations of the different initial states are themselves temperature-dependent. Problem F shows that if an increment in the collision energy is as effective in promoting reaction as an increment in the internal energy of the reactants, then the two sources of T-dependence will be equivalent. As we have noted, such equivalence is **not** necessarily the case.

3.2   Two-body microscopic dynamics of reactive collisions

This section is a simple microscopic approach to reactive collisions. What we want is to determine the rate of those collisions that result in a chemical reaction. The *reaction cross-section* σ_R is a measure of the effective size of the molecules as determined by their propensity to react, at a given collision energy. We will not explicitly indicate that the reaction cross-section is velocity-dependent, but please bear this in mind. In order to keep the view simple in this section we overlook the role of the internal energy states of molecules. That is, we use the discussion of the approach motion of structureless particles, as in Section 2.2, augmented by the notion of the reaction probability. We will be able to make quite a bit of headway, but much will be left for Chapter 5 where the polyatomic nature of the collision is explicitly recognized.

3.2.1   The opacity function

As the reactants collide (at a given energy) we characterize their initial approach in terms of the impact parameter. We define the *reaction probability* or *opacity function*, $P(b)$, as the fraction of collisions with impact parameter b that lead to reaction. Two properties of the opacity function are obvious. Since at most all collisions can lead to reaction, $0 \leq P(b) \leq 1$. Moreover, for a chemical reaction to take place, it is necessary for the reactant molecules to get close to each other so that "chemical (exchange) forces" will operate and the atomic rearrangements that constitute the chemical change can take place. For high-impact-parameter collisions the centrifugal barrier (Section 2.2.2) acts to keep the molecules apart (recall that the distance of closest approach $R_0 \to b$ for large b). We therefore expect that reaction will take place only when b is "small," i.e., of the order of the range of the chemical force, and that the reaction will fail to occur, i.e., $P(b) = 0$, for higher values of b.

3.2.2   The microscopic view of the reaction cross-section

For collisions with an impact parameter in the range b to $b + db$, the reaction cross-section is given in terms of the opacity function by

$$d\sigma_R = 2\pi b P(b) db \qquad (3.12)$$

$2\pi b \, db$ is the area presented to the colliding reactants when their impact parameter is in the range b to $b + db$. $P(b)$ is the fraction of all such collisions that lead to reaction. Due to the possibility of reaction, the total collision cross-section, $2\pi b \, db$, is partitioned between the cross-section for *reactive*, Eq. (3.12), and *non-reactive* collisions given by[6]

$$d\sigma_{NR} = 2\pi b [1 - P(b)] db \qquad (3.13)$$

Here $1 - P(b)$ is the fraction of collisions that are non-reactive at the impact parameter b. One can say that reaction "quenches" the non-reactive scattering.

It is important to note that while the partitioning between reactive and non-reactive collisions depends in an intimate fashion on the collision dynamics, the sum of the two cross-sections is $2\pi b \, db$, irrespective of the details. If several different reaction paths are energetically possible, their cross-sections can increase

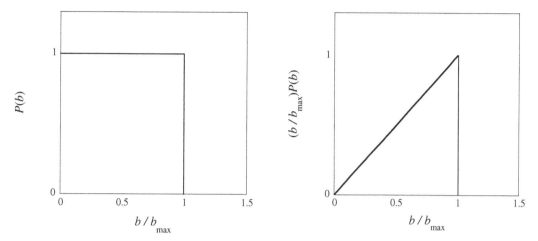

Figure 3.4 The simplest model for a reaction probability: a unit step function. It shows a reaction that occurs with 100% probability whenever the reactants can get near enough to one another and then has an abrupt cutoff at b_{max}. The impact parameter is shown in units of b_{max}. From Chapter 1 you expect that such a model fails to allow for steric effects, but it is realistic for the harpoon-type reactions that we discuss next. Also shown in the figure is the integrand $bP(b)$ of Eq. (3.14) that defines the reaction cross-section, so the contribution of different bs to the reaction cross-section is directly the height of this curve. This shows graphically how the highest bs for which reaction is possible, contribute most.

only at the expense of each other, or of the non-reactive cross-section. There is never more (nor less) than $2\pi b \, db$ to partition among all possible final outcomes of collisions when the impact parameter is in the range b to $b + db$. Recall that the discussion of the measurement of the total cross-section of the experiment in Section 2.1 did not specify what happens as a result of the collision beyond that incoming molecules disappeared as far as the detector of the beam flux is concerned. If a force is applied, leading to a deflection, the molecule in the beam fails to reach the detector. That experiment measures the *total* cross-section irrespective of how the outcome of the collision is partitioned between the different kinds of processes that are possible.

The total reaction cross-section is the sum ($=$ integral) over all collisions irrespective of their impact parameter

$$\sigma_R = 2\pi \int_0^\infty bP(b)\,db \tag{3.14}$$

True to the principle of "he that hath, receiveth," σ_R weights $P(b)$ such that the contributions at higher b values tend to contribute more heavily because of the $2\pi b$ term in the area element. Hence, if a reaction has an opacity function that extends to higher b values, it will have a particularly large reaction cross-section, see Figure 3.4.

3.2.3   A simple opacity function

A determination of the reaction cross-section does not uniquely specify* $P(b)$, but only its b-weighted average, Eq. (3.14). Because of the b-dependent weighting, the reaction cross-section is more sensitive to the value of $P(b)$ at higher bs. Hence, rather than trying to determine $P(b)$ in detail, let us assume a simple functional form with very few parameters and determine these parameters from the observed σ_R or from dynamical models. Since $P(b)$ is non-negative, cannot exceed the value 1, and vanishes at high bs, the simplest approximation is a unit step function as shown in Figure 3.4:

$$P(b) = \begin{cases} 1, & b \le b_{max} \\ 0, & b > b_{max} \end{cases} \tag{3.15}$$

b_{max} is the cutoff impact parameter, the highest b for which a reaction occurs. The reaction cross-section is then simply

$$\sigma_R = 2\pi \int_0^{b_{max}} b \, db = \pi b_{max}^2 \tag{3.16}$$

b_{max} (and its possible energy dependence) can be determined from the measurements of σ_R or from models for the dynamics, as we shall show below.

3.2.4   The harpoon mechanism

An illustration for our simple considerations where the reaction cross-section is well estimated by Eq. (3.16) is the long-range electron transfer process known as the *harpoon mechanism*. This mechanism only applies for reactions of atoms (or molecules) with rather low ionization potential with molecules of rather high electron affinity. It was proposed by Michael Polanyi (1932) to explain the very large (over 100 Å^2) reaction cross-sections of alkali atoms with halogen molecules, see table in Problem H. Since Polanyi's time the idea of a charge transfer during the approach motion of the reactants has become much more widely applicable. What distinguishes the harpoon mechanism is that the charge transfer occurs at quite a large separation of the reactants, well before what one would customarily think of as the range of onset of chemical forces.

Consider a collision between K and Br$_2$. The first stage of the reaction is envisaged as the transfer of the valence electron of the alkali metal atom to the halogen molecule. Such a transfer is shown by Eq. (3.18) below to be possible even when reactants are quite a few Ångstrom apart. Once the transfer takes place and a temporary ion-pair (e.g., K$^+$Br$_2^-$) is formed, the strongly attractive coulomb force accelerates the two ions toward each other. This is followed by formation of

---

* We shall see later (Section 4.4.4) that *angular distribution* studies can provide information on the b-dependence of the opacity function.

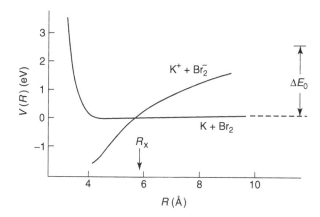

Figure 3.5 Schematic drawing of intersecting potential curves to model the harpoon reaction of K + Br$_2$. The nearly flat curve represents the long-range K–Br$_2$ interaction where both species are neutral. The value of this energy at large separations sets the zero of energy. The upper curve is a coulombic attraction approximating the ionic K$^+$–Br$_2^-$ potential at long range. As R is decreased to the crossing point R_x, the lowest energy state of the KBr$_2$ system switches from the covalent to the ionic form and follows the lower curve until K$^+$Br$^-$ forms, with the ejection of a Br atom. At **very** large separations the potential energy curve for K$^+$ + Br$_2^-$ levels out at the difference, ΔE_0, between the ionization potential of K and the (vertical) electron affinity of Br$_2$. Therefore the asymptotic (= in the reactants' region) separation in energy of the two potential curves is ΔE_0.

the stable KBr and rejection of the Br atom. The metal atom has, in effect, used its valence electron as a "harpoon" in order to pull in the halogen molecule.

A simple estimate of the range of the harpoon can be obtained easily by considering the energetics of the charge transfer. First we realize that the ionization potential of the metal (which may be an alkali or alkaline earth) exceeds the electron affinity of the halogen molecule, so that, at large separations, charge transfer is endoergic and cannot take place in a low-energy collision. However, as the reactants approach one another the formation of the ion-pair **can** occur owing to the gain in energy from the coulomb attraction in the newly formed ion-pair!

The largest separation R_x at which charge transfer can take place on energetic grounds is estimated from the solution of the equation (cf. Figure 3.5) that determines the largest distance at which the coulomb attraction between the two oppositely charged ions is sufficient to provide the required energy ΔE_0

$$-e^2/R_x + \Delta E_0 = 0 \qquad (3.17)$$

where $\Delta E_0 =$ IP(metal) $-$ EA(halogen molecule), the endoergicity, is the difference between the ionization potential of the alkali and the electron affinity of the halogen molecule. The nuclei of the Br$_2$ molecule are at their equilibrium distance apart and so the electron affinity that we need is the so-called *vertical*[7] *electron affinity*. In writing Eq. (3.17) we have kept only the leading terms and neglected the long-range dispersion interaction between the two neutral reactants, which is negligible compared to e^2/R at R_x as well as any polarization interactions in the ion-pair. Thus Eq. (3.17) provides an approximation formula for R_x:

$$R_x \cong \frac{e^2}{\Delta E_0} = \frac{14.4}{\Delta E_0 \,(\mathrm{eV})} \quad (\text{in Å}) \tag{3.18}$$

where the numerical factor is the subject of Problem H.

Owing to the large coulomb attraction, reaction follows "immediately" after the electron transfer. For $b > R_x$ the reactants will never get "close enough" for charge transfer, i.e., to within R_x. The reason is that during their initial approach the neutral reactants are moving under the very weakly attractive long-range potential. So their distance of closest approach is essentially b. When $b < R_x$ charge transfer can occur and then the strong coulomb attraction overcomes the centrifugal repulsion and draws the two ions together, downhill. Reaction between the two ions follows.* Thus $b_{\max} \cong R_x$ or $\sigma_R \cong \pi R_x^2$. See also Problem H.

Equation (3.18) implies that R_x will increase for a decrease in ionization potential of the metal: thus σ_R increases for the sequence Li, Na, . . . , Cs. The trend in σ_R with the electron affinity of the halogen is also qualitatively explained by Eq. (3.18). This simple mechanism thus offers a qualitative explanation for both the large magnitude of the alkali–halogen reaction cross-sections **and** the trends with the metal and halogen involved, Problem H. The harpoon model also provides an understanding of the mechanism of the collisional ionziation reactions, e.g., $K + Br_2 \rightarrow K^+ + Br_2^-$, and explains why the threshold energy for such reactions equals the endoergicity, without an additional barrier.

In this section we have discussed curve crossing along the approach coordinate and we next provide some further examples. However, the model is extensively employed, particularly so when we generalize in Section 5.1 the notion of an approach coordinate to a more general one that is known as the reaction coordinate or reaction path. Then we can treat the more general case; an example of such an oxidation reaction is shown in Figure 5.6, where the reactants begin their chemical interaction by a charge transfer, but one that takes place when they are closer in.

3.2.4.1   A modern variation on an old theme: excimer lasers

A new application of the harpoon mechanism is to the pumping of excimer lasers. These lasers are the workhorses in the UV region and their typically orange-painted housing is easily recognized in the laboratory. Excimer lasers operate by taking advantage of the alkali atomlike character of electronically excited rare gas atoms[8] Rg*. These metastable atoms are efficiently produced by electron impact. Being rare gases, their first electronically excited state is high in energy, which means that these excited states have low ionization potentials. Therefore, they

---

* In Chapter 5 we will regard this as an example of an "early downhill" or an "attractive" potential.

will readily undergo a harpoon reaction, with a large cross-section, with halogen molecules, X_2

$$Rg^* + X_2 \rightarrow Rg^*X + X$$

The chemically bound ionic molecule thus created, Rg^*X (the "excimer"), is electronically excited. It emits UV light and goes down to its ground state RgX. But the ground state is not chemically bound and promptly dissociates. An excess population of the excited state is thereby maintained and the system acts as a laser. Repeating a discharge through the Rg/X_2 mixture excites fresh Rg^* atoms and the laser can be operated again and again. It is capable of producing rather high powers. The usable frequency range of this laser is governed by the choice of rare gas and halogen.

There is much more to the technology of excimer lasers. But two key points are made possible by the special chemical nature of rare gas atoms. First is the unusually high excited state energy that creates a state Rg^* with a rather low ionization potential, thereby making a harpoon reaction with X_2 possible. Second is the efficient draining of the final state after emission due to the Rg–X repulsion for those Rg–X interatomic distances corresponding to the well of Rg^*X.

3.2.4.2 Hardness and electronegativity

The harpoon model uses the ionization potential (IP) and the electron affinity (EA) as the measures of electron donating and accepting. In terms of the energy of the N-electron molecule, the IP and EA are defined by

$$\begin{aligned} IP(N) &\equiv E(N-1) - E(N) \\ EA(N) &\equiv E(N) - E(N+1) \end{aligned} \tag{3.19}$$

For species that are not quite so ready to give or receive an electron it is useful to use the *hardness* (H) and the *electronegativity* (E) defined by (Kohn *et al.*, 1996)

$$\begin{aligned} H &= (IP - EA)/2 \\ E &= (IP + EA)/2 \end{aligned} \tag{3.20}$$

Often it is a useful approximation to use an orbital picture where electrons are assigned to orbitals, as in Section 5.1.5, and these orbitals are taken to be the same for the cation, the neutral, and the anion. Then the energy of adding or subtracting an electron is just the energy of the orbital it goes out of or into. This so-called Koopmans theorem is an approximation that does not always hold.

*3.2.4.3 Dynamics in condensed phases: a simple application of curve crossing

The potentials during the approach motion of the reactants can be much modified when the reactants are not isolated but are coupled to an environment. In the gas phase, if the ionization potential is not low, the energy ΔE_0 needed to form an ion-pair is high and the curve crossing of the ionic and covalent potentials does not occur as the reactants approach (or recede). In water or another polar solvent,

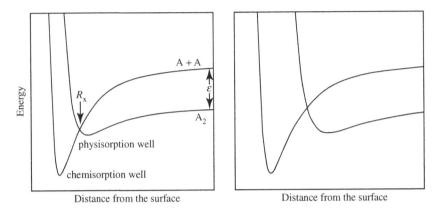

Figure 3.6 The Lennard-Jones curve-crossing model for dissociative chemisorption, left: without and right: with an energy barrier. The undissociated A_2 molecule is physisorbed at the surface. The A atoms are chemisorbed. The energy of the two new metal–A bonds suffices to compensate for the A–A bond energy ε and the depth of the physisorption well. Therefore the interaction potential of the undissociated A_2 molecule with the surface is asymptotically lower, by the A–A bond energy. But near the surface this potential curve is crossed by the interaction of two A atoms with the surface. The limitation of the two-body point of view is evident in this plot. The A–A bond distance, that is surely a key variable, is not represented in this simple view. More on this topic in Chapter 12.

the ions are solvated and, if we continue to use a two-body point of view, the ionic potential curve is lowered compared to the covalent curve. For example, in water, KBr dissociates to ions and not to neutral atoms, as it would in the gas phase. For a quantitative estimate of the solvation energies see Chapter 11. Other processes that are highly endoergic in the gas phase, such as the autoprotolysis of water, $H_2O + H_2O \rightarrow H_3O^+ + HO^-$, also become possible in water.

Lennard-Jones used a curve-crossing model in an early explanation of dissociative adsorption of diatomic molecules (such as N_2 or H_2) on metal surfaces (Fe, Pt, . . .). Far away from the surface, the energy of two atoms, say of nitrogen, is significantly higher than that of the diatomic N_2 molecule (by about 945 kJ mol^{-1}, the triple N_2 bond is extremely strong). But on the metal surface the bond is broken. The required energy is provided by the bonding of the atoms to the metal. If we plot the potential energy as a function of the distance to the surface, then the interaction of the two atoms must, at some distance, cross the weaker interaction of the undissociated molecule and the surface, Figure 3.6. On many metals the weakening of the molecular bond occurs through charge transfer from the metal to the empty, antibonding, orbitals of the molecule.

The two-body point of view is highly oversimplified, particularly in that it fails to describe the structure and the response of the environment, but it does capture certain essential features. Current studies that probe the important role of the dynamics of the surface (or of the solvent) are discussed in Chapters 11 and 12.

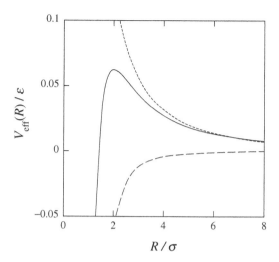

Figure 3.7 The effective potential for an ion–molecule reaction shown only at longer ranges where the potential (in units of ε, dashed line) can be approximated by its long-range form $V(R) = -\varepsilon \, (\sigma/R)^4$. The effective potential, in units of ε, is plotted vs. R/σ for $(E_T/\varepsilon)(b/\sigma)^2 = 0.5$. The centrifugal barrier, $E_T b^2 / R^2$ is shown, in units of ε, as a dotted line and it is seen that at longer range it easily overcomes the ion–molecule attraction. But closer in, the attraction dominates, resulting in a barrier in the effective potential. Problem J shows analytically that the location of the barrier is at $R_{max}/\sigma = (2 \, / \, (E_T/\varepsilon)(b/\sigma)^2)^{1/2}$.

3.2.5 The centrifugal barrier to reaction

A simple quantitative model for the translational energy dependence of b_{max} and hence σ_R is obtained by adopting the following criterion for reaction: reaction occurs with unit probability, i.e., $P(b) = 1$, if the colliding molecules have sufficient translational energy to overcome the energy barrier for reaction and reach the region of chemical forces. Consider first reactions without an energy threshold. The only barrier to the approach motion of the reactants is the centrifugal barrier (Section 2.2.2), and so reaction is always energetically possible at low bs. For a given translational energy E_T and impact parameter the effective potential

$$V_{eff} = V(R) + E_T b^2 / R^2 \tag{3.21}$$

shown in Figure 2.9 may have a local maximum,[9] at some $R = R_{max}$ where R_{max} is, in general, a function of b and E_T. This maximum is due to the opposing influence of the very long-range centrifugal repulsion falling off as R^{-2} and the long-range attractive part of $V(R)$ that falls off faster, as shown in Figure 3.7. Therefore the repulsive centrifugal barrier dominates at large R values. But the attractive $V(R)$ can overcome the centrifugal effect at somewhat smaller Rs.

3.2.5.1 Computing the capture cross-section for reactions with no energy threshold

The capture model assumes that all trajectories that can energetically cross the barrier in the effective potential do cross and proceed to form products. The maximum of the effective potential occurs at the distance R_{max} that is determined by the equation for the location of a stationary point of a function:[10]

$$\frac{\mathrm{d}}{\mathrm{d}R}[V_{\mathrm{eff}}(R)]\bigg|_{\text{evaluated at } R=R_{max}} = 0 \tag{3.22}$$

Only when the molecules can approach to within R_{max} can they cross the barrier in the effective potential and enter the region of chemical forces. Hence, the criterion for reaction is that during their approach motion the molecules can reach $R = R_{max}$ with at least some kinetic energy left, so that they are able to enter the "reaction zone." Making use of Eq. (3.21) we can write this criterion as a quantitative statement about the radial kinetic energy at $R = R_{max}$:

$$\tfrac{1}{2}\mu(\mathrm{d}R/\mathrm{d}t)^2\bigg|_{\text{evaluated at } R=R_{max}} = [E_T - V(R) - E_T b^2/R^2]_{R=R_{max}} \geq 0 \tag{3.23}$$

where R_{max} has been determined by Eq. (3.22). The right-hand side of Eq. (3.23) is decreasing when b is increasing. So we define b_{max} as the value of b such that the right-hand side of Eq. (3.23) comes down to zero,

$$[E_T - V(R_{max}) - E_T b^2/R_{max}^2]\bigg|_{\text{evaluated at } b=b_{max}} = 0 \tag{3.24}$$

If we approximate the intermolecular potential between the reactant molecules by its long-range form*

$$V(R) = -C/R^s \quad (s > 2) \tag{3.25}$$

after some work** using the implicit Eq. (3.22) we get

$$R_{max}^2 = (sC/2E_T b^2)^{2/(s-2)} \tag{3.26}$$

Solving for $V(R_{max})$ using Eq. (3.25) we find b_{max} from Eq. (3.24) and thus, finally, the reaction cross-section

$$\sigma_R = \pi b_{max}^2 = \pi q(s)(C/E_T)^{2/s} \tag{3.27}$$

where $q(s) = (s/2)[(s-2)/2]^{-(s-2)/s}$.

For the particular case of a singly-charged ion–molecule reaction, shown in Figure 3.8, $s = 4$ and $C = \alpha/2$, where α is the polarizability of the molecule

$$\sigma_R = \pi(2\alpha/E_T)^{1/2} \tag{3.28}$$

This is known as the Langevin–Gioumousis–Stevenson form of the reaction cross-section. It implies that the reaction rate constant $v\sigma_R$ is (nearly) independent of the collision energy so that the thermal reaction rate constant is nearly independent of temperature.

---

* This is permissible only if R_{max} is large enough so that it is in the long-range tail of the potential.
** $\mathrm{d}[V_{\mathrm{eff}}(R)]/\mathrm{d}R = sC/R^{s+1} - 2E_T b^2/R^3$ so R_{max} is the solution of $sC/R_{max}^{s-2} = 2E_T b^2$.

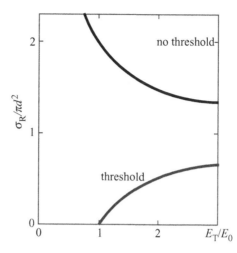

Figure 3.8 The translational energy dependence of the reaction cross-section for reactions with and without an energy threshold [adapted from R. D. Levine and R. B. Bernstein, *J. Chem. Phys.* **56**, 2281 (1972)].

The simple capture model does account for the moderate decrease of σ_R with E_T for reactions without energy threshold. In the particular case of ion–molecule reactions it explains why their rate constant is nearly independent of temperature and it even accounts roughly for the magnitude of σ_R as well, as shown in Figure 3.2.

3.2.6 Reactions with an energy threshold

For reactions with an energy threshold, we must take account of both the centrifugal barrier and the threshold energy barrier (the "permanent" barrier for reaction). We now replace Eq. (3.23) by the condition that at some separation d, the energy available for the motion along R exceeds the threshold energy E_0, or*

$$[E_T - E_T b^2/d^2 - E_0] \geq 0 \tag{3.29}$$

We take b_{max} as the largest value of b for which (3.29) is not negative, i.e., $b_{max} = d(1-E_0/E_T)^{1/2}$ so that

$$\sigma_R = \pi b_{max}^2 = \begin{cases} 0, & E_T \leq E_0 \\ \pi d^2 \left(1 - \dfrac{E_0}{E_T}\right), & E_T > E_0 \end{cases} \tag{3.30}$$

The reaction cross-section vanishes below the threshold E_0 and steeply rises above it, reaching the asymptotic high-energy limit of πd^2, meaning that reaction

---

* In the absence of the potential, $E_T(1-b^2/d^2)$ is the kinetic energy for motion along R at $R = d$. Therefore the reaction criterion (3.29) is sometimes formulated as: reaction occurs if the kinetic energy along the line of centers of two hard spheres exceeds E_0. The criterion and the cross-section (3.30) is then known as the "line of centers" model. This interpretation of (3.29) is correct but not essential. We can think of E_0 as the value of $V(R)$ at $R = d$. Equation (3.29) is then the criterion that the motion of the reactants under the potential $V(R)$ can reach the separation $R = d$ where $V(d) = E_0$ and E_0 is positive.

is possible for every collision that reaches the critical configuration $R = d$. The increase of σ_R with E_T is in qualitative accord with the results often found for reactions with an energy threshold, but deviations are possible at higher energies, see Problem L. Even more critical is the empirical observation that the value of the geometrical cross-section πd^2 obtained by fitting (3.30) to the observed energy dependence of σ_R is typically too small to make good chemical sense. The observed reaction cross-section is smaller than it ought to be and the deviation is more serious when the reactants are polyatomic. We pick up on this point in Section 3.2.7 and again in Section 6.1.4.1.

Figure 3.8 is a schematic summary of the translational energy dependence of σ_R for reaction without and with an energy threshold. In Figure 3.8 the reaction cross-section is far larger if there is no energy threshold. This is indeed the general implication of this section. If there is no energy barrier the reaction can take place at larger impact parameters and the cross-section, because it gives higher weight to larger impact parameters, is far larger.

3.2.7   The steric factor

So far we have discussed the approach motion of the reactants. But there are usually other requirements, besides the closeness of the approach, for a reaction to take place. In particular, there may be *steric requirements*: some configurations of the colliding molecules may be more conducive to reaction. A direct experimental verification of the steric requirement can be obtained using oriented reactant molecules. The experiment allows the determination of the reactive asymmetry or the relative reactivity for the two configurations, say

$$CH_3I + Rb \xrightarrow{f} RbI + CH_3$$
$$Rb + CH_3I \xrightarrow{u} CH_3 + RbI.$$

Here f and u stand for favorable and unfavorable alignment configurations (CH_3I molecules have been oriented by the use of a specially designed configuration of electrical fields). For nearly head-on ($b \approx 0$) collisions the experimental reactive asymmetry is shown schematically in Figure 1.4. See also Figure 10.1.

To examine the notion of steric requirements, consider the reaction $H + D_2 \rightarrow D + HD$. At low energies it is an abstraction reaction, meaning that the potential energy of the three-atom system is such that the low-energy path to reaction is when the approach of H to D_2 is nearly collinear. The impact parameter b is the "miss-distance" from H to the center of mass of the D_2 molecule. At each value of b, all possible orientations of the axis of D_2 with respect to the H–D_2 distance are allowed. Owing to collisions with unfavorable orientations, $P(b)$ will fail to reach the maximal value of unity. To account for such steric (and any other) requirements it is traditional to modify the simple, unit-step function by

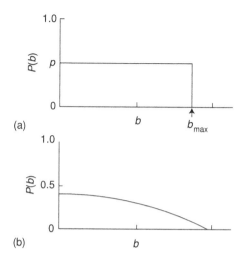

Figure 3.9 Opacity functions $P(b)$, where P is the reaction probability for a collision at given b (for a specified total energy E). (a) The simplest step function, Eq. (3.31) with a constant steric factor p. (b) Computed for the $H + H_2$ exchange reaction at $E = 11\,kcal\,mol^{-1}$ [adaped from M. Karplus, R. N. Porter, and R. D. Sharma, *J. Chem. Phys.* **43**, 3259 (1965)]. This shows that the steric factor p is not necessarily independent of the impact parameter as assumed in case (a) and in Eq. (3.31).

the introduction of a *steric factor* $p < 1$,

$$P(b) = \begin{cases} p, & b \leq b_{max} \\ 0, & b > b_{max} \end{cases} \tag{3.31}$$

This leads to the expected smaller value of the reaction cross-section

$$\sigma_R = 2\pi p \int_0^{b_{max}} b\,db = \pi p b_{max}^2 \tag{3.32}$$

We expect that the value of p is smaller the more constraining are the steric requirements for reaction. However, a measurement of σ_R alone can only determine the value of the product $p b_{max}^2$. Additional information is required to determine the magnitude of p. We next proceed to argue that, as shown in Figure 3.9(b), there are good reasons to assume that the steric factor is both energy- and b-dependent, so that the simple approximation (3.31), as shown in Figure 3.9(a), is not really adequate.

*3.2.7.1   A simple model of steric requirements: the cone of acceptance

The derivation of Eq. (3.30) assumes that the height of the activation barrier is independent of the orientation of the reagents. The experimental results show that at a given energy and impact parameter (e.g., $b \approx 0$ in Figure 1.4), the reaction probability depends very much on the initial orientation angle. Equally, quantum

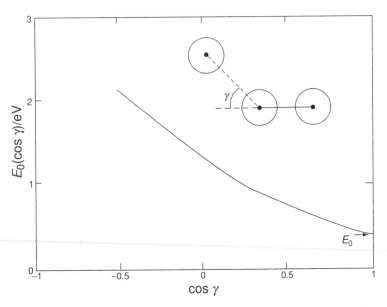

Figure 3.10 Barrier height E_0 vs. $\cos\gamma$ obtained from *ab initio* computations of the potential energy of H_3. Here γ is the "bend angle," defined in the insert. The minimum barrier corresponds to the collinear attack, indicated by the arrow at $E_0 = 0.425$ eV [ab initio computations by P. Siegbahn and B. Liu, *J. Chem. Phys.* **68**, 2457 (1978); parametrized surface by D. J. Truhlar and C. J. Horowitz, *J. Chem. Phys.* **68**, 2466 (1978), **71**, 1514 (1979); adapted from R. D. Levine and R. B. Bernstein, *Chem. Phys. Lett.* **105**, 467 (1984)]. Note that the $H + D_2$ collision occurs on the same potential as that of $H + H_2$. Only the masses are different.

mechanical computations of the barrier height (discussed in detail in Chapter 5) show a significant dependence of the barrier on reagent orientation. The results of such computations for the simplest exchange reaction, $H + H_2$, are shown in Figure 3.10.

To incorporate these results in the previous model, we modify Eq. (3.30) by allowing the barrier height E_0 to depend on the (cosine of the) orientation angle* γ so that E_0 is replaced by $E_0(\cos\gamma)$. The condition for going over the effective barrier is now written as

$$[E_T - E_0(\cos\gamma) - E_T b^2/d^2] \geq 0 \tag{3.33}$$

Reaction is possible when the orientation of the reactants is such that the condition (3.33) is satisfied and, as before, we assume that when crossing of the barrier is possible the reaction occurs with unit probability.[11]

The reaction cross-section for oriented reagents, i.e., for a given value of γ, is obtained by summing ($=$ integrating) $2\pi b\,db$ over the range of impact parameters

---

* For the reactants, at each impact parameter, the entire range of values of γ, from 0 to π, is, in principle, possible. Only such values that are consistent with Eq. (3.31) can reach $R = d$.

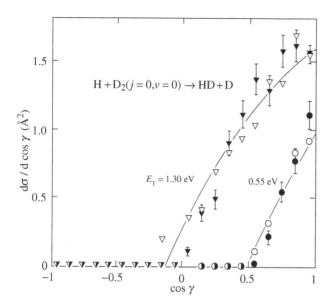

Figure 3.11 Orientation dependence of the cross-section for the reaction $H + D_2(v = j = 0) \rightarrow HD + D$ at the two indicated values of the collision energy E_T. The ordinate is $d\sigma_R/d\cos\gamma = 2\sigma_R(\cos\gamma)$. The solid curves were calculated from the angle-dependent line-of-centers model, Eq. (3.34), and the (open and filled) points represent dynamical computations (these are quasi-classical trajectory results that have statistical error bars as discussed in Chapter 5) on the *ab initio* potential surface referred to in Figure 3.10 [adapted from N. C. Blais, R. B. Bernstein, and R. D. Levine, *J. Phys. Chem.* **89**, 20 (1985)].

consistent with Eq. (3.33). The result looks like Eq. (3.30) but it is for a given orientation,

$$\sigma_R(\cos\gamma) = \pi(b_{max}(\cos\gamma))^2$$
$$= \begin{cases} 0, & E_T \leq E_0(\cos\gamma) \\ \pi d^2\left(1 - \dfrac{E_0(\cos\gamma)}{E_T}\right), & E_T > E_0(\cos\gamma) \end{cases} \qquad (3.34)$$

Equation (3.34) breaks the range of orientations into two regions. The *cone of acceptance* for reaction, defined by the angles γ that satisfy $E_T > E_0(\cos\gamma)$ and the cone of no reaction, $E_T \leq E_0(\cos\gamma)$, where the collision energy is not sufficient to overcome the barrier. The *steric hindrance* familiar from organic chemistry is due to atoms (or groups) that are "in the way," meaning those directions of attack for which $E_0(\cos\gamma)$ is high or even very high.

The result, Eq. (3.34), as a function of γ, is compared to actual dynamical computations for the $H + D_2$ reaction in Figure 3.11. The barrier height used in Eq. (3.34) is that shown in Figure 3.10.

The orientation angle γ is defined in a plane containing the three atoms, cf. insert in Figure 3.10. For reactions of randomly oriented reactants we need to integrate over the range of values of γ. To do so, note that the angular range γ to $\gamma + d\gamma$ defines a ring on a sphere drawn around the center of mass. The ring is generated by allowing for all azimuthal angles ϕ. The area of the ring, for a unit sphere, is $2\pi \sin \gamma \, d\gamma = 2\pi d \cos \gamma$. Then

$$\sigma_R = \int_{-1}^{1} \sigma_R(\cos \gamma) d \cos \gamma \Big/ \int_{-1}^{1} d \cos \gamma \qquad (3.35)$$

Care must be exercised, since $\sigma_R(\cos \gamma)$ is non-zero only for such orientations satisfying $E_T \geq E_0(\gamma)$. In general, the integration needs to be done numerically; however, near the threshold for reaction, one can expand $E_0(\gamma)$ in a Taylor series and retain only the linear term,[12]

$$E_0(\cos \gamma) = E_0 + E_0'(1 - \cos \gamma) \qquad (3.36)$$

In (3.36), E_0' is the negative of the slope of the plot of the barrier vs. $\cos \gamma$ at $\cos \gamma = 1$ (the barrier in Figure 2.21 is lowest for a collinear approach). Using (3.36) in Eq. (3.35) leads to an analytic integral, which in the post-threshold regime, $E_T < E_0 + 2E_0'$, is

$$\sigma_R = \pi d^2 (E_T - E_0)^2 / 2E_0' E_T \qquad (3.37)$$

The labor of performing the analytic integration, Problem M, turns out to be worthwhile: in the post-threshold regime it predicts a concave-up increase of σ_R with the collision energy. This is in contrast to the concave-down functionality of the simpler Arrhenius-like form of Eq. (3.30).

Experimental results often show the concave-up dependence, but near the threshold, where the cross-section is small, the unavoidable finite spread in E_T induces large uncertainties in the σ_R values. Dynamical computations do, however, suggest that the initial rise of σ_R vs. E_T is indeed typically concave-up.

*3.2.7.2   The cone of acceptance can depend on energy and on the impact parameter

What of the steric factor itself? Assume that reaction occurs whenever Eq. (3.33) is satisfied. Then we can determine the probability for reaction at a given b by summing (i.e., integrating) over the range of γ values that allows for the energy along the lines of centers to exceed the barrier height. This leads, in the post-threshold regime, to

$$P(b) = \tfrac{1}{2} \int_{-1}^{1} d \cos \gamma \; H \left[E_T \left(1 - \frac{b^2}{d^2}\right) - E_0(\cos \gamma) \right] = \tfrac{1}{2}(1 - \cos \gamma_{\max}(b)) \qquad (3.38)$$

where $H(x)$ is the unit step function, $H(x) = 1$ for $x \geq 0$ and $H(x) = 0$ for $x < 0$, and $\gamma_{\max}$ is the b- and energy-dependent opening angle of the cone of acceptance, defined as the solution of

$$E_T (1 - b^2/d^2) = E_0(\cos \gamma_{max}(b))   (3.39)$$

It follows that the steric factor, defined historically as how far the reaction probability is below unity, is here derived to equal $(1 - \cos \gamma_{max}(b))/2$ and, as such,
depends on both collision energy and impact parameter. The actual dependence
is determined via the implicit Eq. (3.39) and so requires knowing how the barrier
to reaction varies with the approach angle γ.

3.2.7.3 Steric hindrance

The simplest intuitive idea of an atom or a group being "in the way" of the reaction
corresponds to a barrier that has a constant height E_0 for all approach angles γ
$\leq \gamma_0$, and is infinitely repulsive otherwise. Then $\gamma_{max} = \gamma_0$ and the cone of
acceptance is b-independent. This model for $E_0(\gamma)$ is simpler than Eq. (3.36) but
captures well the idea of a steric hindrance and the reactive asymmetry. It leads
to a steric factor p given by $(1 - \cos \gamma_0)/2$. We expect the size of molecules to
decrease with increasing collision energy and so a rigid and non-yielding steric
hindrance is an idealization. Realistically, the role of steric hindrance should
diminish at higher collision energies.

3.2.8   Two aspects of scattering

The simple concept of centrifugal energy representing the energy of the rotation
of the interparticle distance has thus taken us part way toward an understanding
of reactive molecular collisions. There is one final step that we can still take:
to consider a rather direct manifestation of this rotation, namely the angular
deflection of the particles induced by reactive collisions. We take this step in
Chapter 4. Scattering also means the change in the internal energy states. To
describe this aspect we must develop a polyatomic view of the chemical change.
To do so now, proceed from here to Chapter 5.

3.B   Appendix: Dynamics in strong laser fields – a curve-crossing picture

The curve-crossing model provides a useful view of how strong laser fields can
be used to control the system. In the field of a strong laser there are many photons.
One can therefore regard the molecule as being in a bath of photons. Continuing
with this chemical language, let us solvate the molecule by the light or, in the
customary technical terminology, dress it. To do so we label the potential not only
by the state of the system but also by the number of photons that are present.

Figure B3.1(a) is the conventional representation of the lower state potential
$V_g(R)$ between two ground state Hg atoms. Also shown is an excited state potential
$V_e(R)$ for the excimer Hg*Hg that asymptotically correlates to a ground and an

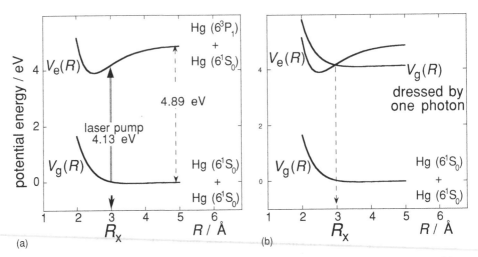

Figure B3.1 (a) The potential energy curves of Hg + Hg and Hg + Hg* [adapted from P. Gross and M. Dantus, *J. Chem. Phys.* **106**, 8013 (1997)]. The long-range attraction and the well (depth 370 cm^{-1}) in the ground state potential, $V_g(R)$, are hardly noticeable on the energy scale shown. The vertical asymptotic separation of the two potential curves is the resonance excitation energy of a Hg atom, corresponding to the $^1S_0 \to {}^3P_1$ transition. When the two Hg atoms are closer, the electronic energy gap (4.89 eV for an isolated atom) is lowered due to the stronger attraction in the excited state. At the relative separation R_x the laser frequency matches the potential gap. In other words, the Franck–Condon "window" is where the two atoms are at the separation R_x apart. A vertical transition at R_x prepares a bound vibrational state of the excited electronic state. This is known as laser-assisted recombination, Problem O. (b) Panel (a) drawn in the dressed states picture.

excited atom. Note that the excimer is more tightly bound than when the two atoms are in their ground state. Also shown in the figure is the Franck–Condon region, see Section 7.0.1, for absorption of light by two Hg atoms during their collision. The laser frequency is tuned slightly to the red (i.e., a shade to lower energies) from the resonance frequency needed to excite an isolated Hg atom. Figure B3.1(b) is the same but drawn in the dressed states picture. The lower state is that of two ground state Hg atoms and n photons. The dressed state is the potential of Hg + Hg and $n - 1$ photons, one photon having been absorbed. In the dressed states picture, the laser field has shifted the energy of the lower state such that there is a curve crossing at the interatomic distance R_x defined by $V_g(R_x) + h\nu = V_e(R_x)$ where ν is the laser frequency. This criterion is used extensively in Section 7.3 where we discuss the absorption (or emission) of light during a collision.

What was the Franck–Condon regime is, in the dressed states picture, replaced by a crossing of two potential curves. The immediate implication of the two ways of looking at the problem is the same: the laser-induced transition is localized about R_x. However, the dressed states picture not only allows us to view the laser as an environment that alters the potential curves of the bare molecule, but also anticipates other effects that are possible using higher-power lasers. The most notable is that, if the density of photons is high, the ground state dressed by n photons can cross higher excited states dressed by $n - 2$, $n - 3$, etc. photons, leading ultimately to ionization.

Problems

A. The exoergicity of the reaction is defined as the change in energy between the ground state of the reactants and that of the products. Why is the $H + D_2$ $(v = 0) \rightarrow D + HD$ just a shade endoergic? Much of our understanding of isotope effects on reactivity stems from the considerations you will make in solving this problem.

B. Chemists are still arguing why the branching between the two possible products in the $Ar^+ + HD$ reaction is not 1:1. Why not make your own reasons? If you do, make your predictions for how the initial rotational state of HD affects the branching. Recall or note from Chapter 2 that the center of mass of HD is nearer to the D atom.

C. One sometimes loosely writes $k(T) = \langle v\sigma_R(v) \rangle \approx \langle v \rangle \langle \sigma_R(v) \rangle$ where the brackets denote an average over a thermal velocity distribution. The exact result, Eq. (3.8), shows that this is really not so. To correctly factor out the mean (relative) velocity from the thermal rate constant we can proceed as follows. One can define not just the number density of molecules with velocity in a given range, it is also possible to determine the flux density of molecules with velocity in a given range. (a) Show that the flux density gives a somewhat higher weight to faster molecules and then (b) verify the comment after Eq. (3.8). Historically, the velocity distribution of molecules was measured by O. Stern in 1911, in a setup similar in principle to Figure 2.1, with a velocity selector but without any scattering cell. The raw experimental results did not fit the Boltzmann distribution. Einstein then pointed out that what the experiment measures is the flux density and not the number density. All fell into place but we still sometimes fail to distinguish between a number and a flux density.

D. Arrhenius temperature dependence and saddle-point integration. Many interesting phenomena are the outcome of two conflicting trends, so knowing how to do a saddle-point integration, at least in simple cases, is handy. For the thermal reaction rate constant, Figure 3.3 or Eq. (3.8), the opposing factors are the Boltzmann term, which is a steeply decreasing function of energy, and the reaction cross-section that rises rapidly in the post-threshold region.* The

---

* Of course, if the reaction has no threshold or if we are at an energy range well past the threshold then our assumed rapid energy increase of the reaction cross-section is not correct. For such cases the reaction rate constant will not be well represented by an Arrhenius form with a positive activation energy. The Tolman expression for the activation energy, Eq. (3.9), is, however, exact and does not depend on any assumptions about the energy dependence of the reaction cross-section.

product of these two will therefore have a sharp peak as shown in Figure 3.3. Let us write the product as $\exp(F(\beta E_T))$, $\beta \equiv 1/k_B T$, where from Eq. (3.8) $F(\beta E_T) = -\beta E_T + \ln(\beta E_T \sigma_R(E_T))$. If the integrand has a sharp maximum, the function $\exp(F(\beta E_T))$ must have a maximum at some energy E^* where this energy may be T-dependent. E^* is determined by the condition for a maximum

$$\partial \exp(F(\beta E_T)) / \partial E_T = 0 \quad \text{at} \quad E_T = E^*$$
$$\partial^2 \exp(F(\beta E_T)) / \partial E_T^2 < 0 \quad \text{at} \quad E_T = E^*$$

It follows that in the vicinity of E^*, $F(\beta E_T)$ must decrease parabolically on either side of the maximum. Replacing the function $F(\beta E_T)$ by its parabolic approximation

$$\int dE_T \exp(F(\beta E_T)) \approx (-2\pi/F''(\beta E^*))^{1/2} \exp(F(\beta E^*))$$

where the double prime denotes the second derivative. Verify that this is an Arrhenius-like form but that E^* and the pre-exponential factor can have a temperature dependence. See also Problem O in Chapter 6.

E. The entropy of activation. If, as is quite often the case, the reaction requires some restrictions on how the reactants come together, then the entropy of the reactive reactants should be lower than that of the reactants. You could then say that there is an entropic barrier to reaction. By examining the temperature dependence of the activation energy, Eq. (3.9), you can come up with a rigorous expression for the entropy of activation. Do so. Later we shall recognize that there can be situations where entropic considerations act in the opposite direction and favor the reaction (this is most common for unimolecular elimination or dissociation processes). Can you suggest an example where the reactive reactants will be less constrained than the reactants?

F. Reaction rate constant for state-selected reactants. Say the internal state of the reactants is selected, as in Appendix 3.A. Then the rate constant is given by Eq. (3.8) except that now the reaction cross-section is for reaction from the particular internal state. To obtain the thermal reaction rate we need to average over the internal states of the reactants as in Eq. (A.3.3). There are then two factors that govern the temperature dependence of the thermal rate, the T-dependence of the reaction rate of the state-selected reactants and the change in the populations of different reactant internal states as the temperature is changed. For reactions with an energy threshold both lead to an increasing contribution at higher T. If the rate constant for reactants in internal state i, Eq. (A.3.1), increases with temperature as $k_i(T) \propto \exp(E_i/k_B T)\kappa(T)$, where $\kappa(T)$ is an (increasing) function only of temperature, then show that (a) the activation energy for internal state i is lowered compared with the ground state exactly by the excess internal energy and (b) the thermal reaction rate has no T-dependence owing to the differing occupations of internal states. In other words, under such circumstances the translational and internal energies are equally effective in promoting reaction. What does it take to have this kind of behavior? Examine Eq. (3.8) and conclude that this requires that the product* $E_T \sigma_{R_i}(E_T)$, not the reaction cross-section itself, be only a function of the total energy, e.g., $E_T \sigma_{R_i}(E_T) \propto Y(E_T + E_i)$. This means that if you plot $E_T \sigma_{R_i}(E_T)$ vs. E_T, the curves for different initial states look identical but shifted apart by their internal energy.

G. The Tolman expression for the activation energy. The derivation of the Tolman result for the activation energy as the difference between the mean energy of the reactive reactants and the mean energy of all reactants, Eq. (3.9), was for the special case that there was only translational energy. In Appendix 3.A we showed how to express the reaction rate constant as an average of contributions from different internal states of the reactants. Show that in this more general case we have

$$E_a \equiv \langle E_T + E_{\text{internal}} \rangle_{\text{reactive reactants}} - \langle E_T + E_{\text{internal}} \rangle_{\text{all reactants}}$$

H. The harpoon mechanism. The table shows the energetic data needed to estimate the radius R_x of the harpoon and the measured values of the reaction cross-sections (in Å^2) between alkali atoms and halogen molecules [adapted from J. Maya and P. Davidovits, *J. Chem. Phys.* **59** (1973)].

	Cl_2	Br_2	I_2	IP (eV)
Na	124	116	97	5.1
K	154	151	127	4.3
Rb	190	197	167	4.2
Cs	196	204	195	3.9
EA (eV)	1.6	1.7	1.2	

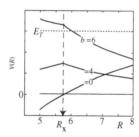

(a) Derive the factor 14.4 in Eq. (3.8). This follows readily from the hydrogen atom where the ionization potential is 13.6 eV (note how high it is compared to that of the alkali metals) and the radius of the 1s orbit, 0.53 Å. (b) Neglecting the weak long-range dispersion potential between the neutral reactants, plot the long-range part of the effective potential as shown above and hence conclude that a harpoon can be cast whenever $b < R_x$. (c) Plot the theoretical estimate for the reaction cross-sections vs. the electron affinity of the halogen molecule, take it to be a continuous variable, for different alkali metals. Are the trends consistent with the entries in the table? (d) Compute the reaction cross-section. Are the absolute values consistent with the entries in the table? (e) Show how to plot all the entries in the table on one common plot of reaction cross-sections vs. a suitable variable. Put all experimental entries in your plot and also draw the theoretical curve. Is there an agreement? If you have access to a curve-fitting program,

---

* We will meet this product in a variety of other contexts, particularly so when detailed balance is discussed. It is the yield function as discussed in Levine (1969) and also called the cumulative rate (Miller, 1998).

see how to get an even better fit. (f) For the harpoon reactions, the approxima-tion $k(T) = \langle v\sigma_R(v)\rangle \approx \langle v\rangle\langle\sigma_R(v)\rangle$ (see Problem C) is not unreasonable. Why? (g) Compute the reaction rate constant.*

*I. The harpoon mechanism for excited states. Excited electronic states have a lower IP and so are generally more likely to react by a harpoon mechanism. But the lowest IPs are for excited electronic states of the alkali metals. (a) The first excited state of Na is about 2.1 eV and of K about 1.56 eV above the ground state. How will the reaction cross-section in the table above change if these atoms are used as reagents? (b) Why stop at the first excited state? How will the reaction cross-section in the table above change if higher excited state atoms are used as reagents? Experimental answer: the reaction cross-section will not continue to grow indefinitely (Bersohn, 1976). Why not? Hint: look at the value of the crossing radius. For more insight see Section 9.3.2.

J. Ion–molecule reactions. The long-range potential between an ion and a molecule is dominated by the polarization potential, Eq. (2.16). (a) Determine the location of the barrier along the effective potential as a function of the collision energy and impact parameter. (b) Examine closely the result in (a) and conclude that our model will have to break down at high enough collision energies. Make an estimate of when this will be the case. To do so requires the value of the polarizability, α, of the molecule. Use the rule of thumb that polarizability is a measure of the volume of a molecule, check that this at least makes sense from the point of view that αe^2 needs to have the dimension of energy length^4 (why?) and hence make a reasoned guess for α. Try to do all of this in scaled (dimensionless) variables. Using $\alpha = \sigma^3$ may simplify your work but it is not essential. (c) From (a) determine the height of the barrier along the effective potential and hence solve for the maximal impact parameter for which reaction is possible at a given collision energy.

K. Steric factor for ion–molecule reactions. Purpose: to compute the cone of acceptance for reaction, Section 3.2.7.1, for the model discussed in Prob-lem H. When the molecule is not spherical, e.g., H_2, the long-range poten-tial depends not only on the ion–molecule distance but also on the orientation angle, γ, between the axis of the molecule and the ion–molecule distance R, see Eq. (2.18). (a) For a non-polar molecule we can take this into account by

---

* You will find rather high reaction rates. Significantly higher than what you computed in Problem E of Chapter 2. In the older literature this is sometimes described as the reaction cross-section being larger than gas kinetic. The statement is correct but, taken superficially, it can give a misleading impression. What we really should say is that the collision cross for $M + X_2$, being larger than the reaction cross-section, is exceptionally large. It is far larger than what we would conclude if we take it to be a mean of the collision cross-sections for $M + M$ and for $X_2 + X_2$ collisions. This is what was meant in Chapter 2 by the statement that the collision cross-section (and even more so, a reaction cross-section) is a joint property of both partners. For example, an alkali atom can be of quite different sizes when colliding with partners of low or of high electron affinity. See the comparison of $K + I_2$ and $K + CH_3I$ in Chapter 4.

allowing the polarizability of the molecule to be different along its axis and perpendicular to it. Using the second Legendre polynomial, a convenient functional representation is $V(R, \gamma) \xrightarrow{\text{long range}} - (C/R^4)(1 + a P_2(\cos \gamma))$, where a is the asymmetry parameter and $C = e^2\alpha/2$ for a singly-charged ion. If you have access to a suitable program, plot the potential on a polar grid where R is the radius and γ is the angle. If not, plot the potential vs. R for $\gamma = 0$ and $90°$. (b) Determine the maximal impact parameter for which reaction is allowed as a function of the approach angle γ. (c) If the molecule is polar, e.g., HCl, it will have a dipole moment. Then the leading terms in the long-range potential are $V(R, \gamma) \xrightarrow{\text{long range}} - (C/R^4) - e\mu \cos \gamma / R^2$. For this case too, determine the maximal impact parameter for which reaction is allowed as a function of the approach angle γ. In the general case, one should also allow for the asymmetry of the R^{-4} term and for the polarization of the ion by the dipole, which gives rise to an R^{-6} term.

L. The line-of-centers model and its limitations. The result (3.30) is often known as the line-of-centers model because it assumes that reaction will take place if the energy along the line of centers, R, exceeds a threshold value, E_0, at $R = d$. (a) Sometimes one adds the assumption that the colliding partners are hard spheres so that there is no potential for $R > d$. This is not necessary, but let's do so in (a) only. Then, compute the cross-section for non-reactive collisions. (b) Experimentally, the line-of-centers energy dependence is found to be reasonable for energies in the post-threshold regime, but at higher energies the reaction cross-section goes down, as shown, rather than up:

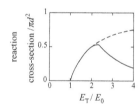

There are several reasons why this can happen. Suggest at least two, and argue that the value of the energy where the deviations begin can be used to distinguish between them. (c) There is a particular reason why the deviation shown in the figure insert above can occur, and it has to do with the breakdown of an assumption made in deriving the model. Which assumption? (Hint: we do not mean a steric requirement because, in general, cones of acceptance for reaction should open up as the energy is increased, Problem M.)

M. The line-of-centers model with a steric requirement. (a) Derive the result (3.35) for the energy dependence of the reaction cross-section when the barrier depends on the approach angle. The point is that you will have to be careful about the limits of integration and this will make you think in detail about the energy dependence of the cone of acceptance. See Smith (1980). (b) Derive the results in two different ways (of course, you should get the same final answer). At a given angle of approach, first integrate over all bs. This will give you the reaction cross-section vs. γ. Plot it. Then integrate over γ. Alternatively, at a given b, first integrate over all γs. This will give you the opacity function. Examine the result to convince yourself that the steric factor is both b- and collision energy-dependent. Plot the result and see how different it is from the simple approximation (3.31).

N. Peripheral collisions. Reactions such as $Cl + CH_4$ or $H + HCl$ may be more likely to occur at higher impact parameters. This is because reaction requires the three atoms, $Cl-H-C$ or $H-H-Cl$, to be nearly in a collinear configuration and this is easier to achieve for an off-center collision. Make a model where the center of mass of HCl is on the Cl atom. Place the bound H atom on a sphere about the center. Assume that reaction occurs when $H-H-Cl$ is in a collinear configuration. Hence compute the opacity function vs. b [adapted from P. M. Aker and J. J. Valentini, *Is. J. Chem.* **30**, 157 (1990)].

O. Laser-assisted recombination. The problem is to compute the cross-section for forming a bound species during the collision.* The key consideration is that the colliding Hg atoms, as described in Figure B3.1, will absorb light when they are at a distance R_x apart. This is the distance when the electronic energy separation between the ground and excited states equals the energy of the laser photon. Say that all collisions absorb light. (a) Show that the cross-section for formation of the bound excited state is $\pi R_x^2 (1 - V_g(R_x)/E_T)$. (b) Plot this cross-section vs. the collision energy and show that it is monotonically decreasing, being largest for very slow collisions. (c) Argue that your result in (a) also provides a small correction for the cross-section of harpoon reactions [see P. Gross and M. Dantus, *J. Chem. Phys.* **106**, 8013 (1997), J. Kohel and J. W. Keto, *J. Chem. Phys* **113**, 10551 (2000), Dantus (2001) and references cited therein for examples of laser-assisted reactions].

Notes

1 Texts of chemical kinetics that also discuss dynamics include Steinfeld *et al.* (1999), Houston (2001).

2 In the past chemists sometimes avoided studying the kinetics of such reactions because their rates were too fast to be able to maintain steady conditions. Often these are reactions between an ion and a molecule. But for many reasons, most notably aeronomy and the

---

* Ordinarily, when two atoms collide they will not recombine to form a stable bound diatomic. Rather, they will emerge as two unbound atoms. This is so even when their potential is attractive and has a deep well corresponding to a stable diatomic. The reason is that the total energy needs to be conserved. In the region where the potential energy is attractive, the kinetic energy increases so as to keep the sum constant. To form a bound state we need some mechanism for removing the energy from the pair of atoms. Typically this is done by a third body that takes the excess energy away. Following Gross and Dantus (1997), Appendix 3.B suggests that we can do something else: use a laser to form a bound, electronically excited, stable molecule.

formation of molecules in space (Duley and Williams, 1984; Herbst, 1995; Klemperer, 1995; Smith and Rowe, 2000), reactions that have no threshold energy (cf. Section 3.2.6) are quite important.

3 Should there not be a barrier to atom exchange? As a rule of thumb, yes, a barrier is to be expected. The trick of ion–molecule reactions is that there is a significant gain of energy when the reactants approach, due to the strong ion–molecule physical attraction, Section 2.1.9. This attraction is often sufficient to overcome the chemical barrier so that, in reference to the reactants that are far apart, there is no threshold energy requirement. By solvating the reactants in a medium of high dielectric constant, one can reduce the electrostatic attraction of the reactants and thereby observe the role of the barrier. More on this in Chapter 11.

4 Reasoning in the other direction needs care. A measured second-order reaction rate constant with a negative activation energy does not necessarily mean that the mechanism is bimolecular with a cross-section that decreases with energy. See Problem M of Chapter 5.

5 The threshold energy is not quite the same as the Arrhenius activation energy. As we have seen, the latter is the average of the energy of those collisions that lead to reaction, minus the average energy of all collisions.

6 Strictly speaking, $d\sigma_{NR}$ also contains an additional term, the "shadow" term, quantal in origin, which is always present, Chapter 4.

7 "Vertical" meaning that the nuclei do not change their position and the anion Br_2^- is formed at the equilibrium bond distance of the neutral Br_2. The additional electron goes into an antibonding orbital so that the anion is expected to be less strongly bound than the neutral molecule. Hence the equilibrium bond length of Br_2^- is larger. The anion is therefore formed with some vibrational excitation. The energy difference between the ground vibrational states of the anion and the neutral is the *adiabatic electron affinity*.

8 The electronically excited rare gas atom is sometimes called a "super-alkali" because its ionization potential is so low. There are also super-halogens and not only super-alkalis (Herschbach, 1966; Bersohn, 1976). The super-halogen has a particularly high electronic affinity; NO_2, with an electron affinity of about 2.4 eV, is an example; so is $(CN)_2$. Excited states of organic molecules are used as effective electron donors. Complexes of transition metals in unusually high oxidation states are keen acceptors and vice versa for complexes where the metal is nominally neutral.

9 The idea that reaction is possible when the system can cross the barrier in the effective potential is sometimes known as a capture model. The image is that the molecules are "captured" by the strong chemical forces that operate at closer range than R_{max}. Recent developments of the model extend it to reactants with internal structure, so that the barrier depends also on the initial vibrational state; see Clary (1990), Troe (1992, 1997). For the experimental situation see Sims and Smith (1995), Smith and Rowe (2000).

10 An exceptional case, for which there is no maximum, is if the long-range part of $V(R)$ has an R-dependence with an inverse power less than or equal to 2. This includes the case of reactions of ions of opposite charge, where $V(R) \propto -R^{-1}$. See B. H. Mahan, *Adv. Chem. Phys.* **23**, 1 (1973) and Problem M.

11 A caveat is needed. Equation (3.31) is the condition at the barrier. Experimentally, however, one cannot easily prepare reactants within a narrow range in γ relative to the interparticle distance **R**. This is because the direction of **R** is rotating during the approach motion, whereas we can only prepare a range of γ with respect to laboratory coordinates.

In the laboratory, the well-defined direction is that of the initial velocity $\mathbf{v}$. Therefore the laboratory orientation angle γ_L is defined as the angle that the axis of the molecule makes with respect to $\mathbf{v}$. For low impact parameters the two angles are essentially the same because $\mathbf{R}$ is essentially in the direction of $\mathbf{v}$ and the collision is nearly head on. Otherwise, a transformation is needed. Implicit in such a transformation and in our entire discussion is the assumption that the axis of the molecule is hardly rotating during the collision. Dynamicists are very used to the idea that rotation of molecules is slow compared with the duration of a collision or a vibrational motion. That is correct, and is why experiments using selected reactants can demonstrate the reactive asymmetry between the two ends of a molecule. On the other hand, the intermolecular forces are not isotropic and can channel reactants preferentially into the cone of reaction or away from it. Only the detailed computations that we discuss in Chapter 5 can fully address such issues.

12 Such an expansion is only valid for $E_T < E_0 + 2E_0'$ because otherwise reaction is possible at all values of γ and there is no cone of no reaction. Furthermore, it need not be the case that the barrier is lowest for a collinear approach. For insertion reactions one expects the transition state to be bent. Detailed electronic structure computations show that, contrary to simple chemical intuition, some abstraction reactions can also have the lowest barrier for a slightly bent configuration.

Chapter 4
Scattering as a probe of collision dynamics

The most direct probe of collision dynamics is the observation of the scattering (i.e., the deflection and energy-state alteration) of the colliding particles as a result of their interaction. A pioneering example is the observation of large deflections in α-particle collisions with atoms by Rutherford, early last century. The results of this experiment led him to propose a model of the atom where the positive charge is concentrated in a small, central nucleus. Ever since, the angular distribution **after** the collision has been a primary diagnostic tool in the attempt to understand the interactions **during** the collision.

We begin with an introduction to this method as applied to elastic (energy-state unchanging) collisions. First we look at scattering using classical mechanics and then we provide a quantum mechanical view.

4.1 Classical scattering of structureless particles

Dynamics means the description of the motion under the action of forces. For our problem of two particles, this means solving Newton's equation of motion for the relative motion

$$\mu \frac{d\mathbf{v}}{dt} = \mu \frac{d^2 \mathbf{R}}{dt^2} = -\mu \frac{dV(R)}{dR} \hat{\mathbf{R}} \qquad (4.1)$$

obtaining $\mathbf{R}$ as a function of time. We wrote the force as derived from the inter-particle potential $V(R)$ and directed along the center-to-center vector $\mathbf{R}$ (the carat denotes a unit vector). Solving Newton's second-order differential equation of motion is what a computer does when it is running a molecular dynamics program. This can be implemented very efficiently. Our task in this chapter is to understand the scattering and the information obtainable from it.

4.1.1 Conservation of angular momentum

Our first task is to examine the time evolution of the vector $\mathbf{R}$. This will uncover a conservation condition that was already used by Newton to explain Kepler's observations about planetary motion. The angular momentum vector $\mathbf{L}$ is defined as the vector product of the position and momentum vectors

$$\mathbf{L} = \mathbf{R} \times \mu \, d\mathbf{R}/dt \qquad (4.2)$$

By the definition of a vector product, $\mathbf{L}$ is a vector perpendicular to the plane defined by the position and momentum vectors.

For a collision of structureless particles, the acceleration is in the direction of the relative position vector $\mathbf{R}$, cf. Eq. (4.1). The chain rule then shows that the vector $\mathbf{L}$ is not changing with time:

$$d\mathbf{L}/dt = d\mathbf{R}/dt \times \mu \, d\mathbf{R}/dt + \mathbf{R} \times \mu \, d^2\mathbf{R}/dt^2 = 0 \qquad (4.3)$$

Both terms in Eq. (4.3) vanish, since, in either term, the two vectors that are vector multiplied are parallel to one another.

A vector has both a magnitude and a direction. That the vector $\mathbf{L}$ is conserved means that the magnitude and also the direction of $\mathbf{L}$ are constant in time.

The conservation of the direction of the vector $\mathbf{L}$ implies that there is a particular, privileged, plane, such that the collision is confined to this special plane. This plane is defined, before the collision, by the two vectors $\mathbf{R}$ and $\mathbf{v}$ where $\mathbf{v}$ is the initial relative velocity. By its definition, Eq. (4.2), as a vector product, $\mathbf{L}$ is normal to this plane. We now want to show why, as time moves on, the collision remains confined to this plane. At some finite time t, consider the plane defined by the two vectors $\mathbf{R}(t)$ and $\mathbf{R}(t + \Delta t)$. Take Δt to be so small that $\mathbf{R}(t + \Delta t) = \mathbf{R}(t) + (d\mathbf{R}/dt)\Delta t$ and hence $\mathbf{L}$ is normal to the plane defined by $\mathbf{R}(t)$ and $\mathbf{R}(t + \Delta t)$. Since the direction of $\mathbf{L}$ is constant in time, for any value of t, $\mathbf{R}(t)$ is confined to the same single plane defined by the initial conditions.

The conservation* of angular momentum puts two conditions at our disposal: the direction of $\mathbf{L}$ is conserved and the magnitude of $\mathbf{L}$ is conserved. The two following subsections make use of these two conditions.

---

* The interaction of real molecules is not necessarily central ($=$ along the direction of $\mathbf{R}$), so that for application to molecular collisions $\mathbf{L}$ need not be conserved because the second term in Eq. (4.3) will not necessarily vanish. Even so, the conservation of $\mathbf{L}$ is often not a poor approximation. The argument is that if we consider two colliding molecules and rotate the entire system as a rigid body, the dynamics are unchanged. This implies that the total angular momentum $\mathbf{J}$ is strictly conserved. Now $\mathbf{J} = \mathbf{L} + \mathbf{j}$ where $\mathbf{j}$ is any internal angular momentum of the colliding molecules (for structureless particles $\mathbf{j} = 0$). If we put realistic values in the classical estimate $L = \mu v b$ we see that the value, L, of the angular momentum is many tens (or even more) of units of $\hbar$, Problem A. So, when $\mathbf{J}$ is only a few $\hbar$ units in magnitude, $\mathbf{J} \approx \mathbf{L}$. Particularly for reactive collisions, where the individual masses of the two particles will change, there are important extensions where one needs to impose the conservation of $\mathbf{J}$ rather than just $\mathbf{L}$. These are quite interesting and we will point them out in detail in Chapter 10. But these should not make us overlook the simpler cases when it is a realistic approximation to conserve $\mathbf{L}$.

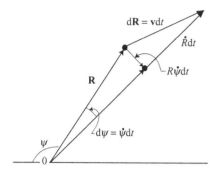

Figure 4.1 In a short time interval d*t*, the vector **R**(*t*) changes to the vector **R**(*t* + d*t*). The figure shows the resolution of the vector change in position, **R̊** d*t* = **R**(*t* + d*t*) − **R**(*t*), into a radial *R̊* d*t* and a tangential *R* d *ψ* = *R* *ψ̇* d*t* component. The dot denotes the time derivative. The radial component, which is in the same direction as **R**(*t*), represents the change in the magnitude of **R**(*t*). The tangential component, which is perpendicular to **R**(*t*), represents the change in direction of **R**(*t*) at a fixed magnitude *R*. The vector **L** is normal to the plane defined by **R**(*t*) and **R**(*t* + d*t*).

4.1.2 The angle of deflection

The conservation of the direction of **L** means that the relative motion of two particles under the influence of a force that depends only upon their mutual separation is confined to a plane. It is therefore sufficient to use two coordinates to specify the relative motion. The relative position vector **R** can then be specified in terms of its length R and its orientation ψ with respect to a fixed direction, say the initial direction of the velocity, as shown in Figure 4.1.

In the course of the collision, both the relative separation R and the orientation angle ψ will vary with time. We use a dot to designate time rate of change (the time derivative). Thus $\dot{R}$ is the velocity with which the colliding particles approach (or recede) from one another. It is often called the velocity along the line of centers. $\dot{\psi}$ is the (angular) velocity for the rotation of **R** (Figure 4.1) during the collision. We will need to know R as a function of time since $R(t)$ describes the approach and the separation of the colliding particles during the collision. Why, however, do we bother with $\psi(t)$?

There are two reasons for our concern with the orientation of the interparticle separation. The first is that the rotation of **R takes up energy**. Using Pythagoras' theorem for the two components of **R̊** d*t* we find two terms in the kinetic energy of the relative motion

$$K = \frac{1}{2}\mu\left(\frac{d\mathbf{R}}{dt}\right)^2 = \frac{1}{2}\mu\left[\left(\frac{dR}{dt}\right)^2 + R^2\left(\frac{d\psi}{dt}\right)^2\right] \tag{4.4}$$

Here we have resolved d**R**/d*t* into its components, Figure 4.1, and the two terms represent the kinetic energy due to the changes in magnitude and in the direction of **R**, respectively. The first term in Eq. (4.4) is the radial kinetic energy, i.e., the

Figure 4.2 The collision trajectory in the c.m. system.* The solid curve represents a trajectory with initial velocity **v**, impact parameter b, and mass μ. The relative separation $\mathbf{R}(t)$ is uniquely defined in terms of the distance R and the orientation angle ψ. The trajectory is symmetric about the apse line, which passes from the origin through R_0 where R_0 is the distance of closest approach. The final deflection angle is $\chi = \pi - 2\psi_0$ where ψ_0 is the value of ψ at the mid-point of the trajectory.

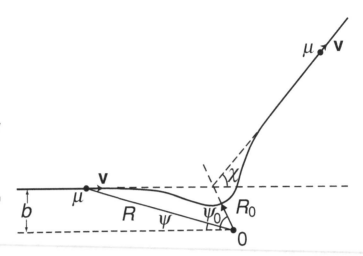

kinetic energy due to the component of the velocity along the line of centers of the collision partners. The second term is the centrifugal energy, i.e., the kinetic energy due to the component of velocity perpendicular to the line of centers of the collision partners.

The centrifugal energy, $(\mu/2)R^2(\dot{\psi})^2$, is the subject of this subsection. It is the kinetic energy due to rotation of the interparticle distance $\mathbf{R}$. It can be written as a rotational energy, namely $(I/2)\omega^2$, where $I = \mu R^2$ is the moment of inertia and $\omega = \dot{\psi}$ is the angular velocity.

The second reason for our concern with $\psi(t)$ is so important to our theme that Section 4.2 is devoted to it: before and after the collision, when the particles are far apart and no forces act between them, they travel in a straight path. Hence, before or after the collision $\dot{\psi}(t)$ is zero; ψ only changes when the particles are "close." By measuring the overall change in ψ due to the collision we can obtain considerable insight into the interparticle forces acting during the collision. We continue with this theme in Section 4.1.4. Here we just reiterate that even in the absence of a force ψ will change by π as the particles pass by one another. What we need to know is by how much less or more will ψ change when a force is acting. Figure 4.2 is an example for a trajectory in the presence of a force. What we want to determine is the net *deflection* χ, the angle that the final velocity makes with respect to the initial direction.

Using the conservation of the magnitude of the angular momentum we can solve for $\dot{\psi}$ as follows. During the collision, the component of $\dot{\mathbf{R}}$ perpendicular

---

* Section 2.2.7 showed that when two particles collide under the action of a potential that depends only on their separation R, it is possible to transform to the center-of-mass (c.m.), system and discuss only the relative motion because the c.m. is at rest. See also Problem C. In the c.m. system, the relative motion is equivalent to the motion of a single particle with a reduced mass $\mu = m_A m_B/(m_A + m_B)$ and position R.

to $\mathbf{R}$ is (cf. Figure 4.1) $R\dot{\psi}$ so that, from the definition of $\mathbf{L}$ as a vector product, $\mathbf{L} = \mathbf{R} \times \mu\dot{\mathbf{R}}$:

$$L = \mu R^2 \dot{\psi} \qquad (4.5)$$

On the other hand, before the collision when the impact parameter b is the component of $\mathbf{R}$ that is perpendicular to $\mathbf{v}$, $L = \mu v b$. These two values of L are necessarily the same, since the value of L is constant throughout the collision. Thus we obtain an explicit solution for the angular velocity of the relative motion

$$\dot{\psi} = bv/R^2 \qquad (4.6)$$

We can now rewrite the kinetic energy, Eq. (4.4), using Eq. (4.6), as

$$K = \frac{1}{2}\mu\left(\frac{\mathrm{d}R}{\mathrm{d}t}\right)^2 + \frac{E_\mathrm{T}b^2}{R^2} \qquad (4.7)$$

In this way, for a given value of the collision energy and impact parameter, the kinetic energy is expressed as a function of $R(t)$ only.

Having obtained Eq. (4.7) we have completed our reduction of the number of coordinates necessary to specify a collision trajectory. In the c.m. system, for given initial values of E_T and b, the trajectory is uniquely specified by a single function of time, $R(t)$, the (magnitude of the) center-to-center distance of the colliding particles. In Section 2.2.2 we derived Eq. (4.7) in another way and indeed found that it provided insight for understanding the approach motion.

4.1.3 The deflection function for hard spheres and for realistic potentials

The potential deflects the colliding molecules from their original, pre-collision paths. The deflection caused by the collision is defined as the angle between the final and initial relative velocity vectors. Section 4.1 showed that the initial collision energy and the impact parameter specify a unique collision trajectory. It follows that the deflection is a unique function of the initial collision energy and impact parameter and can be computed, Section 4.2.3, once we have determined the collision trajectory $R(t)$.

Before we turn to the more quantitative considerations, recall the essential features in the dependence of the deflection upon the impact parameter, Section 2.2.5. What we really need is "an inversion" that provides the ability to determine the impact parameter from the observed deflection. As we shall see, for a realistic potential that has a well, a strict inversion is not simple. For the moment we shall be content with observing the final deflection and inferring what was the impact parameter. The need to do so arises because even in a collision between molecules of well-defined velocities, *all values* of the microscopic impact parameter are possible. Usually, we cannot control the value of this parameter, so the next best thing is to know, a posteriori, what its value was.

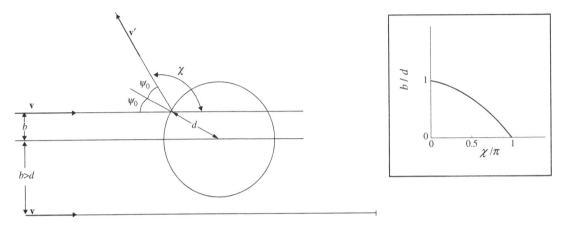

Figure 4.3 Construction of the deflection angle for the collision of rigid spheres with a potential of range d. At the given impact parameter the angle $\chi = \pi - 2\psi_0 = 2\arccos(b/d)$ for all $b \le d$; $\chi = 0$ otherwise. The functional dependence given by Eq. (4.8) is plotted in the right-hand panel. When the impact parameter is higher than d, lower trajectory, there is no deflection.

The interaction between two molecules, $V(R)$, is typically attractive at long range and steeply repulsive at short range. If the potential was purely repulsive we could approximate the scattering as that of hard spheres. Then the trajectory is easy to visualize. If $b > d$, d range of the hard-sphere repulsion, there is no scattering because the centrifugal barrier does not allow the molecules to approach as close as d. For lower impact parameters, Figure 4.3 shows a trajectory for a hard-sphere collision. It is constructed by making a reflection of the trajectory at the point of impact. Defining the deflection function as χ, it is evident from the figure that $\chi = \pi - 2\psi_0$ where $(b/d) = \sin\psi_0$. Thus for the rigid-sphere potential the deflection is

$$\chi = 2\arccos(b/d) \tag{4.8}$$

for $b \le d$, and $\chi = 0$ for $b > d$. As expected, the deflection is fully backwards ($\chi = \pi$) when the collision is "head-on" ($b = 0$) and then it decreases monotonically as b increases. By $b = d$ the two hard spheres just graze one another and there is no deflection.

We next examine the consequences of the presence of both the attractive and repulsive parts of $V(R)$. Figure 4.4 shows schematically collision trajectories for different values of the impact parameter and the resulting deflection. For large b the molecules sample only the long-range attractive force and the trajectories deviate only slightly from the axis, leading to a small (negative) deflection angle χ. As b decreases the trajectory samples a stronger attraction and the deflection angle becomes increasingly negative. As b is decreased further, the influence of the repulsive force begins to be important. Consequently, a most negative

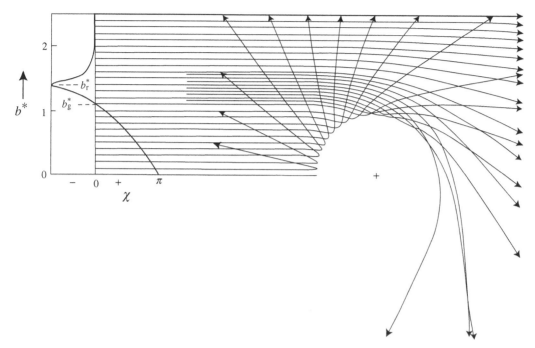

Figure 4.4 Collision trajectories at different impact parameters for a given collision energy. The ordinate is the reduced impact parameter $b^* = b/R_m$ where R_m is the equilibrium distance of the well in the potential. The resulting deflection function is shown on the left-hand side. Note the qualitative difference between hard-sphere scattering and scattering by a realistic potential: for hard spheres there is no attractive part of the potential so going from the deflection $\chi(b)$ to b is single valued; $\chi(b)$ is defined uniquely by the value of b. But in the presence of a well in the potential there can be more than one value of b that results in scattering into a given value of $\chi(b)$.

deflection is reached, at the so-called *rainbow* angle χ_r and a corresponding impact parameter b_r. At the rainbow, the attractive and repulsive forces on the trajectory balance one another and so we expect that $b_r \approx R_m$. For collisions with impact parameters below the rainbow the molecules get sufficiently close that the role of the repulsive force becomes increasingly important. As b decreases further, the deflection angle increases, passes through zero when $b = b_g$, the *glory* impact parameter, and becomes progressively more positive. As b decreases to zero, the collisions become more nearly "head-on," and the molecules rebound, essentially in the backward direction.

Owing to the cylindrical symmetry of the scattering about the $b = 0$ axis (Figure 4.6), the sign of the deflection angle is not experimentally meaningful. The observable deflection angle θ is thus the absolute value of the computed deflection angle, i.e., $\theta = |\chi|$.

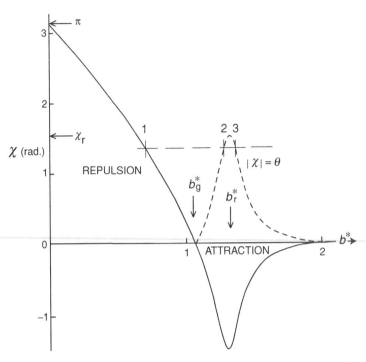

Figure 4.5 Typical deflection function $\chi(b)$ at low energy for a realistic potential (cf. Figure 4.4). b is expressed in units of R_m. At large b the angle of deflection is negative (but very small), arising from the long-range intermolecular attraction. As b is decreased χ becomes more negative, and eventually at $b = b_r$ the greatest negative deflection is obtained. This is the rainbow, already mentioned in Section 2.2.5. As b is further reduced the influence of the repulsive forces becomes increasingly evident and at $b = b_g$, the long-range attraction has been compensated for by the short-range repulsion, so the net (overall) deflection is zero. See also Figure 2.13. This is known as the glory. At still smaller values of b repulsion dominates, and backward scattering results with $\chi \to \pi$ as $b \to 0$. At low collision energies one can also have orbiting collisions where the molecules persistently rotate about one another, as discussed further in Section 4.3.6. Operationally only the absolute value and not the sign of the deflection is meaningful. Where the deflection itself is negative the absolute value is shown as a dashed line. The dashed horizontal line identifies three values of the impact parameter, labeled 1, 2, and 3, that lead to scattering into an angle just below the rainbow at χ_r. Trajectories 2 and 3 sample the well region while trajectory 1 gets into the repulsive region because its impact parameter must be below the glory value b_g.

The quantitative relation between χ and b for a realistic intermolecular potential is summarized in Figure 4.5. For $|\chi| > |\chi_r|$ a one-to-one correspondence exists between b and χ; for $|\chi| < |\chi_r|$ three different values of b correspond to the same deflection angle $\theta = |\chi|$. This multi-valued relation between b and θ is not found for a purely repulsive potential, where $\chi(b)$ is a monotonically decreasing function of b, for example, Eq. (4.8).

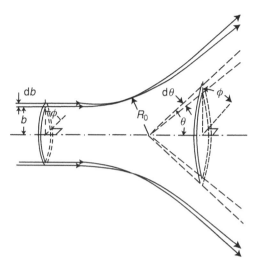

Figure 4.6 A construction showing the equivalence between the trajectories with initial impact parameters in the range b to $b + db$ and those with final deflections in the range θ to $\theta + d\theta$, $\theta = |\chi(b)|$. For a central potential there is no dependence on the azimuthal angle ϕ. When, as is usually the case, all initial values of ϕ are equally probable, the scattering has cylindrical symmetry. Trajectories with the final deflection angle θ will thus form a cone of equi-intensity.

If we could sample the b-dependence of the deflection function we would have a very sensitive probe of the intermolecular potential. In fact we shall see below that, at least for higher bs, the functional dependence of $\chi(b)$ on b is that of $V(b)$, the potential evaluated at $R = b$. In particular, the existence of a minimum in $\chi(b)$ is an indication of a well in the potential. Therefore the rainbow impact parameter should serve as a good indicator of R_m, the equilibrium distance of the potential. This expectation is used in drawing Figure 4.5.

4.1.4 Angular distribution in the c.m. system: the differential cross-section

Our aim is to measure the angular distribution after the collision, that is, the flux of molecules undergoing a deflection into the range θ to $\theta + d\theta$. Consider therefore a bunch of collision trajectories with impact parameters in the range b to $b + db$. As shown in Figure 4.6, the final deflection angle will be in the range θ to $\theta + d\theta$, where $\theta = |\chi(b)|$. For a spherically symmetric or "central"* potential, $V(R)$, there will be no dependence on the azimuthal angle ϕ. Thus all the trajectories that entered through a ring of radius b and width db will exit via an annular cone into a solid angle $d\omega$ with ϕ in the range 0 to 2π and θ in the

---

* Called central because it depends only on the center-to-center distance R.

range θ to $\theta + d\theta$. The final solid angle through which the trajectories exit is given by* $d\omega = 2\pi \sin\theta \, d\theta$.

We can measure** the number of molecules $d\dot{n}(\theta)$ deflected, per unit time, into the solid angle $d\omega = 2\pi \sin\theta \, d\theta$. In the same way that the collision cross-section is a measure of the rate of all collisions, Eq. (2.8), we introduce the differential collision cross-section $d\sigma/d\omega$ as a measure of the rate of collisions leading to deflections in a narrow angular range about θ. Let the incident flux be I. The initial velocity is perpendicular to the plane in which the ring of area $2\pi b \, db$ is drawn. By construction, all the trajectories that lead to deflections in the range θ to $\theta + d\theta$ have impact parameters in the range b to $b + db$ and vice versa. So the number of collisions per unit time with trajectories coming through the ring is $d\dot{n}(\theta) = I 2\pi b \, db$. We define the differential cross-section, $d\sigma/d\omega$, with units of area per steradian, by the condition of conservation of matter, i.e., $d\dot{n}(\theta) = I(d\sigma/d\omega)d\omega$ or $(d\sigma/db)db = (d\sigma/d\omega)d\omega$. Introducing the scattering intensity $I(\theta)$ by $d\sigma = I(\theta)d\omega = 2\pi b \, db$ we can express the differential cross-section as[†]

$$\frac{d\sigma}{d\omega} \equiv I(\theta) = 2\pi b \frac{db}{d\omega} = 2\pi b \bigg/ \frac{d\omega}{db}$$

$$= \frac{b}{\sin\theta (d\theta/db)} = \frac{b}{\sin\theta \, |d\chi/db|} \qquad\qquad b \leftrightarrow \theta \equiv |\chi(b)| \qquad (4.9)$$

(where the absolute sign is needed because $d\chi/db$ can be negative and we here assume a one-to-one relation between b and θ). The total cross-section is the integral of the differential cross-section:

$$\sigma = \int d\sigma \; \left(= 2\pi \int b \, db\right) = \int I(\theta)d\omega = 2\pi \int_0^\pi I(\theta)\sin\theta \, d\theta \qquad (4.10)$$

Given the deflection function $\chi(b)$ at the energy E of the experiment we can calculate (Figures 4.7 and 4.8) $I(\theta)$, the angular distribution of the scattering in the c.m. system, except[‡] where Eq. (4.9) is divergent. At these problematic points

---

* This is a standard result of solid geometry: consider a sphere of a (large) radius R about the origin. The trajectories cross the sphere through a ring (shown in Figure 4.6) of width $R \, d\theta$ and radius $R \sin\theta$. The area of the ring is thus $2\pi R^2 \sin\theta \, d\theta$. The solid angle subtended by the ring is this area for a unit sphere, $d\omega = 2\pi \sin\theta \, d\theta$. The unit of the solid angle is a steradian and the solid angle subtended by an entire sphere is 4π.

** Section 2.2.7.2 discusses the construction required for going from the c.m. angular distribution to the observed scattering in the laboratory system. The inverse problem is harder.

† There is an important refinement when the deflection function (cf. Figure 4.5) is such that there may be more than one value of b corresponding to a given $\theta = |\chi|$. This is discussed in Section 4.2.5.

‡ The divergence of Eq. (4.9) is a defect of the classical mechanical approach to what is really a quantal problem (cf. Figure 4.8). There are really three problems: (1) the divergence for large values of b, where $\theta \to 0$. This is discussed further in Section 4.2.4. (2) The divergence at the "glory impact parameter" where $\chi(b_g) = 0$, so that $\sin\theta = 0$ (also discussed in Section 4.2.4). (3) The divergence at the rainbow impact parameter where $d\chi/db = 0$, discussed

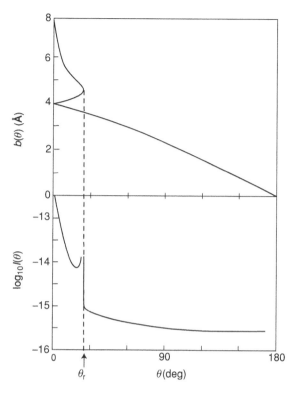

Figure 4.7 Classically calculated (Eq. [4.9]) angular distribution of the elastic scattering $I(\theta)$, for a realistic molecular system. Plotted is log $I(\theta)$ ($cm^2\,sr^{-1}$) and the corresponding $b(\theta)$ (Å), for the same collision energy. Note the rainbow divergence at θ_r and the divergence in the forward direction. In the backward direction the scattering is hard-sphere-like.

quantum mechanics comes to the rescue, as already mentioned in Section 2.2.6 and will be discussed in more detail in Section 4.3.

4.2 Elastic scattering as a probe of the interaction potential

4.2.1 Scattering as a probe of the potential

We are finally ready to use the angular distribution as a probe for the potential. As usual, we begin with the rigid-sphere model. Using Eq. (4.8) for the deflection function needed in Eq. (4.9), Eq. (4.10) yields,* for the hard-sphere scattering,

$$I(\theta) = d^2/4 \qquad\qquad (4.11)$$

a constant (independent of θ) angular distribution. This equation implies not only isotropic scattering, but also that the scattering is independent of collision energy.

in Section 4.2.5. In all three cases the quantal treatment yields a finite value. The physical manifestations of these classical singularities are an important source of information about the intermolecular potential. The classical divergences are all made finite by quantal interference effects and so are also important as an illustration that interference effects do have observable implications. We shall later discuss how this allows us an option of control.

* Introduce the variable $x = \cos(\theta/2)$. Equation (4.8) gives $x = b/d$. Hence compute db^2/dx^2 and use $dx^2/d\cos\theta = 1/2$.

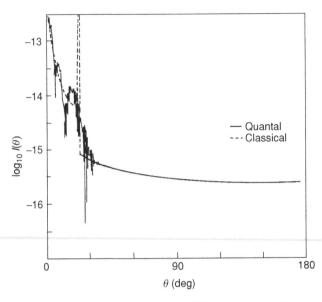

Figure 4.8 Comparison of quantal, solid curve, vs. classical, dashed, calculation of $I(\theta)$, on a logarithmic scale, for the same realistic potential at the same collision energy. The classical differential cross-section diverges at $\theta = 0$ and at the rainbow angle while the oscillatory quantal cross-section is everywhere finite. In the backward direction, where there is only one classical trajectory, there is close agreement between the classical and quantal results. The backward scattering results from low b collisions, and these sample the inner repulsive core of the potential. The backward angular distribution is therefore hard-sphere-like, see Eq. (4.11).

For a more realistic intermolecular potential function, however, the angular distribution will be anisotropic and energy-dependent. We expect the scattering to be concentrated in the low-angle, forward, direction as shown in Figure 4.7. The reason is the large contribution to the scattering from ever larger annular areas ($2\pi b \, db$) at large bs and thus small deflections (cf. Figure 4.5). Thus the low-angle scattering is mainly caused by, and thus tells us about, the long-range part of the potential. The wide-angle, backward scattering comes from the repulsive collisions at small b (note that $\chi(b \to 0) \to \pi$) and is approximately that of the hard-sphere model (to the extent that the realistic deflection function, Figure 4.6, resembles, at low bs, that for the hard-sphere model, Figure 4.3).

In summary, we reiterate: forward scattering originates from large impact parameters and hence is characteristic of the long-range force. Backward scattering originates from small impact parameters and hence is characteristic of the short-range force. The region of the well of the potential induces scattering in the rainbow range. This summary is so important that, below, we shall first derive it in a semi-quantitative fashion. Then we will get exact results in classical mechanics and finally discuss the quantal approach to the angular distribution of the elastic scattering.

4.2.2 The angle of deflection as a measure of the potential

A semi-quantitative route from the force acting on the trajectory to the resulting deflection is obtained by noting that a deflection means that the velocity vector has acquired a component perpendicular to its initial direction. When the initial impact parameter is high so that the deflection is small enough $\tan \chi \approx \chi$. Using $\perp$ to denote the component perpendicular to the initial direction of the velocity

$$\chi \cong \text{momentum}_\perp/\text{initial momentum} \qquad \text{high } bs \qquad (4.12)$$

By definition, before the collision, $\text{momentum}_\perp = 0$. Therefore, the required component of the momentum can be computed as the integral, along the trajectory, of the $\perp$ component of the force

$$\text{momentum}_\perp = -\int_{-\infty}^{\infty} dt \, (dV/dR)_\perp$$

$$= -2 \int_{b}^{\infty} dR \, (dt/dR) \, (dV/dR)_\perp \qquad (4.13)$$

The evaluation of the integral over the trajectory is done by breaking the trajectory into two identical contributions, up to the point of closest approach (which, to be consistent with the small angle approximation, we take to be at $R = b$) and from that point on. The integral $\int dt |(dV/dR)_\perp|$ can be evaluated exactly* for the long-range part of the potential, $-C_s/R^s$. Its approximate value is $V(b)$. Putting $E_T = \mu v^2/2$ we have the simple final result

$$\chi \approx V(b)/E_T \qquad (4.14)$$

All the approximations are forgiven in view of the considerable insight that is gained: the deflection function is the image of the potential. The small deflections arise from the long-range part, the rainbow minimum is the minimum of the potential and there is a steep increase in the deflection for small bs. Equation (4.14) was derived for high bs, but the only range where it is qualitatively wrong is $b \to 0$, where the correct deflection function goes to a finite value, namely π. Furthermore, Eq. (4.14) is a very explicit and simple result as to the role of the potential and the energy of the collision.

---

* Using y as the $\perp$ Cartesian component of R and the straight-line trajectory, $\partial R/\partial y = y/R \cong b/R$ or $(dV/dR)_\perp = (\partial V/\partial y) = (dV/dR)(\partial R/\partial y) \cong (dV/dR)b/R$. Also, from Eq. (4.7) or (4.17), $\partial R/\partial t = v\sqrt{R^2 - b^2}/R$.

*4.2.2.1. The energy and impact parameter dependence of
the angle of deflection*

Equation (4.14) is so useful that it is tempting to seek a better approximation. In
the high-energy limit where $V(b)/E_T < 1$, it can be shown that the variable $E_T\chi$
is a function of b **only** (independent of E_T). Therefore the useful generalization
is to replace $V(b)$ by a quantity $\tau(b)$, defined by

$$\tau(b) = E_T\chi(b, E_T) \tag{4.15}$$

such that for all bs, $\tau(b)$ is a function of b only and at large impact parameters
the b-dependence of τ is essentially that of $V(b)$. In terms of the variable τ, i.e.,
using Eq. (4.15), we can bring Eq. (4.9) to the form

$$\theta\sin\theta I(\theta) = (\tau/2)|db^2/d\tau| \tag{4.16}$$

where τ is only a function of b so that there is no explicit dependence upon E_T.
Hence, in this high-energy approximation, a plot of $\theta\sin\theta\, I(\theta)$ vs. τ is independent of the collision energy.

For a long-range potential of the form $V \propto R^{-s}$, $\theta\sin\theta\, I(\theta) \propto \tau^{-2/s}$. For small
angles ($\sin\theta \cong \theta$), this yields

$$\theta\sin\theta(E_T\theta)^{2/s} I(\theta) \cong E_T^{2/s}\theta^{(2s+2)/s} I(\theta) = \text{const.} \qquad \text{forward scattering} \tag{4.17}$$

For realistic molecular systems the attractive long-range potential is dominated
by an inverse sixth power R-dependence. For $s = 6$, Eq. (4.17) implies that
$I(\theta) \propto \theta^{-7/3}E_T^{-1/3}$, which has been found to accord well with the experimental
low-angle scattering data.

4.2.3   The quantitative route from the potential to the deflection function

Computing the deflection function enables us to go from the potential to the
observed scattering. Ultimately what we want to do is an *inversion*, that is, to
go from the observable angular distribution $I(\theta)$ to the intermolecular potential.
First, however, we discuss the direct route.

We evaluate the deflection function using the equation of motion, Eq. (4.6),
for the angle, ψ, of orientation of the interparticle vector **R**:

$$\dot{\psi} = bv/R^2 \tag{4.18}$$

where v is the initial velocity. To integrate Eq. (4.18) along the collision trajectory,
we need to know the classical path $R(t)$ as a function of time and to select the
initial, pre-collision value of ψ. From Eq. (4.7)

$$\dot{R} = v\left[1 - \frac{V(R)}{E_T} - \frac{b^2}{R^2}\right]^{1/2} \tag{4.19}$$

and so

$$(\mathrm{d}\psi/\mathrm{d}R) = \dot{\psi}/\dot{R} = -(b/R^2)\left[1 - \frac{V(R)}{E_\mathrm{T}} - \frac{b^2}{R^2}\right]^{-1/2} \tag{4.20}$$

The minus sign arises because the value of ψ increases with decreasing R, Figure 4.2. We can now integrate Eq. (4.20) from the pre-collision condition ($R \to \infty$, $\psi \to 0$) to the point of closest approach ($R = R_0$):

$$\psi_0 = \int_\infty^{R_0} (\mathrm{d}\psi/\mathrm{d}R)\mathrm{d}R = b\int_{R_0}^\infty \frac{\mathrm{d}R}{R^2\left[1 - \frac{V(R)}{E_\mathrm{T}} - \frac{b^2}{R^2}\right]^{1/2}} \tag{4.21}$$

From Figure 4.2, $\chi = \pi - 2\psi_0$ so that

$$\chi = \pi - 2b\int_{R_0}^\infty \mathrm{d}R\, R^{-2}[1 - V_\mathrm{eff}(R)/E_\mathrm{T}]^{-1/2} \tag{4.22}$$

For the collision of structureless particles, where the potential depends only on their separation R, Eq. (4.22) is exact within the framework of classical mechanics.

With Eq. (4.22) we have established a quantitative route from the potential $V(R)$ to the deflection function $\chi(b, E_\mathrm{T})$ and thereby to the differential cross-section $I(\theta)$. Next we begin to consider the inverse problem: how can we go from the observation of the angular distribution back to $V(R)$.

4.2.4 The total cross-section and the glory effect

We recall from Section 2.2.4 that the total collision cross-section can be written as the integral

$$\sigma = 2\pi\int_0^{b_\mathrm{c}} b\,\mathrm{d}b = \pi b_\mathrm{c}^2 \tag{4.23}$$

where b_c is the largest value of b that can lead to an observable angular deflection.

For finite-ranged potentials, Eq. (4.23) is very explicit* but for a realistic intermolecular potential while $\chi(b)$ decreases, it remains finite at any large b. Equation (4.23) therefore predicts an infinite total collision cross-section. Quantum mechanics rounds off this singularity and shows that the cross-section is finite.

One way to understand this qualitative failure of classical mechanics is to note that the origin of the problem is our attempt to observe very small deflections as a function of b. As we saw, a deflection arises from momentum transfer in the direction of b. So what we are seeking to do is to measure simultaneously momentum and position in the same direction. The Heisenberg uncertainty principle puts a limit on our inherent ability to do so. When can we no longer

---

* For example, for the rigid-sphere potential, $\chi = 0$ for $b \geq d$, so $b_\mathrm{max} = d$ and $\sigma = \pi d^2$.

resolve a deflection? When the inherent quantal uncertainties become comparable to the quantities themselves. The cutoff is therefore at $b = b_c$: momentum
transfer$_\perp \cdot b_c \cong h$. Dividing both sides by the initial momentum we find that the
cutoff is at $\chi \cdot b_c \cong h/p = \lambda$. Here λ is the de Broglie wavelength of the relative
motion. When the wavelength is λ, what the uncertainty principle tells us is that
the smallest angle that can be resolved is given by $\chi(b_c) = \lambda/b_c$. Below we offer
yet another way of looking at this essentially quantal phenomenon.

Thus a measurement of the total cross-section σ is equivalent to a measurement
of $\chi(b)$ at the *one* value of $b = b_c$. A measurement of the velocity dependence
of σ is therefore equivalent to measuring $\chi(b)$ over a range of b (at large b). In
particular, when b_c is large we can deduce the long-range part of the potential
function from the measured $\sigma(v)$. For the long-range form $V(R) \rightarrow -C_s/R^s$ we
have, using Eq. (4.14) for the small deflection angle, $\chi_c \approx 2C_s/b_c^s \mu v^2$. Equating
this value to the uncertainty principle cutoff, $\chi_c = h/\mu v b_c$, we have the upper limit
on the impact parameter leading to a discernible deflection, $b_c = (2C_s/hv)^{1/(s-1)}$.
Since b_c is larger at low velocities, at thermal energies the total collision cross-
section is determined primarily by the long-range part of the potential. Explicitly

$$\sigma = 2\pi (2C_s/hv)^{2/(s-1)} \tag{4.24}$$

where a more detailed evaluation shows that the fore-factor is only approximately
2π.

Referring to Figure 2.4 we note that a plot of $\log \sigma(v)$ vs. $\log v$ is essentially
linear also at higher velocities. This is also in accord with Eq. (4.24), because at
high collision energies the scattering is sensitive primarily to the repulsive part
of the potential, and this potential also scales roughly as R^{-s} except that for the
repulsive part typically $s > 6$.

Experiments with a good velocity resolution show an undulatory velocity
dependence of $\sigma(v)$ about the mean line predicted by (4.24). These *glory oscillations* are a manifestation of a quantum-mechanical interference phenomenon,
Section 2.2.6, between trajectories that are forward scattered, see Figure 4.5, and
those trajectories that are deemed to be unscattered, because their impact parameter is so large that it exceeds the cutoff, $b > b_c$. This interference depends on
the collision velocity as discussed in Section 2.2.6.

4.2.5 Rainbow scattering as a probe of the potential well

For closer-in collisions and when the potential has a well there can be three values
of b that contribute to the scattering, each corresponding to the same observable
deflection angle $\theta = |\chi|$, see Figure 4.5. For such a case Eq. (4.9) needs to be
written as

$$I(\theta) = \sum_{i=1}^{3} \frac{b_i}{\sin \theta |d\chi/db|_{b=b_i}}, \quad \theta \leq \theta_r \tag{4.25}$$

where the three values b_i are the three solutions of the implicit equation $\theta = |\chi(b_i)|$.

For scattering at angles over the rainbow, $\theta > \theta_r$, there is only one possible value of b, but for $\theta < \theta_r$ there are three. Thus there is a discontinuity in $I(\theta)$ at θ_r. But when θ is very near θ_r we have a range of impact parameters giving rise to scattering at the same angle, i.e., $|d\chi/db|_{b_r} = 0$. This reasoning implies a divergence in $I(\theta)$, but this discontinuity and divergence at θ_r is an artifact of the classical approximation.

In the quantal treatment of Section 4.3, the issue is resolved. The angular distribution exhibits an oscillatory interference pattern with an envelope showing successive humps and valleys for $\theta < \theta_r$ leading to a large maximum at about the rainbow angle θ_r. This oscillatory pattern is followed by a drop in the cross-section on the "dark" side of the rainbow. Beyond this drop the quantal and classical cross-sections are in close agreement with one another and so are roughly constant, as expected for hard-sphere scattering. Figure 4.8 showed a comparison of a quantal vs. classically calculated angular distribution.

Experimental observations of the rainbow effect provided early evidence for the existence (and magnitude) of the shallow, intermolecular potential well, for chemically non-bonding systems.

To sum up, elastic scattering measurements serve as a probe of the collision dynamics and can reveal detailed information on the intermolecular potential. But to do a complete job we cannot overlook quantal effects.

4.3 Elements of quantal scattering theory

4.3.1 Essential quantum mechanics: the superposition principle

In electricity and magnetism, the intensity is given by the square of the corresponding electric field amplitude. In quantum mechanics the probability is expressed as the (absolute) value squared of a corresponding amplitude. We are concerned with the scattering intensity and it is expressed in terms of the scattering amplitude $f(\theta)$ as

$$I(\theta) = |f(\theta)|^2 \tag{4.26}$$

The amplitude is generally a complex number. Computing it is the business of the Schrödinger equation. The essential point is that when a final outcome can be reached in several different ways, the amplitude to realize the outcome is the sum (= superposition) of the amplitudes for the different routes.* For molecular collisions, where classical dynamics is often a realistic description, it is a useful

---

* But in an experiment that is capable of determining which alternative is actually followed, we are
  to add intensities. In such an experiment there is no interference.

approximation* to regard different classical trajectories that lead to the same observation as the different routes. In other words, we assign an amplitude to each classical trajectory. For example, the scattering intensity is written differently on the two sides of the rainbow:

$$I(\theta) = \begin{cases} |f(\theta)|^2, & \theta > \theta_r \\ |f_1(\theta) + f_2(\theta) + f_3(\theta)|^2, & \theta < \theta_r \end{cases} \qquad (4.27)$$

For angles on the dark side of the rainbow there is only one trajectory that leads to scattering into a given angle. Classical mechanics may fail to predict fully quantitatively the scattering intensity, but it is not expected to have qualitative shortcomings. The situation is quite different on the bright side. In addition to the three classical terms, which are the analogs of the three terms in the purely classical result, Eq. (4.25), there are interference terms:

$$I(\theta) = |f_1(\theta)|^2 + |f_2(\theta)|^2 + |f_3(\theta)|^2 + 2|f_1(\theta)f_2(\theta)|\cos\delta_{1,2}$$
$$+ 2|f_1(\theta)f_3(\theta)|\cos\delta_{1,3} + 2|f_2(\theta)f_3(\theta)|\cos\delta_{2,3}, \qquad \theta < \theta_r \quad (4.28)$$

$\delta_{n,m} = \delta_n - \delta_m$ is the difference in the phase of the corresponding amplitudes, where $f_n = |f_n|\exp(i\delta_n)$. The determination of the phase is one of the key subjects of this section.

The intensities, $|f_n(\theta)|^2$, can be reliably estimated from classical mechanics. In other words, each such term is of the form of Eq. (4.9). Hence the moduli of the interference terms are not small and so their importance is to be judged by the phase differences. For a perfect resolution in scattering angle and in collision energy the quantum and classical results can therefore be very different. But the phase difference is large compared to 2π and it is rapidly varying with energy and scattering angle so that the interference terms tend to average out. There are two very important exceptions caused by the phase difference between different amplitudes being slowly varying. Either case corresponds to interference between only two trajectories. These two cases are therefore not only important in their own right, but also as clear examples of the observable role of interference. As much as we love classical mechanics because of its simplicity and intuitive appeal, we live in a quantum world in which trajectories on the microscopic level are not well defined. Not only is the quantum world one that is inherently more wiggly and undulatory, it is also one that removes the discontinuities of classical treatments and replaces them with transitions from oscillatory behavior on the bright side of the rainbow to an exponential falloff on the dark side.

As shown in Figure 4.5, there are three trajectories that lead to scattering into angles just below the rainbow. Trajectories 3 and 2 sample the well region of the

---

* Approximations can go wrong. What we say is almost always correct but there are very interesting exceptions. These are possible because certain outcomes are observable and yet they cannot occur in classical mechanics. Tunneling through a barrier is the most familiar example.

potential and so have comparable impact parameters. From the approximation Eq. (4.14), the bs are just above and just below the equilibrium distance, R_m, of the potential. Trajectory 1, which samples the repulsive forces, has an impact parameter smaller than the range parameter σ, which is the distance of the onset of repulsion, cf. Figure 2.5. Trajectories 3 and 2 are similar and the interference between them leads to the rainbow oscillations in the quantum mechanical cross-section.

Glory oscillations occur for low angle scattering. It is now trajectories 2 and 1 that are similar. Trajectory 3, which samples the long-range attractive forces, has a high impact parameter and so corresponds to a far higher cross-section. Trajectories 2 and 1 correspond to a low angle scattering because they sample the attractive and repulsive forces to about the same extent. It is their interference that leads to glory oscillations.

Computing the amplitudes and their phases is the subject of quantum mechanical scattering theory although, as a practical matter, semiclassical approximations are quite useful.

4.3.2 The quantum mechanical approach to elastic scattering

Our purpose is to discuss the key aspects of the quantum mechanical treatment of elastic scattering rather than to offer detailed derivations.[1] We shall be guided by the classical approach. The essential difference is the superposition principle, unique to quantum mechanics, as introduced in the last subsection. What this means at the technical level is that the Schrödinger equation is linear, so any linear combination of solutions is a physically possible solution. The way we want to use this idea is to find such linear combinations of wave functions that are appropriate for a scattering event. The second guide will be the familiar orbital structure of the hydrogen atom, where the wave function describes the relative motion of the electron and proton under a central (= coulomb) force. We here use the time-independent approach, where a stationary scattering wave function of definite energy is to describe the entire course of the collision. In later chapters we shall introduce the time-dependent approach, often known as a wave-packet picture.

The physics that corresponds to the stationary, time-independent, description is that there is a large number of binary collisions. Some have started a long time ago and the products are already receding, others are yet to enter the range of the force. In such a description the rate of collisions is constant. It is like a crossed molecular beam experiment that is running under steady conditions. Like the experiment, what we want to determine is the flux of products.

To write a stationary scattering wave function we argue by a familiar example. Textbooks write the wave function for the hydrogen atom at a given energy as a product of a radial wave function, the orbital, labeled by the principal quantum number n and the angular momentum quantum number l ($l = 0, 1, 2, \ldots$, often being designated as $s, p, d, \ldots$), and a wave function giving the angular shape of the orbital. This angular part of the wave function is labeled by l and m, where m is the projection of l on the z axis, $-l \leq m \leq l$. m provides the discrete quantum version of the azimuthal angle ϕ that specifies the plane of the collision. One other familiar result about the hydrogen atom is that orbitals of different values of l but the same value of n are degenerate, i.e., they have the same energy.

The form of the stationary scattering wave function is an immediate application of what we know about the hydrogen atom and what we discussed about the classical description of the approach motion. First of all, we typically do not select m and l. We would like to be able to select the plane of the collision and the magnitude of the impact parameter, but very often we do not. Therefore, the scattering wave function will be a superposition of degenerate wave functions of given m and l. As for the hydrogen atom, these wave functions are products of an orbital part, a part that describes the relative motion, and an angular part. Two technical points. First, we need an analog of the principal quantum number n, the quantum number that determines the energy. This extension is required because for our problem the relative motion is unbounded. It is traditional to use the wave number k for this purpose. We define k in terms of the energy $E = \hbar^2 k^2 / 2\mu$ or, as a vector, $\mathbf{p} = \hbar \mathbf{k}$. Second, electrons are lighter than nuclei and so, as we have already seen in Problem A, quite high values of l, of the order of $k\sigma$, σ range of the potential, can contribute to the wave function for heavy particle collisions. It is an important practical point but there is nothing essential about it. For collisions of very cold atoms, for which the value of k is rather low, it can easily be the case that only very few values of l, or even only s-wave ($l = 0$) collisions, are important.

We are ready for the scattering wave function at a given energy, expressed as a superposition of states over all l and m values:

$$\psi_{\mathbf{k}}(\mathbf{R}) = 4\pi \sum_{l=0}^{\infty} \sum_{m=-l}^{l} \mathrm{i}^l \psi_{kl}(R) Y_l^m(\hat{\mathbf{R}}) Y_l^{m*}(\hat{\mathbf{k}}) \tag{4.29}$$

The Y_l^ms are the spherical harmonics. The carat denotes the solid angle in the direction of the vector. The (absolute value) squares of the spherical harmonics are familiar as the angular shape of electronic orbitals. The key point about the superposition Eq. (4.29) is that it gives equal weight to all the orbitals of given l and m. This is because there is no selection of either the magnitude or the direction of the angular momentum of the relative motion of the two particles that collide. This is just as in classical mechanics, where we allowed all values of the azimuthal angle ϕ and all values of the impact parameter. The new feature is that we sum over amplitudes and not over observables. The wave functions $\psi_{kl}(R)$ that describe the relative motion for a given value of l need yet to be determined by requiring that the form Eq. (4.29) solves the Schrödinger equation. For convenience, we took the factor $\left(\sqrt{-1}\right)^l = \exp(\mathrm{i}l\pi/2)$ out of these wave functions.

The collision is also specified by the initial momentum that we write as the wave vector $\mathbf{k}$. We can refer the direction of the vector $\mathbf{R}$ to the initial direction of the vector $\mathbf{k}$. The addition theorem of spherical harmonics (Zare, 1988), where the $P_l(\cos\theta)$s are the Legendre polynomials,

$$P_l(\hat{\mathbf{R}} \cdot \hat{\mathbf{k}}) = \frac{4\pi}{2l+1} \sum_{m=-l}^{l} Y_l^m(\hat{\mathbf{R}}) Y_l^{m*}(\hat{\mathbf{k}}) \qquad (4.30)$$

is the made-to-measure tool for doing so. The resulting form of the wave function is

$$\psi_{\mathbf{k}}(\mathbf{R}) = \sum_{l=0}^{\infty} (2l+1) \, i^l \psi_{kl}(R) \, P_l(\cos\theta) \qquad (4.31)$$

and this form is what you will find in most textbooks. θ is the polar angle of the position vector $\mathbf{R}$ with respect to the incident velocity. The orbitals that appear in the expansion of the wave function are known as partial waves.[2] As for the hydrogen atom and other central potential problems it is convenient to factor out an R^{-1} dependence from the orbital and to define a *radial wave function* $G_l(R) = kR\psi_{kl}(R)$. As shown in Eq. (4.36) below, the radial wave function vanishes at the origin.

Let us check. A scattering wave function must also describe the relative motion of two structureless particles if no force acts between them. Such a motion is described by a *plane wave*, where the momentum is constant along the trajectory. We have the mathematical identity (Levine, 1969; Zare, 1988)

$$\exp(i\mathbf{k}\cdot\mathbf{R}) = \exp(ikR\cos\theta) = \sum_{l=0}^{\infty} (2l+1) \, i^l j_{kl}(R) \, P_l(\cos\theta) \qquad (4.32)$$

Here $j_{kl}(R) = j_l(kR)$ are the (regular) *spherical Bessel functions* and are therefore the orbitals for a free motion. The adjective "regular" refers to an orbital that is finite at the origin, just as in the theory of electronic orbitals where, as here, on physical grounds irregular solutions are excluded.

The essential new feature of scattering problems as compared to bound state problems is that the **boundary conditions**, namely the behavior of the orbitals as $R \to \infty$, need to be explicitly specified.* Common to all elastic scattering problems is that the incident motion is in the direction of decreasing R and the scattered motion is in the direction of increasing R. First, say that there is no potential. Then the orbitals should give equal weight to the incoming and outgoing waves. The mathematical form of the Bessel functions for large argument shows that this is

---

* For bound states, the condition that the wave function is normalized implicitly specifies that it vanishes at large distances. For scattering at a fixed energy, we do not want to impose the condition that the number of collisions is one (or some other finite number). Rather, we want collisions to occur at a steady rate. This requires a constant flux of incoming molecules. The flux is, as usual, the velocity times the density. The density is $|\phi|^2$ where ϕ is the incoming wave. The velocity is constant because the wave vector k is specified.

indeed so:

$$G_l(R) \propto kRj_l(kR) \to \begin{cases} (\text{i}/2)\{\exp\left[-\text{i}(kR - l\pi/2)\right] \\ \quad - \exp\left[\text{i}(kR - l\pi/2)\right]\}, & R \to \infty \\ 0, & R \to 0 \end{cases} \quad (4.33)$$

where $j_l(kR)$ is the regular spherical Bessel function of order l. It is an oscillatory function having essentially its sinusoidal R-dependence all the way back to the point where the first maximum of $\sin(kR - l\pi/2)$ occurs; it then declines to zero. Because $l + 1/2 \approx \mu vb/\hbar = kb$, the innermost maximum of the radial wave function occurs at about the classical turning point, $R = b$, of the free motion.

When there is a potential, it will no longer be the case that the outgoing wave in Eq. (4.33) has the same weight as the incoming wave. Therefore, when R is large enough that the role of the potential is negligible, we expect the wave function $G_l(R)$ to be a superposition, as in Eq. (4.33), but now the two components differ in their amplitude. We write this as

$$G_l(R) \xrightarrow{R \to \infty} (\text{i}/2)\{\exp\left[-\text{i}(kR - l\pi/2)\right] - S_l \exp\left[\text{i}(kR - l\pi/2)\right]\} \quad (4.34)$$

Here S_l is the *scattering amplitude* for the lth partial wave. It is our first encounter with an element of what is known in general as the *scattering matrix*. The conservation of the angular momentum and of the collision energy means that for our case the scattering matrix is diagonal and its diagonal elements are the S_ls. The matrix is defined such that, if there is no scattering it reduces to the unit matrix. Indeed, for $S_l = 1$, Eq. (4.34) reduces to (4.33). In the general case of elastic scattering, where the flux that comes in must equal the flux that goes out, it remains the case that $|S_l| = 1$. We therefore put

$$S_l = \exp(2\text{i}\delta_l) \quad (4.35)$$

and refer to the real number δ_l as the *phase shift*. The origin of the name is as follows. In terms of the shape of the orbitals, what we have learned is that the presence of the potential causes the wave function to have its argument shifted. The radial wave function has a sinusoidal shape so we refer to its argument as the *phase* of the orbital. Owing to the action of the potential, the radial wave function will have a phase somewhat different from that for the free motion. What matters to us is the charge, or *shift*, in the phase. We expect, for example, that if the potential is repulsive, the wave function is pushed out compared to the case of no scattering, see Figure 4.9. Compared to the orbital without a potential, $kRj_l(kR) \to \sin(kR - l\pi/2)$, the general solution of the scattering problem, Eq. (4.34) with (4.35), has the form

$$G_l(R) \to \begin{cases} \exp(\text{i}\delta_l)\sin(kR - l\pi/2 + \delta_l), & R \to \infty \\ 0, & R \to 0 \end{cases} \quad (4.36)$$

where δ_l is the phase shift.

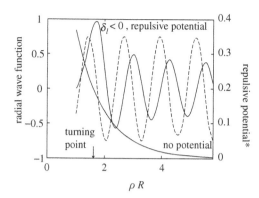

Figure 4.9 The orbital for scattering by a repulsive potential $\exp(-\rho R)$. The figure shows that a repulsive potential keeps the wave function out of the region that is accessible to the motion without a potential. Outside the range of the potential, this orbital must describe a free motion. It therefore differs from an orbital that describes a motion free for all Rs, dashed curve, by having the argument of its trigonometric function (= the phase) shifted.

For elastic scattering the only role played by the potential is to shift the phase of the wave function. The magnitude of this shift is computed by solving* for the radial wave function and determining δ_l from its definition by using the asymptotic form, Eq. (4.36).

What we have concluded is actually rather remarkable. If all that we can do is to probe the system after the collision is over, that is, as $R \to \infty$, then everything about the dynamics of the lth partial wave is contained in one number, the phase shift, δ_l! We turn to some important applications of this result. In addition, this observation points to a direction for further research: to get deeper we need to be able to probe the collision as it is taking place and not only after it is over.

4.3.3 The scattering amplitude

The amplitude $f(\theta)$ of the scattered wave at the angle θ from the direction of the incident wave is defined by comparing the full scattering wave function to $\phi(\mathbf{R})$, the wave function for free motion:

$$\psi_{\mathbf{k}}(\mathbf{R}) \xrightarrow{R \to \infty} \phi(\mathbf{R}) + \frac{\exp(i\mathbf{k} \cdot \mathbf{R})}{R} f(\theta) \tag{4.37}$$

Using Eqs. (4.31), (4.32), and (4.36), we obtain after some rearrangements

$$f(\theta) = (2ik)^{-1} \sum_{l=0}^{\infty} (2l + 1) \left[\exp(2i\delta_l) - 1\right] P_l(\cos\theta) \tag{4.38}$$

---

* For a general potential the solution needs to be done numerically. Even a most modest PC can handle this task. Many mathematics packages offer dedicated routines. The secret is to insure that one generates a regular solution, namely an orbital that vanishes at the origin. Semiclassical approximations for evaluating the phase shift as a definite integral are readily available and have the additional merit of providing insight.

The scattered intensity is $I(\theta) = |f(\theta)|^2$ as in Eq. (4.26), where $I(\theta)$ is the differential cross-section.

Two features of our result for the scattering amplitude, Eq. (4.38), deserve immediate comment. First we need to prove that the infinite sum in Eq. (4.38) is convergent. The key is that it is $S_l - 1$ rather than S_l itself that appears in the scattering amplitude of the lth partial wave. For sufficiently large l, the wave function hardly samples the potential (being kept away by the centrifugal barrier). As l increases, the phase shift must therefore ultimately decline to zero, or $S_l \rightarrow 1$ when $l \gg k\sigma$, where σ is the range parameter of the potential (shown in Figure 2.5). Hence the sum of Eq. (4.38) is effectively a sum over a (typically large, see Problem A, but) finite number of terms and thus convergent! The classical divergences of the differential cross-section have been eliminated.*

Still, the sum of Eq. (4.38) will, in general, contain many terms. The reason is that at all but the very lowest relative velocities, atoms and molecules are heavy enough that their de Broglie wavelength λ is significantly shorter than the range σ of the potential. But if $\lambda = 2\pi/k < \sigma$, then $k\sigma > 1$ and we must retain many terms, up to $l \gg k\sigma$, in the sum of Eq. (4.38). Because the Legendre polynomials are oscillatory functions of θ, there will be very many highly oscillatory terms when we compute $|f(\theta)|^2$, as shown in Figure 4.8. At higher relative velocities, as the de Broglie wavelength gets shorter (and more terms need to be included), the oscillations become more rapid and tend increasingly to average out to the classical scattering angle dependence for a detector with finite angular resolution.

Much as the deflection function serves as an intermediate construct between the potential and the observed differential cross-section in classical mechanics, so does the set of phase shifts in quantum mechanics. Indeed, the relation between the two is even closer than this formal similarity might suggest. A sketchy but correct relation is provided in the footnote, leading to the conclusion that while all the partial waves contribute to the scattering at a given angle θ only a narrow range of l values contributes significantly. This range is determined by**

$$\theta = 2\frac{\partial \delta_l}{\partial l} \tag{4.39}$$

---

* The secret is the factor $S_l - 1$ that appears in Eq. (4.38). The "1" comes from the presence of the incident wave, $\phi(\mathbf{R})$, in the scattering wave function, Eq. (4.37). It is the superposition of the incident and scattered waves that eliminates the classical divergences. This is also the origin of the shadow contribution that we mention later.

** To derive Eq. (4.39) take l to be a continuous variable and regard the partial wave sum in Eq. (4.38) as an integral over l. In principle all the partial waves contribute to the scattering at a given angle θ. In practice, because the Legendre polynomials at a given θ are a rapidly varying function of l, only a narrow range of l values contributes significantly. This range is determined as follows. When the partial wave sum is written as an integral, the integrand rapidly oscillates about the value zero. It therefore averages out. Why is the integral not zero? Because there is a narrow range in l where the integrand is not oscillatory. This is the range where the oscillation of $P_l(\cos \theta)$ exactly cancels the oscillation of $\exp(2i\delta_l)$. At a given θ, $P_l(\cos \theta)$ varies with l as $\sin(l\theta)$. The range in l where the integrand is a stationary function of l is where its phase, $\exp(i(2\delta_l - l\theta))$, regarded as a function of l, is constant. The point of stationary phase is at $\partial(2\delta_l - l\theta)/\partial l = 0$.

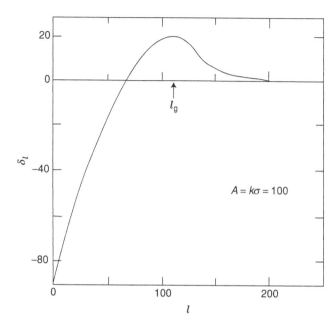

Figure 4.10 The phase shift δ_l computed for a realistic interatomic potential vs. l. The computation is for the realistic value $A = k\sigma = 100$, meaning that many partial waves contribute to the scattering (σ is the range of the potential and $k = p/\hbar$ is the wave vector). Note the steep variation of the phase shift at lower ls that is due to the repulsive core of the potential (the initial decline is with a slope of $\pi/2$, which is what we expect for a hard-sphere scattering). The stationary point occurs at the glory impact parameter, $l_g = k b_g$.

Equation (4.39) is a "semiclassical" correspondence between the l-dependence of the phase shift as shown in Figure 4.10, and the deflection angle at the impact parameter $b \cong l/k$. Conversely, for a given θ, Eq. (4.39) determines the dominant l value that leads to scattering into that given angle.

4.3.4 The cross-section and the random phase approximation

By integrating the scattering intensity over all scattering angles θ and over all collision planes specified by the azimuthal angle ϕ we obtain the quantum mechanical total cross-section as a sum over the contributions of partial waves

$$\sigma = \iint d\cos\theta \, d\phi \, |f(\theta)|^2 = \frac{4\pi}{k^2} \sum_{l=0}^{\infty} (2l+1) \sin^2 \delta_l \tag{4.40}$$

The infinite sum effectively terminates when the phase shift becomes small enough. When many terms contribute to the sum, it follows from Eq. (4.39) that before the phase shift declines to zero it falls through many values of π. Therefore $\sin^2 \delta_l$ can be approximated by the typical value of $1/2$. This limiting behavior, that the phase shift is large enough that its restriction to the interval 0 to π

yields effectively a random variable, is a useful approximation in many aspects of the scattering of heavy particles. It does have its limitations, for example, when the phase shift is not a rapidly varying function of l. Such will be the case near the glory condition, where the deflection is zero. The random phase approximation for the cross-section, Eq. (4.41) below, therefore will not exhibit the glory oscillations.

Ultimately the phase shift stops changing rapidly. We denote by l_c the highest value of l at which $\sin^2 \delta_l = 1/2$. Neglecting the contribution from higher l values to the sum (which is an error of about 10–20%), the cross-section is given by

$$\sigma \cong \frac{4\pi}{k^2} \sum_{l=0}^{l_c} (2l+1)\frac{1}{2} \cong \frac{2\pi l_c^2}{k^2} \cong 2\pi b_c^2 \tag{4.41}$$

where $b_c \equiv l_c/k$ is the maximal impact parameter that contributes to the cross-section. The result looks classical but, on closer inspection, it is twice as large! This quantal effect is real and it is known as *shadow scattering*.

4.3.5 Time delay and resonances

The phase shift is a function of the collision energy and of the angular momentum. The l-dependence gives rise to the angular deflection as explicitly seen in Eq. (4.39). Here we show that the energy dependence of the phase shift measures the time deflection of the scattered wave. We define the temporal deflection just as we did for the angular deflection, that is, with reference to the unscattered wave.

Recall that as a result of the scattering the wave function acquires a phase shift compared with the unperturbed wave. The time-dependent form of the outgoing wave is $\exp[i(2\delta_l + kR - Et/\hbar - l\pi/2)]$, where we multiplied the stationary outgoing wave, Eq. (4.34), by the time dependence of a stationary state, $\exp(-iEt/\hbar)$.

To proceed correctly we should form a wave-packet as a linear superposition of waves of different energies. Taking a liberty, we look at one component and ask when will the wave function reach a point R, where R is large and so the products are out of the range of the potential. The wave function is delocalized, but for such R values that the rapidly varying exponent of $\exp[i(kR - Et/\hbar - l\pi/2)]$ will be nearly stationary the wave function is maximal. (About that point the wave function is not oscillatory and so there is a high probability of locating the system in that region.) Because $E = \hbar^2 k^2/2\mu$, taking a derivative with respect to k shows that the stationary phase is at $R = (\hbar k/\mu)t = vt$, which is just what we would expect from classical mechanics. If there is a potential, the stationary phase of the outgoing wave function is at $R = (\hbar k/\mu)t + 2\partial\delta_l/\partial k = v(t + \tau)$ where τ

$$\tau = \frac{2}{v}\left(\frac{\partial\delta_l}{\partial k}\right)_l = 2\hbar\left(\frac{\partial\delta_l}{\partial E}\right)_l \tag{4.42}$$

is the additional *time delay* arising from the scattering.

The formalism does not promise that the time delay is positive. Indeed, why should it necessarily be positive? If the potential is repulsive, the scattering wave function cannot penetrate into the repulsive core while the wave function in the absence of the potential can get further in. So the scattered wave will come out ahead. We can almost guess from the shift $2\partial\delta_l/\partial k$ in the location of the wave front that for low ls, $\partial\delta_l/\partial k$ is the range, d, of the repulsive core so that $\delta_l \approx -kd + l\pi/2$. The second term is to insure the backward scattering for low impact parameters, cf. Eq. (4.39). A well in the potential will also lead to a negative delay because the kinetic energy over the well is higher than it is in its absence. Structureless particles appear to exit rather promptly from the region of their interaction. We speak of direct collisions in this case.

In a situation where the time delay is positive, the colliding particles spend a longer time in the region of their interaction, longer than the time necessary to transverse the region. What can cause such a delay? We shall encounter many examples when the colliding particles do have an internal structure. Then the kinetic energy of the relative motion can be converted to internal excitation, leaving the relative motion bound in the attractive well of the potential $V(R)$. Eventually the energy will reflow into the relative motion and the particles will separate, after being delayed. For structureless particles, a recognized route to a delay is by quantal tunneling through the centrifugal barrier, as discussed in Section 4.3.6.

Unlike the case of a direct collision, a delay requires that the phase shift increases with energy.[3] A long delay requires a steep increase and this behavior is often represented by a Breit–Wigner form

$$\delta = \delta_{\text{direct}} + \arctan\left(\frac{\Gamma/2}{E_0 - E}\right) \tag{4.43}$$

shown in Figure 4.11. Here δ_{direct} is the *background* or *direct* contribution, which is present as a term that decreases moderately with energy. E_0 is the center energy of the resonance and Γ is the range in energy over which the contribution of the resonance is significant. For a number of good reasons we refer to Γ as the *width of the resonance*. One such reason is the delay time associated with a resonance located at E_0 is $\hbar/\Gamma$. From the time–energy uncertainty principle we therefore expect that the energy of the resonance is defined only up to Γ. An equivalent way to understand this behavior is to plot the contribution of the resonance to the cross-section as also shown in Figure 4.11. From Eq. (4.40), the cross-section has an energy dependence known as a *Fano profile*

$$\sin^2\delta = \sin^2\left[\delta_{\text{direct}} + \arctan\left(\frac{\Gamma/2}{E_0 - E}\right)\right] = \frac{(\varepsilon + q)^2}{1 + \varepsilon^2}\sin^2\delta_{\text{direct}}$$

$$\xrightarrow{\delta_{\text{direct}} = 0} \frac{(\Gamma/2)^2}{(E - E_0)^2 + (\Gamma/2)^2} \tag{4.44}$$

The two line profile parameters are the index, $q = \cot\delta_{\text{direct}}$, which measures the asymmetry of the line arising from the interference[4] between the direct and

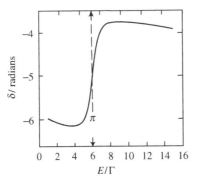
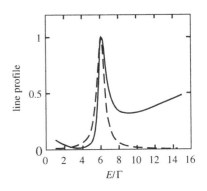

Figure 4.11 A resonance in elastic scattering. Left: the phase shift in the energy vicinity of a resonance, Eq. (4.43), vs. the energy in units of the width, Γ, of the resonance (the resonance is at the energy E_0; in the plot, $E_0/\Gamma = 6$). Outside the range of the resonance the scattering is direct and the phase shift decreases with increasing energy. The dashed line is to show that the resonant part of the phase shift increases by π as the energy increases about E_0. Right: the contribution of such a resonance to the cross-section, Eq. (4.44). The role of the direct scattering is measured by the asymmetry parameter q. If there is no direct background, the line profile is a symmetric Lorentzian about E_0 with a width Γ, shown as a dashed line.

compound scattering and the reduced energy, $\varepsilon = (E_0 - E)/(\Gamma/2)$. If there is no direct scattering, we find, using $\sin \arctan(1/\varepsilon) = 1/(\varepsilon^2 + 1)^2$, the symmetric *Breit–Wigner profile*, which has a Lorentzian shape.

The increment in the cross-section arising from a resonance as discussed above occurs in a particular partial wave. Under usual scattering conditions, when many partial waves contribute, the observable role will be small. To observe a resonance we need a large width, or a mechanism where many partial waves are affected, or the experimental ability to select one or a few partial waves. As mentioned in the next section, "half collisions" are one way toward such a selection.

4.3.6   Low-energy collisions: classical orbiting and quantal resonances

Figure 2.10(a) shows the effective potential for a collision with a finite impact parameter. Coming from far apart and moving classically, the two particles can only approach up to the distance of closest approach where their kinetic energy equals the effective potential, Eq. (2.31). But quantum mechanically, the radial motion can tunnel through the barrier and then stay in the classical regime in the well to the left of the barrier. Eventually, the particles will tunnel back out and escape. This situation is possible whenever there are multiple solutions for the classical turning point, R_{tp}, defined by analogy to Eq. (2.31):

$$E_T = V_{\text{eff}}(R_{\text{tp}}) = V(R_{\text{tp}}) + \hbar^2 l(l+1)/2\mu R_{\text{tp}}^2 \tag{4.45}$$

The classical signature of these resonances is the orbiting; that is, the rather large (negative) deflection angle χ when the collision energy equals the top of the barrier in the effective potential (Ford and Wheeler, 1959). For an exactly orbiting collision, the two particles can rotate around their center of mass for any number of times before they separate again. The height of the barrier in the effective potential is not high. We have already computed it, with another motivation, in Chapter 2, but for realistic potentials it is typically below the value of the well depth. So the orbiting region and hence the tunneling resonances are characteristic of low-energy collisions, unless there is a strong chemical interaction between the particles.

Historically, tunneling resonances and their line profiles were first seen in so-called half collisions. That is, the quasi-bound state of a given value of l is prepared by optical excitation from a bound lower state. The absorption line to such a state will have a width in frequency arising from the inherent uncertainty in the energy of a state that is only temporarily bound. When the dissociation is by tunneling through a centrifugal barrier, hydrides offer the best examples because the width will not be too small. The reason is that, by the uncertainty principle, the width in frequency is the inverse lifetime. As a rough guide, the width is given by the frequency of the bound state motion reaching the left side of the barrier, $\hbar\omega/2\pi$, times the (small) probability p of tunneling through the barrier:[5]

$$\Gamma = (1/\text{lifetime}) \approx (1/\tau_{\text{vib}}) \text{ probability of tunneling} \qquad (4.46)$$
$$= (\hbar\omega/2\pi)p$$

Semiclassically, $p \approx \exp(-\int dR\sqrt{2\mu(V(R)-E)})$ where the integral spans the region where the system moves under the barrier. We therefore expect p to scale exponentially down with the reduced mass. This type of "momentum gap" estimate provides a useful guide also in other examples of what we call predissociation. For tunneling, the *momentum gap rule* also tells us that states at energies just below the barrier height have the largest widths and so will be most amenable to observation in the frequency domain.

4.4 Angular distribution for reactive molecular collisions

Our agenda in this section is quite rich. We start with direct reactive collisions and the physically complementary case of compound collisions where the molecules stay for a while in the region of the chemical forces. Consistently with earlier sections of this chapter, we look at the angular distribution from a two-body point of view. Only later in the book will we leave the two-body simplicity and recognize that the products have different internal states that can be populated. The discussion will not be as quantitative as in earlier sections, and will blend physical

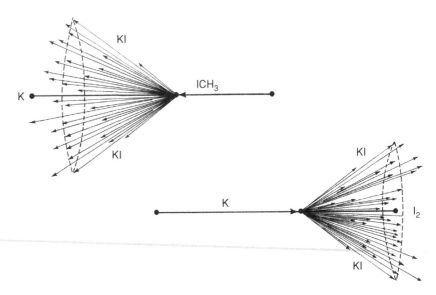

Figure 4.12 The "cones" containing most of the products' intensity (in the c.m. system). For the $CH_3I + K$ reaction (above) the KI product appears primarily in the backward ("rebound") direction; not so for the KI from $K + I_2$ (below), which shows typical stripping behavior.

insights with what we have learned about elastic scattering. This approach is sometimes known as the *optical model*.[6]

4.4.1 The angular distribution as a probe of direct vs. compound collisions

The observation of the very specific energy disposal in exoergic reactions (Section 1.2) is often accompanied by a strong preferential angular disposition of the products. For example, in reactions such as

$$CH_3I + K \rightarrow KI + CH_3$$

The KI product molecules were observed in the "backward" hemisphere, with respect to the incident K atom. This is a *rebound mode*. On the other hand, for reactions such as

$$K + I_2 \rightarrow KI + I$$

the KI is scattered strongly "forward," i.e., in the direction of the incident K atom. This forward scattering is typical of *stripping mode* behavior (the spectator limit), discussed already in Chapter 1. A generalization will be discussed in Section 4.4.3 below and then in Section 10.2.2.

Figure 4.12 shows the predominant "cones" of scattered products for two KI-forming reactions. Such anisotropic angular distributions, when first observed

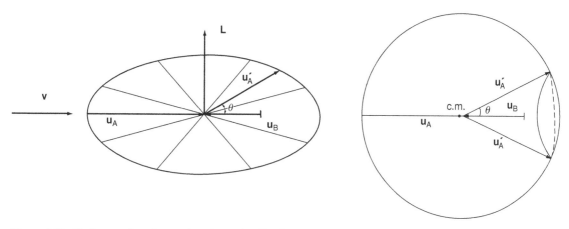

Figure 4.13 Understanding the products' angular distribution in complex-forming reactions. Left: A + B collision in the center-of-mass system. The collision takes place in a plane perpendicular to the direction of the angular momentum **L**. A long-living A–B "diatomic" complex is formed. The complex dissociates along its axis. If A–B rotated for a while prior to dissociation, the products would come out in the plane of the collision, uniformly in all directions. This is called, pictorially, the *sprinkler model*. Right: construction for the observed angular distribution. The direction of **L** is not selected. It is therefore necessary to average over all possible orientations. When we rotate **L** about the axis formed by the initial (c.m.) velocities the products come out distributed on a sphere, as shown. In particular, all products that dissociate into the angular range θ to $\theta + d\theta$ come out through a ring. The scattering intensity through a ring is constant, if the complex is long living. The area of the ring, $2\pi \sin\theta \, d\theta$, is minimal near the two poles, where ($\theta \to 0$), and maximal near the equator. The scattering intensity is defined as the flux of products into the angular range $\sin\theta \, d\theta \, d\phi$. Hence the scattering intensity is maximal near the two poles.

in the early molecular beam scattering studies of the early 1960s, were considered rather unexpected. The preferential scattering indicates that the process of reaction (i.e., the atomic rearrangement) must be over quickly. Chemists expected the reactants to form a "complex" that lasts for a while before forming the products. Say that this mechanism is correct. A two-body view of two particles forming a complex is shown in Figure 4.13. The conservation of angular momentum does not depend on how sticky the collision is, so the two-particle complex rotates in the plane of the collision. It acts just like a bound diatomic molecule with the rotational angular momentum $l \cong \mu v b / \hbar$. Eventually the complex will dissociate and the two particles will fly apart. If the complex survived for more than a few classical rotation periods the products would fly off in a random direction and not in such a concentrated, "directed" cone disposed so specifically with respect to the initial relative velocity vector.

Reactions that proceed through a compound mechanism and where the complex is long living (compared to a rotational period of the interparticle axis) will have an angular distribution that is uniform in the plane of the collision. Here again we remember that typically the orientation of this plane is not selected and so we must allow for all azimuthal angles ϕ. When we do so we recognize that not all scattering angles θ are equivalent. Products that exit in the original direction or opposite to it, i.e., products that exit near the poles, $\theta \approx 0$ or π, all come out in a small solid angle. Products that exit near the equator, $\theta \approx \pi/2$, all come out in a wide solid angle. Therefore, for compound collisions, the angular distribution in space is not uniform. Rather, it is maximal near the poles and minimal near the equator:

$$\frac{d\sigma}{d\omega} = \frac{d\sigma}{2\pi \sin\theta \, d\theta} \propto \frac{1}{\sin\theta} \qquad \text{sprinkler model} \qquad (4.47)$$

The key point is that, for a long-living complex, the angular distribution has a forward–backward symmetry. The dynamics appears to have forgotten the initial direction. The only thing that it remembers is that the angular momentum (and, of course, energy) is conserved. Problem D and Section 10.3.1.1 provide a quantum version of this result.

Thus we have an experimental or "operational" definition of a *direct reaction*, i.e., whenever the angular distribution is not symmetrical with respect to the forward vs. backward hemisphere. This establishes a time base, such that if the "time of reaction" is less than, say, one rotational period of the combined system (typically 10^{-12} s), we can say the reaction is of the "direct" type.

4.4.2 Direct reactions: forward vs. backward scattering

We have seen in Section 2.2.5 how the differential elastic cross-section $I(\theta)$ serves as a sensitive probe of the dynamics of the collision of structureless particles. In a similar fashion one can introduce the differential reactive cross-section $I_R(\theta)$, except that here we mean the number of product molecules scattered at the (center-of-mass) angle θ (per unit time and unit solid angle) divided by the incident flux of reactant molecules. In other words, we write Eq. (3.5) as

$$\frac{d^2 \dot{n}_R(\theta, \phi)}{d^2\omega} = (I_A n_B) I_R(\theta) \qquad (4.48)$$

where $\dot{n}_R(\theta, \phi)$ is the number of reactive collisions per unit time, with the products scattered into a given angular range $d^2\omega = \sin\theta \, d\theta \, d\phi$ in the direction specified by θ and ϕ. If the *differential reaction cross-section* $I_R(\theta)$ is integrated over all solid angles it gives us the *total reaction cross-section*, σ_R.

We shall defer a simple but quantitative analysis of $I_R(\theta)$ to Section 4.4.3. Here we are interested in a qualitative picture, comparable to the interpretation of σ_R in terms of the opacity function as presented in Section 3.2.

Figure 4.14 shows a comparison of the two angular distributions corresponding to the two KI-forming reactions. We note that the entire backscattered cross-section for the KI from the rebound reaction of $K + ICH_3$ fits under the backward

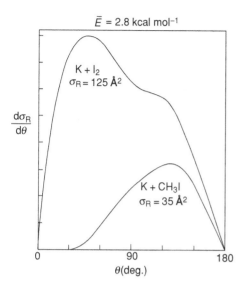

Figure 4.14 The complete angular distribution of KI for the K + I_2 and K + CH_3I reactions. Plotted is the polar differential reaction cross-section $2\pi \sin\theta \, I_R(\theta)$ vs. θ. When integrated over θ, this gives directly the reaction cross-section σ_R and the result is quoted in the figure [adapted from the experimental results of K. T. Gillen, A. M. Rulis, and R. B. Bernstein, *J. Chem. Phys.* **54**, 2831 (1971); A. M. Rulis and R. B. Bernstein, *J. Chem. Phys.* **57**, 5497 (1972)].

"tail" of the angular distribution for the predominantly forward-scattered KI from the reaction of K + I_2.

We interpret this in terms of the reaction probability as a function of impact parameter, the opacity function $P(b)$ (Section 3.2.1). Low-impact-parameter collisions (nearly "head-on") usually lead to reaction and because they are mainly head-on, the product diatomic rebounds backwards; K reacts with I_2 by the harpoon mechanism, Section 3.2.4. Therefore, encounters at long range with very large impact parameters are also effective in yielding reaction, i.e., $P(b)$ is near unity out to very large b (roughly $b_{max} \approx 7\text{Å}$). We know that this is so because the reaction cross-section is large, $\sigma_R \approx 125 \text{ Å}^2$. For these large-$b$ encounters the K^+ ion, remember the harpoon, simply "picks up" an I^- ion and carries it forward with very little deflection, and the remaining I atom must sadly recoil in the opposite direction with a velocity determined by momentum conservation. This is the so-called stripping mechanism, which is characterized by the forward scattering of the product molecule.

Of course we have oversimplified the picture of direct reactions; there is obviously a range of angles involved and for each reaction the degree to which the scattered product concentrates in a given angular region will be different. In some cases we find preferential *sideways* scattering in certain direction reactions. In general, the empirical trend is clear: the more "forward" the reaction, the larger the magnitude of the reaction cross-section. For the K + CH_3I reaction, which is predominantly "backward," the reaction cross-section is much smaller (by a factor of c. 3) in comparison with the magnitudes of the stripping reaction cross-sections, Figure 4.14.

We turn now to a more quantitative application of the concept of the reaction probability to the analysis of scattering in direct reactions.

4.4.3 Scattering in direct reactions

The optical model provides a unified framework, based upon general principles, for a simplistic analysis of collision dynamics. We begin, as usual, with the form of the reaction collision cross-section for reactants with an impact parameter in the range b to $b + db$,

$$d\sigma_R = 2\pi b P(b)\, db \tag{4.49}$$

To get the angular distribution we need to relate the impact parameter and scattering angle. For elastic collisions, this is possible because at a given collision energy, the impact parameter specifies a unique classical trajectory that describes the collision. This is not true for reactive processes. There are more variables that are necessary, such as the initial orientation, etc. We either have to compute many classical trajectories, all having the same initial impact parameter, which we will do in Chapter 5, or we have to simplify the situation by a physical assumption.

We shall assume that direct reactions are sudden-like; that there is a critical configuration at which the reactants instantly become products. Up to that configuration, the reactants approach.* Beyond that configuration, the products recede. For direct reactions it is therefore reasonable to assume that since the actual rearrangement is so rapid, the net angle of deflection is the sum of the angle of deflection of the reactants coming in and that of the products going out. Specifically, we replace the elastic scattering result for the deflection angle $\chi = \pi - 2\psi_0$ by $\chi = \pi - (\psi_{in} + \psi_{out})$ where the two ψs refer to the angle of rotation of R on the incoming and outgoing parts of the reactive trajectory.

For explicit results we need, however, to adopt specific models. One such model is that of rebound reactions. Here one assumes that reaction only occurs on close collisions when the reactants are subject to the short-range repulsive part of the intermolecular potential. The rearrangement thus takes place "at close quarters" and the newly formed products recede under the influence of the short-range repulsion. Hence, the net deflection is that typical of hard-sphere scattering.

For hard-sphere scattering we have seen that the angular distribution is a constant, independent of the collision energy, $(2\pi b\, db/d\omega) = d^2/4$. If d is the radius at which reaction takes place, we therefore have for rebound reactions

$$d\sigma_R/d\omega = (d^2/4)\, P(b(\theta)) \tag{4.50}$$

---

* In Chapter 10 we apply this idea to vector quantities. Problem E is a quantal version of the model.

Here, the only dependence on θ is through the b-dependence of the opacity function and the higher the impact parameter the more forward is the scattering. Our model predicts, therefore, backward scattering of the products, since for rebound reactions $P(b)$ contributes only at low b values. As we increase the collision energy, a rebound character at low energy changes to a more forward scattering at higher energy. There are two factors that govern this behavior. First, for reactions with an energy threshold, the range of b values that lead to reaction increases with increasing E_T, Section 3.2.7. Then, a strict hard-sphere scattering is an idealization. There are forces on the trajectory en route to the reaction shell of radius d. So the deflection angle is expected to scale inversely with the collision energy. The general statement we made earlier is that $E_T \chi(b) = \tau(b)$, Eq. (4.15), so that collisions at a given impact parameter, and hence given $\tau(b)$, will lead to decreasing deflections as the collision energy is increased. In conclusion, the angular distribution for direct reactive scattering lends itself to the approximate two-body representation

$$I_R(\theta) = P[b(\theta)]I_R^{\circ}(\theta) \tag{4.51}$$

where the reference angular distribution $I_R^{\circ}(\theta) = b/\sin\theta\,|d\chi/db|$ is to be computed from the motion into and out of the configuration at which reaction occurs.

*4.4.4   Information gained from non-reactive scattering

The cross-section for non-reactive collisions can be expressed in terms of the fraction, $1 - P(b)$, of collisions that do not lead to reaction

$$d\sigma_{NR} = 2\pi b[1 - P(b)]db \tag{4.52}$$

The non-reactive cross-section is necessarily smaller than it would have been had the reaction not taken place, i.e., if $P(b) = 0$, $d\sigma_{NR}^{\circ} = 2\pi b\,db$. Here $d\sigma_{NR}^{\circ}$ is the cross-section if $P(b)$ were zero for all b. In picturesque terms we can say that the non-reactive cross-section is quenched by the occurrence of the reaction. The closer the reaction probability $P(b)$ to unity the more severe the quenching.

While we cannot measure the differential cross-section $d\sigma_{NR}/db$, we can measure the angular distribution $I_{NR}(\theta)$ of the non-reactively scattered molecules. We expect that $P(b)$ will contribute primarily at low b values and hence that the quenching of $d\sigma_{NR}$ will correspond to the (near) absence of non-reactively scattered molecules in the backward direction. In quantitative terms we thus have that

$$I_{NR}(\theta) = [1 - P(b(\theta))]I_{NR}^{\circ}(\theta) \tag{4.53}$$

The qualitative implications of Eq. (4.53) are clear: $I_{NR}(\theta)$ is quenched in the presence of reaction. The greater the range of impact parameters over which reaction is possible, the wider the angular range over which $I_{NR}(\theta)$ is quenched. But what can we use for the relation of b to θ? Here too we need a model assumption, namely a reasonable estimate of $\chi(b)$. Given that information, we can determine $P(b)$ from Eq. (4.53). Intuitively, the "reference" angular distribution $I_{NR}^0(\theta)$ should not differ much from the angular distribution for a closely related but inert system. For example, a reasonable guess is to use the angular distribution of K + Xe as the "reference" one for K + CH_3I. Using such techniques one can obtain an estimate of $P(b)$ from Eq. (4.53).

4.4.5 Summary

We can summarize as follows: those direct reactions that occur primarily at low impact parameter are characterized by a small reaction cross-section; there is predominantly backward scattering of the products and the non-reactive angular distribution is quenched in the backward direction. Reactions that occur with a high probability over a wide range of impact parameters will have a large reaction cross-section, the products will appear mostly in the forward direction, and the non-reactive cross-section will be severely quenched except in the very forward directions. As the energy is increased the tendency towards forward scattering of the products should be favored.

4.4.6 On to polyatomics

Thus far we mainly used a two-body point of view. From now on the discussion will emphasize the polyatomic nature of the dynamics of chemical reactions. This is the same transition that is made in books on spectroscopy. These go from bound AB to bound ABC, while we go from unbound AB to unbound ABC. There is more than one vibrational coordinate in ABC. Which one is to be unbound? Well, this is very much part of the discussion of Chapter 5. Nor is it only the stretch vibrations that are of interest. The bending vibration of ABC is the carrier of the steric preference during the collision.

Problems

A. How large is the angular momentum in molecular collisions? The angular momentum L has the same dimensions, length $\cdot$ momentum, as Planck's constant. The quantum number l is a measure of L in units of $\hbar$, $L = l\hbar$. Classically, $L = \mu v b = \hbar k b$, where k is the wave number, $k = \mu v/\hbar$. For molecular collisions the reduced mass μ can vary by over two orders of magnitude, but even for H + H collision it is three orders of magnitude bigger than the mass of the electron. For reasonable values of the impact parameter and collision velocity, and remembering that for reactions with a barrier we need that $E_T = \mu v^2/2 > E_0$, compute a reasonable range for L values. Conclude that typically $l \gg 1$ and that it requires very low velocities to be in the fully quantal regime where only a few values of l are important. The dimensionless parameter $A = k\sigma$, where σ is the range parameter of the potential, is a useful guide to the range of l values that contribute.

B. Use the higher velocity data shown in Figure 2.4 and Eq. (4.24) to draw conclusions about the R-dependence of the short-range repulsive potential. What can you conclude from the low-velocity end of the plot?

C. The relative position vector of two particles is $\mathbf{R} = \mathbf{R}_1 - \mathbf{R}_2$. Assume that the potential between them depends only on the distance R. Derive Eq. (4.1) by writing equations of motion for $\mathbf{R}_1$ and $\mathbf{R}_2$.

D. Angular distribution in resonance scattering. If there is a resonance in the lth partial wave then one term in the sum (Eq. (4.38)) dominates. Therefore $I(\theta) \propto |P_l(\cos\theta)|^2$. Averaging over the fast oscillations of the Legendre function, derive Eq. (4.47). See also Section 10.3.1.

E. Direct chemical reactions are fast. Suppose the two reactants approach one another and suddenly switch into products that then depart. By analogy with the scattering amplitude for elastic collisions, Eq. (4.36), we can now write the amplitude for reaction as $\exp(i\delta_{out})\,|S_l|\,\exp(i\delta_{in})$, where $P(b) = |S_l|^2$. Show that this amplitude accounts not only for the deflection angle having the form in the text but also for the fact that there is no time delay due to the reaction itself. The delay is entirely made up from any excess time it takes the reactants to move in, and similarly for the products.

F. Back to chemistry. (a) Suggest a plausible interpretation for all three observations: K atoms scattered in a collision with Kr are found preferentially in the forward direction with respect to the incident K atoms; the intensity in the backward direction is lower and is almost independent of angle. K atoms scattered in a collision with HBr are found very preferentially in the forward direction; there is almost no scattering intensity in the backward direction. K atoms scattered in a collision with NaBr are found somewhat preferentially in the forward direction; the intensity in the backwards direction is significantly higher than that for K + Kr and it increases as $\theta \to 180°$. *(b) K atoms scattered in a low-energy collision with SO_2 are found preferentially in the forward direction. The intensity in the backward direction is however also high and it increases as $\theta \to 180°$; see D. O. Ham and J. L. Kinsey, *J. Chem. Phys.* **48**, 939 (1968).

G. The K + $(CN)_2$ and K + CH_3CN reactions. In many ways the CN radical acts as a halogen atom (Bersohn, 1976). Compare and contrast the two KCN-forming reactions. Chapter 5 will allow you to draw conclusions regarding the vibrational excitation of the nascent KCN. Another similarity is the stereochemistry, with KCN being formed preferentially for a K approach from the CN end of the CH_3CN molecule, see Figure 1.5. A collision of fast (hyperthermal) K atoms with CH_3CN molecules leads to the formation of separated ions K^+ and CN^-. The CN^- ions are produced predominantly by attack of the K atom on the CH_3-end of CH_3CN [S. A. Harris, P. W. Harland, and P. R. Brooks, *Phys. Chem. Chem.*

Phys. **2**, 787 (2000)]. The electron apparently enters the lowest unfilled π*CN orbital to form an unstable linear molecular negative ion that then breaks up as it attempts to bend into the geometry of the stable products, KCN and CH_3.

Notes

1 Classical, semiclassical, and quantum elastic scattering are thoroughly covered in Child (1974). The same material but within a more mathematical approach can be found in Newton (1982). The classic review is Bernstein (1966). See also Pauly (1979). The seminal paper on semiclassical scattering is Ford and Wheeler (1959). The venerable scattering theory text is Mott and Massey (1965). A more accessible text is Johnson (1982). It is nowadays straightforward to generate numerical tables of, say, the deflection function vs. impact parameter for a range of collision energies. Still, the discussion and extensive tables in Hirschfelder *et al.* (1954) provide much insight. An introduction to quantum mechanics with an emphasis on the superposition principle is Feynman *et al.* (1966).

   The theoretical discussion of this section should not make us forget that the theory is to guide and interpret real experiments. Four books stand out in making this connection: Bernstein (1979, 1982), Scoles (1986), Pauly (2000).

   We deal only with collisions of structureless particles. But there is a highly developed quantum scattering theory for both inelastic and reactive collisions. For general background see Levine (1969). Inelastic collisions are simpler to handle because the need to change coordinates in going from reactants to products, discussed in Appendix B of Chapter 5, does not arise. See, for example, Gianturco (1979), Murrell and Bosanac (1989). There are many sources for reactive collisions, including Baer (1985), Kouri (1985), Clary (1986), Schatz (1988), Miller (1990), Bowman and Schatz (1995), Schatz (1996), Clary (1998), Miller (1998), Nyman and Yu (2000), Althorpe and Clay (2003). Quantum theory is nowadays reaching all the way to enzyme kinetics, see Gao and Truhlar (2002).

2 We shall not make use of it, but to complete the analogy with the electronic problem, we point out that the partial waves can be determined by the solution of the Schrödinger equation. $(E - H)\psi(\mathbf{R}) = 0$ is a differential equation in three variables, the three components of $\mathbf{R}$. The conservation of angular momentum allows us to rewrite it as many separate Schrödinger equations, one for each partial wave. These have the form of a radial equation where the name refers to the differential equation depending only on the radial (i.e., scalar) distance R:

$$\left(k^2 + \frac{d^2}{dR^2} - \frac{l(l+1)}{R^2} - \frac{2\mu}{\hbar^2} V(R) \right) G_l(R) = 0, \quad l = 0, 1, 2, \ldots$$

   with $\psi_l(R) = G_l(R)/kR$. The collision energy E is expressed in terms of the wave number k, $E = \hbar^2 k^2/2\mu$, so that $k = p/\hbar = 2\pi/\lambda$, where λ is the de Broglie wavelength. Note the effective potential energy for radial motion, $V_{\text{eff}}(R) = V(R) + \hbar^2 l(l+1)/2\mu R^2$ that we already met in the classical scattering.

3 It is not obvious, but perhaps not unintuitive, that the second virial coefficient, Section 2.1.10.1, is closely related to the time delay. The two regimes where $B(T)$ is respectively positive and negative correspond to negative and positive time delays.

4 Interference between different ways of reaching the same product state will be a key element in our discussion of control in later chapters. The Fano line shape is a special case where the alternative routes have a clear physical interpretation, one being direct scattering

and one scattering through a resonance. Another example of an interference is the rainbow scattering. For measuring the resonance contribution to the phase shift, $\delta - \delta_{\text{direct}}$, see Gordon *et al.* (2001).

5 This estimate shows that the widths of two adjacent tunneling resonances are significantly smaller than their spacing, $\hbar\omega$. Such resonances are said to be non-overlapping.

6 For a detailed application to alkali atoms and alkyl iodide reactions, see J. L. Kinsey, G. H. Kwei, and D. R. Herschbach, *J. Chem. Phys.* **64**, 1914 (1976).

Chapter 5
Introduction to polyatomic dynamics

In this chapter we recognize that reactants (and hence products) have internal structure. We need to describe the potential energy that gives rise to the forces that act during the collision and to determine the dynamics. With this background we examine what features of the potential energy play a special role in the dynamics and how the internal states of the reactants participate. The **rate** of chemical reactions for polyatomic reactants is discussed in Sections 6.1 and 6.2. The derivation therein makes essential use of two features of the potential that we examine in this chapter. These are the potential energy barrier that separates reactants and products and the not unusual possibility that there are two (or more) barriers, between which there must be a hollow. To fully appreciate the approximations that are made in deriving expressions for the reaction rate that are based only on structural information, it is necessary to examine the dynamics, as we do in this chapter.* We begin by reminding you about the Born–Oppenheimer separation of electronic and nuclear motions that underlies our work in this chapter,[1] and the possibility of its failure.

5.0.1 The Born–Oppenheimer separation: a caveat

This chapter is based on generalizing the concept of an interatomic potential to the polyatomic case. Specifically, we are interested in many-atom systems that have enough energy to undergo a chemical change. It is therefore important to point out that the concept of a unique potential function governing the motion of the atoms is an approximation. It requires that, throughout its time evolution, the system remains in the same electronic state. When this is the case then the

---

* In this chapter we emphasize energetic aspects and these are scalar quantities. If we want to understand more about reaction directionality then we need to examine vectorial aspects, which are treated in Chapter 10. In this chapter we treat the dynamics by classical mechanics. When we want to understand coherence effects we need to examine quantal superposition of states. We pick up on this theme from Chapter 7 onwards. There we will also extend the discussion to reactions that do not necessarily proceed for reactants in their ground electronic state.

electronic energy is the potential for the nuclear motion.* Born and Oppenheimer showed that this can be so because the electrons are much lighter and hence move much faster than the nuclei. Thus in the time it takes the nuclei to move a very small distance the electrons can thoroughly sample their accessible space. This enables us to treat the two motions separately. First, we freeze the nuclei in some configuration and diagonalize the electronic Hamiltonian to determine the lowest eigenvalue and the stationary wave function for the electrons. The nuclei are then moved to another configuration for the atoms where the diagonalization is repeated. Continuing in this manner one eventually has the (lowest) electronic energy for an entire range of configurations of the nuclei. This electronic energy is the desired potential function and it can be used to determine the dynamics of the nuclei, assuming that as the heavy atoms move, the light electrons "instantaneously" adjust to the new position. In Chapter 9 we shall come to call this rapid and smooth adjustment of the electrons to the new circumstances an *adiabatic* behavior. The maintaining of the same quantum state despite a perturbation (= an adiabatic limit) is typical when the time scale of the perturbation is long compared to the time scale over which the system can adjust. In the present context, the perturbation is the motion of the heavy nuclei and we take it to be slow on the time scale of electronic reorganization.

The *Born–Oppenheimer separation* is a most useful approximation because it allows us to understand the geometrical structure of molecules at equilibrium and also for small vibrations about equilibrium. Another advantage is that the potential remains the same under isotopic substitution. For our purpose we need a potential that describes large displacements, large enough that bonds can be broken and formed. Furthermore, we often need to deal with open-shell systems. All of this means that we need to be more careful than colleagues dealing with stable species. The ground electronic state of the system is often sufficiently

---

* Some details. Write the (non-relativistic) Hamiltonian as $H = T + H_{el}$, where T is the kinetic energy operator for the relative motion of the nuclei. For a diatomic molecule, $T = -(\hbar^2/2\mu)\nabla_{\mathbf{R}}^2$. The "electronic Hamiltonian" (Hel) is a sum of the kinetic energy operator of the electrons and potential terms, all electrostatic. We list them so as to emphasize that a number of these terms depend on the distance R between the two nuclei: the repulsion between the two nuclei, the attraction of the electrons to the two nuclei, and the repulsion between the electrons. In the Born–Oppenheimer separation we first hold R constant and diagonalize the electronic Hamiltonian $H_{el}\psi(r;R) = E_{el}(R)\psi(r;R)$. We write $\psi(r;R)$ as a reminder that while the electronic wave function is a function of the coordinates, r, of the electrons, the function has been determined at a given value of R. Diagonalizing the electronic problem provides a number of eigenvalues, the lowest one being the ground state. The diagonalization is now repeated at other values of R. This yields the electronic states and their energies as a function of R. For any particular electronic state the wave function for the full Hamiltonian $H\Psi = E\Psi$ is approximated as $\Psi = \chi(R)\psi(r;R)$. The nature of the approximation is that we do not allow T to operate on the electronic state $\psi(r;R)$. Then, from $\langle\psi(r;R)|H|\Psi\rangle = E\langle\psi(r;R)|\Psi\rangle$ we get $(T + E_{el}(R))\chi(R) = E\chi(R)$. In words, the electronic energy plays the role of the potential for the motion of the nuclei. Different electronic states give rise to different potentials for the nuclei. For excited states some of the potentials can be repulsive. The approximation fails when the matrix element of T between two different electronic states is not small, Problem A.

lower in energy than electronically excited states that the Born–Oppenheimer separation of fast-moving electrons and slow-moving nuclei is valid. The reason is that, as we shall show in Chapter 7, the "time" available for the electrons to adjust to the motion of the nuclei is Planck's constant h divided by the gap, where the *gap* is the difference between the electronic energy of a given electronic state and its near neighbors, see Problem A. The gap can be large, so large that the electrons adjust before the nuclei appreciably move and then all is well. But the gap depends on the configuration of the nuclei and while it can be large in the reactants or products region, it need not necessarily stay large when the system is in the transition state region. In fact, in the polyatomic case the gap can outright vanish, leading to a so-called *conical intersection* of two electronic states.[2]

How can one tell if the Born–Oppenheimer separation is valid during the reactive event? We have no easy answers. We provide qualitative reasoning about when the separation is likely to fail, but only quantum chemistry can tell for sure. We need to know the electronic energies and even then we need to know more, as discussed in Chapter 7. A limited guidance is available from experiment. If reactants in a given state can give rise to products in more than one electronic state, or vice versa, then the separation has certainly failed. Unfortunately, the converse does not follow. Entrance and exit can be on a single electronic state and yet a second electronic state can be populated during the collision. Life is simpler when we have the concept of a unique potential for the motion of the nuclei. Pragmatically therefore one tends to assume that, in the absence of evidence to the contrary, the Born–Oppenheimer separation is valid. We shall do so in this chapter, but the caveat must be kept in mind. This warning is particularly important when we discuss photochemistry, Chapter 7, where the system is promoted to an excited electronic state whose dynamics are of interest.

5.1 Potential energy functions and chemical reactions

5.1.1 Potential energy surfaces

We are about to generalize the concept of an inter*atomic* potential. We need to know the interaction energy as a function of the configuration of the system throughout the rearrangement from reactants to products.[3] Even in the simplest case of an atom–diatom collision, A + BC, the potential energy is a function of three coordinates, e.g., the three interatomic distances.* However, a function of even three variables is difficult to visualize. It is therefore customary to begin by examining the potential for the "collinear" (A–B–C) configuration, when the

---

* The interaction potential is a function of the positions of the three nuclei and hence, in principle, a function of nine coordinates. Imagine now the three atoms as the vertices of a triangle in an otherwise empty space. The interparticle energy will not be changed if the triangle is rotated or translated as a rigid body. Hence the potential is a function only of the three coordinates required to specify the atomic triangle. These can be the three interatomic distances, but they can also be two distances and the angle between them or even one distance and two angles.

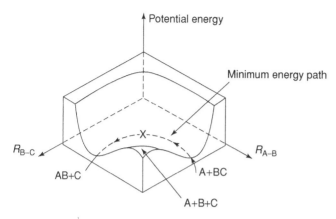

Figure 5.1 A potential energy surface for the A + BC → AB + C reaction in the collinear configuration (schematic). The *minimum energy path* leading from reactants to products is shown as a dashed line. It is also referred to as the *reaction coordinate*. The highest point along the minimal energy path is the *saddle point* (X). It is a saddle because the energy increases if we deviate from the reaction coordinate in either direction but along the path it is a maximum. In the vicinity of the saddle point the motion along the minimum energy route, where one bond shrinks and the other bond extends, is what, in a stable molecule, we call the asymmetric stretch motion of A–B–C. For a direct reaction as shown here this is an unbounded motion that takes the system from reactants to products. The motion in the direction perpendicular to the minimum energy path is bound. As we cross the saddle point region this motion changes in character from the A–B vibration to the symmetric vibration of A–B–C to the vibration of B–C. The plateau corresponds to the three separated atoms A+B+C. The inner repulsive wall is the region where two or all three atoms get close to one another.

three atoms are confined to a straight line so that there are only two independent interatomic distances, the A–C distances being constrained to equal the sum of the A–B and B–C distances. In this case we can plot the potential as a function of the two coordinates as a *potential contour map*, showing equipotential lines in a two-dimensional plot in a manner made familiar by topographical maps.*

Chemical intuition can help us identify the main topographical features expected when the potential is plotted as a function of the two bond distances: there is the "old" A–B bond distance, limited to motion near equilibrium in the reactants region when atom C is far away, and there is the "new" B–C bond distance, limited to motion near equilibrium in the products region when atom A is far away. Chemistry is what happens in the middle and we expect that the plot looks schematically as Figure 5.1. It shows a stylized potential energy function

---

* A topographical contour map is a 2D plot of the height as a function of longitude and latitude. We plot the potential as a function of the two bond distances. Computers easily allow you to make a three-dimensional perspective plot of the potential as a function of two variables. Even for the triatomic ABC system what we really want is to view the potential as a function of three variables. We know of no easy way for doing so.

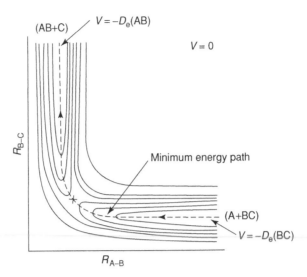

Figure 5.2 A contour map for the potential energy surface of Figure 5.1. The solid lines are contours of given potential energy. The minimum reaction path is shown as a dashed line. The entrance valley has a depth $-V = D_e(BC)$, the dissociation energy of BC (as measured from the bottom of the valley) and similarly for the exit valley. At the saddle point (X) the potential energy is above that of either valley but below that of the dissociation plateau, where both bonds are broken and the three atoms are far apart from one another. It is customary to take the zero of energy as the energy of the plateau.

drawn in a perspective three-dimensional form. The same function is displayed as a contour map in Figure 5.2.

The "mountain pass" en route from the reactants to products (along the minimum energy route) is a dominant feature of such a surface, explaining why the energy threshold for reaction is often much smaller than a bond dissociation energy. As pointed out by Eyring, Polanyi, and Evans in the early 1930s, a chemical reaction, say

$$D + H\text{—}H \rightarrow H + H\text{—}D$$

proceeds not by breaking the "old" bond and subsequent formation of the "new" bond, but via the concerted motion of the nuclei in a continuous transformation from the reactants' to the products' valley. The new bond forms as the old bond is broken. A potential energy surface thus serves to mediate between the reactants' and products' configuration.

Because the reaction path passes through the local minima of the surface, the potential energy increases when we deviate sideways from the path. Hence, near the barrier the potential surface has the form of a saddle. The location of the barrier is thus referred to as the *saddle point* of the surface (sometimes as the *col*). The configurations about the saddle point are the *transition state region*.

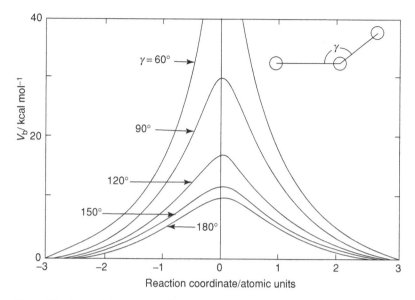

Figure 5.3 Potential energy profile along the reaction path for the collinear and for a number of bent configurations of H_3. Each one of these profiles is for a fixed value of the approach angle as in the insert. From an *ab initio* computation with an accuracy of better than 1 kcal mol^{-1} [adapted from P. Siegbahn and B. Liu, *J. Chem. Phys.* **68**, 2457 (1978) and the analytical representation[4] by D. G. Truhlar and C. J. Horowitz, *J. Chem. Phys.* **68**, 2466 (1978)]. Note how the barrier height increases for a sideways approach of the H atom to the H–H axis. To interpret this, note that H_3 has three electrons. In Section 5.1.4 we show that two can be placed in a strongly bonding orbital extending over the three atoms. The third electron goes into an orbital that becomes increasingly antibonding as the bend angle decreases.

The height of the barrier along the minimum reaction path is the lowest maximum of the potential between the reactants' and products' valleys. In classical mechanics this height is therefore the minimal energy for a trajectory to go over. Such a trajectory represents a possible motion of the nuclei during a reactive collision. Of course, in an actual collision the three atoms need not be confined to move on a line. The trajectory needs to be computed using the potential when all three bond distances are allowed to move independently, subject of course to the forces.

In Figure 5.3 we read that the lowest barrier in the potential for the $H + H_2$ reaction is at the collinear configuration and its height is about 10 kcal mol^{-1}. For thermal reactants at room temperature only an exponentially small fraction of collisions will have enough energy to cross the barrier (recall Figure 3.3). Reaching the three-atom plateau is even less likely because it is about 100 kcal mol^{-1} higher than the reactants' valley.

It is not invariably the case that the reaction is concerted with the new bond forming as the old bond is being broken. Organic chemists are familiar with

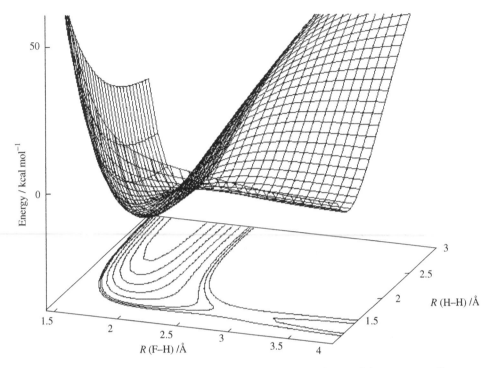

Figure 5.4 *Ab initio* computed potential energy surface and the corresponding contour map for the collinear configuration of FH_2.[5] Energy in kcal mol^{-1} and the zero of energy is taken here to be at the reactants' well. The contour extending from the reactants' to the products' region is at 3 kcal mol^{-1}. Below it are contours at 1.5, 0, −5, −10, −15, −20 and −25 kcal mol^{-1} [drawn from the results of K. Stark and H.-J. Werner, *J. Chem. Phys.* **104**, 6515 (1996)].

so-called S_N1 reactions,[6] where the old bond is entirely or largely broken before the new bond forms. We expect such reactions to have higher barriers because energy is required to loosen the old bond before there is any gain in energy from the formation of the new bond.

One of the most significant recent trends in quantum chemistry is the development of computationally tractable schemes for the evaluation of three-atom potential surfaces that can reach a kJ mol^{-1} accuracy. Such variational calculations must be better than Hartree–Fock (i.e., they must take into account electron correlation) in order to be usable in the kind of dynamical problems in which we are interested, where bonds rearrange.

5.1.2   The reaction path

To better understand the origin of a threshold energy for reaction consider the "cost" in energy required to cross over from the reactants' to the products' regions. We define a *reaction path* as the line of minimal energy from the reactant to the

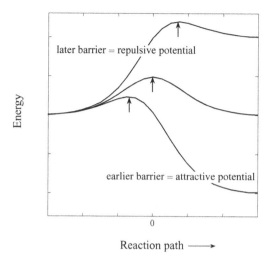

Figure 5.5 Potential energy profile along the reaction coordinate for an early, midway, and late barrier. We show the two asymmetric cases as corresponding to an exoergic and endoergic reaction. The correlation between location of the barrier, shown by an arrow, and relative stability of reactants and products can be understood on the basis of the *Evans–Polanyi model* that we will discuss. The correlation shown is a useful rule of thumb, known in organic chemistry as *Hammond's postulate*. In structural terms this is sometimes stated as: the more exoergic the reaction, the more the configuration at the barrier will be reactant-like.

product valley. Figure 5.3 shows the potential energy along such a minimum energy path for the H_3 system in the collinear configuration (which is the one of lowest energy) and for several bent configurations. We note that the barrier along this "energy profile" is relatively small, $E_b \cong 10$ kcal mol^{-1}, in the collinear configuration. This barrier is less than 10% of the dissociation energy of H_2. It is the presence of a low-energy mountain pass between the reactants' and products' valleys that favors the concerted (i.e., bimolecular) mechanism of atom (or group) transfer.

Figure 5.4 shows an *ab initio* calculated potential surface for a highly reactive system: $F + H_2 \rightarrow HF + H$. For this very exoergic reaction we see that the barrier is significantly smaller than for H_3 and it occurs at a rather "early" stage of reaction. Immediately beyond the barrier the potential energy drops rather steeply. For such a reaction, the exoergicity is released early along the reaction path and is available to be pumped into the vibration of the emerging new bond. A surface with an early release of the exoergicity is often referred to as an *attractive potential energy surface*. A surface of the opposite type, i.e., a late release, is termed *repulsive*, see Figure 5.5. (Of course, if a surface is attractive for the forward reaction it is repulsive for the reverse reaction.)

The $F + H_2 \rightarrow HF + H$ potential shown in Figure 5.4 is an example of a quite early barrier for a very exoergic reaction. As can be seen from the location of the barrier in terms of the two bond distances, the transition state region resembles the reactants: the H—H bond is hardly stretched and the F atom is still far from the H atom it will depart with. We expect this to be the more common situation; exoergic reactions have an early barrier as shown schematically in Figure 5.5. We do not require a rule for endoergic reactions because their potential is the same as that for the reversed exoergic reaction. So the more endoergic the reaction, the later along the reaction path is the barrier. For the thermoneutral $H + H_2 \rightarrow H_2 + H$, Figure 5.3, the barrier is in the middle.

There is a hierarchy in what one can expect from quantum chemistry.[7] At most we want full, accurate, computations of potential surfaces for many-atom systems. This is still not easy to do, but when it can be done the computation provides not only the energy but also its gradient, namely the force.[8] What is currently realistic is to reduce the labor by restricting attention to the potential along the reaction path. What is definitely possible is to examine only the stationary points of the potential along this path. The results of such a computation are shown in Figure 5.6 for the important *combustion* reaction[9]

$$C_2H_5 + O_2 \rightarrow \begin{cases} C_2H_4 + HO_2 \\ c\text{-}CH_2CH_2O + OH \\ CH_3CHO + OH \end{cases}$$

As the reagents approach, some charge transfer from the hydrocarbon radical to molecular oxygen takes place. The higher ionization potential of the radical means that the transfer is only partial and occurs when the reactants are already close in.

One new feature revealed in Figure 5.6 is that there can be more than one barrier along the reaction path. Between two barriers there has to be a well, as shown. Such a well corresponds to an intermediate. When this intermediate is reached from the reactants then it is not stable because it has enough energy to cross the barrier back to the reactants' valley. The intermediate will be stable if its energy is lowered so that it cannot dissociate.* Also seen in Figure 5.6 are precursor states that are en route from the reactants to the barrier. These too will be stable if drained of the energy that is available when they are formed from the reactants.

In a large biological molecule such as a protein, the *potential energy landscape*[10] for its intramolecular motions is far richer than we can easily visualize or even characterize. The interplay of chemical forces between bonded atoms and longer-range forces between spatially adjacent atoms leads to a potential function that is replete with local minima separated by both low and high barriers. It is expected that there is a *funnel region* about the lowest energy state so that a protein (unlike, say, a glass) can reasonably quickly fold to its natural

---

* This is not possible for an isolated collision but can be achieved if we intervene from the outside and cool the intermediate.

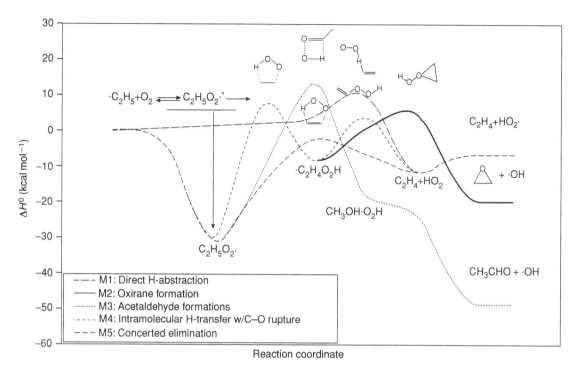

Figure 5.6 Five possible mechanisms for the $C_2H_5+O_2$ reaction. The mechanisms are distinguished by the different stationary points (both minima and maxima) along the reaction paths [adapted from J. C. Rienstra-Kiracofe, D. Allen, and H. F. Schaefer III, *J. Phys. Chem. A* **104**, 9823 (2000)]. Higher-level quantum chemistry computations can identify the configurations at the stationary points along these paths of steepest descent and determine not only their energies but also the forces that act on the atoms (Pulay, 1995; Pulay and Baker, 2001). Density functional methods can then be used to determine the potentials (and forces) along the minimum energy paths connecting the stationary points and even map the entire potential energy surface, for use in dynamical computations [see, for example, Kohn *et al.* (1996), Tse (2002)].

state. But there are also numerous (exponentially many in the number of amino acids) secondary minima.[11] We will come back to this when we discuss *molecular machines* in Section 6.2.4.

*5.1.2.1   Input from spectroscopy of large-amplitude motions

For stable molecules, the potential energy in the vicinity of the well is determined experimentally by spectroscopic means. Increasingly, spectroscopy is providing access to outer reaches. Overtone spectroscopy directly pumps higher vibrational states. The method relies on deviations from the harmonic approximation, but the higher up we go the weaker is the transition strength. Overcoming this limitation is possible by promoting the molecule to an excited electronic state that has a different equilibrium configuration. This state will emit light and one can

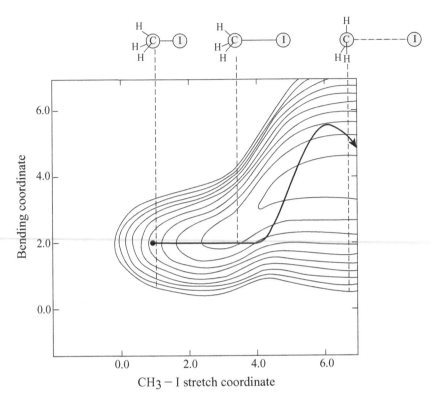

Figure 5.7 Extracting information about the potential energy surface from the light emitted by electronically excited CH_3I as it is falling apart following laser excitation from the ground state. An interpretation of the emission spectrum is provided using a contour plot of the excited state potential as a function of the C–I stretch coordinate and the umbrella-bending mode of CH_3. The photoexcitation promotes the molecule to the spot shown. From this region the (mean) motion traces a trajectory represented as a heavier line. The trajectory shows that early in the photodissociation process, $CH_3I^* \rightarrow CH_3 + I^*$, it is the C–I bond that elongates. Later, the CH_3 group that is initially pyramidal flattens to a planar configuration. The different configurations of the molecule are shown as snapshots above the plot. The sequence of events is inferred* from the nature of the emission and the Franck–Condon principle as discussed in Section 7.01 [adapted from Imre *et al.* (1984); B. R. Johnson *et al., J. Phys. Chem.* **100**, 7743 (1996)].

enhance the signal by stimulating the downward transition, see Chapter 7. It is not necessary that the excited state be stable. Figure 5.7 shows the potential inferred from the emission of light from a CH_3I molecule as it is falling apart on an excited repulsive state.

---

* In doing so it should be remembered that the emission depends not only on the potentials but also on the variation of the electronic transition dipole with the position of the nuclei.

The analysis of the light emission experiment provides a trace of the mean path (heavy line) followed by the dynamics. The trajectory shown in Figure 5.7 takes a wide oscillation about the umbrella-bending mode of CH_3, indicating that this mode is born highly excited. There is an important lesson: the reaction path is a key idea but the actual motion is under no obligation to follow it. We will return to this lesson.

5.1.3 Semi-empirical potential surfaces

Quantum chemical computations can provide the main topological features of the surface for systems of chemical interest and can even map the complete surface for the simpler, fewer-atom systems. There is still a computational barrier to the generation of purely theoretical potentials for larger systems and there is another, seemingly trivial but all the same real, problem. The potential is a function of the bond distances. The computations provide the potential (and its gradients, namely the forces) for different configurations. It is not easy to store and recall a function defined over discrete points. It would be much better to fit the computations to a functional form. Getting a flexible enough form is sometimes a challenge.*

A simple empirical functional form is the use of "switching functions." It is based on the realization that the surface in effect switches off the old bond and switches on the new one. For the simple case of a surface for the $A + BC \rightarrow AB + C$ collinear reaction

$$V(R_{BC}, R_{AB}) = V_{BC}(R_{BC})f_1(R_{AB}) + V_{AB}(R_{AB})f_2(R_{BC}) \qquad \text{incomplete} \qquad (5.1)$$

Here V_{BC} and V_{AB} are the spectroscopically known potentials of the stable diatomics, and the switching functions f_1 and f_2 range from zero to one, increasing to unity for large values of their argument. Such a form correctly describes the asymptotic form of the potential (and hence the correct overall energetics) and brings out explicitly the concept of the potential surface as the mediator between reactants and products. We have to guess at the switching functions but more than that, Eq. (5.1) misses a three-body term, sometimes called a mixing term, that insures the potential has a central barrier and has the correct stereochemistry.

The use of known diatomic potentials to estimate the three-atom potential function is at the heart of the so-called London–Eyring–Polanyi(–Sato) (LEP(S)) method. This is a semi-empirical scheme based on the London equation, originally intended to deal with four one-electron S-state atoms. In its most primitive form, we begin by writing the potential between two atoms as a sum of a coulomb (Q)

---

* As an alternative there is the approach that gives up trying to have a global form of the potential. Instead, it computes the potential at those configurations where it is of interest or where it is called for during a dynamic computation. The latter is known as *computing on the fly* because the potential is determined as we are flying over the potential landscape. (We are **over** rather than **on** the potential because our height is the total energy. It is the sum of the kinetic and potential energies.)

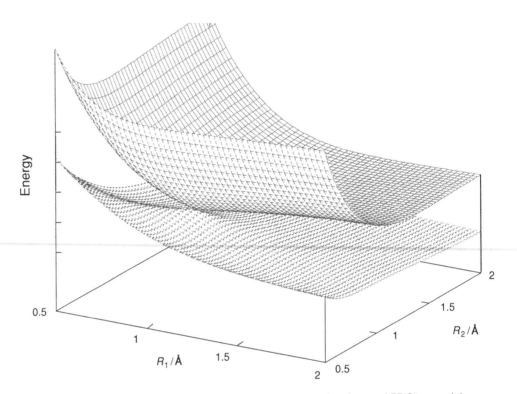

Figure 5.8 A conical intersection in H_3 plotted using the two LEP(S) potential functions, Eq. (5.3). The lower surface corresponds to the ground state reactants. The upper surface is the other solution of the LEP(S) potential. The plot is for H_3 in a triangular configuration. The H–H distance that is the base of the triangle is held constant at 1 Å while the other two distances are allowed to vary. The intersection occurs where all three bond distances are equal. In this manner of plotting, every displacement leads away from the intersection and, for small displacements, the energy varies linearly thereby generating a conelike structure. A conical intersection provides an efficient funnel out of the excited and into the ground electronic state. Passage between the two potentials is very much aided by a conical intersection because transitions between two electronic states are (exponentially, see Chapter 9) more likely the smaller is the energetic gap between the two potentials. Therefore, a system prepared on the upper state is funneled by a conical intersection into a localized region of the lower potential energy surface, in the very vicinity of the apex. Entry into the lower state via a conical intersection offers a way for starting the subsequent dynamics on that state with a rather restricted range of initial conditions.

and an exchange (J) term:[12]

$$V(R_{AB}) = Q_{AB} \pm J_{AB} \tag{5.2}$$

Both Q and J are functions of R that tend to zero for very large R. We need two equations so as to separately determine Q and J. When the singlet, ground state, potential, and repulsive triplet potential (that correlates to ground state atoms) are known,[13] we assume that both known potentials can be expressed as Eq. (5.2) taking the positive and negative sign respectively. This determines J and Q.

When three atoms are in close proximity, the LEP(S) potential function is

$$V_{ABC} = Q_{AB} + Q_{BC} + Q_{CA}$$
$$\pm (1/2)^{1/2} [(J_{AB} - J_{BC})^2 + (J_{BC} - J_{CA})^2 + (J_{CA} - J_{AB})^2]^{1/2} \quad (5.3)$$

When one atom is removed to infinity, the coulomb and exchange terms for the two broken bonds vanish and Eq. (5.3) reduces to (5.2).

5.1.3.1 The conical intersection for the LEP(S) potential

The LEP(S) functional form provides an explicit example of a *conical intersection*. Generally, the two surfaces represented by Eq. (5.3) are quite far apart. The lower surface has a barrier for atom exchange. The upper surface has a well corresponding to a bound but electronically excited ABC molecule. The conical intersection is when the barrier reaches all the way up to the bottom of the well, Figure 5.8.

To examine the intersection, consider for simplicity the case of three identical atoms. Then the coulomb and exchange integrals for all three pairs have the same functional dependence on the distance. At any equilateral configuration of the atoms the square root term in (5.3) must therefore vanish and the upper and lower roots merge. Now make a small displacement from the equilateral configuration. If the displacement is small, the exchange integrals differ by a term linear in the displacement.* Therefore, about the intersection point, the two surfaces decrease and increase linearly. This linear divergence is like two cones joined at their apex:

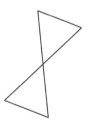

Of course, further away from the intersection the divergence need not be linear. Also, please do not be misled. The intersection need not be a point. It occurs along the entire surface defined by the condition that the square root term vanishes. For three atoms, the configuration space is three-dimensional so the intersection occurs along a two-dimensional surface.

---

* Keep the AB and BC distances at the same value R and increase the AC distance by δR. Of the three terms in the square root in Eq. (5.3), the first vanishes and the other two are equal to one another and to $(J'(R)\delta R)^2$, where the prime denotes the derivative. Hence the two values of Eq. (5.3) differ by $J'(R)\delta R$.

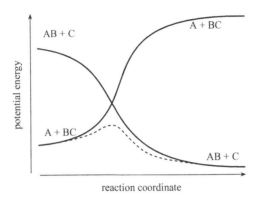

Figure 5.9 A curve-crossing model for a potential energy along the reaction coordinate, for a somewhat exoergic reaction. Each solid curve is the potential when the electronic state is constrained to be either that of the reactants or that of the products. So if we fix the state as that of A + BC but bring the nuclei to AB being near and BC being far apart, the energy is high because we broke the BC bond but did not compensate by the formation of the AB bond. The (ground) adiabatic potential is shown as a dashed line.* It changes smoothly from the reactant BC being bound to the bound AB product. Within this Evans–Polanyi type model, it is easy to visualize how the location of the barrier shifts if the reaction is made less exoergic. See Figure 5.5 and Problem G.

5.1.4 The Evans–Polanyi model

A considerable amount of work has been done on the properties of semi-empirical surfaces. There is evidence to suggest that in a family of related reactions of decreasing barrier heights, the location of the barrier occurs correspondingly earlier along the reaction coordinate, as is summarized by the Hammond postulate. Moreover, in a family of related reactions, the barrier height decreases with an increase in the reaction exoergicity. The curve-crossing model as discussed in Section 3.2.4 has been used by Evans and Polanyi to rationalize these correlations. It has found many applications in organic chemistry,[14] and we will return to it also for a unified approach to the role of solvent and solute interactions in Chapter 11. The feature that we emphasize here is the crossing taking place along the reaction coordinate, Figure 5.9. As in any curve-crossing model, we start with potential curves for the system in a **frozen** *diabatic* electronic state. These are not necessarily the Born–Oppenheimer (or adiabatic) electronic states because we do not allow the electrons to fully adjust to the motion of the nuclei. Instead, we restrict the states accessible to the electrons. In the *Evans–Polanyi model* we take the two states as the electronic states corresponding to the reactants

---

* The variational principle of quantum mechanics tells us that when the wave function is made more flexible, the energy of the ground state goes down (or stays the same). The dashed line in Figure 5.9 is an example of this general result because it results from allowing the electrons to fully adjust to the position of the nuclei.

A + BC and the products AB + C. In other words, as we move along the reaction coordinate we respectively do not allow the new bond to be formed or, starting with the state of the products, we do not allow the old bond to be formed. The adiabatic potential, the one where we freely allow the electrons to adjust to where the nuclei are, is different, and it is shown in Figure 5.9 as a dashed line. More on this model in Chapters 6 and 7.

5.1.5 The cone of acceptance: qualitative considerations

We have already discussed the origin of the *steric effect* in chemical reactions as due to preferential orientation at the barrier to reaction. In Figure 5.3 we show that the barrier of the H + H_2 reaction is lowest for the collinear configuration. Thus, at collision energies only slightly higher than the barrier, the preferred orientation for reaction is expected to be essentially collinear. As the collision energy increases, the range of "acceptance angles" for reaction increases, that is, the *cone of acceptance* opens up. The line-of-centers model introduced in Section 3.2.7 interprets this as showing that a greater range of impact parameters can then contribute to reaction at a higher energy.

The results for the angle-dependent barrier to reaction shown in Figure 5.3 are based on accurate *ab initio* computations. Can one, however, understand the origin of the orientation dependence from simple ideas of chemical bonding? This is important not only for its own sake but also for demonstrating that electronic structure theory provides a unified approach encompassing chemical reactivity.

A simple interpretation of the bonding "during" a reactive collision can be obtained via the standard molecular-orbital approach. The molecular orbitals (MOs) are constructed as a linear combination of atomic orbitals (AOs). The number of independent MOs is equal to the number of AOs used. To arrange these MOs in order of increasing energy, we note that when an MO acquires another node, its energy is higher. Once the order of the orbitals is established, the electronic configuration is determined by assigning two electrons (of opposite spin) to each MO, starting with the one lowest in energy, until all electrons are assigned.

To discuss the H + H_2 system, consider first the H_2 reagent. Using a $1s$ AO on each H atom (denoted by a sphere), we have two independent linear combinations:

$\sigma^*(1s)$

$\sigma(1s)$

The dots represent the nuclei and the shading defines the sign (sometimes known as the phase) of the atomic orbitals. In the σ orbital, the two $1s$ functions have the same sign. Therefore this MO has no node and hence it is a low-energy bonding orbital. In the σ^* MO, the two atomic functions have opposite signs. There is a node midway between the two nuclei so the energy of the σ^* orbital is higher and it is called an antibonding orbital. The ground state configuration of H_2 is $\sigma(1s)^2$.

Now let an H atom (with its $1s$ orbital) collinearly approach H—H. The $1s$ orbital can add, with the same sign or with an opposite one, to either σ or σ^*:

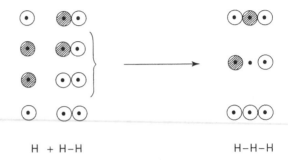

$$H + H\text{--}H \qquad\qquad H\text{--}H\text{--}H$$

The in-phase addition to σ gives rise to a lowest (node-free) MO of H_3. The out-of-phase addition to σ^* (producing a node at each bond) gives rise to a strongly antibonding MO, one with two nodes. The two middle "alternatives" shown in the left column are really the same: a node at one H—H bond, no node at the other. An alternative description is to add them together so as to have the orbital picture on the right, which shows that this orbital is somewhat antibonding: it has a node midway between the two end nuclei.

Three electrons are to be assigned: two go to the lowest, bonding orbital; the third is assigned to the higher-energy, somewhat antibonding orbital. We expect a small energy barrier for the collinear approach of H to H_2.

Next, let the H atom approach at an angle ($<180°$) with respect to the H—H bond. The corresponding orbitals for H_3 are pictured as follows:

The energy of the bonding orbital is not greatly affected by the change in the H_3 bond angle. But the middle, singly occupied orbital, that was a somewhat antibonding orbital, is now more strongly antibonding, because its node is between two nuclei that are now nearer to one another and so it screens them less. The electron assigned to that orbital will thus have a higher energy. Therefore, the barrier to a sideways approach of H to H—H is higher and should increase monotonically as the bond is bent further. The detailed computations (e.g., see Figure 5.3) verify this qualitative conclusion.

A computation-free test of our considerations is the case of the H_3^+ ion, where only two electrons need be assigned, and they go into the lowest bonding orbital. We thus expect that H_3^+ will be a stable species and, moreover, that H^+ can approach H_2 from all directions. Such is indeed the case. Are the dynamics of the $H^+ + H_2$ going to be simpler than $H + H_2$? Unfortunately, no. There is a second potential that correlates asymptotically to $H + H_2^+$. The two surfaces can cross and the dynamics is not electronically adiabatic.

*5.1.5.1 From structure to reactivity: on orbital steering

Let us get further into chemistry. The purpose is to show how considerations of electronic structure that were tested for the geometry of stable triatomic molecules can be used to understand the configuration of the transition state. We will want even more, namely how the orbitals steer the approach to the barrier. As an example consider the approach of a hydrogen atom to a halogen molecule, beginning with a homonuclear one, say Cl_2. Again, let the initial approach be collinear, say along the z axis. The $1s$, $2s$, $2p$, and $3s$ Cl orbitals are far too low-lying in energy to interact with the $1s$ H orbital. In our simplistic approach only the $3p$ orbitals are relevant. Of the $3p$ orbitals of each Cl atom, two are orthogonal to the Cl_2 bond and hence, by symmetry, do not interact with the spherically symmetrical $1s$ H orbital, which approaches along the z direction. Only the $\sigma(3p_z)$ and $\sigma^*(3p_z)$ orbitals can be combined with the $1s$ H. To construct the HClCl orbitals, consider the available Cl_2 orbitals

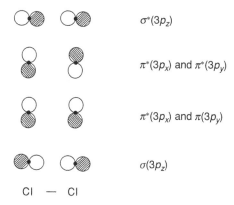

Next add the $1s$ H orbital in- and out-of-phase, using only the $\sigma(3p_z)$ and $\sigma^*(3p_z)$ orbitals:

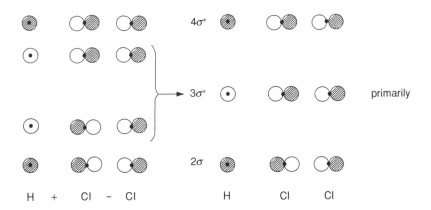

As in H_3, we have three independent MOs made out of three atomic ones, $1s$ on H and $3p_z$ on each of the Cl atoms. In contrast to the H_3 case, here it is found that the middle orbital is primarily bonding on H—Cl and antibonding on Cl—Cl, rather than "equal mixture" as in H_3. The rationale is, of course, that the H—Cl bond is stronger than the Cl—Cl bond.

Each Cl atom has five $3p$ electrons and H brings one for a total of 11: two are in the bonding HClCl orbital, four are in the bonding $\pi(3p_x)$ and $\pi(3p_y)$ Cl_2 orbitals, and another four in the antibonding ones. Thus 10 electrons have been assigned. As in H_3, the remaining odd electron is assigned to the middle orbital of HClCl, that is, the orbital that is antibonding on Cl—Cl but bonding on H—Cl. It is a singly occupied molecular orbital (SOMO), just as in H_3. Therefore there should not be a large barrier for collinear approach and the energy release upon reaction should be as a *repulsion* between the two Cl atoms.*

Say now that the H atom approaches Cl_2 sideways, in the x–z plane. The $\pi(3p_y)$ MO remains orthogonal to the plane and hence by symmetry, is indifferent to the approaching $1s$ H atom. Not so for the $\pi(3p_x)$ and $\pi^*(3p_x)$ MOs of Cl_2. In particular, the $\pi^*(3p_x)$ acquires strong H—Cl bonding character and hence is considerably stabilized. The stabilization of the $\pi^*(3p_x)$ orbital is countered, however, by the less efficient overlap in the 2σ orbital. Will a collinear or a bent H + XY intermediate be more stable? Here we require real quantitative computations. An example of the results of such computations is shown in Figure 5.10 for the simpler case of the F + H_2 reaction. It is seen that the energies of several MOs change with the approach angle.

---

* The HClCl complex will thus dissociate along the Cl_2 bond axis. The products' scattering angle is then the angle that the Cl_2 bond makes with the initial relative velocity. Since the H atom is light and usually fast moving, the Cl_2 molecule will not rotate much during the interaction time. For H + Cl_2 the angle of scattering of the products is therefore indicative of the preferred approach geometry. Experimentally, the HCl product is found to be backscattered, that is, the HCl rebounds, consistent with the prediction of a nearly collinear preferred approach.

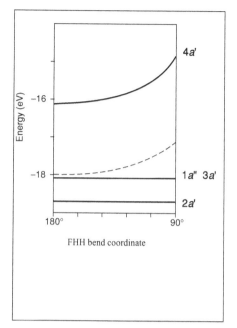

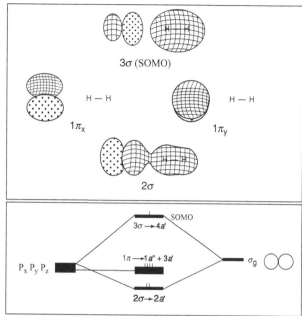

Figure 5.10 Understanding the barrier and its orientation dependence for the F + H$_2$ reaction. Right: molecular orbital interaction diagram. The 1s and 2s orbitals of F are too low down in energy and are essentially uninvolved and not shown. For a collinear approach along the z axis, the p_x and p_y orbitals of F are orthogonal to the σ orbital of H$_2$. In terms of their symmetry they are labeled as 1a'' and 3a'. The p_z orbital of F interacts with the σ orbital of H$_2$ forming a bonding, 2a' orbital and antibonding, 4a' orbitals. The latter is the SOMO and its energy increases when the molecule is bent. The variation with bond angle of the energy of the MOs is shown on the left. Also shown (dashed line) is the sum of the energies of the MO showing a preference for a collinear attack. Note however that there is a delicate balance because if the atom is a shade more electronegative, e.g., Ne$^+$, then the energy of the bonding, 2a', orbital will decrease faster with bending and the lowest barrier can be for a somewhat off-collinear configuration [adapted from D. M. Proserpio, R. Hoffmann, and R. D. Levine, *JACS*, **113**, 3217 (1991)]. Walsh diagrams, as shown on the left, are useful for understanding the geometry of bound molecules (Walsh, 1953). They can also be used to advantage in understanding the origin of steric effects in reaction dynamics. See Herschbach (1973), Mahan (1975). This application of Walsh diagrams reminds us that the same forces operate in bound and unbound molecular systems.

One aspect of the general H + XY case that still allows for a qualitative discussion is which "end" of the heteronuclear molecule will the H atom attack preferentially? To answer that, we note that in a heteronuclear diatomic the bonding MO is more localized on the more electronegative atom, whereas the antibonding MO is more localized on the atom that is less electronegative. Now the crucial orbital (3σ^*), which governs the approach geometry, is essentially an in-phase linear combination of the 1s H orbital with the σ^*(3p) antibonding XY orbital, localized on the less electronegative atom. The H atom will therefore preferentially approach the less electronegative halogen atom. Thus, for the reaction of H and ICl, HI will be preferred over HCl as the principal product. (This is despite the obvious fact that the reaction to form HCl is more exoergic!)

Qualitative considerations cannot replace detailed computations of the height of the barrier vs. the approach angle; however, such arguments do provide valuable insight and also must serve as guides for more complex systems. For such many-atom systems one also must not overlook the role of atoms or groups that are "in the way" and block the entrance. These repulsive interactions can, at zeroth order, be well anticipated on the basis of molecular models, cf. Figure 1.5, or, more quantitatively, using *molecular mechanics* (Davidson, 1993), a method that assigns sizes to atoms and bonds. At the same time we must note a recurring motif in many-atom systems: molecules are not rigid and they can and do adjust to the approach of a reactive reactant. During a reaction, substrates are more flexible than our conventional chemical upbringing, with its emphasis on rigid structures, prepares us for. New bonds form during the reaction and these can modify the preferred geometry.

5.1.6 The steric effect: the polar map representation

An often-used representation of the steric effect is by a *polar map*, showing equipotential contours as a function of the **angle** of orientation of the approaching atom with respect to the molecule. Figure 5.11 shows such a plot for the approach of Cl to HI (based on a semi-empirical surface for the Cl + HI → I + HCl reaction). The bond distance in HI is held constant (at its equilibrium value) as we vary the distance and orientation. The contours show clearly the steric hindrance to the H-atom abstraction by the bulky iodine atom, leaving only a narrow *cone of approach* along which the chlorine can approach the small hydrogen atom and abstract it. In the course of an actual reaction, the HI distance gradually increases as the Cl approaches, so such a polar map is not quite a complete indication of the steric requirements. In general one expects that the cone of acceptance will open up when the old bond is stretched.

5.1.7 Stable and unstable polyatomics

The potential energy function for a stable polyatomic molecule must exhibit a well. The bottom of the well is the equilibrium geometry of the bound molecule. At low vibrational energies the motion of the atoms is confined to be near the bottom of the well. As the energy is increased the molecule vibrates more vigorously and can access higher-up regions. At sufficiently high energy a molecule can dissociate. It follows that the well must connect to one or more exit valleys. If there is a barrier between the well and the exit valley then the threshold

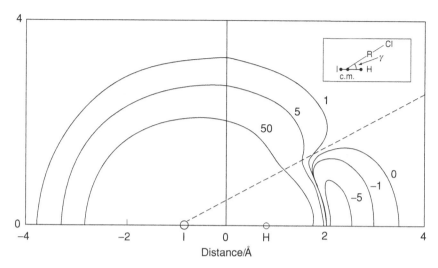

Figure 5.11 A polar representation of a potential contour map for the ClHI system, based on a semi-empirical LEP(S) computation. The potential energy contours (energy in kcal mol⁻¹) show the dependence of the potential on the approach angle γ of the Cl, at fixed H–I separation. The dashed line shows the cone of approach within which the chlorine can abstract the H atom [adapted from C. A. Parr, J. C. Polanyi, and W. H. Wong, *J. Chem. Phys.* **58**, 5 (1973)].

energy for dissociation is higher than the bond energy. Otherwise, the two are equal.

The wells along several potential energy profiles as shown in Figure 5.6 are therefore not unexpected. We should anticipate that, in general, more than one barrier occurs along the reaction coordinate. Between any two such barriers there is necessarily a well. If the total energy of the system is above the barrier, its motion is not confined to the well region. But at a lower energy the system must stay in the well; it is bound. Of course, such a truly bound system cannot be formed in an isolated collision starting from either the reactants' or products' side. Such starting conditions, plus the requirement that we reach the well, necessarily imply that the energy is higher than the barrier. But once we reach the well region the system can be stabilized, say by collisions with a third body. Stable bound states are therefore routinely well known.

We shall come to think that direct reactions are typical for a single barrier along the reaction coordinate. A well along the reaction coordinate means *complex dynamics*. For direct reactions we needed considerable ingenuity to initiate the reactants in the middle of a collision. Not so when there is a well. A suitable starting configuration is readily available. What we need to do is to supply the energy required so that the stable bound state can dissociate. This is what we usually call a *unimolecular reaction*. No wonder there is a rich literature on this and related subjects, as we discuss throughout Chapters 6 and 7.

5.1.8   Collision-induced dissociation

At higher translational energies it becomes possible for the actual reaction trajectories to deviate increasingly from the minimum energy path, and eventually it is energetically unnecessary that the system exits along the product valley at all. Provided that the available energy exceeds the thermochemical requirement (the dissociation energy of the reactant molecule) the atoms can separate, i.e., the "system" can exit to the plateau corresponding to the free atoms.

One well-studied example is the reaction of H_2^+ with He, which for $E < 2$ eV proceeds mainly to HeH$^+$ + H but for $E > \Delta E_0 \cong 2$ eV goes almost exclusively into dissociation:

$$H_2^+ + He \rightarrow H^+ + H + He$$

This is also an illustration of our rough correlation of energetic effects: as the translational energy is increased, newly allowed reaction paths take over at the expense of the old ones.

5.1.9   On to energy requirements and energy disposal of chemical reactions

We have learned to recognize energy to be a factor of central importance in the dynamics of chemical change. Energy is needed to drive the reaction forward; energy determines the magnitude of the cross-section or reaction rate and the branching between different possible final states; energy governs the details of the collision trajectory and energy is what we can get out of the products. To make such a discussion quantitative we need to know how one calculates the collision dynamics on a given potential. At the same time we continue to seek to correlate qualitative surface features (such as a well) with important dynamic effects (such as a sticky collision). We can thereby infer the major features of the surface from the observed dynamics.

5.2   The classical trajectory approach to reaction dynamics

5.2.1   From the potential surface to the dynamics

Classical mechanics provides a direct route from the potential energy surface to the dynamics of the collision, namely, the (numerical) solution of the classical equations of motion for the atoms. The solution uses Newton's law of motion to determine the position of each atom as a function of time. This output is known as a trajectory. It allows us to visualize how each atom moves as the reaction is taking place. Trajectory computations are carried out for two purposes. First, as a diagnostic of trends, i.e., features of the dynamics arising from different features of the surface or from changes in reactants' energies, masses, and so

Experiment   QCT

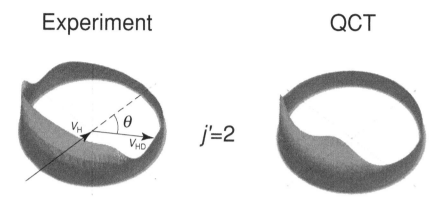

$j'=2$

Figure 5.12 Comparing experiment and a classical trajectory simulation for the angular distribution of state-selected HD molecules from the $H + D_2 \rightarrow HD(v',j') + D$ reaction at an initial translational energy of 1.64 eV [adapted from Fernandez-Alonso and Zare (2002)]. The experiment determines the angular distribution of HD in a definite final vibrational and rotational state (shown to be backward scattered for $v' = 3$, $j' = 2$) as a function of scattering angle with respect to the incident H atom. It is the higher initial translational energy that allows higher rovibrational states of the HD product to be populated. The classical trajectory computations are seen to mimic the observed distribution of final outcomes. This is because certain classical variables, such as the initial impact parameter, have a range of possible values. Each classical trajectory has definite initial conditions. We therefore need to run many trajectories to generate the required distribution, as is discussed further in Section 5.2.2.

on. Second, to investigate particular reactions for the purpose of accounting for observed chemical dynamical behavior or reaction rate or for anticipating experimental results. Before discussing the main results of diagnostic studies we need to elaborate below on the point that a classical trajectory describing the collision has definite starting conditions and these lead to a definite outcome. To generate the distributions, say over an angle, that are measured in an experiment one must generate an ensemble of trajectories that mimics the experimental distribution in initial conditions. In this way we can simulate all the non-quantal features of the reaction dynamics[15] as shown, for example, in Figure 5.12. This section is not a manual on how to generate such plots, but a discussion of a few essential principles.

The essence of the procedure is to choose a set of initial conditions and solve the classical equations of motion for each atom. In other words, one computes the time-development of the coordinates of each particle by solving the Newton second-order-in-time differential equation of motion. To do so we need the force acting on each atom. The force is computed as the change in the potential when that atom is displaced. Here is where we need to know the potential as a function of the positions of the atoms. An equivalent method for generating a trajectory is to solve Hamilton's equation for the position and the momentum. There are

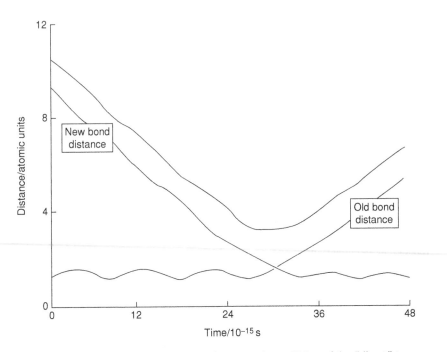

Figure 5.13 Plot of a collision trajectory for a reactive collision of the "direct" type (the time scale is that for the H + H$_2$ reaction). The collision is direct because of the fast switchover between the old and new bonds. Note also the vibration of the reactant BC molecule prior to the encounter with the reactant atom and the low-amplitude oscillation of the newly formed AB bond, indicating only a minimal product vibrational excitation. To interpret the dynamics recall that the slope of a plot of distance vs. time is the velocity. From Newton's first law, a change in the slope in such a plot indicates that a force is acting.

twice as many equations, but the equations are first order in time so we need just as many initial conditions as when we use Newton's second-order differential equations.

Computers allow us to exhibit a single trajectory as a movie showing the actual atoms as they move. One can also have a poor person's movie, namely a set of snapshots of the atoms at different times.* A simple and instructive alternative is to plot the interatomic distance $R_{ij}(t)$ for a pair of atoms i, j vs. time, as shown in Figure 5.13 for a three-particle system.

The idea of a direct reaction, one that is over in a vibrational period as seen in Figure 5.13, was not quite what chemists expected. It was somehow felt that chemical rearrangement is more protracted. The interpretation of the early experiments probing the dynamics of direct reactions, such as those discussed in Chapter 1, received further support from trajectory computations.

---

* At the other extreme, if you have access to a virtual reality room you can view the reaction taking place in the space around you.

Along the trajectory one knows the position and velocity of each atom. These completely specify the mechanical state of the system, so it is possible to compute all the observables of interest. But this determinism carries with it an implication. A classical trajectory has a completely definite outcome. The collision is either reactive (meaning that the new bond distance is short whereas the old bond distance is long) or it is not reactive. The products come out in a definite angle. Their internal energy is sharply defined, and so on. What gives rise to all the distributions that we have been talking about?

5.2.2 The need for averaging trajectory results

The trajectory shown in Figure 5.13 corresponds to a particular choice of initial values for the coordinates and velocities of the participating atoms. To obtain quantities of physical interest (such as cross-sections, rates, etc.) it is necessary to average over some or all of these initial conditions. For example, to determine the opacity function, $P(b)$, at fixed energy E, it is necessary to average over all possible initial **orientations** of the reactants at the given b. We have already seen (Section 3.2.7) how this averaging leads to the concept of a steric factor, that is, $P(b) < 1$.

The sampling of initial conditions for the purpose of executing an average over them is often done by a so-called Monte Carlo (i.e., a random selection) procedure. Thus the computation of the observable dynamical quantities by the use of averages over classical trajectories is often referred to as the *Monte Carlo method*. We describe the technical essence in Appendix 5.A and here we discuss the reason why averaging over the initial conditions is **inherently** required.

An essential difference between classical and quantal mechanics is the number of initial conditions that need to be specified if the initial state is to be fully defined. In classical mechanics one must specify both the position and the momentum for each degree of freedom. In quantum mechanics the uncertainty principle implies that if, say, the momentum is well specified, the value of the position can be anywhere within its possible range. Since molecules are inherently quantal, a complete specification of initial conditions for a collision in a system of n degrees of freedom consists of n quantum numbers. In contrast, a classical trajectory for the system requires $2n$ initial conditions. The method of classical trajectories **mimics** this quantal aspect by running many classical trajectories where, of the $2n$ initial conditions, n are held constant (the same n that correspond to the quantal case) while the other n are allowed to vary. The final outcome is determined by averaging over those initial conditions that are varied. We refer to these initial conditions that need to be inherently averaged over as the *phases*.[16]

In conclusion, there are two separate reasons why we need to run many trajectories and then to average over the results. One is inherent. A classical trajectory is necessarily specified beyond what quantum mechanics allows and therefore corresponds to an initial state that cannot be realized in the laboratory. We can

mimic the quantum uncertainty by the so-called *quasi-classical method* of generating many trajectories and only looking at the averaged results. The second and additional reason for averaging is practical. In the laboratory one seldom prepares a fully (quantal) state-selected initial state. We would like to, but it is often just not practical to realize this wish. A clear example is that we typically do not select the angular momentum quantum number *l* or its projection, *m*. In classical language, we sample all values of the impact parameter *b* and all orientations of the plane of the collision. As we discussed, this experimental reality requires us to average over those initial conditions that could be selected, but were not. Fortunately, both needs to average can be combined in one procedure and the Monte Carlo method, Appendix 5.A, offers a practical procedure for doing so. Thereby, classical trajectories can provide the most exquisite details about the dynamics, as was shown in Figure 5.12 for the highly resolved final state distribution in the reaction $H + D_2 \rightarrow HD + D$.

*5.2.1 Chaos and longer time evolution of the quasi-classical trajectory method

Even within classical mechanics there are a number of limitations on the description of the dynamics using numerically generated classical trajectories. Surprisingly, the sheer number of atoms is not, in itself, a limitation. One can, given a very modest workstation, keep track of the motion of a very large number of atoms, enough even for the description of biological processes. Experts can handle millions of atoms so that, e.g., the behavior of bulk matter under extreme disequilibrium (fracture, turbulent flow, and so on) can be simulated. What is computationally challenging is not the solution of a large number of differential equations of motion that are coupled (because the force on one atom depends on where the other atoms are). What often is the slowest step is the computation of the forces.[17]

There are several important limitations on what one can do with classical trajectories. An important one is technically known as *deterministic chaos*.[18] The characteristic of chaos that is relevant here is the *sensitivity to initial conditions*. This means that two trajectories that start with rather similar initial conditions (= positions and velocities of the atoms) end up after the collision with rather different outcomes, up to that, one trajectory is reactive and another, with a rather similar starting point, is not.

On the one hand, chaos is good news. It explains the sense in which a classical mechanical system can forget where it came from.[19] It implies that unless the initial conditions are fully tightly specified, the longer-time outcome is not totally determined. Initial classical conditions can never be truly fully specified because this calls for keeping an infinite number of digits for each number (position and velocity) and, in any case, quantum mechanics implies a necessary fuzziness in the classical initial conditions. Further, a real experiment will have even more averaging on initial conditions. The more averaging, the more statistical is the result likely to be. We shall come to find that this is an important consideration. Averaging over initial conditions averages over details in the outcome and makes the results look less specific.

Chaos is bad news if one wants numerically to integrate the equations of motion. Two aspects contribute. One is the necessarily finite precision with which the initial conditions are specified and the other is the inevitable numerical error (due to the finite precision of the algorithm and the computer round-off of digits) in any time propagation step. After some propagation time it is no longer certain that the trajectory that is being propagated is indeed the one starting from the specified initial conditions. The only check is to reverse the directions of all the velocities and to integrate back in time. If one (approximately) recovers the initial conditions, all is well. Typically, if the reaction is direct, this check will work. But if the dynamics is not direct and, in general, if the integration time is long, the trajectory will not integrate back. Hence the problem, in many-atom systems the method of classical trajectories is only useful for relatively not too long times: times that are definitely long on the inherent scale of a vibrational motion but short on the scale of bulk interest such as the time interval between collisions.

In the face of chaos the option that we recommend[20] is to initiate the dynamics at the configuration of interest, which need not be that of the separated reactants. For example, if we have a barrier with wells on either side, as in Figure 5.6, and one needs to understand the dynamics, start the integration at the barrier. Similarly, if one wants to simulate the role of the solvent in barrier crossing, start the (solvated) reactants in the transition state region.

For long-time evolution of a many-atom system it is probably best to recognize that a large subset of all degrees of freedom is not strongly coupled to motion along the reaction path and to other primary degrees of freedom. They can therefore be thought of as "secondary" or "solvent" modes, meaning that they act as a bath for the primary degrees of freedom. Many-atom systems of biological interest are in a real sense solvated. Similarly, for example, for a chemical reaction taking place on a surface. The role of the bulk of the solvent (the part of the environment that is not strongly coupled as well as effects due to other intramolecular degrees of freedom) can be mimicked by *Brownian* (or *Langevin*) *dynamics*.[21] These are equations of motion for the primary degrees of freedom that allow for the role of the other degrees of freedom (= the environment) through introducing a *frictional* or *dissipative force* that drains energy out of the primary degrees of freedom. But if this is uncompensated for then the motion in the primary degrees of freedom will eventually come to a stop. So, in addition to the persistent friction, the *Langevin equation of motion* includes a *fluctuating force* due to the environment that can restore energy to the primary system. The combination of a frictional and fluctuating force insures that at long times the system reaches thermal equilibrium.[22]

5.A Appendix: Monte Carlo sampling

The advantages of Monte Carlo sampling are familiar to us because we are constantly exposed to the results of polls of public opinion. In a tightly contested election, pollsters approach just over 500 representatively chosen persons to get a reasonably reliable (reported as <5% uncertainty) prediction. What is the secret?[23] How can such a small sample reliably predict the outcome of a country-wide election? The point is that the pollsters ask a single yes/no question where the answers are about equally likely. They do not seek to check if a yes answer correlates with the gender of the person nor with her/his income level, etc. Nor do they ask about fringe candidates. If they did any of this, they would need a much larger poll. The empirical observation is therefore that a simple yes/no question needs a relatively small sample to provide an approximate but reliable prediction. Rather than the major effort of asking everybody, we accept a small sampling error and ask far, far fewer people.

How does this apply to our problem? What we want to do is to properly sample the initial phases. Take the simplest case, A + BC. Excluding the center of mass, we have six degrees of freedom. Therefore, six phases need to be sampled.[24] Say we plan to do so systematically. A phase angle spans the range 0 to 2π, so ten different values of the initial phase is a reasonable minimal sample for each degree of freedom. But there are six different coordinates and a full specification of initial conditions means choosing a value for each phase. So the sample has to have 10^6 systematically chosen different initial conditions. If all we want is an answer to "what fraction of collisions are reactive?" then we suspect that this is too large a sample. Just like the opinion pollsters we should be able to make do with about 500 different trajectories if, like them, we know how to sample in a representative manner. We shall show that this is possible by an explicit example. Our demonstration will be that a relatively small sample is enough when we seek a simple answer. The more resolution that we want, the more trajectories we need to run to be able to generate a representative sample where enough trajectories end up with the desired outcome.

The simplistic discussion above is really the heart of the issue. The method of quasi-classical trajectories is most suitable when the degree of resolution required is not very high. Otherwise, the more correlations we need between different attributes of the products, the more trajectories need to be computed. For example, generating results as shown in Figure 5.12, where the product HD vibrational and rotational states are determined as a function of scattering angle, requires very many trajectories. Determining the probability of rare events, such as the crossing of a high barrier by thermal reactants, will require so many trajectories that fine-tuning of the sampling is inevitable.* Classical trajectories with Monte Carlo selection of initial conditions is the standard workhorse.[25]

---

* After we discuss transition state theory we will be in a better position to handle this issue. Pun intended.

*5.A.1 An example of Monte Carlo sampling

The Monte Carlo sampling of initial conditions in the method of classical trajectories is illustrated here for the particular problem of computing the reaction
cross-section σ_R given in terms of the opacity function $P(b)$ by, Eq. (3.14),

$$\sigma_R = 2\pi \int_0^\infty bP(b)\,\mathrm{d}b \qquad\qquad (\text{A.5.1})$$

Now $P(b)$ is the probability of reaction at impact parameter b. Let us therefore
run $N(b)$ classical trajectories, all of which have the same initial parameter b.
Not all of these trajectories will necessarily lead to reaction. The reason is that
they will differ in some other initial conditions (e.g., the phase of vibration or
rotation of the diatomic reagent). Let $N_R(b)$, $N_R(b) \le N(b)$, be the number of
trajectories that do exit in the products' valley. Then if $N(b)$ is large enough,
$P(b) = N_R(b)/N(b)$, and so we need to evaluate the integral

$$\sigma_R = 2\pi \int_0^B b\,\frac{N_R(b)}{N(b)}\,\mathrm{d}b \qquad\qquad (\text{A.5.2})$$

where B is a sufficiently large impact parameter such that no reaction takes
place for $b > B$ (i.e., $N_R(b) \cong 0$ beyond B^*). We could replace the continuous
integration over b by a sum over a large number of discrete b values (i.e., generate
a histogramic representation of $P(b)$). At each b value we would need to run $N(b)$
trajectories, determine $N_R(b)$, and then compute the sum. This is feasible but
requires running trajectories at a number of different b values. However, what we
are after is not $P(b)$ but the cross-section σ_R. Say that we are willing not to know
$P(b)$, will it simplify the task?

Let us sample initial b values with some attention to the physics. In other
words, like the public opinion pollsters let us prepare a representative sample.
Higher b values are more heavily weighted in the cross-section, via the annulus
$2\pi b\,\mathrm{d}b$. Therefore, if we plan to run a grand total of N trajectories, let us allot the
number $N(b)\Delta b$, where

$$N(b) = (2\pi b/\pi B^2)\,N \qquad\qquad (\text{A.5.3})$$

to initial values of the impact parameter in the range b to $b + \Delta b$. Note that in
doing so we run more trajectories at higher values of b.

Integrating Eq. (A.5.3), over b from 0 to B, we verify that N is the total number
of trajectories

$$\int_0^B N(b)\,\mathrm{d}b = N \qquad\qquad (\text{A.5.4})$$

---

[*] In Section 3.2 we saw how to estimate such a value. To be on the safe side, add 10%. Taking a
larger value of B requires running a shade too many trajectories but will not change the computed
value of the reaction cross-section, Problem J.

Let $N_R(b)\Delta b$ be the number of trajectories in the range b to $b + \Delta b$ that did lead to reaction. Then, for the particular choice of $N(b)$ given in (A.5.3), we have, by substituting in Eq. (A.5.2),

$$\sigma_R = \pi B^2 \frac{\int_0^B N_R(b)\,\mathrm{d}b}{N} \tag{A.5.5}$$

We now introduce N_R as the grand total of trajectories that did lead to reaction,

$$N_R = \int_0^B N_R(b)\,\mathrm{d}b \tag{A.5.6}$$

Since $N_R(b) \le N(b)$, $N_R \le N$. The number N_R of reactive trajectories is obtained by sampling according to Eq. (A.5.3) and counting how many trajectories are reactive. Using Eqs. (A.5.5) and (A.5.6), the reaction cross-section can now be written as

$$\sigma_R = \pi B^2 \frac{N_R}{N} \tag{A.5.7}$$

What we have gained is that far fewer trajectories are now needed. Why? Because the percentage error in a sampling of a probability like $P = N_R/N$ is $((1 - P)/N_R)^{1/2}$. So it scales as $1/\sqrt{N}$. (If we want to estimate the cross-section to, say, 10% accuracy and we take $P = 1/2$ we need, roughly, $(0.5/N_R)^{1/2} = 0.1$ or $N_R/P = N = 100$ trajectories.) Roughly the same number of trajectories is needed to compute $P(b) = N_R(b)/N(b)$ at a given value of b. Therefore, a much higher number of trajectories is needed if we first compute $P(b)$ and then compute the cross-section. By giving up the ability to provide an accurate sampling for $P(b)$ we have achieved a computational advantage.* Conversely, the more resolution we need to generate, the higher the computational effort. To know everything that one possibly can about even an atom–diatom reaction in a given initial state requires running millions of trajectories.

The Monte Carlo sampling is equally useful for other types of averages, for example the thermal average that defines the thermal reaction rate constant, $k(T) = \langle v\sigma_R \rangle$. Problem J asks for a sampling procedure that will generate $k(T)$ directly without computing σ_R first and shows that the computational effort is comparable to that required for Eq. (A.5.7). This is the strength and the weakness of the method. There is a price to pay if higher resolution is needed but more averaged quantities can be efficiently determined. By all means, look at the results of a single trajectory but know that no reaction can be represented by only one trajectory. Instead, we must average over all initial conditions.

---

* Technically, we have achieved this advantage by using a more representative or a "biased" sampling: we gave a higher weight in the sample to higher values of b.

5.3 Energy and dynamics of the chemical change

Aided by the ability to run and visualize classical trajectories we discuss the role of energy in driving chemical reactions and the complementary question of energy disposal in reactions.

5.3.1 Energy disposal in direct exoergic reactions

Our journey begins with the groundbreaking "atomic flame" experiments of M. Polanyi. These suggested that in such harpoon reactions as

$$K + Br_2 \rightarrow KBr^\dagger + Br$$

the newly-formed salt molecule is highly vibrationally excited (as denoted by the dagger), so much so that in a subsequent collision with an alkali atom it can cause electronic excitation of the alkali:

$$KBr^\dagger + K \rightarrow KBr + \quad K^*$$
$$\downarrow$$
$$K + h\nu$$

The electronically excited atom decays by fluorescence leading to a visible "flame," with a color typical of the alkali metal used.

At the same time Eyring, Polanyi, and others used a form of the London equation to generate the first "realistic" LEP(S)-type potential energy surfaces, Eq. (5.3). Then they sought to interpret the dynamics of such reactive collisions (including the vibrational excitation of the products) using these potential functions. Ever since, energy disposal in exoergic reactions has been interpreted and discussed in terms of the analysis of classical trajectories.

One of the correlations that has emerged is as follows: for given reactant masses (and at not too high energies) the conversion of the reaction exoergicity into the product vibration occurs while the attacking atom A is still on its approach motion toward the BC molecule, see Figure 5.14. An early release of the exoergicity is manifested as a repulsion between B and C and this channels the exoergicity into the relative motion of A and B, that is, into product vibration. The efficacy of the conversion will be greater the earlier is the release of the exoergicity. A potential where the attacking atom is being accelerated toward the other reactant is known as an attractive potential energy surface, as shown in Figure 5.14.

An alternative representation of the same phenomenon is shown in Figure 5.15, where the trajectory is superimposed on the (attractive) potential energy surface. Recall that, by the Hammond postulate, we expect that the more exoergic the reaction, the more attractive will the potential be. As a rough guide, we expect that exoergic reactions channel the energy into the vibration of the newly formed bond.

The early release of the exoergicity appears initially as kinetic energy in the AB direction. This leads to the so-called *bobsled effect*. As the potential contours

Figure 5.14 Schematic interpretation of the correlation between product vibration and the timing of the release of the exoergicity along the reaction path. Upper: an attractive (early downhill) surface. The progress of the collision, as shown in the adjacent panel, illustrates an early release of the exoergicity (designated by *) during the approach of A. This accelerates A and B toward one another and leads to a vibrationally excited AB product. For the repulsive surface (lower) the late release of the exoergicity, while the new molecule AB is already receding from C, ejects AB with high translational energy but with low vibrational excitation [adapted from Polanyi (1972)].

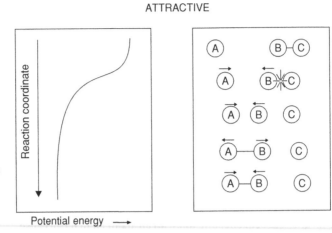

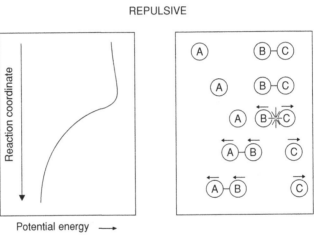

bend round (because of AB and BC repulsions) the trajectory cannot follow the minimum-energy path but tends to continue in a straight line (owing to the large momentum in the AB direction) leading to a "bobsled" motion. Ultimately, the AB repulsion converts the exoergicity into vibration of the AB product. This is in accord with our expectations of an early release of exoergicity in alkali–halogen reactions based on the harpoon model. The early barrier for the $F + H_2$ reaction, Figure 5.4, similarly leads to highly vibrationally excited HF. Abstraction reactions of F with organic molecules RH also release much of the energy into the vibration of HF, meaning that the large organic radical R is a spectator and remains internally cold.

All of our discussion thus far on the relations between the topology of the potential surface and the dynamics of the system has so far ignored the important question of the influence of the *masses* of the particles upon the dynamical

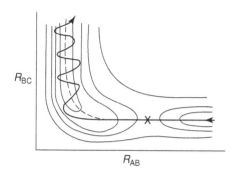

Figure 5.15 Schematic representation of efficient channeling of the reaction exoergicity into product vibrational excitation, for an attractive potential energy surface. The drawing is generated from a trajectory, as shown in Figure 5.13, by plotting successive values of the two bond distances. The early release of the exoergicity (past the early saddle point) leads to a deviation of the actual collision trajectory (solid line) from the minimum-energy path (broken line). The early acceleration enables the trajectory to climb to the side of the potential and thereby initiates an A–B vibration. Had the trajectory followed the minimum-energy path the exoergicity would be released as relative translational energy of the products.

behavior of the system. It was recognized early on that there would be strong mass effects on any given potential surface (since, after all, Newton's laws involve the masses rather directly!). In order to see these effects clearly a *mass-weighted* coordinate system was devised upon which to plot the potential energy surface.

In this representation the system can be characterized by a point particle moving on an actual potential energy landscape, as further discussed in Appendix 5.B. By imagining the motion of a mass point on such a potential one can more easily understand the influence of initial translational energy, or of initial vibrational energy of the reactants upon the trajectories and the exoergicity disposal. The effect of mass-weighting of the axes in such a plot can be such that it is possible to have high product vibrational excitation even on a repulsive, late downhill surface, when the attacking atom is very light. Problem E describes a simple potential for which mass effects can easily be visualized.

As an illustration of the application of the classical trajectory method to the problem of the energy disposal in exoergic reactions, Figure 5.16 compares the computed and the measured HF vibrational energy distribution for the H + FCl → HF + Cl reaction.

Hydrides have a very high vibrational frequency and also a low moment of inertia (a high rotational constant). Extensive product vibrational excitation is not unique to hydrides, as already shown in the atomic flame experiments and as found for such well-studied exoergic reactions as O + CS → CO + S, Figure 5.17, and F + I$_2$ → IF + I. Nor is it essential for the reaction to be exoergic. One can rapidly approach the bend along the reaction path by initiating

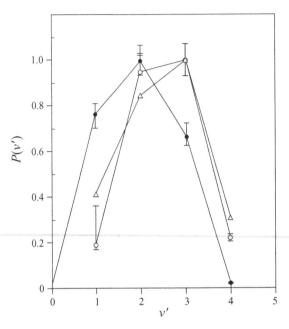

Figure 5.16 Comparison of the experimental HF vibrational state distribution, points (normalized to unity at the most probable final state), and the results of trajectory computations (based on an *ab initio* potential energy surface) for the very exoergic reaction H + FCl → HF + Cl, open points. The error bars shown for the trajectory results are due to the Monte Carlo sampling [adapted from R. Sayós *et al.*, *Phys. Chem. Chem. Phys.* **2**, 523 (2000); experimental results o from D. Brandt and J. C. Polanyi, *Chem. Phys.* **35**, 23 (1978) and △ from K. Tamagake and D. W. Setser, *J. Phys. Chem.* **83**, 1000 (1979)]. Trajectory computations for D + FCl → DF + Cl, not shown in the figure, indicate that the DF vibrational distribution is essentially the same as that of HF when the results are plotted vs. f_v, the fraction of the available energy in the vibration of the product. As discussed in Section 5.1.5.1, the reaction is expected to proceed by two mechanisms, a direct abstraction and a migratory route where the H atom approaches from the direction of Cl. The migratory route is favored by higher-impact-parameter collisions and so leads to more forward scattering [for earlier results for the H + F_2 → HF + F reaction see N. Jonathan, S. Okuda, and D. Timlin, *Mol. Phys.* **24**, 1143 (1972) and J. C. Polanyi, J. L. Schreiber, and J. J. Sloan, *Chem. Phys.* **9**, 403 (1975)].

a collision with high translational energy. The bobsled effect will equally take place and some of that energy will be converted to products' vibration, as shown for H + D_2 in Figure 1.3, but the conversion is not as efficient as when exoergicity is released early on along the reaction path.

5.3.2   Energy requirements for reactions with a barrier

For endoergic reactions, and also for exoergic reactions with a barrier located along the reaction path, we seek to obtain information on the energy requirements

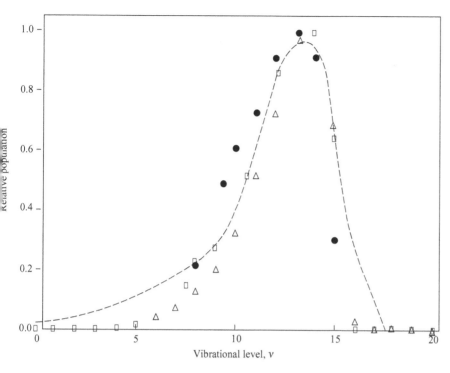

Figure 5.17 CO vibrational state distribution following the highly exoergic $O(^3P)$ + CS → CO + $S(^3P)$ reaction. Dots: experimental [adapted from G. Hancock, B. A. Ridley, and I. W. M. Smith, *J. Chem. Soc. Faraday Trans* 2 **68**, 2117 (1972)]. Open symbols: trajectory computations for thermal (△) and translationally hot (□) atoms [adapted from D. Summerfield *et al.*, *J. Chem. Phys.* **108**, 1391 (1997)]. Dashed line: a fit of the distribution by a linear vibrational surprisal as discussed in Section 6.4 [adapted from H. Kaplan, R. D. Levine, and J. Manz, *Chem. Phys.* **12**, 447 (1976)].

for reaction. Obviously we require a total energy E in excess of the barrier height E_b to achieve a reactive trajectory. But must we supply this energy, say, as initial translation?

On the basis of microscopic reversibility and the discussion of Section 5.3.1 on energy release, we expect a correlation between the location of the barrier along the reaction path and the form of initial energy most conducive to reaction. *Translational* energy is most effective for passage across an *early* barrier while *vibrational* energy of the reactant molecule is more efficient for surmounting a *late* barrier (i.e., in the exit valley).

These requirements can be interpreted, as shown in Figure 5.18, in terms of the availability of kinetic energy in the proper coordinate. For an early barrier, one requires momentum along R_{AB}, while to overcome a late barrier the energy is needed in the R_{BC} coordinate.

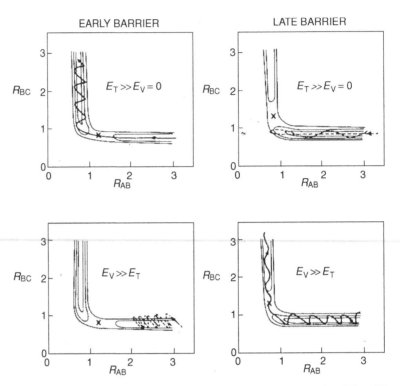

Figure 5.18 Influence of reactant energy for the thermoneutral $A + BC \rightarrow AB + C$ reaction on a LEP(S) surface with a barrier (X). In all four cases $E > E_b$ so that reaction is allowed on energetic grounds. When the relative translational energy (E_T) is high (top row) there is ample kinetic energy for motion along R_{AB}. Reaction thus occurs for a barrier in the entrance valley (i.e., along R_{AB}, left column), but fails if the barrier is in the exit valley (i.e., along R_{BC}, right column). The opposite is found when the reactant diatomic has high vibrational excitation (E_V) but E_T is low, bottom row. The reactant vibrational energy (which means that kinetic energy is available along R_{BC}) helps in surmounting a late barrier [adapted from J. C. Polanyi and W. H. Wong, *J. Chem. Phys.* **51**, 1439 (1969)].

The energy requirements of reactions will obviously become less restrictive at higher collision energies. Even so, experiments, simulations (Figure 5.18), and considerations of detailed balance all agree that at low collision energies (and particularly so in the thermal regime) the selective energy requirements can significantly influence the reaction rate.

5.3.3   Direct vs. compound collisions

For a given potential surface (and mass), classical mechanics can be used to illustrate the progress of the reaction through plotting the interatomic distances as a function of time. Typically, for a *direct* reaction the switchover between the

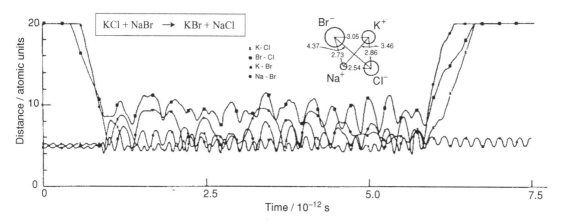

Figure 5.19 Bond distances (in atomic units) vs. time from a classical trajectory computation for the KCl + NaBr → KBr + NaCl reaction [P. Brumer, Ph.D. thesis (1972)]. The most stable structure of the tetratomic ionic intermediate is shown as an insert.

two bonds occurs within a very short time interval, corresponding to about one vibrational period, as shown in Figure 5.13. By way of qualitative contrast, let us have a look at another trajectory calculation for the (gas-phase) reaction

$$KCl + NaBr \rightarrow KBr + NaCl$$

where the experimental evidence suggests that there is a long-lived intermediate. The potential energy surface has a deep well and the equilibrium structure of the ionic complex is shown in Figure 5.19. The trajectory shown in this figure is very "snarled" and it is quite reasonable that it would look very much the same if we started it at the KBr + NaCl configuration. By the time the trajectory gets out of the well it hardly remembers from which side of the well it entered.* By running many trajectories at different initial conditions we verify that the exit to products or back to the reactants is essentially independent of how the collision was initiated. This is an extreme example of the compound or *complex* mode of reaction.

Many reactions are more similar to one mode or the other but it is quite possible for the same chemical transformation to proceed by both a direct and a complex mode. Early on, Figure 1.4, we noted that the insertion reaction of $O(^1D)$ into H_2, that is, primarily a complex mode, also has a direct component that proceeds by abstraction and is detected as backward scattering. Even the $H + H_2$ reaction

---

* How can an accurately computed trajectory forget? All that we need to do is to run the trajectory backwards in time and eventually it will show clearly which side of the well it entered on. Correct. But as discussed in Section 5.2.2.1, if we make even a very tiny change in the initial conditions of the trajectory then, owing to the required long integration time, it may well not exit in the same valley. So what we really mean is that a sample of trajectories, all quite similar, forgets. Some members go out as reactants, others as products.

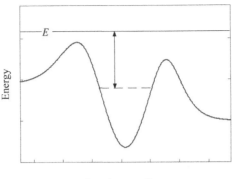

Reaction coordinate

Figure 5.20 Potential energy along the reaction coordinate with a well between two barriers, for an exoergic reaction. Reactants starting at the left with the total energy E being mostly in translation can cross the first barrier and proceed promptly to form products. However, once the trajectory enters the well region, there is another option: some fraction of the total energy, indicated by the double arrow, can flow into the other degrees of freedom, the internal degrees of freedom of the system. In classical mechanics the total energy must always be above the potential energy so this scenario is only possible because of the presence of the well in the potential. Once this energy redistribution occurs, the energy along the reaction coordinate, dashed line in the figure, is insufficient for the molecule to escape across either barrier. The system behaves as if it is bound even though it has enough energy to dissociate. What happens next depends on the size of the reactants, a central theme of Section 6.2. Sooner, if the reactants are small, or later, energy will flow back to the reaction coordinate and the energy-rich molecule will "dissociate," meaning that it will cross a barrier and exit.

has a minor contribution from an insertion mechanism, particularly at higher energies (Fernandez-Alonso and Zare, 2002). Another example that we already encountered is the $C_2H_5 + O_2 \rightarrow C_2H_4 + HO_2$ reaction, Figure 5.6, where at lower collision energies the reaction will proceed by a complex mode because the direct path has a higher barrier.

5.3.3.1 Complex mode trajectories and unimolecular reactions

Trajectories that proceed over a well along the reaction coordinate can be delayed and the reaction occurs by a complex mode.[26] Why should that be? At low energy the well represents a stable molecule, but a trajectory that comes from the reactants' region has enough energy to cross over the well and directly proceed to the products' side. The "molecule" that is formed is energy-rich and can immediately dissociate. Figure 5.20 shows schematically why the presence of a well enables a trajectory to spend some time seemingly as a stable molecule before dissociation takes place.

When the trajectory rattles many times within the well before exiting, the dynamics can be broken into two stages: capture into the well and the delayed dissociation of an energy-rich molecule. The second stage is what we know as a unimolecular dissociation. To cleanly separate the two stages we need the system to spend a long time over the well, long enough that it forgets how it got into the well. As we discuss in Section 6.2, the more internal degrees of freedom there are, the far more ways there are to distribute the energy E and the much longer it will take for it to return to the reaction coordinate. For few-atom systems the delay can be but a few vibrational periods, which is quite short.

*5.3.3.2 Prompt trajectories

The majority of trajectories are delayed by a passage over the well because they will lose energy to other degrees of freedom. Yet a small but finite fraction will exit promptly. One can see that this can be so by the very argument that suggests that the energy is partitioned amongst all degrees of freedom. Let us accept that the energy is distributed. Even then there is a small probability that energy in excess of the dissociation energy remains localized in the reaction coordinate. This fraction of molecules dissociates promptly because it has the energy in the mode where it is needed. Furthermore, the energy redistribution depends on the coupling between the motion along the reaction coordinate and the other, internal, modes.

Another sense in which an intermediate situation arises is to note that we have contrasted two limits. Direct reactions, where the trajectory crosses over immediately from the reactants' to the products' valley and compound collisions, where the trajectory rattles many times within the well. So many times that it can lose the memory of where it came from. It is reasonable to expect that there are intermediate situations where the trajectory goes through the well region more than once but it still does so for only a limited number of times before it exits.

5.3.4 Stereodynamics

Exploring the dynamics using classical trajectories can provide more than a validation of our intuitive view of the cone of acceptance for reaction, Figure 1.5. Nor are they limited to the simple A + BC problem. A much-studied example is the prototype S_N2 reaction

$$Cl^- + CH_3Cl \rightarrow ClCH_3 + Cl^-$$

The energy along the reaction path has a double-well shape, with two symmetrical minima that are due to the strong ion–molecule polarization force and a central barrier to atom exchange. The structure at the barrier corresponds to a planar CH_3 group. The reaction therefore proceeds through a Walden inversion, Figure 5.21.

An important point clearly brought forth by trajectory computations is that reactants accommodate one another en route to the reaction. The steric requirements are then not simply those of the isolated reactants. A simple example,

Figure 5.21 The configuration of the reactants, transition state, and products along the reaction path for the S_N2 reaction of Cl^- with CH_3Cl. The approach (or retreat) motion to the transition state can deviate from a collinear Cl–C–Cl configuration.

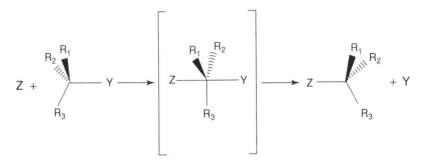

observed also in atom–diatom reactions, is how the elongation of the old bond opens up the cone of acceptance for reaction. The reason is that at the transition state, particularly if it is not early, the old bond distance is extended well above its value at equilibrium for the reactant. Therefore, the barrier height as a function of approach angle is lower when the old bond is extended and this allows for a wider range of approach angles to react. Figure 5.22 shows the effect for the H_3 energy surface.

Another aspect is that atoms or groups that could offer a steric hindrance move out of the way. This can be as simple as the transition to planarity of CH_3 during the S_N2 reaction shown in Figure 5.21 to the "gate opening" when a ligand approaches a biological receptor. It is a speciality of proteins that their internal adaptation can be limited and yet the resulting change in binding can be considerable.* A well-known example is when hemoglobin adjusts to admit oxygen, then the second oxygen molecule is admitted more easily.** But the conformational change of biopolymers can also be substantial, as in the unstacking of adjacent base pairs in the double helix required for the intercalation (of drugs or other effectors) in DNA, Figure 5.23.

The theme of structural adaptation during reaction is not limited to biological substrates. In Chapter 12 we discuss surface chemical reactions and catalysis. We shall then draw attention to the flexible nature of the surface during a chemical reaction taking place on it.[27]

Steering is another stereochemical concept that spans the range from quite simple reactions to the association of biological molecules. Steering originates from the anisotropic intermolcular force[28] that can pull molecules into or deflect

---

* This is, of course, determined by the features designed by nature into the potential energy surface. Unlike the more rigid molecules where the potential has one dominant well, the potential energy landscape of a protein is far richer in minima with the result that the protein may be ready for some action yet this does not take place until it is triggered by a binding of, typically, an ion, in some site possibly quite remote. Proteins can communicate small local structural changes over long distances. More on *molecular machines* in Section 6.2.4.

** This cooperative effect means that the fraction of bound oxygen has a sigmoid dependence on the pressure of oxygen with the well-known result that a modest drop in oxygen pressure results in oxygen being released.

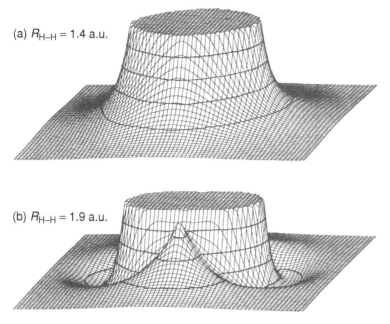

(a) $R_{H-H} = 1.4$ a.u.

(b) $R_{H-H} = 1.9$ a.u.

Figure 5.22 Polar representation of the H_3 potential energy. Shown is the potential, for a fixed reactant bond distance, as a function of the distance of the approaching H atom from the center of H_2 and of the approach angle, cf. Figure 5.3. The two plots are for H_2 at its equilibrium separation and for a higher bond distance. Note how a cone of acceptance opens up when the reactant bond is stretched [adapted from I. Schechter, R. Kosloff, and R. D. Levine, *Chem. Phys. Lett.* **121**, 297 (1985)].

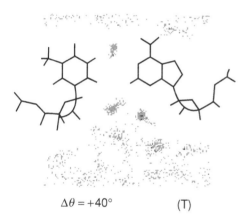

$\Delta\theta = +40°$ (T)

Figure 5.23 A model study of the organization of water during the unstacking of DNA base pairs (adenine and thymine). Each dot is a time-averaged position of the O atom of a water molecule. The conformation shown corresponds to an opening of thymine toward the major groove [adapted from E. Giudice, P. Várnai, and R. Lavery, *Chem. Phys. Chem.* **2**, 673 (2001)].

them out of the cone of acceptance. Such effects are more noticeable for rotationally excited reagents and trajectories have examined this role.[29] The simplest effect is that rotation can either mask or expose a reactive site, for example, in $K + HCl \rightarrow KCl + H$, where the center of mass of HCl is essentially on the Cl atom. Rotational excitation of HCl makes the H atom rapidly orbit around Cl and shield it from reacting with the K atom.

To go beyond the simpler considerations above we must recognize that stereo-dynamics is about the direction of attack and other quantities that are vectors. Chapter 10 takes up this theme.

5.3.5 On to the specificity of energy disposal and selectivity of energy requirements

What we have learned in this section is that certain dominant features of the potential energy surface govern the chemical reactivity. Can we use this knowledge to develop some simpler (albeit more approximate) theoretical approach and thus obviate the need for extensive trajectory calculations? Reactions for which the saddle-point region of the potential surface plays the major role would be the most likely to yield to such an attack. We shall explore this possibility in Chapter 6.

5.B Appendix: Mass-weighted coordinate systems

The solution of the classical equations of motion to yield trajectories requires a digital computer. However, we can obtain considerable physical insight into the dynamics simply by examining the potential-energy contour map drawn using a *mass-weighted* coordinate system. We shall illustrate this approach for the case of a collinear collision where one can exactly simulate the actual trajectory by the process of "rolling a ball" on the actual potential energy surface. The Monte Carlo averaging is then performed by varying the initial conditions for the motion of the ball. We examine first the mathematical transformation to the mass-weighted coordinates and then the physical implications.

Consider the collinear reactive collision $A + BC \rightarrow AB + C$. $V(R_{AB}, R_{BC})$ is then a function of only two bond distances, $R_{AB} = R_B - R_A$ and $R_{BC} = R_C - R_B$, and these are the natural variables for thinking about the potential. The issue is that the kinetic energy (in the center-of-mass system) is not a simple function of these two coordinates. It is a simple function if, instead, we use the old bond distance, R_{BC}, and the distance, R, of the attacking atom A to the center of mass of BC, $R = R_{AB} + (m_C/(m_B + m_C))R_{BC}$. Explicitly, using a dot to denote the derivative with respect to time and denoting the reduced masses by μ,

$$T = \frac{\mu_{BC}}{2}\,\dot{R}_{BC}^2 + \frac{\mu_{A-BC}}{2}\,\dot{R}^2$$

In a similar fashion, using a prime to denote products, $R' = R_{BC} + (m_A/(m_B + m_A))R_{AB}$

$$T = \frac{\mu_{AB}}{2}\,\dot{R}_{AB}^2 + \frac{\mu_{C-AB}}{2}\,\dot{R}'^2$$

So what we want is to plot the potential as a function of the two Cartesian coordinates, R_{BC} and R or, equivalently, the pair R_{AB} and R', Figure B5.1.

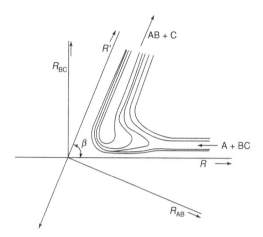

Figure B5.1 The two (equivalent) pairs of Cartesian coordinates, R_{BC} and R, or, R_{AB} and R', that allow the kinetic energy to be written as a sum of squares [adapted from F. T. Smith, *J. Chem. Phys.* **31**, 1352 (1959)]. The reactants' and products' valleys are along the new and old physical bond distances, R_{AB} and R_{BC}, and so appear to be at a skew angle β with respect to one another. The mass-scaled coordinates Q are defined in the text.

In terms of either pair of Cartesian coordinates the motion is simple because each coordinate is only coupled to the other by the force and not by the kinetic energy. This is really all there is to it except that different masses will lead to different motions because of the mass factors. We can eliminate these by scaling the coordinates by the (square roots) of the reduced masses μ. To do so introduce two new coordinates Q_1 and Q_2 by the transformation

$$Q_1 = aR_{AB} + bR_{BC}\cos\beta = aR$$
$$Q_2 = bR_{BC}\sin\beta = R_{BC}\sqrt{\mu_{BC}} \tag{B.5.1}$$

Here a, b, and $\cos\beta$ depend on the masses only, $M = m_A + m_B + m_C$ is the total mass and the μs are the reduced masses in the reactants' and products' channels:

$$a = [m_A(m_B + m_C)/M]^{1/2} = \sqrt{\mu_{A-BC}}$$
$$b = [m_C(m_B + m_A)/M]^{1/2} = \sqrt{\mu_{C-AB}} \tag{B.5.2}$$
$$\cos^2\beta = m_A m_C/(m_B + m_C)(m_A + m_B) \tag{B.5.3}$$

The point of the coordinate transformation from the bond distances to the Qs is that the kinetic energy for the collision (in the c.m. system) has the form

$$T = \tfrac{1}{2}\left(\dot{Q}_1^2 + \dot{Q}_2^2\right) \tag{B.5.4}$$

Equation (B.5.4) has the interpretation of the kinetic energy of a point particle (often referred to as the system point) of unit mass, whose position is specified by the two Cartesian coordinates Q_1 and Q_2.

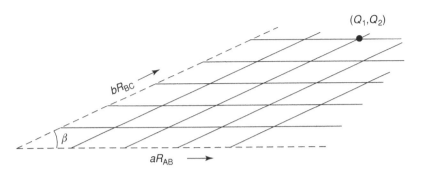

Figure B5.2 The construction of the Cartesian coordinates Q_1 and Q_2 in terms of the physical internuclear distances R_{AB} and R_{BC}, Eqs. (B.5.1) and (B.5.2) [adapted from J. O. Hirschfelder, *Int. J. Quant. Chem. IIIS* **17** (1969); see also B. H. Mahan, *J. Chem. Educ.* **51**, 308 (1974)]. When the coordinates Q_1, Q_2 are used as Cartesian axes the physical distances are skewed by the angle β, as shown. Grid lines for the skewed coordinates are shown as dashed lines. Why is the potential skewed? Because the atoms repel when they are close in so that neither bond distance can be negative and, by Eq. (B.5.1), neither can Q_1 nor Q_2. The angle between the physical bond distances is not 90° because we want to express the dynamics as that of a single ball rolling. This confines the motion to the sector shown. See also Figure B5.1.

When we express the potential surface as a function of Q_1 and Q_2, the solution of the classical equations of motion for Q_1 and Q_2, i.e., $Q_1(t)$ and $Q_2(t)$, are identical to the solution for the motion of the point particle on the potential $V(Q_1, Q_2)$. Rather than solve numerically the classical equation of motion, we can simulate the solution by letting a ball (of unit mass) roll without friction along a surface whose height at any point Q_1, Q_2 is $V(Q_1, Q_2)$.[30]

The geometrical significance of the change of variables to Q_1, Q_2 is illustrated in Figure B5.2. If we use Q_1 and Q_2 as two Cartesian axes, the effect of the transformation is to *skew* the two bond distances at the angle β to one another. Hence, if we regard the potential energy as a function of Q_1 and Q_2 and draw it in the Q_1, Q_2 plane, the entrance and exit valleys will be at an angle β to one another. Such a contour map is known as a (mass-weighted) *skewed axis* representation.

Such effects as the efficient conversion of the exoergicity into product vibration on an attractive surface can readily be visualized in the skewed coordinate system. On an attractive surface the ball is rolling "downhill" as Q_1 decreases and hence it enters the "bend" along the reaction path with a high speed. The ball will then fail to make its exit along the products' valley and instead it will climb the shoulder of the potential (the bobsled effect) and thereby convert much of the exoergicity to product vibration.

The lighter is the transferred atom, the larger is $\cos^2 \beta$ and the sharper is the "bend" between the reactants' and products' valley. You can imagine that the rolling ball will try to cut the corner. In so doing it will hit the repulsive shoulder of the products' valley and be reflected back to the reactants' valley. It may chatter in this way for a few times between the two heavy partners, an effect that can be discerned when we initiate the system near that region, as discussed in Section 1.2.6 for Cl + HI. The lighter is the attacking (or retreating) atom, the smaller is $\cos^2 \beta$ and the more is the products' valley stretched compared to the reactants' valley or, technically, the larger is the scale factor b as compared to a (or vice versa). Hence a light atom approach on a repulsive surface can still lead to significant products' vibrational excitation, as is observed for H + XY.

Problems

*A. Failure of the Born–Oppenheimer approximation. Let g and e denote the ground and excited electronic states. The Born–Oppenheimer separation fails when the matrix element $\langle \psi_g(r; R)|T|\psi_e(r; R)\rangle$ is not negligible, where T is the kinetic energy operator for the nuclei. Take the case of a diatomic molecule, examine the commutator $[H, \partial/\partial R]$, and hence show that $(E_g(R) - E_e(R))\langle \psi_g (r; R)|\partial/\partial R|\psi_e(r; R)\rangle = \langle \psi_g(r; R)|\partial H_{el}/\partial R|\psi_e(r; R)\rangle$. The "gap" between two different electronic states is $(E_g(R) - E_e(R))$ and depends on the nuclear coordinate. Conclude that the Born–Oppenheimer approximation is most liable to fail when the gap is small.

B. The qualitative topography of the potential energy surface. Using chemical knowledge (a) draw potential energy surfaces for a collinear configuration of the atoms for the following collisions: Ar^+ + HI, H + OH, Cl + K_2 (recall Problem B of Chapter 2), Br + HCl. For each surface locate the regions where the potential energy is high and also where it is low, the relation between the well depth of reactants and products, etc. (b) Plot also the potential along the reaction coordinate and state whether the barrier is early or late. (c) For the reaction K_2 + Cl you should be able to make a drawing in (a) and (b) where at least one distance is given a magnitude.* Do so. (d) How will your drawings change if we use a D rather than an H atom? If we use an Ar atom instead of an Ar^+ ion? (e) Discuss the energy requirements and energy disposal in these reactions along the lines introduced in Section 5.3. Next, we consider steric effects. (f) Taking the CH_3 group as an "atom" draw potential energy surfaces for the collinear approach K + CH_3I and the collinear approach K + ICH_3. (g) How will the polar map for the K + I_2 reaction vary with the angle of attack?

C. Qualitative considerations of energy disposal. The CO vibrational state distribution from the reaction of O + CN is found to be bimodal; there is significant population of quite high vibrational states, say $v = 10$ up to the exoergicity limit but also lots of rather vibrationally cool molecules with a thermal-like distribution

---

* The K_2 dimer has a relatively low ionization potential and a bond energy that is low by chemical standards. (Use MO theory to explain why the ionization potential of the weakly bound dimer is only a bit higher than that of the atom.)

[J. Wolfrum, *Ber. Bunsenges. Phys. Chem.* **81**, 114 (1977)]. It is proposed that this highly exoergic reaction releases sufficient energy to form an N atom in the excited ^{2}D state. On the potential energy surface for the reaction leading to N*, NCO is an intermediate. The reaction producing ground state N atoms is direct. Discuss the suggested explanation and propose one or more tests for it. (Hint: recall from Figure 5.12 that we can measure the angular distribution for different internal states of the products.) In a further experiment the reaction was initiated with vibrationally hot CN. How will the CO vibrational distribution change?

D. Hard-sphere models can be extended for reactive collisions [e.g., B. H. Mahan, *J. Chem. Educ.* **51**, 308 and 377 (1974)]. Since the hard-sphere potential is purely repulsive, the one refinement that is needed is to modify the hard-sphere potential so as to mimic a chemical bond. We do so by adding a square well, of depth $-D$ and finite range, to the atom–atom potential. It will then look like (i) below. For an A + BC collision the hard-sphere potential energy surface will then look like (ii).

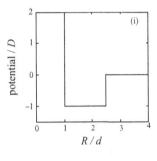
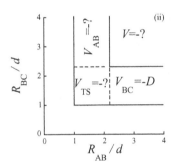

Panel (ii) lets you design your own potential energy surface by choosing suitable values for two of the question marks. (a) Explain the labels on the potentials and choose such values that the reaction is exoergic with a low barrier and that the reaction is endoergic with a barrier that is/is not higher than the endoergicity. (b) For either case in (a), what is the dissociation energy of AB? (c) The third question mark corresponds to a value that is not arbitrary but is dictated by the problem. What is it? There is also the question of what is the value of the potential for short distances. (d) Specify values that will make the reaction proceed over a well. (e) What is the threshold energy for reaction for the two potentials that you specified in (a). (f) What is the threshold for collision-induced dissociation on this potential?

*E. Mass scaled coordinates. Statics. Using Appendix 5.B, scale the potential surface of problem D for the case when (a) the attacking atom is light/heavy compared to the two others, (b) atom B is far heavier than the two others, (c) all three masses are equal. (d) Increase all masses by the same factor. How will the scaled potential surface change?

*F. Mass scaled coordinates. Dynamics. Using Appendix 5.B, discuss the dynamics for the potential energies you generated in (b) and (c) above. (a) Start with a simple case – thermoneutral reaction with no barrier – and contrast the role of reagent translation for cases (b) and (c). (b) How will you drive the reaction in case (b)? (c) Say the exchanged atom is very light, conclude that it will rebound a couple of times between its two heavy partners before the products exit. (d) As a different example consider the (roughly) thermoneutral reaction of H with I_2. (e) Trajectory computations show that T, the heavy (mass 3) isotope of hydrogen, has a significantly higher reaction cross-section. Explain. *(f) Assuming that there is no barrier to reaction show that the cross-section is higher by about $\sqrt{3}$.

*G. The Evans–Polanyi model. Go to the classic text Glasstone *et al.* (1941) and see the derivation that leads to the diabatic potentials for the reactants. In the notation of Section 5.1.3, this diabatic surface has the functional form

$$V_{A+BC} = Q_{AB} + Q_{BC} + Q_{CA} + \left[J_{BC} - \tfrac{1}{2}\left(J_{AB} + J_{AC}\right)\right]$$

(a) Write, by symmetry, the diabatic potential that correlates to the products. (b) Draw the two diabatic potentials for a collinear collision for a very exoergic, a mildly exoergic, and a thermoneutral reaction. You will need to infer the J and Q terms as discussed in the text. The drawing can be done using even a hand-held graphic calculator but is best done on a computer using a graphic program. (c) Infer the Hammond postulate from your work in (b). (d) Get another result: the height of the barrier to reaction correlates with the exoergicity.

*H. How high in energy is the conical intersection in H_3? Use spectroscopic data to obtain the J and Q terms for H_2 and hence determine the potential energy at the intersection of the LEP(S) surfaces. The numerical value will depend on the length of the side of the triangle, but what can be said in general? If the system is NaH_2 what will be the geometry at the intersection?

*I. Chaos results in a rapid erasure of details of the starting point. We consider a very simple example and for brevity do not discuss the kinetic context in which this example arises. Let θ be an angle, measured in units of 2π. Say we write θ in a binary notation. Then trigonometric functions depend only on the fractional part of θ. Let successive values of θ be generated by the rule $\theta_{n+1} = 2\theta_n$. Imagine implementing this iteration on a computer. Let θ_0 be the initial value. On a computer θ_0 is specified as a sequence of digits of finite length, say 16 figures. Show that with each iteration you lose one digit in accuracy, θ_1 is known only to 15 digits, θ_2 is known only to 14 digits, etc. Implication: an expression can be mathematically an identity and yet, when implemented on a digital computer, it can rapidly lose numerical accuracy.

J. Monte Carlo sampling of initial conditions. (a) Discuss why the reaction cross-section computed using Eq. (A.5.7) need not change in value if the value, B, of the maximal impact parameter for reaction is taken larger than it has to be. (b) The reaction rate constant for thermal reactants is obtained, Section 3.1, as an average over the reaction cross-section. Determine a sampling of the relative velocity of thermal reactants such that the reaction rate constant is computed directly without the need to explicitly perform the integral over the distribution of the velocity, Eq. (3.7). The result is $k(T) = \pi B^2 \langle v \rangle N_r/N$. Derive it, specify how you will compute N_r, and discuss the implications. Will you or will you not use this method if you want to compute the rate constant over a wide temperature range? If not, what is a reasonable alternative? (Hint: the alternative will, for a comparable computational effort, tell you more about the dynamics.)

K. Limitations on the Monte Carlo sampling of initial conditions. Say you want to compute the reaction rate constant when there is a rather small steric factor and/or a rather high activation energy. (a) Will you use a Monte Carlo sampling of initial conditions? Discuss in detail why we think that getting an accurate result will require a considerable consumption of computer time. (b) Can you try to suggest at least two different options for getting the result while computing far fewer trajectories? For a quick start see R. L. Jaffe, J. M. Henry, and J. B. Anderson, *J. Chem. Phys.* **59**, 1128 (1973).

L. The Lindemann mechanism. Textbooks usually invoke the steady-state approximation for solving this kinetic scheme, see Section 6.2.1. But the use of this approximation is not essential. Write $[A^*] = a_1 \exp(-k_1 t) + a_2 \exp(-k_2 t)$ and show that you can solve for the reaction rate using the kinetic scheme that Lindemann proposed and initial values for the concentrations, e.g., $[A^*] = 0$ at $t = 0$. Recover the steady-state result as an approximation. What have we gained? That there is also a prompt decay component! Conclude that it is smaller in magnitude than the delayed decay but it is not necessarily zero.

M. Reactions with a negative activation energy. Reactions that proceed via formation of a complex and are measured at higher pressures can show a negative activation energy. As an example consider the reaction of C_2H_5 with O_2 as shown in Figure 5.6. The first step is the formation of the $C_2H_5 \cdot O_2$ complex. At higher pressure this complex is stabilized by collisions and is brought to thermal equilibrium with the reactants. The rate-determining step is the isomerization of the complex to $CH_2=CH_2 \cdot O_2H$ as shown in the concerted elimination pathway. (a) Examine Figure 5.6 and explain why the rate of the net chemical change can manifest a negative activation energy. The measured [O. Dobis and S. W. Benson, *JACS* **115**, 8798 (1993)] high-pressure value is about $-20 \, \text{kJ mol}^{-1}$. (b) Suggest why the Arrhenius A factor for the barrier crossing isomerization will be low. (c) The reaction can also proceed by a direct route as shown in Figure 5.6. Discuss if this route is favored at low or at high temperatures.

Notes

1 The Herzberg treatise on spectroscopy of molecules is in three volumes – one on diatomics and two on polyatomics. This chapter is in the spirit of the first polyatomic volume, with the essential difference that we deal with large-amplitude motions. The second Herzberg polyatomic volume is on electronically excited states.

2 Conical intersections is a topic that we will discuss in detail. See Michl (1972), Michl and Bonacic-Koutecky (1990), Bernardi *et al.* (1996), Yarkony (1996, 1998), Worth and Cederbaum (2004).

3 There is a rich literature on the topic of potential energy surfaces, e.g., "Potential energy surfaces." *Faraday Discussions of the Chemical Society* **62** (1977), Truhlar (1981), Hirst (1985). This is because the potential is the structural underpinning of chemical reactivity. Nor is the concept limited to reactive collisions. Energy transfer collisions, Chapter 9, are possible because the potential depends not only on the relative position but also on the internal coordinates. By Chapters 11 and 12 we will further extend the concept to interaction with the environment. When it comes to rational drug design or catalysis, such an extension is essential.

4 Fitting a potential energy function to an analytical form is not a simple task. See, for example, Hirst (1985), Schatz (1989), Schlegel (1995), Hollebeek *et al.* (1999).

5 For the history leading to this accurate computation and for the implications to the dynamics of the $F + H_2$ reaction, see D. E. Manolopoulos, *J. Chem. Soc. Faraday Trans.* **95**, 673 (1997).

6 See, for example, Rauk (1994), Pross (1995).

7 The quantum chemical software industry has produced quite a number of packaged programs that use methods of increasing sophistication to compute at least the stationary points along the reaction path and often much more. When using such packages it is always advisable to assess the accuracy of the results by checking a few configurations at a higher level of computational accuracy to know that the results are about converged. Such a check is necessary not only for the barrier height (or well depth) but also for geometrical features such as the configuration at the barrier and for the forces, see note 8.

8 The force is a derivative of the potential. But the potential is the electronic energy so the force can be computed from knowing the electronic wave functions at just one point by using the *Hellmann–Feynman theorem*. We sketch a proof because it brings out the essential point that the theorem is exactly correct only when we have exact wave functions. Using the notation of Problem A, $E_g(R) = \langle \psi_g(r; R)|H_{el}|\psi_g(r; R)\rangle$. Using a prime to denote a derivative, $\partial E_g(R)/\partial R = \langle \psi_g'(r; R)|H_{el}|\psi_g(r; R)\rangle + \langle \psi_g(r; R)|\partial H_{el}/\partial R|\psi_g(r; R)\rangle + \langle \psi_g(r; R)|H_{el}|\psi_g'(r; R)\rangle$. If we have exact (and normalized) wave functions, then $H_{el}|\psi_g(r; R)\rangle = E_g(R)|\psi_g(r; R)\rangle$ and the first and third terms vanish because $\partial \langle \psi_g(r; R)|\psi_g(r; R)\rangle/\partial R = \partial 1/\partial R = 0$. The electronic Hamiltonian is a sum of coulomb terms and so its derivative is easy to compute.

9 For more on combustion see Miller *et al.* (1990), Pilling *et al.* (1995), Pilling (1996), Pilling and Robertson (2003).

10 Elber and Karplus (1987), Davidson (1993), Onuchic *et al.* (1997), Dobson *et al.* (1998).

11 As a last resort one can draw a *connectivity diagram* (Becker and Karplus, 1997), and this is useful not only for proteins but also for other highly fluxional species such as clusters (Berry, 1993; Wales *et al.*, 1998; Wales, 2001). The diagram shows the lowest barrier height along the path connecting one basin to another. Any more detailed description requires a choice of the coordinate(s) along which the potential energy is to be displayed; see, for example, Hayward and Go (1995).

12 The *Sato modification* allows for a finite overlap S between the orbitals on atoms A and B. Then Eq. (5.2) needs to be modified to $V(R_{AB}) = (Q_{AB} \pm J_{AB})/(1 \pm S^2)^{17/2}$.

13 Consider a Morse potential, $V(R) = D_e(\exp(-2(R - R_m)/\rho) - 2\exp(-(R - R_m)/\rho))$, with the zero of energy such that the potential vanishes at infinity. An often-used device is to fit this potential to the ground-state potential and approximate the triplet potential by an *anti-Morse* functional form, which is an everywhere repulsive potential generated by replacing the central minus sign on the right-hand side by a plus. The same procedure can also be used for a Lennard-Jones potential. Either way, J is negative and $|J| > Q$ near the equilibrium separation of the diatomic. The Sato modification is equivalent to taking a different value of D_e for the Morse and the anti-Morse functions.

14 See, for example, Pross and Shaik (1983), Pross (1995).

15 There can be important differences between classical and quantal dynamics results (cf. Sections 2.2.6 and 4.3). Even so, classical mechanics is very useful in that it provides an easily visualized representation of the dynamics. It also provides the correct overall trends and average behavior even when it errs in the details. An important limitation is that the method can only mimic but not really reproduce the quantization of internal degrees of freedom. The proper handling of zero-point energy is a case in point. Quantal wave-packet methods, Section 8.0.1, have the advantage that like a classical trajectory they provide an image of the unfolding in time of the dynamics.

16 For example, in the simplest case of the A + BC reaction, one must average over the phases of the vibrational and rotational motions of BC. What are the phases and why do we regard the phases as randomly distributed? Because the *phase angle* is the classical variable that is conjugate to the *classical action*. The semiclassical correspondence is that a classical action, in units of Planck's constant, is the quantum number. (We have seen this already in the correspondence $L \leftrightarrow \hbar l$ for the angular momentum.) If the quantum number is well specified the corresponding phase angle is unknown or, in the quasi-classical mimic, it is averaged over. As an example, for a harmonic oscillator, the position and momentum are related to the classical action J (energy $E = J\nu = J(\omega/2\pi)$) and the phase φ by Goldstein (1950); Child (1991)

$$x = (2E/k)^{1/2}\cos(\omega t + \varphi)$$
$$p = (2mE)^{1/2}\sin(\omega t + \varphi)$$

Therefore the initial conditions for the trajectory can just as well be specified by the action and angle variables. The classical actions are constants of the motion of the unperturbed molecule and are therefore the natural variables if the state of the separated reactants is stationary. When we come to discuss time-resolved studies we will allow for *coherent states* where the phase angle is not random. (Yes, for such states, the quantum number will not have a sharp value but only a mean value, see Figure 1.7.)

17 This is the case even if the potential function is given in an analytical form. An exception is if the potential is extremely simple such as including only two-body terms. But such a simple potential is not realistic when a chemical reaction is possible. If one needs to compute the potential by quantum chemical methods then this becomes the bottleneck. This is true even though so-called gradient methods compute not only the potential but also the forces. For examples of computing the potential at each point along the trajectory where it is needed, see Michalak and Ziegler (2001). This is often known as computing the potential "on the fly."

18 See Brumer (1981), Rice (1981), Brumer and Shapiro (1988), Uzer (1991), Wyatt *et al.* (1995) for applications to intramolecular dynamics and Sagdeev *et al.* (1988), Schuster

(1988) for background. Deterministic chaos is the chaos in the solution of mechanical equations of motion, that is, equations that strictly speaking allow a reversal of the direction of time. (Checking that time reversal is satisfied for the trajectory is a good test of the validity of the numerical integration.)

19 So why not give up dynamics altogether and only do statistics? The technical problem is that it need not be the case that all possible initial conditions lead to chaotic time evolution. A trivial example is if the impact parameter is higher so that the forces are weaker because the reactants do not get close in. The problem is to delineate the boundary of the chaotic regimes.

20 Stated very crudely, another option is to say that a trajectory is a trajectory. It therefore does not matter if it originated in the initial conditions that were given or from similar but not identical values. The problem with this is that when errors accumulate it is no longer certain that the trajectory has "similar" initial conditions. If such an approach is used it is necessary to check that energy and other constants of the motion are conserved. Engineers, who need to follow the properties of matter over longer times, are actively developing integration schemes where larger time steps can be taken. To insure convergence time is discretised not directly in the equations of motion themselves but in a variational principle. See also 'Tackling time-scale problems' in Frenkel and Smit (2002).

21 For more on dynamics in large systems, see Adelman and Doll (1977), Adelman (1980), Tully (1981), McCammon and Harvey (1983), McCammon and Karplus (1983), Hynes (1985), Warshel (1991), Warshel and Parson (2001).

22 The *Langevin equation* can be derived from the mechanical equations of motion and, as such, the exact equations are reversible in time. However, there are commonly made approximations, for example that the energy spectrum of the secondary (or "bath") degrees of freedom is continuous (rather than a very dense but discrete quasi-continuum). These, physically quite realistic, approximations insure the dissipative behavior. In general, it is physically reasonable to mimic a dense manifold of states by a continuum if the mean spacing, D, of states is such that h/D is a significantly longer time scale than is relevant to the experiment, see Chapter 7. The exception is if the quasi-continuum of states does not have a smooth density of states but has a coarse-grained structure (it is a *bumpy quasi-continuum*).

23 The technical answer is the Monte Carlo theorem. We will not state it but paraphrase to say that if there are more than four variables, a systematic sampling is not the most economic route. See Porter and Raff (1976), Truhlar and Muckerman (1979).

24 For the expert, actually five will do but this does not change the argument.

25 There are important problems that we do not discuss in detail because they are of a more technical nature. These include the need to assign quantum numbers to the products. This is typically done by *binning*, that is by collecting all products' states in a given energy range into the same box and assigning the quantum number that corresponds to the mean energy. Note that this goes against microscopic reversibility because we do not treat reactants and products on an equal footing. There is the problem of zero-point energy that was already mentioned. This is particularly serious in many-atom systems where the energy "hidden" in zero-point motion is, classically, high.

26 Unimolecular reactions are extensively discussed in texts of chemical kinetics. In addition to Steinfeld *et al.* (1999), Houston (2001), see the more specialized sources of Forst (1973), Tardy and Rabinovitch (1977), Beynon and Gilbert (1984), Pritchard (1984),

Reisler and Wittig (1986), Manz and Parmenter (1989), Gilbert and Smith (1990), Green *et al.* (1992), Baer and Hase (1996), Holbrook *et al.* (1996), Schinke *et al.* (1997).

27 The mobile surface is discussed by Somorjai (1994), Somorjai and Rupprechter (1998).

28 The strongest anisotropy is due to an asymmetrical charge distribution or to electronic orbital steering as discussed in Section 5.1.5.1.

29 For the role of reagent rotation, see Sathyamurthy (1983), Levine (1990), Loesch (1995).

30 The technique of solving equations of motion by constructing a mechanical system that is moving under the same equations as the real problem is known as *analog* (as opposed to digital) computation. Note, however, that here this is only useful for collinear collisions and cannot describe, for example, *migratory dynamics* such as in H + ICl where the attack is on one atom, iodine, as discussed in Section 5.1.5.1, but the final product is HCl.

Chapter 6
Structural considerations in the calculation of reaction rates

We address the question that every chemist asks: given thermal reactants, how do you compute the rate of crossing the barrier toward the products? We seek to cast the answer using traditional tools and, specifically, the structure of the system at the barrier. Using just one, physically realistic, approximation, *transition state theory* enables us to do that. The theory identifies a *bottleneck* for the reaction and computes the rate of passage through it.

The success of transition state theory inspires us to do more. We shall, but we require additional assumptions to be made at each point where we seek a generalization. The most pressing reasons for doing this are that there may be more than one barrier separating the reactants and products and that there can be multiple reaction paths. The case of the $O_2 + C_2H_5$ reaction, shown in Figure 5.6, represents the norm rather than the exception. Transition state theory allows us to compute the rate of barrier crossing, but to get to the products we may need to cross several barriers and/or take different paths. It is for this reason that quantum chemists have grown proficient in computing the structures at each barrier (and each hollow). But we still need to know how to compound the effects of multiple bottlenecks to obtain the overall reaction rate – and this task calls for either dynamical computations as in Chapter 5 or for an additional assumption as introduced in Section 6.2. Another reason for wanting to go beyond transition state theory is that we often measure the reaction rates for specific internal states of the products or for selective initial states of the reactants. Even more, already in Chapter 4 and in detail in Chapter 10 we want not just the final populations, which are scalar quantities, but final vector quantities such as the orientation of the products.

We divide the chapter into three sections. First, transition state theory (TST) proper, where we compute the rate of barrier crossing for reactants in equilibrium. Next, computing the reaction rate when more than one barrier is present, and finally, state-resolved reactivities. As much as possible, we emphasize the close link between structure and dynamics.

6.1 Transition state theory: the rate of barrier crossing

Transition state theory is a method for predicting the rate of chemical reactions. Technically, what the theory provides is the rate of crossing of a barrier. If there is only one barrier between reactants and products, then transition state theory specifies how to compute the reaction rate constant. Transition state theory assumes the validity of only one condition, but a cardinal one, namely that on one side of the barrier, the states of the system are in equilibrium. If there is only one barrier between reactants and products, then it is the reactants that should be kept at equilibrium. The simplicity of transition state theory is lost if the reactants are state-selected.[1]

The equilibrium condition required by the theory makes one sometimes think that transition state theory is a theory like that described in Section 6.2 where there is a collision complex. This impression is categorically false. If anything, transition state theory is at the opposite extreme. It is a theory of direct crossing. In one and only one go, the system is to cross the barrier.* The statistical assumption made by transition state theory is not about the dynamics of the reaction; instead, it is about the equilibrium nature of the reactants that are located to one side of the barrier. A very important implication of this assumption is that it allows the theory to be cast in structural terms, terms that are familiar to chemists.

The statistical assumption made by transition state theory is a specification of the reactants to which the theory can be applied. It is not an approximation. Either the reactants that you are interested in are in equilibrium and the theory is applicable or the reactants are not in equilibrium and you should look for another way to compute the reaction rate. Transition state theory does make one simple and physically clear approximation to which we now turn.

6.1.1 The point of no return and the transition state

Transition state theory assumes that there is a *configuration of no return* such that, starting from the reactants, when and if the system reaches this critical spatial configuration, it will necessarily proceed to form products. If such a configuration can be identified then it must be crossed by any reactive trajectory and transition state theory assumes that it is only crossed once. The approximation of a single crossing en route from reactants to products is the key to the simplicity of the results. It implies that the reaction rate is given by the rate at which the colliding

---

* The expression for the rate of barrier crossing, as obtained by transition state theory, can also be usefully applied when there is a well, with barriers on either side, separating reactants and products. Under such circumstances, the theory can be applied but it is applied to the crossing of each barrier separately, as discussed in detail in Section 6.2. Such an application requires that the motion of the system rattles many times in the well region between the two barriers, see Section 5.3.3. Transition state theory itself computes the rate of barrier crossing, and this rate equals the reaction rate only for direct reactions.

molecules reach (and thus pass through) this configuration. We do not need to know how the system approaches this configuration, nor what dynamics it undergoes beyond this configuration. All that is needed to compute the reaction rate is to stand at the configuration of no return and to count the number of reactants passing over the barrier to products, per unit time. Dynamics is thereby reduced to counting, which is a real stroke of genius.

It is clear, however, that the existence of such a point of no return is an approximation. Strictly speaking, it is only when the nascent products are significantly separated from each other that we can be **sure** that they will indeed proceed along the exit valley to form stable products. Even so, everything that we know thus far supports the view that when there is a barrier along the reaction path, the location of the barrier, i.e., the saddle point, is the sensible choice for a point of no return. For reactions without an energy threshold, as we have seen in Section 3.2.6, the centrifugal barrier can be used to determine the location of the point of no return, which is identified with the maximum of the effective potential.

Dealing with thermal reactants and with barriers that are high compared to thermal energies, the assumption is quite reasonable that if the barrier has been crossed, the motion downhill to the products will not reverse upon itself. At higher energies, when the barrier becomes less of a handicap, the barrier can be recrossed. But transition state theory is intended to be useful at ordinary temperatures when, due to the Boltzmann factor, there is not much excess energy available for crossing the barrier. Even if the theory errs, note that it must always provide an upper bound. Trajectories that start from the reactants and recross the barrier may* fail to form products. But the theory counts them as reactive because they crossed the barrier.

The term transition state is often applied to the *configuration of no return*.** This configuration is most certainly the divider (dividing surface) between reactants and products, and so the name is appropriate.† We have already encountered experimental techniques, and we will find more that initiate the dynamics in the

---

* Not will, but may. Trajectories that originate from the reactants and cross the barrier an odd number of times will form products. The theory counts each such crossing as an independent trajectory and also for this reason it overestimates the rate, i.e., it provides an upper bound.

** The transition state in the sense that we use it here is exactly the opposite of any intermediate species between reactants and products. An intermediate complex exists over a well in the potential energy and so the trajectory samples it many times, as in Figure 5.19. But for its excess energy, an intermediate complex is a stable species and even with the extra energy it survives for a while. The transition state is the least stable configuration. As an example, you can visualize it as a trajectory moving periodically along the bisector of the H_3 potential energy surface shown in Figure 5.2. It is a trajectory tracking the bound symmetric stretch of H_3. But the smallest change in the initial conditions of such a trajectory will make it roll down either to the reactants' or to the products' region. With experimental ingenuity the transition state in our strict sense can be discerned, for example Figure 8.3.

† We can be fanciful and think of the configuration of no return as the surface of a black hole. Any trajectory that crosses the surface is assumed not to come back.

region of the barrier but not necessarily on it. We shall then speak of the *transition state region*. It certainly makes quantum mechanical sense and even in classical mechanics it surely takes at least a vibrational period to be sure that having crossed the barrier the trajectory will proceed to form products.

6.1.2   The statistical condition

For chemical kinetics, transition state theory is most useful in the form that starts from reactants in thermal equilibrium. For our purpose we want a more detailed version, that of reactants with a total energy in the range E to $E + dE$. If we know how to do that, we can and will average over a Boltzmann distribution in E to obtain the thermal results. The first task at hand is to define what is meant by reactants at equilibrium at a total energy within the range (and at given values of any other conserved quantum numbers). It is the foundation of statistical mechanics that equilibrium under such conditions means that all possible quantum states of the reactants are equally probable.*

Now comes a point that needs attention. In principle, the condition of the reactants at equilibrium does not identify which states of the reactants will proceed to form products. But we are armed with a strong assumption: there is a configuration of no return. Trajectories that originate at the reactants' side of the barrier either fail to reach this point or they do reach it and then they proceed to form products, without turning back. Every trajectory that goes through to products has originated in the reactant region. Therefore, all quantum states at the transition state are equally probable because each such state can be traced back to a state of the reactants[2] and at equilibrium within a narrow energy range, the states of the reactants are equally probable.

The physical meaning of the conclusion that all quantum states at the transition state are equally probable is as follows. There are various ways of partitioning a given total energy between the internal (bound**) degrees of freedom of the transition state and the translational motion that corresponds to the formation of the products, i.e., the motion along the reaction coordinate. Clearly if there is a lot of translational energy in, and thus velocity along, the reaction coordinate, the passage of that trajectory across the transition state configuration will be

---

* In the technical language of statistical mechanics, reactants in equilibrium within a narrow energy range are said to have a microcanonical distribution. Reactants in thermal equilibrium have a canonical distribution.

** The motion along the reaction coordinate is unbounded because we are crossing a barrier. But the motion along other degrees of freedom is bound. For A + BC these other modes include the symmetric stretch of ABC (this stretch is orthogonal to the reaction coordinate, which at the saddle point is the unbound asymmetric stretch motion), the bending motion of ABC (the origin of the steric factor), and the overall rotation of ABC. For more atoms there are more bound vibrations. For a nonlinear molecule of n atoms there are a total of $3n - 6$ vibrations ($3n$ degrees of freedom minus 3 to account for c.m. coordinates and 3 for rotations). Of these, one vibration is the motion along the reaction coordinate, leaving $3n - 7$ other vibrations.

swift, whereas if most of the energy is elsewhere, the crossing will be slow. For computing the reaction rate it is thus of importance to know how the total energy is partitioned in the transition state configuration. From the above arguments we conclude that all manners of partitioning are equally probable.

Trajectory computations can validate the assumption of no return. When the available energy is only barely above the barrier, which is typically the case for thermal reactants and chemical-sized barriers, a crossing trajectory seldom returns. This is because the potential on the opposite side of the barrier falls quite steeply so that the trajectory is accelerated toward the products. It is only at higher energies that recrossing becomes more common.[3]

6.1.3   Computing the rate for direct reactions

Our purpose in this section is to obtain an expression for the rate of a direct chemical reaction at a given total energy E.* We shall derive this result in two distinct stages. The discussion is divided into separate parts in order to spell out clearly the successive stages.

First, we must consider the energy balance at the transition state, Figure 6.1. Let E_0 be the minimal energy necessary to reach the transition state configuration measured from ground-state reagents.** At a given total energy E (measured also from the zero level of the reactants) the translational energy for motion along the reaction coordinate, ε_T, can be at most $E - E_0$. In general, the available energy, $E - E_0$, will be distributed between ε_T and energy in the other degrees of freedom of the transition state. These other degrees of freedom correspond to bound motions as in an ordinary stable molecule. We shall refer to them as the internal degrees of freedom of the transition state and assume that the energy $E - E_0$ is the sum of ε_T and that internal energy ε_I.

Our first task is to calculate the rate of passage across the point of no return along the reaction coordinate q. Let us consider a subset of transition states, namely, an ensemble of them, each with its translational energy (along the reaction

---

* Later we shall arrive at our other goals, the (direct) reaction rate at a given temperature T. Then we discuss what to do for the case of complex-forming reactions, where there is more than one barrier.

** E_0 is not quite the height of the potential barrier to reaction. It is the energy of the ground state of the transition state above the ground state of the reactants. The internal modes of the transition state are bound and therefore discrete. The ground state is higher than the barrier top by the zero point energy (zpe) of these discrete vibrations. Similarly, the ground state of the reactants is higher than the barrier bottom by their zpe. These two corrections enter with different signs, but need not cancel, $E_0 =$ barrier height plus zpe of the transition state minus zpe of the reactants. The quantum chemist needs to provide us not only with the height of the barrier but also with the forces at the barrier. From the forces we compute the frequencies for the vibrations at the transition state and hence the zpe. We will need these frequencies in an essential way, so computing the zpe is a necessary labor for quantum chemistry. As already mentioned, computing the forces for a given configuration is nowadays possible.

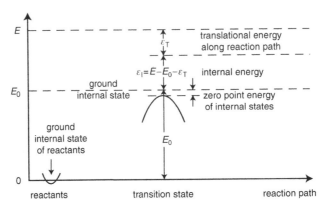

Figure 6.1 Energetics at the transition state: schematic energy profile along the reaction coordinate for a potential with a barrier. E_0 is the ground state energy of the *internal* motions of the supermolecule at the transition state (measured from the zero point internal state level of the reactants). E is the total energy; $\varepsilon_I = E - E_0 - \varepsilon_T$ is the energy available for the internal states of the supermolecule at the transition state [adapted from R. A. Marcus, *J. Chem. Phys.* **43**, 2658 (1965)]. Note that because these bound internal states are quantized, correctly speaking, E_0 is not simply the height of the barrier. Rather, E_0 is the height of the potential barrier minus the zero point energy of the reactants and plus the zero point energy of the internal modes at the transition state. Because certain modes loosen up at the barrier and their frequency goes down, the difference between the two zpes need not be small compared with the net height of the barrier.

coordinate q) in the range ε_T to $\varepsilon_T + d\varepsilon_T$. To compute the rate of crossing for all possible partitioning of the energy E we will later integrate over ε_T. For a given ε_T, there is a corresponding linear speed and linear momentum along q, which we denote by $v^+(= \dot{q})$ and $p^+ = m^+ v^+ = (2m^+ \varepsilon_T)^{1/2}$. Here we use a superscript $+$ to denote properties at the transition state; m^+ is a mass appropriate to the one-dimensional (1D) translation of the system, along q. (It disappears from the final result, however!) The range of ε_T can be written in terms of p^+:

$$d\varepsilon_T = p^+ \, dp^+ / m^+ = v^+ \, dp^+ \tag{6.1}$$

To calculate the reaction rate constant we first compute the rate of barrier crossing. Then we extract the rate constant by dividing by the concentration of the reactants. We determine that concentration by taking one pair of reactants in each initial quantum state. We could take two or any other number. Because the initial quantum states are equally populated, the rate will scale with the number that we choose.

    The rate of barrier crossing (the number of crossings per unit time) for transition states in the specified ε_T range is $dr^+ = v^+ \, dN^+$, where dN^+ is the number of systems per unit length along q (i.e., a 1D number density). Because all available states are assumed equiprobable and occupied, and because the system is in a definite state (of energy ε_I) of the other degrees of freedom, what we need is the number of states for the one-dimensional motion, from left to right, along the reaction coordinate.

The quantum mechanical result* for the number of such 1D translational states in a length dq is $dp\,dq/h$, so the 1D number density per unit length is $dN^+ = dp^+/h$. This gives for the rate of passage across the barrier

$$dr^+ = v^+\frac{dp^+}{h} = \frac{d\varepsilon_T}{h} \tag{6.2}$$

Thus, the rate of passage through the point of no return along the reaction coordinate is $d\varepsilon_T/h$, irrespective of the value of ε_T itself. This result is really most remarkable. Irrespective of any details about the reactants, when the total energy is in the range E, $E + dE$ and the internal degrees of freedom are in a given state, the rate of crossing is universal, namely $1/h$ per unit translational energy. The only assumption made is that of a point of no return; all systems reaching it proceed further on and ultimately become products.

A given total energy E may be partitioned in many ways between ε_I and ε_T, i.e., between the internal modes of the transition state and the motion along the reaction coordinate. We now invoke the result that at equilibrium all these partitionings are equally probable. Each internal state of the transition state contributes $d\varepsilon_T/h$ to the rate of passage. Therefore, when the total energy is in the range E to $E + dE$, the rate of passage is the sum over internal states of all the rates of crossing:

$$\sum_{\substack{\text{internal}\\\text{states}}} dr^+ = \sum_{\substack{\text{internal}\\\text{states}}} \frac{d\varepsilon_T}{h} = \sum_{\substack{\text{internal}\\\text{states}}} \frac{dE}{h} = \frac{dE}{h} \sum_{\substack{\text{internal}\\\text{states}}} 1 \equiv \frac{dE}{h} N^{\ddagger}(E - E_0) \tag{6.3}$$

The sum is over all the internal states of the supermolecule whose energy ε_I is in the allowed range from 0 to $E - E_0$, where, for a given ε_I, we have $d\varepsilon_T = dE$. Hence, for reactants having a total energy in the range E to $E + dE$ the rate of passage is determined by the number $N^{\ddagger}(E - E_0)$ of internal states of the supermolecule at the transition state whose energy ε_I is in the allowed range, $0 \leq \varepsilon_I \leq E - E_0$. The superscript $\ddagger$ is a reminder that $N^{\ddagger}(E - E_0)$, defined by

---

* We require a quantal result because of an inherent flaw in the classical mechanical counting of translational states, which was first realized in connection with black-body radiation density, and which led to the introduction of Planck's constant h. The same failing was subsequently realized in connection with the contribution of the translational motion to the entropy (the Sackur–Tetrode equation) and hence to the other thermodynamic properties of gases. We will need to use this result again in Section 6.4, so here is a short proof using semiclassical considerations. Take an interval of length L. Quantization means that an integral number of (de Broglie) wavelengths ($\lambda = h/p$) fit into L. The nth quantum state is therefore defined by $n\lambda = L$. The same condition written in terms of the momentum is $p = nh/L$. The number of states when the momentum is in the range p to $p + dp$ is $dn = L\,dp/h$, which, per unit length, is dp/h. If you apply this expression to a particle inside an infinitely deep well you will get twice as many states. This doubling occurs because the states in the well can be even or odd, so that linear combinations of them correspond to particles moving from right to left or vice versa. In 3D the result for the density of translational states is $d\mathbf{p}/h^3$. A short proof is to multiply the contributions of the three directions for a box of side L, (dp_x/h) $(dp_y/h)(dp_z/h) = d\mathbf{p}/h^3$.

the sum in Eq. (6.3), represents only the bound internal states at the point of no return, that is, those having a real as opposed to an imaginary frequency. Thus, this sum **excludes** the unbound motion along the reaction coordinate.

Given a knowledge of the potential energy surface near the saddle point we can calculate the energies of the allowed internal states and simply count up how many are in the range $0 < \varepsilon_1 \leq E - E_0$.

With the result shown in Eq. (6.3) we have reduced computing the rate of barrier crossing to a counting problem! Given the structure of the transition state we simply need to count how many internal states are in the allowed energy range. When we use thermal reactants the counting problem can be recast in even more familiar terms and we do so shortly. In the meantime we can either be satisfied by the cumulative rate (per unit energy) of crossing the barrier[4]

$$Y(E) = \frac{1}{h} N^{\ddagger}(E - E_0) \tag{6.4}$$

or we can get a reaction rate constant by dividing the rate of crossing by the concentration of the reactants. Because we have one set of reactants per quantum state, the desired concentration is the number of reactant states per unit volume for energy in the range E to $E + dE$. This number can be written as $\rho(E)dE$, where $\rho(E)$ is defined as the density of states of the reactants, that is, how many quantum states per unit energy. The concept of a density of states, a smooth function for which energy is a continuous variable, serves to simplify the form of several expressions that we derive and also helps in understanding the relation between structure and reactivity. The concept comes from the quantum mechanics of stationary states. By definition, a density of states is the number of quantum states per unit energy interval, and it is discussed in Appendix 6.A.

Dividing the rate by the concentration we get the rate constant

$$k(E) = \frac{N^{\ddagger}(E - E_0)}{h\rho(E)} \tag{6.5}$$

The third, and final, stage, which goes from $k(E)$ to $k(T)$, is carried out in the next section. It is useful however to note that the concept of a differential reaction rate, $k(E)$, does not require the approximation of a configuration of no return and can be derived in complete generality. This follows from our discussion of the reaction rate for state-selected reactants in Appendix 3.A. This allows us to write the rate constant for thermal reactants as an average of the state-selected rates over the internal states of the reactants. By examining this expression, Problem A shows that if $\sigma_R(i)$ is the reaction cross-section for reactants in the initial internal state i (internal energy E_i) then it is an exact result that

$$Y(E) = \frac{1}{h} \sum_i \pi^{-1} k_i^2 \sigma_R(i) \tag{6.6}$$

where $(h^2 k_i^2 / 2\mu) = E_T = E - E_i$ and $k(E) = Y(E)/h\rho(E)$. We can therefore use a reaction cross-section from any source, even including a classical trajectory computation, to compute $k(E)$. Problem A also shows that both the exact expression Eq. (6.6), for the yield and also the transition state approximation Eq. (6.4) satisfy microscopic reversibility. Note that the $1/h$ factor in Eq. (6.3) remains in the final results. The origin of this factor is the failure of classical mechanics to count translational states properly.

6.1.4 From $k(E)$ to $k(T)$

Transition state theory for thermal reactants[5] is readily derived from the results at a given energy. First we note that when the reactants are in thermal equilibrium,[6] the general expression for the thermal reaction rate can be written as[7]

$$k(T) = Q^{-1} \int_0^\infty Y(E) \exp(-\beta E) \mathrm{d}E = Q^{-1} \int_0^\infty k(E) \rho(E) \exp(-\beta E) \mathrm{d}E \qquad (6.7)$$

where $\beta = 1/(k_{\mathrm{B}} T)$ (k_{B} is the Boltzmann constant) and Q is the partition function of the reactants. As discussed in Appendix 6.A, Q is defined in terms of the density of states such that the thermal average is over a normalized distribution:

$$Q \equiv \int_0^\infty \exp(-\beta E) \rho(E) \mathrm{d}E \qquad (6.8)$$

From the definition Eq. (6.8) of the partition function Q of the reactants, the thermal rate constant as given by Eq. (6.7) is the thermal average of $k(E)$, $k(E) = Y(E)/h\rho(E)$. In transition state theory,*

$$h Y(E) = N^{\ddagger}(E - E_0) = N^{\ddagger}(E') \qquad (6.9)$$

where E_0 is the threshold energy and $E' = E - E_0$ (so that $\mathrm{d}E = \mathrm{d}E'$). We can integrate Eq. (6.7) by parts, if we know the value of $\mathrm{d}Y(E)/\mathrm{d}E$. Again we appeal to the concept of a density of states, where what we need here is the density $\rho^{\ddagger}(E')$ of internal states at the transition state

$$\frac{\mathrm{d}[h Y(E)]}{\mathrm{d}E} = \frac{\mathrm{d}}{\mathrm{d}E'}[N^{\ddagger}(E')] \equiv \rho^{\ddagger}(E') \qquad (6.10)$$

Integrating by parts,

$$
\begin{aligned}
k(T) &= Q^{-1} \int_0^\infty Y(E) \exp(-\beta E) \mathrm{d}E = -(\beta Q)^{-1} \int_0^\infty Y(E) \frac{\mathrm{d}}{\mathrm{d}E}[\exp(-\beta E)] \mathrm{d}E \\
&= -(\beta Q)^{-1} \left\{ Y(E) \exp(-\beta E) \Big|_0^\infty - \int_0^\infty \exp(-\beta E) \frac{\mathrm{d}Y(E)}{\mathrm{d}E} \mathrm{d}E \right\} \\
&= \frac{k_{\mathrm{B}} T}{h Q} \int_0^\infty \exp(-\beta E_0) \exp(-\beta E') \frac{\mathrm{d}}{\mathrm{d}E'}[N^{\ddagger}(E')] \mathrm{d}E' \\
&= \frac{k_{\mathrm{B}} T}{h} \frac{\int_0^\infty \exp(-\beta E') \rho^{\ddagger}(E') \mathrm{d}E'}{Q} \exp(-\beta E_0) \equiv \frac{k_{\mathrm{B}} T}{h} \frac{Q^{\ddagger}}{Q} \exp(-\beta E_0) \quad (6.11)
\end{aligned}
$$

---

* Equation (6.7) is equally valid if we use $Y(E)$ as determined by a dynamical computation Eq. (6.6).

The final integral is the partition function $Q^{\ddagger}$ for the internal states (the degrees of freedom excluding the reaction coordinate) at the transition state configuration,

$$Q^{\ddagger} \equiv \int_0^{\infty} \exp(-\beta E')\rho^{\ddagger}(E')\mathrm{d}E' \qquad (6.12)$$

With Eq. (6.11) that we here reproduce

$$k(T) = \frac{k_B T}{h} \frac{Q^{\ddagger}}{Q} \exp(-E_0/k_B T)$$

we have completed the line of reasoning that started in Section 6.1. The computation of the thermal rate constant has been reduced to the computation of partition functions, which involves the structure of the transition state and requires no dynamical knowledge whatsoever. It does require knowing the potential energy surface in the immediate vicinity of the transition state. Fortunately, quantum chemistry can provide this information even for many-atom reactants. When this information is not available, it becomes necessary to assume a structure for the transition state to calculate a value of $Q^{\ddagger}$. We do an example in Section 6.1.4.1. When E_0 and $Q^{\ddagger}$ are computed from a well-defined potential surface, transition state theory in the form of Eq. (6.11) provides quite reasonable results when there is a single barrier separating reactants and products. If such is not the case, more work is needed, as will be addressed in the following discussion.

*6.1.4.1   Transition state theory and the steric factor

One of the immediate successes of transition state theory is that it both explains and offers a quantitative estimate for the steric factor. In particular, it readily shows why the steric factor is smaller when the reactants have a more complex structure. Although our detailed demonstration is almost a caricature of the true nature of the system, it captures the essence of the phenomenon, namely, that the steric factor originates from modes of motion that are unconstrained in the reactants and become hindered in the transition state.[8]

Statistical mechanics is that branch of physical chemistry dealing with partition functions. We appeal to statistical mechanics for two results, discussed in Appendix 6.A. First, a partition function of a molecule can be expressed as a product of the partition function for the motion of the center of mass and an internal partition function. Second, the internal partition function can be approximately represented as a product of contributions from each bound mode. To apply the theory we need to compute partition functions (or to look them up in the JANAF tables*). In what follows, we are extremely cavalier and take all rotational partition functions to be equal, and larger by an order of magnitude than all vibrational partition functions that we also take to be equal to one another. This provides a reasonable order of magnitude but is otherwise not a recommended procedure.

---

* JANAF = Joint Army Navy Air Force, a useful tabulation of thermodynamic data published by the (US) National Institute of Standards and Technology (see http://webbook.nist.gov/chemistry).

First, we inquire under what condition is the steric factor unity, meaning that there are no steric requirements. We already know from Section 3.2 that this condition is met when the two colliding particles are structureless. The partition function for the reactants is then $Q = Q_T^3 Q_T^3$, one translational partition function (for a 3D motion) for each reactant. The point of no return has been identified in Section 3.2.7: the two particles are at a distance d apart and the barrier height is E_0. The transition state is here a diatomic. It has six degrees of freedom. Three are the motion of its center of mass. Of the other three, one is the vibrational motion, which is the reaction coordinate. Therefore, it should not be counted as an internal coordinate of the transition state. The other degrees of freedom are the two planes of rotation of a diatomic molecule. Therefore $Q^{\ddagger} = Q_T^3 Q_R^2$. So the transition theory result for the reaction rate constant, in the absence of any steric effect, is

$$k(T) = \frac{k_B T}{h} \frac{Q^{\ddagger}}{Q} \exp(-\beta E_0) = \frac{k_B T}{h} \frac{Q_R^2}{Q_T^3} \exp(-\beta E_0) \quad \begin{array}{l} \text{transition state theory,} \\ \text{no steric requirements} \end{array} \quad (6.13)$$

Next, consider a reaction between an atom and a diatomic molecule, A + BC. Reactions can differ also in their energetic requirements, but to focus attention on the steric requirements with the energetic effects being equal we take the barrier height E_0 to be the same as in the previous reaction. To have minimal steric requirements let us take the transition state, ABC, to be bent. This choice allows A to approach BC within a cone. Because BC has an internal structure, the partition function for the reactants becomes $Q = Q_T^3 Q_T^3 Q_v Q_R^2$. The transition state is a bent triatomic. It has three vibrations, one of which is the reaction coordinate. (As we saw in Section 5.1, this is the asymmetric stretch vibration.) The bent transition state has three planes of rotation, $Q^{\ddagger} = Q_T^3 Q_v^2 Q_R^3$. Accordingly, for reasons that will become immediately apparent, we write $k(T)$ as

$$k(T) = \frac{k_B T}{h} \frac{Q_v^2 Q_R^3}{Q_T^3 Q_v Q_R^2} \exp(-\beta E_0)$$

$$= \left(\frac{Q_v}{Q_R} \right) \frac{k_B T}{h} \frac{Q_R^2}{Q_T^3} \exp(-\beta E_0) \quad \begin{array}{c} \text{A + BC} \\ \text{TST} \\ \text{bent ABC} \end{array} \quad (6.14)$$

Comparing Eq. (6.14) with Eq. (6.13), the steric factor for a bent ABC transition state is Q_v/Q_R, which results in a reduction of the rate compared with Eq. (6.13) by an order of magnitude. Consider the A + BC reaction again but with a linear transition state. A linear triatomic has four vibrations (it can bend in two independent planes) and two rotations. Therefore, for this, more constrained, transition state

$$k(T) = \frac{k_B T}{h} \frac{Q_v^3 Q_R^2}{Q_T^3 Q_v Q_R^2} \exp(-\beta E_0)$$

$$= \left(\frac{Q_v^2}{Q_R^2} \right) \frac{k_B T}{h} \frac{Q_R^2}{Q_T^3} \exp(-\beta E_0) \quad \begin{array}{c} \text{A + BC} \\ \text{TST} \\ \text{linear ABC} \end{array} \quad (6.15)$$

The steric factor for a linear ABC transition state is $(Q_v/Q_R)^2$, which yields a reduction of the rate by two orders of magnitude. Problem F shows that the reduction is determined by how many free motions in the reactants become vibrations in the transition state. In the most general case of reaction between two nonlinear polyatomics,* the steric factor is $(Q_v/Q_R)^5$, which is quite a reduction!

*6.1.4.2 Variational transition state theory

One possible improvement to transition state theory, as introduced by Wigner, is to find the optimal location of the transition state along the reaction coordinate. This approach is called *variational transition state theory*, and it offers a better quantitative prediction of the reaction rate. Here, however, we are interested in the reasoning behind the theory. At first blush, it may seem obvious that the transition state must be located at the barrier in the potential energy surface. A moment's reflection shows that while this assumption is reasonable it is not necessarily the optimal choice. The true transition state is the *bottleneck to reaction*. The rate through the transition state is always equal to or larger than the true reaction rate. Therefore, the transition state is most aptly located where the flux of trajectories that cross it is minimal. The variational approach seeks to determine that least upper bound to the reaction rate. The barrier height and the entropy** of the other modes, Problem E, determine the flux. Generally, the barrier is more important, but the entropy also matters, particularly at higher temperatures.

Variational transition state theory is particularly useful when the potential energy surface itself has no barrier. In Section 3.2 we used the barrier of the effective potential as the point of no return. Variational theory shows that this is the optimal choice, excluding the role of entropy. Variational theory is also needed when there can be *transition state switching*, by which we mean that a region of the potential that was not a bottleneck at a low energy becomes rate-limiting at a higher energy. It is then necessary to switch the location of the transition state, which then becomes a function of energy.

---

* If the two reactants have n_A and n_B atoms, respectively, there are $3n_A - 6 + 3n_B - 6$ vibrations of the reactants. There are $3(n_A + n_B) - 6$ vibrations of the transition state, one of which is the reaction coordinate. Five vibrations are gained for the other degrees of freedom at the transition state. This is at the expense of translations and rotations of the free reactants. See Problem F.

** How does entropy come in? Because there are the internal degrees of freedom of the transition state. The more of these that can be populated, the higher is the reaction rate, as shown explicitly by Eqs. (6.3) or (6.5). Entropy is a measure of the number of accessible states. If the potential energy surface at the saddle is shallow, meaning that the mountain pass is wide, many internal states are accessible. If the forces in the directions perpendicular to the reaction coordinate are strong, the pass is narrow. For thermal reactants we show in Appendix 6.A that the partition function counts how many states are effectively accessible. A reaction is said to be *entropy controlled* when the rate is unusually large (or small) because of the width of the bottleneck. How many states are accessible is determined by the energy (or temperature), so it is possible for a reaction to be energy controlled at low temperatures and entropy controlled at high temperatures, particularly so if the barrier is low.

6.A   Appendix: Density of states and partition functions

We begin with the density of internal states. First, we need the number, $N_I(E_I)$, of internal states, whose energy is less than or equal to E_I. If the reactants are an atom in its ground electronic state and a diatomic molecule, then the only internal energy is the rovibrational energy of the diatom. We need to count how many states have a rovibrational energy less than or equal to E_I. We can formally write this counting as

$$N_I(E_I) = \sum_{v,j} H(E_I - E_{v,j}) \tag{A.6.1}$$

Here $H(x)$ is a step function,* namely, a function $H(E - E')$ that is zero for $E < E'$ and equals unity otherwise. So, strictly speaking, the number of states is a staircase function of E_I, increasing by unity any time we cross a new threshold. When the system has more degrees of freedom, the summation in (A.6.1) is over more quantum numbers, but the meaning remains the same: make a spectrum of energies of the internal states that are arranged as a stick diagram in order of increasing energy and count how many states fall below a cutoff at E_I. Do not forget that more than one internal state can have the same energy and so each energy level must be counted with its correct degeneracy.

The density of internal states is defined such that $\rho_I(E_I)dE_I$ is the number of internal states in the narrow energy range E_I to $E_I + dE_I$. For the systems of interest to us, the number of states per energy interval is sufficiently high so that the number of states is essentially a smooth function. Therefore the density of states is also a continuous function and is to be understood as

$$\rho_I(E_I) = (N(E_I + \Delta E) - N(E_I - \Delta E)) / 2\Delta E \tag{A.6.2}$$

---

* For all values of E, the step function $H(E - E')$ is a constant. Consequently, its derivative vanishes. At $E = E'$ the step function suddenly changes value and we might imagine that its derivative does not exist because it is infinite at that point and zero elsewhere. But we argue instead that the derivative is an infinitely narrow sharp spike that has a unit area. We offer the following proof for this assertion: integrate the derivative over a short interval in E. If E' is within the interval, you get unity, otherwise you get zero:

$$\int_A^B (dH(E - E')/dE)\, dE = H(B - E') - H(A - E') = \begin{cases} 1, & A < E' < B \\ 0, & \text{otherwise} \end{cases}$$

Following Dirac we call the derivative "the delta function"

$$dH(E - E')/dE = \delta(E - E')$$

To practice handling delta functions see Problem C. Taking the derivative of (A.6.1) with respect to energy, we get the formal definition of the density of states

$$\rho_I(E_I) = \sum_{v,j} \delta(E_I - E_{v,j})$$

The practical definition is (A.6.2).

where even though ΔE is a small energy interval there are many quantum states in it.

The molecular partition function $Q(T)$ is defined as the normalization factor for the Boltzmann distribution at the temperature T. Specifically, the probability p_i of a single internal state with energy E_i is

$$p_i = \exp(-E_i/k_B T)/Q(T) \tag{A.6.3}$$

where $Q(T)$ is the function of temperature given by

$$Q(T) = \sum_i \exp(-E_i/k_B T) \tag{A.6.4}$$

Equation (A.6.4) readily explains the old name for Q as "the sum over states." A rough and ready estimate for Q is to divide all the states into those where $E_i < k_B T$, for which we may approximate $\exp(-E_i/k_B T)$ as unity. For all other states we take the exponent to be too small to contribute to the sum. Q thereby becomes the number of states that are significantly populated at the temperature T.

In the text we used two properties of the partition function. One, Eq. (6.8), is a rewriting of the definition Eq. (A.6.4) as an integral over the density of states. For example, for the internal states, the density of states $\rho_I(E_I)dE_I$ is the number of internal states in the narrow energy range E_I to $E_I + dE_I$, so that

$$Q_I = \int_0^\infty \exp(-E_I/k_B T)\rho_I(E_I)dE_I \tag{A.6.5}$$

If the spectrum of states is sparse, then the integral in Eq. (A.6.5) reduces to a sum as in (A.6.4) because the density of states is a sum of delta functions. Often the spectrum is so dense that ρ_I can be approximated as a continuous function and an ordinary integration can be carried out.

The other important property of the partition function is its factorizability: the partition function of disjoint degrees of freedom is a product of individual partition functions. This factorization can be exact. For example, the energy of a molecule is exactly the sum of its internal energy and the translational energy of the center of mass. Then the partition function of the molecule is exactly the product of a translational partition function and an internal one. Sometimes we use factorization as an approximation. For example, the rovibrational energy is not strictly the sum of vibrational and rotational terms because the vibration stretches the molecule so that its moment of inertia depends, albeit weakly, on the vibrational state. Making the assumption of separability, we have from Eqs. (A.6.4) and (A.6.1)

$$Q_I = \sum_{v,j} \exp(-E_{v,j}/k_B T) \cong \sum_{v,j} \exp(-(E_v + E_j)/k_B T)$$

$$= \left[\sum_v \exp(-E_v/k_B T)\right]\left[\sum_j \exp(-E_j/k_B T)\right] = Q_v Q_R \tag{A.6.6}$$

In Section 3.1.2.4 we have implicitly used a result for the mean energy of the system:

$$\langle E \rangle \equiv \sum_i E_i \, p_i = -k_B \partial \ln Q / \partial \left(\frac{1}{T} \right) = k_B T^2 \partial \ln Q / \partial T \qquad \text{(A.6.7)}$$

This result is needed to do Problem D and others.

6.2 RRKM theory and the rate of unimolecular reactions

In this section we discuss the rate of formation and dissociation of energy-rich molecules. There are two key ingredients. One is transition state theory as discussed in Section 6.1. If there is a single barrier separating reactants and products, transition state theory allows us to compute the rate of the chemical reaction. But as discussed in Section 5.3.3, we are here interested in reactions where there is more than one barrier along the reaction coordinate. Between any two such barriers there must be a hollow in the potential. This may slow the passage. The second ingredient that we need stems from an examination of the nature of the complex dynamics over such potentials.

6.2.1 Unimolecular reactions: the Lindemann and the RRKM hypotheses

Figure 5.20 showed schematically why a well along the reaction coordinate enables a trajectory to spend some time seemingly as a stable molecule before escaping out of the well region. For chemists, this is very old news. They have long understood how to initiate a complex mode reaction in the interaction region, that is, with the molecule whose total energy is on top of the well. Begin with a stable molecule that is at the bottom of the well and drive its energy up. Ever since the proposal by Lindemann, chemists also understood that energy-rich polyatomic molecules survive for a while before they dissociate. Indeed, the very success of the *Lindemann mechanism* allowed them to **clock** the dissociation and specifically to show that the lifetime of an energy-rich polyatomic molecule can be comparable to or longer than the time interval between collisions in the bulk.

Lindemann sought to explain why polyatomic molecules in the bulk gas phase dissociate by a unimolecular rate law, that is, seemingly spontaneously. The question was where the energy required for the bond breaking came from. In the simplest form we assume collisional activation and deactivation of the energy-rich species A* by molecules of the bath gas M, and a competing unimolecular dissociation of A*:

$$A \underset{k'[M]}{\overset{k[M]}{\rightleftarrows}} A^*, \quad A^* \xrightarrow{k_d} \text{products}$$

This is an approximate description because it overlooks the – as we show, quite steep – energy dependence of the dissociation rate constant k_d. In Problem G it is done better. Here we proceed with this simplified picture. The rate of formation of products is $k_d[A^*]$. Since the concentration $[A^*]$ of the energy-rich molecule is low, its rate of change is also low and can be approximately put equal to zero. This is known as the *steady-state approximation*:

$$\frac{d[A^*]}{dt} = k[M][A] - k'[M][A^*] - k_d[A^*] \underset{\substack{\text{steady-state} \\ \text{approximation}}}{=} 0$$

It yields an immediate solution for $[A^*]$:

$$[A^*] = k[M][A]/(k'[M] + k_d) \rightarrow \begin{cases} [A]k/k', & \text{high pressure} \\ k[M][A]/k_d, & \text{low pressure} \end{cases}$$

$$\frac{d[\text{products}]}{dt} = k_d[A^*] \underset{\substack{\text{steady-state} \\ \text{approximation}}}{\longrightarrow} \begin{cases} k_d[A]k/k', & \text{high pressure} \\ k[M][A], & \text{low pressure} \end{cases}$$

If the molecule lives longer than the time between deactivating collisions, $k'[M] \gg k_d$, most of the collisionally activated molecules will also be deactivated by collisions. Then the concentration of energy-rich molecules and hence the rate of formation of products is independent of the pressure of the bath gas.

This is known as the *high pressure limit* because the pressure is the measure of the concentration, $[M]$, of the bath gas. In this limit the rate of dissociation is unimolecular. At low enough pressures, $k'[M] \ll k_d$, the rate of reaction is determined by the rate of activating collisions with an apparent rate constant, $k[M]$, that is proportional to the pressure.

It was Lindemann's point that the more degrees of freedom there are in the energy-rich polyatomic the longer it will stay bound in the well. The reason is that it requires energy in that particular vibration that is the reaction path to cross the saddle point to products. The other $3n - 7$ vibrations of the molecule, which are not the reaction path, provide a sink for the excess energy. Once the energy flows into these many modes it takes a rare event for the required energy to localize in the reaction path. By computing the probability for this event we will show that the lifetime increases exponentially as the number of other modes increases.

The observed rate of dissociation does decline with pressure from high-pressure unimolecular kinetics to a low-pressure bimolecular regime. The Lindemann mechanism interprets this behavior as being due to the long lifetime, low k_d, of energy-rich molecules, a lifetime that is long compared with the time interval between collisions at higher pressures. To securely verify the implication from chemical kinetics that the lifetime is long, it is necessary to directly demonstrate the competition between dissociation and collisional deactivation.

One can prepare energy-rich molecules not only by collisional activation but also by collisions of suitable precursors. Chemists know this as *chemical activation* because the energy is made available by the formation of a new chemical bond. What can then be measured is the competition between dissociation and

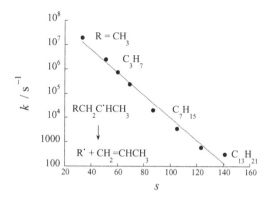

Figure 6.2 The measured (points) dissociation rate constant, logarithmic scale, vs. the number, $s = 3n - 6$, of vibrational modes in a series of energy-rich alkyl radicals of n atoms [data of E. A. Hardwidge, B. S. Rabinovitch, and R. C. Ireton, *J. Chem. Phys.* **58**, 340 (1973)]. The straight line is a fit. All these radicals have about the same excess energy (measured with respect to the bottom of the well), namely the energy of the newly formed primary C—H bond, roughly about 100 kcal mol^{-1}. The energy-rich adducts dissociate primarily by a radical loss, as shown, because the bond that breaks is weaker. We shall indeed show that one signature of energy partitioning is that it is the weak link that breaks, irrespective of which coordinate was initially energy-rich.*

deactivation by collisions, where the rate of deactivation can be varied by varying the pressure of the buffer gas. This allows clocking the dissociation.** Figure 6.2 shows the measured dissociation rate constant in an experiment designed to demonstrate the decrease in the rate in a series of chemically activated alkyl radicals of increasing size:

$$H + CH_2 = CHCH_2R \longrightarrow CH_3\dot{C}HCH_2R^* \xrightarrow{k_d} CH_3CH = CH_2 + \dot{R}$$

The alkyl radical R is identified in the figure. The abscissa is the number, $s = 3n - 6$, of vibrational modes in the energy-rich alkyl of n atoms.

---

* An energy-very-rich large polyatomic molecule has a quasi-continuum of vibrational states, Section 6.2.2.2. Therefore its behavior is almost that of a macroscopic body. Remember that the direction of spontaneous change in a bulk system is primarily governed by energy, but that entropy is also relevant. The same ideas are also effective for a large isolated molecule. So here is that mountain pass again. It is a rule of thumb that it is the weak bond that will break, but entropic aspects must also be considered. If the mountain pass is narrow and constrained it will lower the rate of dissociation, see Section 6.2.2.

** At atmospheric pressure the time interval between collisions is, Chapter 2, in the nanosecond range. Decreasing the pressure increases this time. So clocking by collisions can be used to measure times of the order of nanoseconds or longer. We will discuss clocking by fluorescence and this has a similar range. To measure even shorter times we use the pump–probe technique as already introduced in Section 1.2.5, where the delay between the pump and probe pulses can be shorter or even much shorter than the time interval between collisions. See Chapter 8.

Figure 6.2 also makes another key quantitative point: the lifetimes of energy-rich larger polyatomics are exceedingly long compared with the time scale of vibrational motion. As is consistent with the Lindemann hypothesis, the lifetimes are also longer than the time between collisions at ordinary pressures. This raises a question: for a molecule that lives for so many vibrational periods,[9] will it remember* how it was formed?

The assumption that quasi-bound energy-rich polyatomics have forgotten where they came from is known as the RRKM hypothesis[10] (Rice, Ramsperger, Kassel, Marcus). One in a classic series of experiments by Rabinovitch designed to test the RRKM hypothesis is the formation of a bicyclo compound

$$
CF_2\text{—}CF\text{—}CF = CF_2 + :CD_2 \rightarrow CF_2\text{—}CF\text{—}CF\text{—}CF_2
$$
$$
\backslash \ / \qquad\qquad\qquad \backslash \ / \quad \backslash \ /
$$
$$
CH_2 \qquad\qquad\qquad\qquad CH_2 \quad CD_2
$$

where we have styled the initially vibrationally excited bonds thicker than the rest.

The bicyclic molecule decomposes preferentially by elimination of $:CF_2$. If the energy is rapidly redistributed over all degrees of freedom, elimination of $:CF_2$ can occur from either ring, as observed. This and other experiments have provided the evidence that those molecules that are long-lived have sampled the range of available states. The conclusion that, prior to dissociation, energy is partitioned amongst all the degrees of freedom allows us to considerably simplify the treatment of the dynamics. We do not need to run trajectories. However we prepare the molecule, it forgets how it was formed. But in another sense this *vibrational energy redistribution* presents a challenge because it implies that we cannot control the outcome. The search is therefore on for ways of beating the tendency of molecules to spread the energy over all available modes. Do we stand a chance? One direction is to try and catch those molecules that dissociate early on. The Lindemann mechanism suggests that these are the molecules that dissociate even when the pressure is high. When the rate of collisional deactivation is high enough, Figure 6.2 already showed that the lifetime can be comparable to the time between collisions. The early experimental result[11] was that at a high enough pressure, $:CF_2$ eliminates preferentially from the ring that was initially excited. From that observation the search for non-RRKM behavior was on! We will have more to say but the available evidence is that the onset of energy redistribution is

---

* Strictly speaking, a single classical trajectory does not forget. Indeed, a good test of the numerical integration is to stop the trajectory, reverse the sign of all velocities, and check that the trajectory integrates back to where it started from. But a real experiment is mimicked by an ensemble of trajectories. As we saw in the discussion of the Monte Carlo method, averages computed for an ensemble are not sensitive to the details of initial conditions if these are chosen in an unbiased manner. The ensemble is said to "forget" if you can bias the selection of trajectories and still end up with the same average values for observables of chemical interest.

fairly prompt. So control of the dynamics requires a rather short-time excitation or other means for depositing the energy in a manner that makes the molecule move immediately in the direction of the exit valley.* One tool at our disposal that we are yet to take advantage of, Chapter 7, is electronic excitation because, so far, everything that we have discussed is taking place on the ground electronic state.

6.2.2 The (RRKM) dissociation rate of an energy-rich polyatomic molecule

Imagine an energy-rich polyatomic molecule that resides over a well in the potential energy surface. Typically, a barrier separates the well and the exit valley leading to the dissociation products.** Transition state theory, with one further assumption, allows us to compute the rate of dissociation. Transition state theory assumes that the states to the reactant side of the barrier are in equilibrium and it uses this assumption in an essential manner. If an energy-rich polyatomic lives long enough that it forgets where it came from, then it is in equilibrium and transition state theory can be used. This is the starting point for the RRKM treatment.

What we want to compute is $k(E)$, the rate constant for unimolecular dissociation at the total energy E. The quickest route to this result is to use detailed balance.[12] We equate the rate of association of the products to the rate of dissociation into products when both the energy-rich molecule and the products are at equilibrium. Let $\overleftarrow{k}(E)$ be the rate constant for crossing the barrier from the products' valley into the well. $\overleftarrow{k}(E)$ is known to us from transition state theory, Eq. (6.5), $\overleftarrow{k}(E) = N^{\ddagger}(E - E_0)/h\rho_p(E)$, where $\rho_p(E)$ is the density of states of the products. Detailed balance requires that[13] $\rho_p(E)\overleftarrow{k}(E) = \rho(E)k(E)$, where $\rho(E)$ is the density of states of the energy-rich polyatomic molecule. Hence we obtain the celebrated RRKM result

$$k(E) = \frac{N^{\ddagger}(E - E_0)}{h\rho(E)} \tag{6.16}$$

Equation (6.16) resembles the transition theory reaction rate, Eq. (6.5), and indeed here too $\rho(E)$ is the density of states on the reactant side of the barrier. But here the reactant is the energy-rich polyatomic molecule, and we must **assume** (on the

---

* The non-selective behavior is due, at least in part, to the extensive averaging over initial conditions when the energy-rich molecule is formed by chemical activation. In general we expect that the tighter is the preparation of the ensemble of initial conditions, the longer will it retain its memory.
** An alternative fate for an energy-rich molecule is to isomerize. Then there is a barrier separating the two wells, that of the molecule and that of the isomer. Barriers to isomerization can be significantly lower than those for bond breaking. Since transition state theory does not ask what happens beyond the barrier, the very same approach allows us to compute the reaction rate whether the molecule dissociates or it isomerizes. But the rate of barrier crossing toward an isomer can be fast and care must be exercised before the RRKM approximation is made.

basis of experimental evidence and theoretical considerations) that the molecule has reached equilibrium.

6.2.2.1 Reactions in the bulk

A limit where the equilibrium assumption for the energy-rich molecule is realistic are reactions in the bulk at the limit of high pressure. This is the limit where energy transfer collisions are far more frequent than dissociation. The concentration of the energy-rich molecules is that at thermal equilibrium at the temperature of the gas

$$[A^*(E)] = [A]\frac{\rho(E)\exp(-E/k_B T)}{Q}$$

Here [A] is the total concentration of A molecules and the partition function Q is that of A molecules. The same result is obtained if we form energy-rich A molecules by chemical activation and then let them thermalize by collisions.

The rate of dissociation is, from Eq. (6.16), plus the rule, Appendix 3.A, of averaging over initial states:

$$\begin{aligned}
\text{Reaction rate} &= \int [A^*(E)]k(E)\mathrm{d}E \\
&= [A]\frac{1}{hQ}\int N^\ddagger(E - E_0)\exp(-E/k_B T)\mathrm{d}E \qquad \text{high pressure limit} \\
&= [A]\frac{k_B T}{h}\frac{Q^\ddagger}{Q}\exp(-E_0/k_B T)
\end{aligned}$$

where the last line is by integration by parts as in Eq. (6.11). The reaction is unimolecular with a transition state theory expression for the reaction rate constant. In Problem G the derivation above is extended to lower pressures. In Chapter 11 we ask what happens when collisions are so frequent that they can catch the molecule during the very act of crossing the barrier.

*6.2.2.2 Vibrational state counting: a simplified treatment

The M (for Marcus) in RRKM is also to remind us to count states at the transition state taking into account conservation of angular momentum. For the sake of simplicity, we neglect this constraint. We further assume that all bound vibrations are harmonic (which is manifestly not the case) and make a quasi-classical approximation where the number of states is a continuous rather than a staircase function of the energy, cf. Appendix 6.A. Suppose we have just one harmonic oscillator of frequency ν. Then the number of quantum states below the energy E is $E/h\nu$. Of course, this overlooks the zero point energy of the oscillator, etc., but if the energy is high enough, the order of magnitude is surely right. If there are two oscillators, of frequencies ν_1 and ν_2, the number of quantum states below the energy E is $(E/h\nu_1)(E/h\nu_2)/2$. This is the area of a right-angle triangle whose two sides are $(e/h\nu_1)$ and $(e/h\nu_2)$ with $e \le E$. Each unit area corresponds to a quantum state. The factor $s!$ in Eq. (6.17) is a generalization of the factor 2, which

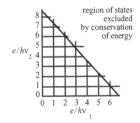

region of states excluded by conservation of energy

is the ratio of the area of the triangle to that of a square. For a sphere, one octant is $1/8$ in volume, etc. If there are s oscillators and the (geometric) mean frequency is $\bar{\nu}$, then

$$N(E) = \frac{1}{s!}\left(\frac{E}{h\bar{\nu}}\right)^s$$

$$\rho(E) \equiv \frac{dN(E)}{dE} = \frac{1}{(s-1)!}\left(\frac{E}{h\bar{\nu}}\right)^{s-1}\bigg/h\bar{\nu}$$

(6.17)

Let the energy-rich molecule have n atoms. Then it has $s = 3n - 6$ vibrations. In the transition state one of these vibrations becomes the reaction coordinate so there are $s - 1 = 3n - 7$ other vibrations. The rate constant for dissociation is, from Eq. (6.16),

$$k(E) = \nu\frac{(E - E_0)^{s-1}}{E^{s-1}} = \nu\left(1 - \frac{E_0}{E}\right)^{s-1}$$

(6.18)

Here $\nu = \bar{\nu}^s/\bar{\nu}^{\ddagger(s-1)}$ is the mean vibrational frequency. Consequently, ν represents the rate of a typical molecular vibration. If $k(E) = \nu$ the molecule will dissociate in one vibrational period. But generally, $k(E) \ll \nu$ because of the second factor in Eq. (6.18), which is typically much smaller than one because $((E - E_0)/E) < 1$ and $s > 1$. This factor is a simple quantitative statement of the physics that we already know from Section 5.3.3.1. First of all, for the unimolecular rate to be slow we need a deep well, cf. Figure 5.20, so that E is just slightly bigger than E_0 and $(1 - E_0/E)$ is significantly smaller than unity. Particularly so for larger polyatomics for which $s \gg 1$. Indeed, the rate of unimolecular dissociation is exponentially small in the number, s, of vibrational modes;

$$k(E) = \nu\left(1 - E_0/E\right)^{s-1} = \nu\exp\left(-(s-1)\ln(1 - E_0/E)\right)$$

$$\approx \nu\exp\left(-(s-1)(E_0/E)\right)$$

See Figure 6.2. At high energies, when the energy per vibrational mode, E/s, becomes comparable to the barrier height, E_0, the system will no longer be confined by the barriers and will exit after one or a few vibrations within the well.

For quantitative purposes one should count states properly; efficient computer programs for doing so are available. The reason why one needs them is that a real molecule is not a collection of s identical modes. The differences between low-frequency bends and high-frequency stretch motions are large, and particularly the C—H stretch vibrations tend not to be significantly populated until fairly high excitation energies.

*6.2.2.3 The vibrational quasi-continuum

At low vibrational excitations polyatomic molecules exhibit a well-defined vibrational spectrum (particularly so if it is initially quite cold). We discuss in this section at what energy a large molecule changes its behavior to become de facto its own heat bath. This change involves a sudden increase in the density of vibrational states as the energy of the molecule approaches the dissociation threshold. Even

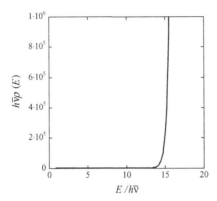

Figure 6.3 The number of vibrational states per mean vibrational spacing, $\rho(E)\Delta E = (E/h\overline{\nu})^{s-1}/(s-1)!$, $\Delta E = h\overline{\nu}$ plotted vs. $E/h\overline{\nu}$ for $s = 42$, which is the number of vibrational modes of the bicyclo species discussed in Section 6.2.1. At about $(E/h\overline{\nu})/(s-1) = 1/e$ the density takes off and reaches millions of vibrational states per mean spacing. At the energy $(E/h\overline{\nu})/(s-1) = 1/2$ the number of such states is in the trillions yet, on average, only $1/2$ the vibrational levels are populated. A more accurate counting of the states does not change the qualitative results. At even higher excitation there are so many states available to store the excess energy that it might seem amazing that states that have at least the energy E_0 in the reaction coordinate have any chance. Indeed, for a large molecule it requires a large excess energy for the rate of dissociation to be measurable.

a bound polyatomic molecule can behave almost classically in that its density of vibrational states is roughly continuous, hence the term *quasi-continuum*.

A short proof of this quasi-continuous density of states is based on the Stirling approximation for the factorial function, $s! \approx s^{s+1/2}\exp(-s) = \sqrt{s}(s/e)^s$. Therefore,

$$\rho(E) = \frac{1}{(s-1)!}\left(\frac{E}{h\overline{\nu}}\right)^{s-1} \bigg/ h\overline{\nu}$$
$$\approx \frac{1}{h\overline{\nu}\sqrt{(s-1)}}\left(\frac{E}{h\overline{\nu}}\bigg/\frac{s-1}{e}\right)^{s-1} \tag{6.19}$$

We find two quite different regimes. When $(E/h\overline{\nu})/((s-1)/e) < 1$, since $s \gg 1$ the density of states is exponentially small, yielding the familiar regime of molecular spectroscopy. The number of quantum states per (mean) vibrational spacing, $h\overline{\nu}\rho(E)$, is limited. But when $(E/h\overline{\nu})/((s-1)/e) > 1$, the density of vibrational states is exponentially large. The transition between the two regimes occurs when the number of vibrational quanta per vibrational mode, $(E/h\overline{\nu})/(s-1)$, is larger than $1/e \approx 1/3$. In other words, when one in every three vibrational modes is singly populated, the molecule is already at the threshold of the quasi-continuum. This onset is quite sharp as a function of energy, which can be seen in Figure 6.3.

6.2.2.4 Do energy-rich polyatomic molecules behave statistically?

In one sense this is no longer an essential question. Why? Increasingly we are able to create polyatomic molecules with a rather localized initial excitation, with details introduced in Chapters 7 and 8. We are also able to probe the products with considerable resolution. Current computational capabilities are such that high-level simulations will accompany this type of experiment. If the dynamics proceed mainly on one potential energy surface, these will be classical trajectories. Very often the quantum chemical computations of the potential will be done at each configuration reached by the trajectory. Density functional theory is the favorite current quantum chemical method for doing so. When different electronic states are involved,[14] one can do quantum dynamics for many nuclear degrees of freedom (Ben-Nun *et al.*, 2000; Hammes-Schiffer, 2001) or one can adopt the method of classical trajectories with an important modification, *trajectory surface hopping* (Tully, 1998), that requires a classical trajectory to bifurcate into two separate trajectories, each propagating on a separate potential. We will have more to say when we discuss particular experiments in Chapters 7 and 8. Here we just anticipate a very useful feature of quantum mechanics that is discussed later. It is that early time dynamics are dominated by a few vibrational modes that are strongly coupled. The weaker coupling terms that eventually will spread the energy do not come into play until later. Early time dynamics in general and also smaller polyatomic molecules are therefore examples where selectivity can override statistics. In particular, the frequency mismatch between modes tends to slow down the initial energy redistribution. In Section 7.2.4.1 we discuss the *dissociation of van der Waals adducts* as an extreme example of such mismatch (Jortner and Levine, 1990): the intramolecular chemical bonds have far higher frequencies than the vibrations in the weak van der Waals wells. Energy redistribution is then the rate-determining step.

In another more chemical sense, the RRKM approach is still very relevant. There are many systems where the energy-rich molecules are not prepared in a very selective manner. Collisional excitation, as in the Lindemann mechanism, is an example where energy content is the dominant variable, as postulated in the RRKM approach. As our discussion of chaos in Chapter 5 noted, the less selective the preparation, the faster will details be erased. For the Lindemann mechanism, in the high-pressure limit, the energy-rich molecules have that distribution in energy that is expected at thermal equilibrium at the temperature of the buffer gas. When the collisional up and down pumping are fast compared to the rate of dissociation, the molecules have fully equilibrated. The dissociation rate constant is then just the RRKM energy resolved rate averaged over a thermal distribution of the excess energy as discussed in Section 6.2.2.1. Thinking about the high-pressure limit also explains why long-living collision complexes that are formed in bulk systems behave statistically. The initial complex may have been selectively formed in an isolated binary collision, but then it suffered so many collisions with the surrounding molecules that the complex is thermalized prior to its undergoing any dissociation or isomerization. The environment can very much enhance a statistical behavior.

As we turn attention to ever larger molecules we enter a regime where, unless the excess energy is high indeed, the RRKM rate of dissociation is exceedingly slow (this is because the density of states of the bound energy-rich molecule is so enormous). Under such circumstances other modes of dissociation, that we otherwise overlook because they are of secondary importance, can become dominant. One option that we have already mentioned is prompt dissociation. This occurs when the molecule does not sample states where the energy is spread over many modes. We discuss this aspect in the RRKM point of view where the states of the molecule are divided into two sets. The majority, that do not have enough energy in the reaction coordinate in order to dissociate, and the minority, that do and hence dissociate promptly. The dissociation rate constant is $\bar{\nu}P$ where $\bar{\nu}$ is a mean vibrational frequency and P is the probability of being in a promptly dissociating state. This assumes that energy redistribution over all possible states is complete prior to any onset of dissociation. The assumption can fail if the initial excitation also samples promptly dissociating states or states effectively coupled to them.[15] In a large molecule the fraction of promptly dissociating states is small, but the absolute number of such states increases with the size of the molecule. Prompt decay is thereby of increasing importance.

In Section 6.2.4 we discuss why understanding energy pathways in large molecules is a key issue with implications that extend beyond the question we discussed so far.

6.2.3   The reaction rate for a complex-forming collision

We are ready to compute the reaction rate for a reaction that proceeds via the formation of a long-lived intermediate. This process requires two or more barrier crossings. The trajectory needs to cross from the reactants' valley into the well and then it needs to cross from the well to the products' valley. When the trajectory rattles enough times between the two barriers, the reaction rate is simply the rate of crossing from the reactants into the well multiplied by the probability that the energy-rich molecule dissociates to products (because the first and second crossings are uncorrelated). Let k_p and k_r be the RRKM rates for dissociation of the complex to products and reactants, respectively. The rate of the initial crossing is given by transition state theory, whereas the probability of dissociating to products is the ratio of k_p to the total rate of dissociation, $k_p + k_r$. Using Eqs. (6.16) and (6.17),

$$k(E) = \frac{N^{\ddagger}(E - E_{0i})}{h\rho(E)} \frac{N^{\ddagger}(E - E_{0f})}{\left(N^{\ddagger}(E - E_{0i}) + N^{\ddagger}(E - E_{0f})\right)} \tag{6.20}$$

Here $\rho(E)$ is the density of states of the reactants before the collision and the entrance and exit barrier heights are designated i and f, respectively.

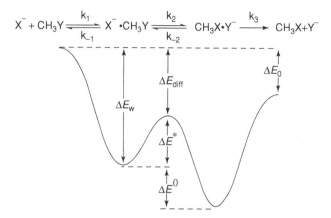

Figure 6.4 Schematic of an energy profile along the reaction coordinate for an exoergic ion–molecule reaction. ΔE_0 is the exoergicity of the reaction. ΔE_w is the well depth on the reactants' side. When the wells on the reactants' and products' side are deep enough, transition state theory can be used to compute the rate of crossing of the central barrier, as if it is an isomerization with an exoergicity ΔE^0 and a barrier height ΔE^*.

6.2.3.1 A case study: ion–molecule reactions

Ions are extensively used as reagents and as catalysts because they are highly reactive. In solution, ions have a strong interaction with the solvent and, as we discuss in Chapter 11, this behavior leads to essential modifications of the dynamics. Therefore, to study their intrinsic behavior we examine here ion–molecule reactions in the gas phase, continuing from the discussion in Section 3.2.6. Specifically, we consider nucleophilic displacement reactions[16] of the type $X^- + RY \rightarrow XR + Y^-$. Such gas-phase S_N2 ion–molecule reactions proceed with reaction rate constants that vary from almost capture-controlled to so slow that they can barely be detected. Moreover, the rate constant exhibits an inverse temperature dependence, that is, the rate of nucleophilic displacement slows down with increasing temperature. This behavior is in marked contrast to the predictions of the Arrhenius equation.

The potential energy profile for $R=CH_3$ is shown in Figure 6.4. It is a double-well potential. The reactants are captured in a physical polarization well as discussed in Section 2.1.9. To chemically rearrange and reach the polarization well of the products, the system needs to cross a barrier. Transition state theory can be used to compute the rate constant for this crossing. The structure of the transition state differs from that of the complex by the loss of rotation of the reactants with respect to one another. We emphasize again the difference between the complex, that is an entity that will be stable if it is drained of its excess energy,[17] and the transition state configuration that has but a fleeting existence. Because ion–molecule reactions can have observed rate constants that are very fast – being essentially determined by the rate of capture into the complex – the barrier height

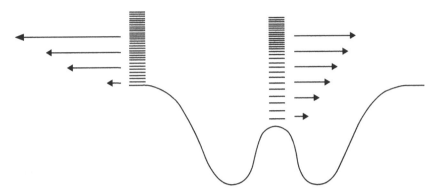

Figure 6.5 Modeling potential energy surfaces using RRKM rate theory, Eq. (6.20). The lengths of the arrows are proportional to the density of states. As the energy increases, the density of states in the direction of dissociation back to reactants becomes higher than the density of states at the barrier separating the reactants' and products' wells.

can be below the energy of the reactants, as shown in Figure 6.4. In the general case we expect that at a given energy above the entrance asymptote, the density of energy levels at the entrance to the complex is lower than that of the barrier region, see Figure 6.5. Because the complex becomes more loosely bound as the energy increases, the densities of energy levels of the two ladders approach one another. Dissociation of the complex to reactants is therefore increasingly more likely than to products, and the rate of the displacement reaction decreases. This behavior is often observed for thermal ion–molecule reactions.

For the asymmetric ion–molecule substitution reaction, $Cl^- + CH_3Br \rightarrow ClCH_3 + Br^-$, trajectory calculations[18] show that the lifetime of the initially formed ion-dipole stabilized adduct $Cl^- \cdot CH_3Br$ is too short to permit randomization of the internal energy so that the internal state distribution is not in equilibrium. Consequently, this reaction is predicted to deviate from RRKM theory. In the RRKM limit, one expects that the reaction rate depends only on the total energy and hence is independent of the initial state of the reactants.[19] The other extreme is marked sensitivity to the state preparation of the reagents. Not surprisingly, most reactions fall somewhere between these limits, with the tendency for statistical behavior to increase markedly with the size of the system. Again as expected, if the well depth of the adduct is deep in the entrance channel, then statistical behavior is expected to ensue because the adduct lives long enough for energy to be efficiently redistributed prior to crossing the transition state. But reaction time scales on the order of a few tens of picoseconds or less are too short to permit efficient energy transfer because they are comparable to or shorter than periods of molecular motions. We can not then use RRKM theory. We shall also see that solvation of the reactants reduces or altogether eliminates the well in front of the barrier so that transition state theory itself can be a suitable tool for computing the rate of the reaction in solution.

6.2.4 Toward molecular machines

In a *molecular machine* the atomic length scale forces that act during a chemical reaction are harnessed to deliver a motion at another region of the molecule. Two examples from nature that we will just touch upon are the *motor proteins* and the active and receptor sites of *enzymes*. We discuss it here because an essential part of the problem is the very many, seemingly superfluous, degrees of freedom of biological molecules.[20] The point of view of Section 6.2 is that in an energy-rich large molecule energy is first equilibrated amongst many degrees of freedom. Yet in biochemistry it is clearly possible for chemical energy produced at one site to be channeled to another site. Distal kinetics, meaning the transfer of localized potential and kinetic energy to a relevant distant functional site, suggests that it is worthwhile to look for effective and selective pathways for *energy transduction*.[21,22] Proteins and other biopolymers differ, however, from strongly bound and rigid molecules. Small conformational changes at one site can affect the entire molecule. The coordinates that describe the deformation are collective and not local and their potential energy landscape is replete with shallow minima. The energies that drive biological function (host–guest interactions) are unusually low, often being non-covalent. The strongest are when charge is involved and *charge-directed action* is sometimes used to explain why nature chose phosphate.* Myosin,[23] is a protein that walks in the familiar hand-over-hand fashion on actin with a step size of about 37 nm.** It does so for many consecutive steps while carrying a cargo on its head. The distinction between forward and backward is made possible because the feet of myosin and also the actin are asymmetric. At low ATP (adenosine triphosphate) concentration both feet are bound by ADP (adenosine diphosphate) to special sites on the actin filament. The front leg is strained. At higher concentrations of ATP, ADP is released from the back leg and ATP binds. ATP cannot bind to the front leg because the strain closes the binding site. When the ATP is bound to the rear leg it hydrolyzes to ADP and this frees the back leg and the strain in the front leg makes the back leg move forward to the next site on the actin. There the phosphate from ATP is released.

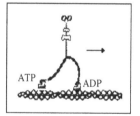

To provide a caricature of a forward-moving process, Figure 6.6 is a sketch of the *ratchet potential* model. It shows how a combination of thermal motion and weak binding can generate a forward motion. The coordinate is the position of

---

* Which can exist in several charge states.
** The distance can be measured by putting a fluorescent probe high up on one of the legs of the protein [A. Yildiz *et al.*, *Science* **300**, 2061 (2003)].

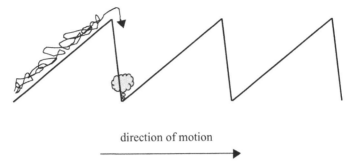

direction of motion

Figure 6.6 The rachet potential model for forward motion (energy of the center of mass of the protein vs. its lateral position). The protein moves up the barrier by slowly gaining energy from the fluctuating thermal motion, shown as a random walk in energy content. Once it reaches the peak the protein falls to the bottom of the (shallow) well where it is weakly bound, shown in gray. The release of the binding repeats the process. The methodology for determining a one-dimensional potential in a system with very many degrees of freedom is discussed in Section 11.2.1.

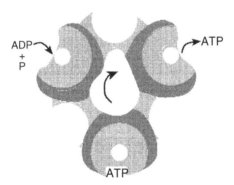

Figure 6.7 The part of ATP synthase that is out of the membrane and rotates to form ATP from ADP and free phosphate. Rotation is driven by a proton gradient in the membrane-bound part of the enzyme [adapted from Boyer (1999)].

the center of mass of the protein. A force due to the environment is acting and its net effect is to move the protein forward. The ordinate is the energy of the center of mass.

Another well-studied molecular machine is the *ATP synthase motor*.[24] It catalyses the formation of ATP from ADP and inorganic phosphate in a process that uses energy from oxidation to create a gradient of protons across a membrane. As the protons move across, protonation and deprotonation of a carboxyl group results in conformational changes in the membrane-embedded part of the enzyme. This drives a rotational motion. Figure 6.7 shows the part of the enzyme that is out of the membrane, where the rotation of the motor in one direction catalyses the formation of ATP while the rotation in the other direction hydrolyzes ATP. A rotation of 120° changes the unit that binds ADP and P to a form that strongly

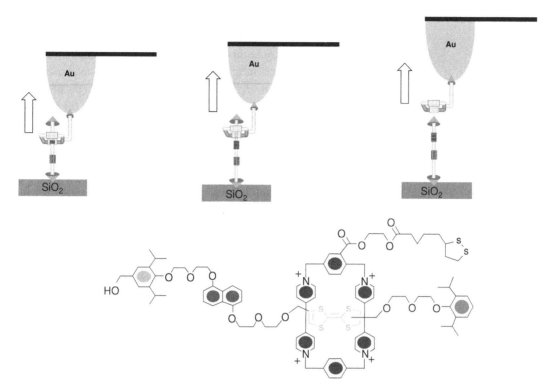

Figure 6.8 The atomic force microscope (top) measures using the gold tip the force necessary to move the dotted pattern wheel along the molecular axle and to extrude (de-slip) it through the stopper at the upper end. The other end of the axle is tethered to the solid support. The vertical and horizontal hatching regions of the axle are used to control the motion of the axle through the wheel. The "axle" and the "wheel" and its "arm" are shown at the bottom. [Figure courtesy of B. Northrop, K. N. Houk, and J. F. Stoddart (2003).]

binds ATP. The unit with tightly bound ATP then changes to a form that releases ATP and the third unit prepares to bind ADP and P.

Progress in understanding molecular machines owes much to the ability to probe single molecules[25] and the ability, Section 11.2.1, to construct suitable potential energy profiles. An example of such a mechanical probing of the force is shown in Figure 6.8. The molecule is a rotaxane* where a bipyridinium unit is encircled by an oxyarene-based (charged) ring with an arm that can attach to the gold tip. The point of the experiment is to probe two motions that are relevant. First, to prepare the interlocked molecule, the temperature is raised so that the ring dilates and it can slip over the "stopper" at the end of the backbone. Once

---

* Latin origin. Rota meaning wheel and axis meaning axle. The wheel is mechanically bound to the axle because the stoppers prevent its extrusion. The wheel can shuttle between two positions along the axle.

mechanically bound, the ring can be made to slide along between different stations (shown in vertical hatching and dotted pattern in Figure 6.8) along the chain. The molecule is a machine because this linear motion can be activated by chemical, electrical, or photochemical means.[26] Computations[27] show the hindrance to sliding is primarily through π–π interactions with a barrier height of about 17 kcal mol^{-1}. Extrusion through a stopper has a far higher barrier, of the order of 70 kcal mol^{-1}. The computations also reproduce the force on the ring as measured by the atomic force microscope (AFM) shown schematically in Figure 6.8.

With the understanding of the forces, simulations of the motion of molecular machines are beginning. These suggest that distant conformational changes are gradually driving the active site through a series of local shallow minima.[28] Much more is forthcoming.[29]

6.3 Resolving final states and populations

Our purpose is to characterize the population distribution of the final states after the collision. We begin with a highly-resolved distribution, the so-called *flux–velocity contour map*, which corresponds to identifying the outcome of the collision both by the state of the relative translation of the products and by the internal energy state.[30] We use the conservation of energy to show that the velocity vector of the separating products specifies this final outcome. The magnitude of the exit velocity reflects how the total energy is partitioned between the internal energy of the products and their translational motion, while the direction of the exit velocity tells us the scattering angle. In Chapter 7 we resolve final states not by measuring the relative velocity vector but by spectroscopic means. The same ideas remain applicable because given sufficient spectral resolution it is possible to determine not only the state of the molecule but also the direction of motion of its center of mass with respect to the laser beam in the laboratory frame. In this section we use the concept of resolving the outcomes in terms of final velocity vectors to characterize the reaction in the context of the products' structure.

6.3.1 Scattering in velocity space: the Newton sphere

Scattering is the correlation of the velocity after the collision $\mathbf{v}'$ with the initial velocity $\mathbf{v}$. For elastic scattering the magnitude of the velocity does not change, only its direction does. Not so for molecular scattering where, as a result of the collision, the internal energy of the molecule(s) can change.

First and foremost, we need to know the total energy. We define the zero for the energy scale by measuring the energy from the ground state of the reactants. Suppose that we know the initial internal energy of the reactants, E_I, and the initial relative translational energy, E_T. Suppose also that the reaction is endoergic with a known standard reaction energy change ΔE_0, i.e., the usual zero-point to zero-point energy change of reaction (cf. Figure 1.1). Conservation of the total energy

E dictates the energy of the products and thereby imposes a relation between the products' internal energy E'_I and their relative translational energy E'_T:

$$E = E'_I + E'_T = E_I + E_T - \Delta E_0 \tag{6.21}$$

Thus, if we can measure the products' relative velocity we know the final relative translational energy and therefore their internal energy E'_I. The total energy is equally known in a photodissociation experiment starting from a molecule in a given internal state of energy E_I. In the equation for conservation of energy, the energy $h\nu$ brought in by the photon replaces the initial translational energy and ΔE_0, the bond-breaking energy, is the energy difference between the ground states of the dissociation products and the parent molecule:

$$E = E'_I + E'_T = E_I + h\nu - \Delta E_0 \tag{6.22}$$

To each internal state of the products there corresponds a velocity whose magnitude is determined from conservation of energy and whose direction is to be measured. Both experiment and theory can provide such resolution, as shown in Figure 5.12. But this requires a separate plot for each final quantum state. Instead, we represent this information on a single plot in such a way that the major dynamical features of the reaction are immediately evident. We do so in the form of a product flux contour map, a map showing the distribution of the final velocity vectors. We develop the concept of a Newton sphere by means of an example. Consider the elementary reaction

$$K + I_2 \rightarrow KI(v', j') + I$$

Each particular rovibrational state of KI has its own internal energy E'_I and hence corresponds to a different final relative translational energy E'_T. But $E'_T = E - E'_I$ and the final relative velocity corresponding to a particular final internal state is therefore known,* $v' = (2E'_T/\mu')^{1/2}$. Here $\mu' = m_{KI}m_I/M$ is the reduced mass of the products where $M = m_K + m_I + m_I$ is the total mass (Figure 6.9).

Next we consider the product distribution in velocity space. A sphere of radius u'_{KI} with the center of mass as the origin is the locus of the "tips" of the velocity vectors of all those KI molecules formed in a particular internal state (and hence given E'_I), for all possible angles of scattering. Different KI internal states would each correspond to a sphere of different radius in velocity space. The one with the

---

* Because measurements are done in the laboratory, we must relate the measured velocity of, say, the KI molecule in the lab system, v'_{KI}, with the desired final relative velocity in the center-of-mass system, v', see Figure 6.9. This follows our discussion in Section *2.2.7, Eq. (2.39') in particular. To repeat the essence, first of all note that the c.m. final relative velocity is the difference $\mathbf{v}' = \mathbf{u}'_{KI} - \mathbf{u}'_I$. Conservation of momentum requires that the velocities in the c.m. system are related by $m_{KI}\mathbf{u}'_{KI} + m_I\mathbf{u}'_I = 0$. Therefore, only one final product needs to be detected and $\mathbf{u}'_{KI} = \mathbf{v}'(m_I/M)$ and $mu'_I = v'(m_{KI}/M)$. The laboratory velocity of the KI is its velocity with respect to the c.m. plus the velocity of the c.m. itself (that is known from the initial velocities $m_K\mathbf{v}_K + m_{I_2}\mathbf{v}_{I_2} = M\mathbf{v}_{c.m.}$): $\mathbf{v}'_{KI} = \mathbf{u}'_{KI} + \mathbf{v}_{c.m.}$.

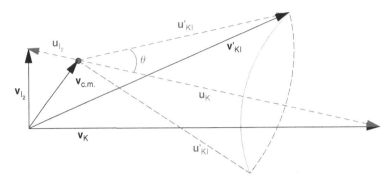

Figure 6.9 Velocity vector diagram (Newton diagram, Section 2.2.7.2) showing the relation between the lab velocity of the product KI, $\mathbf{v}'_{KI}$, and its recoil velocity with respect to the center of mass, $\mathbf{u}'_{KI}$. The initial relative velocity $\mathbf{v} = \mathbf{v}_K - \mathbf{v}_{I_2} = \mathbf{u}_K - \mathbf{u}_{I_2}$ forms the hypotenuse of the Newton triangle. The center of mass is as indicated. $\mathbf{u}'_{KI}$ is the recoil velocity of the KI scattered (shown here in the plane defined by the initial velocities) at a center-of-mass (c.m.) angle θ with a given E'_T. The cone of isointensity (given θ and variable ϕ), see Figures 4.6 and 6.10, is indicated by the dotted circle. All lab velocities $\mathbf{v}$ are solid lines, center-of-mass velocities $\mathbf{u}$ are dashed.

largest radius represents those KI molecules with the maximum recoil velocity allowed by energy conservation, i.e., $(E'_T)_{max} = E$, so those would be ground-state KI molecules. All other spheres represent internally excited KI products. The smaller the radius the larger is the internal energy.

We can regard the spheres as providing a spherical polar coordinate representation of the final-state angular distribution. The radius (fixed for given E'_T) is the recoil velocity and the polar angle θ is the angle of scattering (with respect to $\mathbf{v}$) in the c.m. system. This is the *Newton sphere*. In the absence of external fields the scattering is cylindrically symmetrical about $\mathbf{v}$, so that there is no ϕ-dependence and a given θ gives rise to a cone of isointense scattering.

Taking advantage of the cylindrical symmetry in the absence of fields (i.e., for randomly oriented reactants) allows the representation of the scattering in the plane defined by the velocities of the reactants in the center of mass, Figure 6.9. The origin is taken as the c.m. with the positive x axis along the direction of the velocity $\mathbf{u}_K$ of the K atom. All product KI scattered at a given θ in the c.m. will appear on a given "ray" from the origin while all those with a given E'_T (and thus a given E'_I) would be on a given circle.

For most diatomics, the rotational energy spacings are so small (compared to the energy resolution), i.e., levels so closely spaced, that the final translational energy (and thus $\mathbf{u}'_{KI}$) can be treated as a continuous variable. The circles corresponding to all the possible E'_I and thus E'_T are so very closely spaced they are practically a continuum. Thus, instead of trying to determine the differential solid angle cross-section into a definite internal energy state E'_I, we are satisfied to work with a continuous distribution of product energy states.

6.3.1.1 Flux–velocity maps: qualitative aspects

Figure 6.10 shows a contour map for KI from the $K + I_2$ reaction. The zero of θ is taken in the direction of the initial relative velocity $\mathbf{u}_K$ of the atomic reactant. It is seen from the contours that the KI molecules are scattered predominantly in the forward direction and that the most probable recoil velocity is about 0.5 of

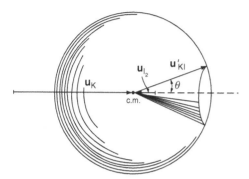

Figure 6.10 Sphere corresponding to locus of recoil velocities $\mathbf{u}'_{KI}$ of scattered KI from the $K + I_2$ reaction for a given translational energy release (constant E'_T and thus $\mathbf{u}'_{KI}$). The vectors $\mathbf{u}_K$ and $\mathbf{u}_{I_2}$ are the initial relative velocities of the reactants with respect to the center of mass (c.m.). The cone corresponds to KI products that recoil at an angle θ with respect to the incident relative velocity of K. The recoiling partner, I, which in the c.m. is always directed opposite to the KI, is not shown. Below we show the projection of the sphere onto a plane, taking advantage of the cylindrical symmetry about the relative velocity axis:

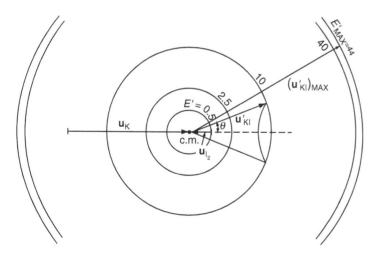

The maximal available energy for the KI + I, at a collision energy of $2.7\,\mathrm{kcal\,mol^{-1}}$, is $44\,\mathrm{kcal\,mol^{-1}}$ and this determines the radius of the outer circle. The upper panel corresponds to a final translational energy of $10\,\mathrm{kcal\,mol^{-1}}$. The experimental results, qualitatively shown in Figure 4.12, are quantitatively summarized in the next panel:

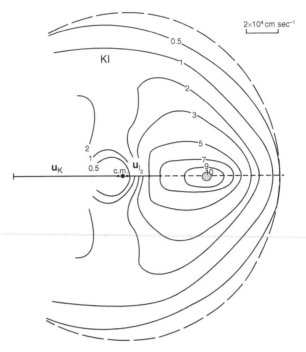

Figure 6.10 (*cont.*) Contour map: the flux (velocity–angle) distribution for the KI product of the K + I$_2$ reaction. The initial velocities are shown in the center-of-mass system where the velocity of the heavy I$_2$ is necessarily small compared to that of K. All KI molecules with the same final velocity $\mathbf{u}'_{KI}$ will lie on a circle centered at the c.m. (•). The dashed circle corresponds to the highest recoil velocity allowed by conservation of energy, and hence to internally cold KI molecules. The closer the point is to the c.m., the higher is the internal energy of the KI product. KI scattered at a given direction θ with respect to the direction of the incident K atom will appear on a given ray (from the c.m.). The maximal scattering is into the contour (arbitrarily) labeled 10. Contrast with the location of the same contour in Figure 6.11. [Adapted from K. T. Gillen, A. M. Rulis, and R. B. Bernstein, *J. Chem. Phys.* **54**, 2831 (1971).]

its maximal value. The recoil translational energy is thus about 0.25 of the total energy. Hence most of the product KI molecules must be significantly internally excited. These findings are in accord with our expectations based on the harpoon mechanism, Section 3.2.4.

We examine another KI-forming reaction:

$$K + ICH_3 \rightarrow KI + CH_3$$

Figure 6.11 displays the contour map, which shows predominantly backward scattering of the KI (a rebound reaction, Section 4.4.3, with a repulsive, or late, energy release).

Figure 6.12 is a map for DI from the reaction

$$D + I_2 \rightarrow DI + I$$

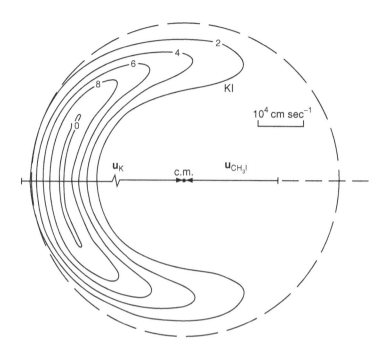

Figure 6.11 Contours of the flux (velocity–angle) distribution for the KI product of the K + CH_3I reaction. Compare with Figure 6.9 where the KI is primarily forward scattered but with a much lower recoil velocity. [Adapted from A. M. Rulis and R. B. Bernstein, *J. Chem. Phys.* **57**, 5497 (1972).]

showing side-scattering of the DI product. Qualitatively similar results are obtained when I_2 is replaced by IBr or ICl. Such a distribution is to be expected on the basis of the electronic considerations of Section 5.1.5. During the approach of the light atom the two heavy atoms hardly move.* We can thus regard the D + IX reaction as a direct photodissociation of IX (with the light atom playing the role of the photon), in which case the heavy atoms will recede along the direction of their bond axis.[31] Owing to the small mass of the D atom, the DI behaves like an I atom and takes off in the direction of the axis of IX at the moment of the bond rupture. The direction of recoil of the DI product with respect to the direction of the incident atom tells us the angle of approach of the D atom for those collisions that favor reaction. Thus Figure 6.12 supports a preferentially "sideways" attack of the D atom on the IX molecule, as discussed in Section 5.1.5.1.

Crossed-beam experiments naturally produce flux velocity–angle contour maps, which can be measured with considerable detail. Applications[32] include a variety of atom–diatom reactions, ion–molecule reactions, complex mode reactions, diatom–diatom reactions, etc. Examples are to be found throughout the text. For the special case when the reaction is photoinitiated we return to this

---

* Quantitatively, this follows from the use of mass-scaled coordinates, Appendix 5.B. For a light atom attack the entrance valley is quite compressed compared with the products' valley.

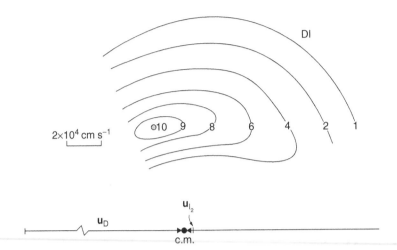

Figure 6.12 Flux (velocity–angle) contour map (upper half only) for the DI product of the D + I₂ reaction, for $E_T \cong 9\,\mathrm{kcal\,mol^{-1}}$. See also Problem P. The extensive "sideward" scattering has been interpreted in terms of a preferred bent geometry for the transition state. This behavior is consistent with molecular orbital considerations (Section 5.1.5.1). [Adapted from J. D. McDonald *et al.*, *J. Chem. Phys.* **56**, 769 (1972).]

topic in Chapter 7. The maps can also be used to display the results of inelastic (non-reactive) scattering experiments. For example the contour of the observed maximum intensity of NO^+ scattered by He:

$$NO^+ + He \rightarrow NO^+(v') + He$$

In the forward direction the contour coalesces with the "elastic" $Q = E_T' - E_T = 0$ circle (i.e., elastic scattering of NO^+). However, there is a significant energy transfer for backward scattering.[33] In Chapter 9 we will have more to say on why this behavior is to be expected.

6.B   Appendix: The quantitative representation of flux contour maps

Returning to our KI example, consider the total number of molecules scattered per unit time into a solid angle $d^2\omega$ with final translational energy in the range E_T' to $E_T' + dE_T'$:

$$d\dot{n}_{KI}(\theta, E_T') = \frac{d^3\sigma(\theta, E_T')}{d^2\omega\,dE_T'}d^2\omega I_K n_{I_2} \tag{B.6.1}$$

As previously, e.g., Eq. (3.5), we define the cross-section through the rate. Here, as in Eq. (4.48), it is the rate of reactive collisions with products scattered into a narrow range in both energy and direction. I_K is the flux of K atoms and n_{I_2} is the number of I₂ molecules. We write the element of a solid angle $d^2\omega$ as a reminder that it is a double differential, $d^2\omega = d\cos\theta\,d\phi$.

The differential cross-section defined by (B.6.1) is related to the state-resolved differential cross-section as follows. First we go over from discrete final internal state indices to a continuous translational energy variable

$$\frac{d^3\sigma(\theta, E_T')}{d^2\omega\, dE_T'} = \sum_{v',j'} \frac{d^2\sigma(vj \to v'j'; \theta)}{d^2\omega}\delta(E - E_T' - E_I') \qquad (B.6.2)$$

The summation is over all those final states that correspond to a final kinetic energy E_T' in the range E_T', $E_T' + dE_T'$. Of course, dE_T' needs to be a small but finite interval for the equation to make physical sense: unless the total energy is defined more precisely than the spacing of rotational states, the discrete quantal nature of the function is smeared out. The velocity analysis experiment itself is usually reported in terms of the scattering with recoil velocity in the corresponding range u' to $u' + du'$ (where $E_T' \propto u'^2$). The corresponding cross-section is

$$\frac{d^3\sigma(\theta, u')}{d^2\omega\, du'} = \frac{d^3\sigma(\theta, E_T')}{d^2\omega\, dE_T'}\left(\frac{dE_T'}{du'}\right) \qquad (B.6.3)$$

The usual flux velocity–angle contour map is of this quantity in a (θ, u') polar coordinate system. Experimentally, it is easier to determine the relative values of the cross-section for different velocities. The results are then presented as a distribution $P(\theta, u') \propto \sigma(\theta, u')$ that can be normalized by integration over all scattering angles and velocities, $\int du' \int d^2\omega'\, P(\theta, u') = 1$.

Cartesian contour maps represent the *velocity-space* distribution, $d^3\sigma(\theta, w')/du'$, where $d\mathbf{u}' = (u')^2\, du' d^2\omega$. Cartesian maps are invariant under a transformation to the laboratory system of coordinates

$$d^3\sigma(\theta,\, u')/d\mathbf{u}' = d^3\sigma(\theta,\, v')/d\mathbf{v}' \qquad (B.6.4)$$

This invariance is invoked in the transformation between the laboratory and the c.m. system. The invariance of the velocity-space distribution to a change in coordinate system is also one of the arguments for the use of this flux map in Section 6.4.

*6.B.1   Reduced distributions: translational energy release and angular distribution

There is a wealth of information in the full flux map. Reduced descriptions are obtained by summing over either the scattering angle or the velocity of the products. For example, integration over all angles yields the cross-section for scattering into a given narrow range of final translational energy,

$$\frac{d\sigma}{dE_T'} = \iint \frac{d^3\sigma(\theta, E_T')}{d^2\omega\, dE_T'} d^2\omega = \iint \frac{d^3\sigma}{d^2\omega\, dE_T'} \sin\theta\, d\theta\, d\varphi$$

$$= 2\pi \int_0^\pi \frac{d^2\sigma}{d\theta\, dE_T'} d\theta \qquad (B.6.5)$$

This is the "continuum" analog of the state-to-state total cross-section,

$$\frac{d\sigma}{dE_T'} = \sum_{v'j'} \sigma(vj \to v'j')\delta(E - E_T' - E_I') \tag{B.6.6}$$

The final integration, over E_T', yields the reaction cross-section for given reactants,

$$\sigma_R = \int_0^E \frac{d\sigma}{dE_T'} dE_T' \equiv \sum_{v'j'} \sigma(vj \to v'j') \tag{B.6.7}$$

When measurements are made from different initial conditions, one can determine the dependence of σ_R on the states of the reactants and the energy requirements of the reaction.

When we can determine only the relative values of the cross-section the result of integration over all scattering angles is the distribution of final translational energy $P(E_T') \propto d\sigma/dE_T'$. This is the continuum analog of the cross-section into final internal states whose energy matches E_T', $E_I' = E - E_T'$. It is normalized to unity by integrating over all final energies,

$$P(E_T') = \int d^2\omega P(\omega, E_T'), \qquad \int_0^E dE_T' P(E_T') = 1 \tag{B.6.8}$$

This is often known as KERD (kinetic energy release distribution).

An alternative route to the cross-section is to integrate first over the distribution of E_T'

$$\frac{d^2\sigma}{d^2\omega} = \int_0^E dE_T' \frac{d^3\sigma}{d^2\omega\, dE_T'} \equiv \sum_{v'j'} \frac{d^2\sigma(vj \to v'j')}{d^2\omega} \tag{B.6.9}$$

Equation (B.6.9) can be converted to the differential polar form, i.e., integrating over the azimuthal angle ϕ to take into account all the "out-of-plane" scattering, at given θ,

$$\frac{d\sigma}{d\theta} = \sin\theta \int d\varphi \frac{d^2\sigma}{d^2\omega} \tag{B.6.10}$$

such that

$$\sigma_R = \int_0^\pi \frac{d\sigma}{d\theta} d\theta' \tag{B.6.11}$$

The differential polar form, $d\sigma/d\theta$, is the fractional contribution to σ_R from any final polar angle θ. It differs from $d^2\sigma/d^2\omega$ by the integration over the azimuthal angle ϕ as in (B.6.10) and by the factor $\sin\theta$. The different character of the reactions of the halogens and the alkyl halides with the alkali metals is also evident in the shape of the polar differential cross-section, Figure 4.14.

6.4 Characterization of energy disposal and energy requirements of chemical reactions

Our purpose is to characterize the population distribution of the final states after the collision. We begin with the *prior distribution*, a method where simplicity is achieved by failing to properly conserve angular momentum. Next, the very same considerations are applied but with conservation followed by a refinement of the description, where the rotational angular momentum of the reactants (or products) is explicitly treated, which brings us to the level of the phase-space theory.

The prior distribution does not seek to match experimental results. Rather, it provides a reference against which to compare. The (logarithmic) deviation of the actual (observed or computed) distribution from the prior one is known as the *surprisal*. Then, armed with detailed balance, we will also seek to characterize the energy requirements of chemical reactions. Consistent with the arguments already presented, the measure of specificity of energy disposal in the forward reaction turns out to be equal to the measure of selectivity of energy requirement in the reverse reaction.

6.4.1 The prior distribution

Reactions differ in the distribution of final states because of different dynamics. If there are no dynamical constraints we take it that the flux into any final quantum state is the same. In other words, in the absence of a dynamical bias and at a given energy all final quantum states are formed with the same rate. This assumption provides an immediate relation between the rate of formation and the structure of the products. To compare the rate into different possible groups of final states what we need to do is to count how many final quantum states are in each group. The relative rates are the ratio of the two numbers. In a perfectly resolved experiment there is no grouping of quantum states. Experiments seldom fully resolve the quantum states of the products. Consequently, as a practical matter, it is useful to have working results for the more common ways of grouping together final states.

We begin with a vibrational state distribution of the reaction product as measured by the chemiluminescence experiment of Section 1.2.1. This is an atom–diatom reaction at a total energy E and all final quantum states where the product diatomic molecule is in a given vibrational state v are grouped together. We show below that the number of products' quantum states that belong to a particular group is proportional to $(E - E_v)^{3/2}$, where E_v is the vibrational energy of the diatomic in the given vibrational state. This distribution is not uniform in the final vibrational energy E_v. Rather, this distribution strongly decreases as $E_v \to E$, as seen in Figure 6.13 for the HF product of the exoergic $F + H_2 \to HF + H$ reaction. The prior distribution is not uniform because different groups can have a different number of quantum states as members. In the case of the group specified by a product diatomic in a given vibrational state, we can easily visualize why. At a given total energy, the more energy is in the vibration the less energy remains for the other degrees of freedom. So fewer other (e.g., rotation of the diatomic product) quantum states are energetically accessible.

Because the prior distribution is not uniform, the characterization of a dynamical bias is a distribution of final states that is deviant from the prior distribution. This, by itself, is already an insightful point. Because the total energy is conserved, when we specify, say, the vibrational state of the product we at the same time implicitly specify limitations on what other states are accessible. This is particularly the case for the vibration because it can take up large chunks of the total energy, but it is equally true if we were to specify, say, a rotation. There are $2j + 1$ quantum states of a given j, but in the absence of a dynamical bias the rotational state distribution will not quite go as $2j + 1$ because of the implications of the conservation of energy.

For reasons discussed in Section 6.4.2 the deviation from the prior distribution is best represented in a logarithmic form as shown in the right panel of Figure 6.13. This logarithmic measure of deviation is known as the *surprisal*. The experimentally observed population distribution is due to the combined role of the dynamics and the statistics. The surprisal factors out the role of the statistics from the observed result. What is shown in Figure 6.13 is that the surprisal monotonically, in fact, linearly, varies as we go up in energy. Why the surprisal can have such a simple form, a linear dependence on E_v, is briefly discussed in Section 6.4.2. In this section the key concept is that of the prior distribution. The point is to examine the role of grouping of quantum states that is the result of an experiment that provides only a partial resolution. Even when the role of the dynamics is fully averaged out, such an experiment does not yield a uniform population distribution. The population of a group can be higher simply because the group has more final quantum states assigned to it. The prior distribution is the distribution when all energetically allowed final quantum states are populated with the same rate. Computing the prior distribution is then the counting of quantum states at a given total energy and grouping them according to the resolution implied by the experiment.

In Section 6.4.1.1 we count the quantum states so as to derive the prior distribution for flux contour maps. Measurement of a flux map provides a relatively high resolution but it too implies grouping of final states: all internal states of the products with energies in a narrow interval are counted together. The complementary experiment is to measure the internal state population of the nascent products. The prior distribution for this case is obtained in Section 6.4.1.2.

*6.4.1.1   The prior flux distribution

The essential technical tool is the result, from Chapter 5, that the density of translational quantum states in the three-dimensional space of the experiment is $d\mathbf{p}/h^3$,

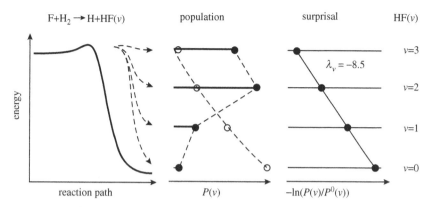

Figure 6.13 Surprisal analysis of HF vibrational excitation in the $F + H_2 \rightarrow H + HF$ reaction. The ordinate in this figure is energy. The energies of HF vibrational states are shown on the right. Because of anharmonicity they are not quite equally spaced. Left panel: energetics. Shown is the energy release along the reaction coordinate. The exoergic reaction can populate up to the $v = 3$ state of HF. Middle panel: the observed distribution of HF vibrations, $P(v)$, solid bars [adapted from M. J. Berry, *J. Chem. Phys.* **59**, 6229 (1973)], and the normalized prior distribution, $P^0(v) = (E - E_v)^{3/2}/\sum_v (E - E_v)^{3/2}$, open circles. On prior grounds it is preferable to populate the $v = 0$ state of HF because this choice leaves so much energy that can be distributed over the rotation of HF and the relative translation of HF and H, which means that many final quantum states can be energetically accessed. Experimentally, this behavior is not what is observed. The dynamics strongly favors channeling energy into HF vibration. Even so, the highest HF vibration is not the most populated. Right panel: the surprisal, the logarithmic measure of deviation, plotted vs. energy. It is seen to be linear in the HF vibrational energy [with a slope of -8.5 when plotted vs. E_v/E, see Eq. (6.35)]. In words, the **observed** HF vibrational state distribution is a **compromise** between the dynamics that favors channeling as much energy as possible into the HF vibration and the statistics (as determined by the structure) that discriminates against higher vibrational states because fewer quantum states are then possible.

where h is Planck's constant and $\mathbf{p}$ is the momentum vector. The density $d\mathbf{p}/h^3$ means that different intervals in momentum $\mathbf{p}, \mathbf{p} + d\mathbf{p}$ contain the same number of translational quantum states. As a result of the scattering the momentum of the relative motion of the products can vary in both magnitude and in direction, so that $d\mathbf{p} = p^2 dp \sin\theta\, d\theta\, d\phi$; θ and ϕ are the direction of the scattering and the flux map, Section 6.3, is the intensity of scattering into a given momentum interval. Still, measuring such a map should not, on prior grounds, yield a constant. The reason is that, by conservation of the total energy, different values of the relative momentum imply that the products have different internal energies. There can be different numbers of products' internal states in different energy intervals. The density of the internal state $\rho_I(E_I)$ was defined in Appendix 6.A to measure this variation. Because energy must be conserved, the number of products' internal quantum states in the interval where the translational energy is in the range E_T

to $E_T + dE_T$ is $\rho_I(E - E_T)dE_T$. All the states in that energy range have the same momentum so the (prior) flux into products' final states is $(p/\mu)\rho_I(E - E_T)dE_T$. But since $E_T = p^2/2\mu$, $p\, dE_T \propto p^2\, dp$ the flux into internal states of the products when the momentum is in the range $\mathbf{p}, \mathbf{p} + d\mathbf{p}$ is proportional to $\rho_I(E - E_T)d\mathbf{p}$. Using a superscript 0 to denote the distribution in the absence of any constraints apart from the conservation of energy,

$$(\mathrm{d}^3\sigma/\mathrm{d}\mathbf{p})^0 \propto \rho_I(E - E_T) \qquad (6.23)$$

Note that all final internal states with the same energy have, according to (6.23), the same flux. In Section 6.4.5 we also impose the conservation of angular momentum. This will restrict the range of final rotational states that need to be counted.

The density of translational states (per unit volume) is defined so as to enable us to go from a momentum to an energy scale. The definitions

$$\rho_T(E_T)\, \mathrm{d}E_T\, \mathrm{d}^2\omega \equiv \mathrm{d}\mathbf{p}/h^3 \qquad (6.24)$$

and $\mathrm{d}\mathbf{p} = p^2\, dp\, \mathrm{d}^2\omega = \mu p\, dE_T\, \mathrm{d}^2\omega$ give $\rho_T(E_T) = \mu p/h^3$. Using this conversion we can rewrite (6.23) in an alternative form,

$$(\mathrm{d}^3\sigma/\mathrm{d}E_T\, \mathrm{d}^2\omega)^0 = (\mathrm{d}^3\sigma/\mathrm{d}\mathbf{p})^0(\mathrm{d}\mathbf{p}/\mathrm{d}E_T\, \mathrm{d}^2\omega)$$
$$\propto \rho_T(E_T)\rho_I(E - E_T) \propto E_T^{1/2}\rho_I(E - E_T) \qquad (6.25)$$

This gives us the prior distribution for the usual way of expressing the flux map.

The densities of states that appear on the right of Eq. (6.25) originate from our choice of using an energy scale to display the distribution of the products. Consider an experiment that resolves the products' internal quantum states and determines their angular distribution as well (e.g., using a Doppler-based detection of the final states or another imaging technique, see Chapter 7; Figure 5.12 is an example). Under such circumstances, the flux from reactants in the energy range of E to $E + dE$ into products in a particular final quantum state is constant.

The prior distribution is not chosen to agree with experiment. It is instead a reference against which experimental (or computational) results are to be evaluated. We discuss such applications in Section 6.4.2. Then we go to measures of selectivity of energy requirements. In Section 6.4.5 we modify the prior distribution by taking into consideration that not only energy but also angular momentum need to be conserved. Explicit results for prior distributions are given in Section *6.4.1.2.

*6.4.1.2 Products' internal state distribution in the prior limit

The prior distribution populates any group of energetically allowed final states according to the number of quantum states in the group, each state with the same weight. To compare the (observed or computed) products' state distribution to the prior one it is necessary to count how many quantum states are in the group.[34] We do so here for the simple case of an atom–diatom reaction at a given total energy for the more common manners of grouping states. Detailed results at a given temperature and/or for products with more degrees of freedom are available elsewhere (Kinsey and Levine, 1979).

The most common grouping is one that integrates over all directions of the final velocity leading to populations of different (groups of) states. As already discussed this approach brings in the density (per unit volume) of translational states

$$\rho_T(E_T) = A_T E_T^{1/2} = A_T (E - E_I)^{1/2} \tag{6.26}$$

where A_T is a collection of constants that cancels out when we normalize the prior distribution to unity. Often, one also does not resolve the $2j + 1$ degeneracy of the rotational states of the diatom. So the number of final states when the diatom is in the (v, j) rovibrational state is

$$\rho(v, j; E) = (2j + 1)\rho_T(E_T) = A_T(2j + 1)(E - E_{v,j})^{1/2} \tag{6.27}$$

The total density of states at the energy E is the sum over all (v, j) rovibrational states that are allowed by conservation of energy*

$$\rho(E) = \sum_{v=0}^{v_{max}} \sum_{j=0}^{j_{max}(v)} \rho(v, j; E) = A_T \sum_{v=0}^{v_{max}} \sum_{j=0}^{j_{max}(v)} (2j + 1)(E - E_{v,j})^{1/2} \tag{6.28}$$

For convenience the summation is written first over all states j that are energetically allowed for a given v and then over all v. The partial sum over j only is the density $\rho(v; E)$ of vibrational states.

The (normalized) prior distributions are therefore

$$P^0(v, j) = \rho(v, j; E)/\rho(E)$$
$$P^0(v) = \rho(v; E)/\rho(E) = \sum_{j=0}^{j_{max}(v)} P^0(v, j) \tag{6.29}$$
$$P^0(E_T) = \rho(E_T; E)/\rho(E)$$

Explicit expressions for the prior distributions require the expressions for the energies of the rovibrational states. For most diatomics, it is sufficient to use the rigid rotor approximation $E_{v,j} = E_v + B_v j(j + 1)$, where B_v is the rotational constant in the vibrational manifold v. It is however necessary to recognize the anharmonicity of the vibrational levels because near the energy cutoff, the translational density of states $\propto (E - E_{v,j})^{1/2}$ varies quite rapidly.

---

* Another route to $\rho(E)$ is to sum first over all internal states that lead to products in a narrow translational energy range,

$$\rho(E_T; E) = \sum_{v=0}^{v_{max}} \sum_{j=0}^{j_{max}(v)} \rho(v, j; E)\delta(E - E_T - E_{v,j})$$

and then to integrate over E_T.

For many purposes it is useful to use reduced energy variables, which are defined by

$$f_v \equiv E_v/E, \qquad g_R \equiv E_j/(E - E_v) \tag{6.30}$$

Note that f_v and g_R classically span the range 0 to 1. Then, normalizing by integration over the reduced variables,[35,36]

$$
\begin{aligned}
P^0(f_v, g_R) &= \frac{15}{4}(1 - f_v)^{1/2}(1 - g_R)^{1/2} \\
P^0(f_v) &= \frac{5}{2}(1 - f_v)^{3/2} \\
P^0(g_R|f_v) &\equiv \frac{P^0(f_v, g_R)}{P^0(f_v)} = \frac{3}{2}(1 - g_R)^{1/2}/(1 - f_v) \\
P^0(f_T) &= \frac{15}{4}(f_T)^{1/2}(1 - f_T)
\end{aligned}
\tag{6.31}
$$

One feature emphasized by these results is that, on prior grounds, there is no isotope effect when the final state distribution is represented on an energy scale. For example, the DF product from the $F + D_2$ reaction has, on prior grounds, the very same vibrational energy distribution as HF. The experimental result is that this behavior is essentially so. The vibrational population distribution for HF and DF looks quite different when examined as a function of the vibrational quantum number. This is necessarily so because at a given total energy the range of the accessible final vibrational states is different. But the two distributions are very similar when the observed distribution is considered as a function of f_v. This is shown for the $Cl + H(D)I$ reaction in Figure 6.14. The reason is, as shown in Eq. (6.31), $P^0(f_v) = (5/2)(1 - f_v)^{3/2}$, that equal intervals of f_v contain the same number of final quantum states. There can be isotope effects due to the dynamics, e.g., an H atom will tunnel far more readily than a D atom. But in the example above as well as in many other cases the isotope effect largely disappears when you compare on an equal basis.

6.4.2   Surprisal analysis

The technical concept of *surprisal* arises when we try to put a quantitative measure on how surprised we are when a particular event occurs.[37] We begin by asking what would we expect such a measure to look like. An obvious requirement is that the more certain is the outcome, the less surprised we are by the event actually taking place. Hence, the surprisal I should be a monotonically decreasing function of the probability P of the outcome. To make the relation between I and P unique we add one more condition: for two independent and unrelated events a and b, if both are observed to occur, and provided that neither event changes our opinion about the other, the surprisal at observing both events should be

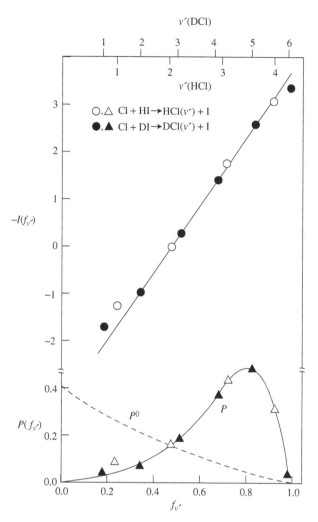

Figure 6.14 Surprisal of energy requirements of the endoergic I + HCl(v) → HI + Cl reaction and the corresponding results for I + DCl(v) as determined, using detailed balance, from the measured energy disposal in the reversed HI + Cl → I + HCl(v) exoergic reaction. Bottom panel: the observed (points) and the prior distribution for the vibrational energy disposal in the exoergic direction. Upper panel: the surprisal. For the energy disposal, left ordinate, the surprisal is $-\ln(P(v)/P^0(v))$. For the energy requirement, right ordinate, the surprisal is the [relative, see Eq. (6.40)] rate constant for I + H(D)Cl(v) → H(D)I + Cl when the fraction $f_v = E_v /E$ of the total energy is in the reactant vibration.

additive, $I(a\&b) = I(a) + I(b)$. The probability $P(a\&b)$ of the combined event is $P(a\&b) = P(a)P(b)$. These two conditions together suffice to prove that there is only one choice:

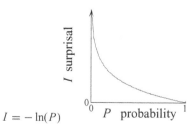

$I = -\ln(P)$

It is easier to verify the converse result: that our choice satisfies the two conditions.

To proceed we must be careful because our events are not elementary outcomes but instead are groups of outcomes. For example, when you toss two coins you are not surprised if the outcome one head, one tail is more probable than when both coins show heads. This is because the outcome one head, one tail can be realized in two ways, first coin head, second coin tail, or the other way around, whereas there is only one elementary event in the outcome both heads.[38] To apply this idea we need to specify what we consider to be the elementary outcomes of the collision. This we have already done. We take it that at a given total energy, all other things being equal, the flux into all final quantum states that are accessible is the same. So we need to divide the observed probability of a group of events by the number of events in the group. The normalized number of events is the probability when all elementary events are equally probable, and is given by the prior distribution. Hence:[39]

$$I(f) = -\ln(P(f)/P^0(f)) \tag{6.32}$$

where $P^0(f)$ is the prior probability of the final outcome f. If what is observed is the prior distribution, the surprisal is zero. Please note again the technical sense in which we are using the concept of being surprised. Whereas you might be astonished that the actual distribution contained no effects from the dynamics that led from the reactants to the products, technically you are not surprised in the least bit.

Figure 6.13 showed an example of applying Eq. (6.32) to the final vibrational state distribution in the $F + H_2$ reaction. The observed distribution is qualitatively different from the prior, as is only to be expected for a direct reaction. Yet the surprisal has a simple (linear) dependence on the fraction of the available energy that is in the vibration. The simplicity of a *linear surprisal* is not caused by the semilogarithmic nature of the plot. Figure 1.3 showed the actual rotational state populations of HD in several vibrational manifolds from the $H + D_2$ reaction. The fit to a linear surprisal is close.

This raises the question of why should the surprisal be a simple function. One answer is the following. For a statistical distribution of final states we take it that all final quantum states are equally probable. The practical version of this expectation is the prior distribution. Because of the dynamics the final quantum states may not be equally probable. What one then needs is a quantitative way to impose the inequality of states as enforced by the dynamics. How to do it?

If all quantum states are equally probable, then one can say that the distribution of final states has a maximal entropy. So far, the maximal value of the entropy is just a restatement of the result that the distribution is as uniform as possible. If the final states are not equally probable, we ask that the entropy of the distribution of states be as large as is allowed by the dynamical constraints. In other words, the working hypothesis is that all states are as probable as possible under the constraints imposed by the dynamics.

*6.4.2.1 The distribution of maximal entropy

The details of determining a distribution of *maximal entropy* are spelled out in a number of reviews.[40] Here we just give an example: suppose the *dynamic constraint* is the final vibrational energy. This means that the dynamics imposes on the distribution of final states a mean value of the final vibrational energy

$$\sum_f E_v(f)P(f) = \langle E_v \rangle \tag{6.33}$$

where $E_v(f)$ is the value of the vibrational energy in the final state f. Then the surprisal of the final state f has the functional form

$$I(f) = \lambda_0 + \lambda_v E_v(f) \quad \text{or} \quad P(f) = \exp(-\lambda_0 - \lambda_v E_v(f)) \tag{6.34}$$

This means that all final states that have a given value of the vibrational energy are equally probable.

Subject to the constraint Eq. (6.33), the surprisal is a linear function with two coefficients, λ_0 and λ_v. There are two coefficients because there are actually two constraints: the dynamical one, Eq. (6.33), and the condition that the distribution of final states is normalized $\sum_f P(f) = 1$. The coefficient λ_0 insures the normalization and when normalization is enforced it becomes a function[41] of λ_v. So a linear surprisal plot is specified by one coefficient. It can be determined from the data by plotting the surprisal, what is known as a *surprisal analysis*, or it can be determined theoretically, if the value of $\langle E_v \rangle$ is given, by regarding Eq. (6.33) as an implicit equation for λ_v.

If we group together those states that have the same energy for the diatomic product we obtain the distribution of vibrational states. Using Eq. (6.34)

$$P(v) = \sum_{\substack{\text{all states } f \text{ such that} \\ E_v(f) = E_v}} P(f) = P^0(v)\exp(-\lambda_0' - \lambda_v E_v) \tag{6.35}$$

The surprisal is the same for all the states in the sum. So the sum is the number of such states. The prior distribution is just this number, divided by a normalization (the total number of states) to render the number into a probability. If there is no dynamic constraint the numerical value of λ_v is zero and the distribution is the prior one. We reiterate that the prior distribution is not the same as a uniform distribution. Rather, it depends on how many products' quantum states fall into the group of states of interest. For example, in Figure 6.13 the prior vibrational distribution falls rapidly with increasing vibrational excitation and so looks thermal-like. Problem H shows that in the limit where the products have many atoms so that the fraction of energy in any particular vibrational mode is likely to be small, the prior distribution is exactly thermal.[42] But in an A + BC reaction there is only one vibrational mode in the products and so the correct form of the prior distribution is required.

It is not necessarily the case that there is only one constraint imposed by the dynamics. If there is more than one constraint, then the surprisal is a linear combination of terms.[43]

6.4.3 Measure of selectivity in energy requirements of chemical reactions

We have just discussed a measure of specificity of energy disposal of chemical reactions. Already in Chapter 1 we drew attention to the implication of microscopic reversibility that energy disposal in a forward reaction and energy requirements of the reversed reaction are one and the same. Here we show that both are characterized by the same surprisal.

For the pair of reactions

$$(A + B)_{\text{state } i} \xrightarrow{k(i \to f)} (C + D)_{\text{state } f}$$

$$(A + B)_{\text{state } i} \xleftarrow{k(f \to i)} (C + D)_{\text{state } f}$$

detailed balance is the statement that

$$[\text{reactants}(i)]_{\text{eq}} \, k(i \to f) = [\text{products}(f)]_{\text{eq}} \, k(f \to i) \tag{6.36}$$

The square brackets denote concentrations and the subscript eq refers to the concentrations at equilibrium. The indices i and f are states or groups of states. For the special case where the initial and final translational energies are sharply defined we prove that detailed balance in the form of Eq. (6.36) is equivalent to the version that can be found in books on scattering theory, namely

$$g_i \left(k_i^2 / \pi \right) \sigma(i \to f) = g_f \left(k_f^2 / \pi \right) \sigma(f \to i) \tag{6.37}$$

The most important point about this equation is that both sides are to be evaluated at the same total energy E. To insure a sharp value of the translational energy, i and f are indices of quantum levels and g is the degeneracy of the level. At a given total energy, the equilibrium condition is that all quantum states are equally probable. So the equilibrium concentrations are proportional to the number of quantum states. There are $g_i \rho_T(E_T)$, $E_T = E - E_i$, states of the reactants and similarly for the products. The reaction rate constants are related to the cross-section as $k(i \to f) = v_i \sigma(i \to f)$, where v_i is the relative velocity of the reactants and similarly for the products. Since $dE_T = v_i \, dp_i$ we have that $v_i \rho_T(E_T) = h^{-1} k_i^2 / (2\pi)^2$ where, as usual, $\rho = \hbar k$. For our immediate purpose, it is more convenient to write detailed balance in its chemical form, Eq. (6.36). At a given total energy we have from (6.37)

$$g_i \rho_T(E - E_i) k(i \to f) = g_f \rho_T(E - E_f) k(f \to i) \tag{6.38}$$

The probability $P(f)$ of the final states is defined as the relative rate into that level

$$P(f) \equiv k(i \to f) / k(i \to), \quad k(i \to) \equiv \sum_f{}' k(i \to f) \tag{6.39}$$

where the prime on the sum is to remind us that summation is only over the states of the products. Therefore[44] we can write that

$$\frac{P(f)}{P^0(f)} = \frac{k(f \to i)}{\sum_f{}' P^0(f)k(f \to i)} \qquad (6.40)$$

The left-hand side is the surprisal of the *specificity of the energy disposal* (for reactants in state i). The right-hand side is the relative rate of reaction (to products in state i) from state f. It is the relative rate because it is the rate constant out of state f normalized by the rate constant averaged over all states (of the reactants of the reversed reaction) at the energy E. In other words, the prior distribution for energy requirement is that, at a given energy E, all initial states react with the same rate. This need not be the observed behavior and Figure 6.14 shows the *selectivity of energy requirements* of the endoergic reaction

$$I + HCl(v) \to HI + Cl$$

determined from the observed energy disposal in the exoergic reversed reaction. Because the surprisal is linear in E_v, vibrational excitation of $HCl(v)$ or $DCl(v)$ leads, at the same total energy, to an exponential enhancement of the reaction rate.

6.4.4   There are deviations from statistics

Much of the material in this section has concerned expected outcomes if the reaction system behaves statistically. If maximal entropy without constraints is really the case, the situation is actually quite dull – all angular distributions are isotropic and featureless, all internal-state distributions approach the form of temperatures, although for molecules with a limited number of degrees of freedom the differences from thermal distributrions can be marked, Problem I, and all the reaction rates are readily calculated. Reality is far stranger and more interesting. It is by comparing observations and calculations against prior expectations that we realize how important or unimportant the dynamics of the process is in altering the outcome. Often there will be even qualitative deviations and we expressed them in terms of constraints. What we next need is the ability to identify the constraints imposed by the dynamics and not only determine them by an analysis of measured (or computed) final-state distributions.

6.4.5   Phase space theory

Here we examine what modifications are required in the definition of the prior distribution when we explicitly take into account the conservation of angular momentum in addition to the conservation of energy. These modifications are required when, say, high final rotational states are being populated, which requires conversion of much angular momentum of initial relative motion to internal angular momentum of the products. As discussed in Section 10.2.2, when an attacking or departing atom is light, one also needs to pay attention to this point. With this modification in hand it is possible to derive *phase space theory*.[45] This is a theory that provides an actual prediction for the distribution of products for reactions where all memory of the initial conditions is lost except for the conservation laws. This is the limit where the trajectories rattle many times across the transition state region before exiting as reactants or products.

To introduce the conservation of angular momentum, consider first the simple case where the particles carry no internal angular momentum. Then the orbital angular momentum $\mathbf{L}$ is conserved and the collision takes place in a plane. For the flux distribution in this plane we use the same assumption as for the discussion of the prior: the flux into any final quantum state is the same. The quantitative change is that now the final velocity is confined to the plane and so the volume element is $d\mathbf{v}' = v' \, d\theta \, dv'$ rather than that in three dimensions, $d\mathbf{v}' = v'^2 \sin\theta \, d\theta \, d\phi \, dv'$. We see two differences. The density of translational states is different in two and three dimensions and the angular distribution is different. In the plane all scattering angles are equally probable. This is the picture that in Section 4.4.1 we called the *sprinkler model*. Next come two points. First, the orientation of $\mathbf{L}$ often is neither selected nor constrained by the dynamics. Therefore, to get the observable scattered flux we need to rotate the direction of $\mathbf{L}$, which means rotating the scattering plane about the $b = 0$ axis. The flux into a given solid angle $d^2\omega = \sin\theta \, d\theta \, d\phi$ will therefore not be constant but will scale as $1/\sin\theta$. Also, if the orientation of $\mathbf{L}$ is not selected it has the weight of $2l + 1$ where $L = \hbar l$. Equation (6.25) is therefore to be replaced by

$$(d^3\sigma/dE_T \, d^2\omega)^0 \propto \frac{1}{\sin\theta} \sum_{l=0}^{l_m} (2l + 1)\rho_l(E - E_T) \tag{6.41}$$

where l_m is the highest angular momentum for which the two particles can form a complex.

Phase space theory does not assume that the particles are structureless and so it is the total angular momentum $\mathbf{J}$, $\mathbf{J} = \mathbf{j} + \mathbf{L}$, that is conserved. The angular factor is no longer a simple $1/\sin\theta$ but retains the essential characteristic that it has forward–backward symmetry. This is further discussed in Chapter 10. The counting of states also needs some additional attention because the need to conserve $\mathbf{J}$ means that $\mathbf{L}$ and $\mathbf{j}$ are now correlated. So the density of internal states is L-dependent and the summation in Eq. (6.41) is not trivial but easily programmable. Figure 6.15 shows the principle of counting states under the restriction that $\mathbf{J}$ is conserved.

Phase space theory seeks to determine the actual cross-section in the limit where the trajectories spend much time in the transition state region before exiting as reactants or products. We therefore factor the cross-section from reactants to products as two contributions, the cross-section starting from reactants to form the collision complex multiplied by the probability that the complex dissociates to products. The statistical assumption enters into the approximation that the

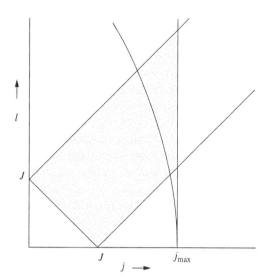

Figure 6.15 The correlation between the rotational and orbital angular momentum quantum numbers at a given total J. Here J can assume all values from $|l - j|$ to $l + j$. This restriction determines the three sides that bound the region of allowed values of l and j, which is shown shaded. The upper range is determined by conservation of energy, $j \le j_{max}$.

only characteristic of the complex is the energy and the total angular momentum J and its (integer valued) projection M, $-J \le M \le J$. Therefore using i and f for initial and final states,

$$\sigma(i \to f) = \sum_{J,M} \sigma(i \to J, M) P(f | J, M) \qquad (6.42)$$

Here $\sigma(i \to J, M)$ is the cross-section for complex formation starting from reactants in state i (and total energy E). $P(f | J, M)$ is the probability that the complex, of given J and M, dissociates to the products in final state f. Note how Eq. (6.42) allows only J and M (and the total energy) to correlate between reactants and products. Section *6.4.5.1 provides the explicit results for computing the cross-section.

*6.4.5.1 Phase space theory: quantitative formulation

We consider an atom–diatom collision where j is the rotational quantum number of the diatom and there are $2j + 1$ (degenerate, in the absence of a field) states of given j. Because J and M are conserved, detailed balance in the form of Eq. (6.37) can be applied to each term in the sum Eq. (6.42). By separating the terms that depend only on i or on f it follows that

$$\sigma(i \to J, M) = \left(\pi / g_i k_i^2 \right) N(J, M) P(f | J, M) \qquad (6.43)$$

where $N(J, M)$ depends only on J and M (and the total energy) but not on the initial state. It is specified by our assumption that the flux into all states (of given J, M, and total energy) should be the same:

$$P(f | J, M) = N(f | J, M) / N(J, M) \qquad (6.44)$$

where $N(f|J, M)$ is the number of states of the complex (of given J, M, and total energy) that dissociate to the final state f. By normalization, $N(J, M) \equiv \sum_f N(f|J, M)$ is the total number of states of the complex and the summation is over states of the reactants as well as those of the products. The final expression for the cross-section is

$$\sigma(i \rightarrow f) = \left(\pi/g_i k_i^2\right) \sum_{J,M} \frac{N(i|J, M)N(f|J, M)}{N(J, M)} \tag{6.45}$$

and it manifestly satisfies detailed balance.

The counting that is required in phase space theory is that of $N(i|J, M)$, the number of states of the complex that are possible starting from the reactants (or products) in a given state. Figure 6.13 showed what is involved. First we specify not only the internal state i and its degeneracy g_i ($= 2j + 1$ if it is an atom–diatom collision where j is the rotational quantum number of the diatom) but also the orbital angular momentum l. There are $2l + 1$ states of given l so, all together, there are $(2j + 1)(2l + 1)$ quantum states under consideration. Out of these we form $N(i, l|J, M)$ quantum states of given J and M. (No states are lost because there are $2J + 1$ states of given J, and J ranges in value from $|l - j|$ to $l + j$.) The number of states that we need for use in (6.44) is obtained by summing over all values of l that can form a complex, $N(i|J, M) = \sum_{l=0}^{l_m} N(i, l|J, M)$. As before, we need to determine the upper limit l_m on the sum.[46]

*6.4.5.2 Locating the bottleneck: variational transition state theory and the statistical adiabatic channel model

When a polyatomic molecule dissociates, most vibrational modes correlate with modes of vibrational character in the products while other modes "disappear," meaning that they correlate with rotation or translation of the separated fragments. As we have seen in Section 6.1.4.1 (and in Problems F and G), the transitory modes are those that determine the shape of the barrier. Taking as an example the dissociation of C_2H_6 to two CH_3 radicals, there are 18 vibrations of the parent. One vibration, the C–C stretch, becomes the reaction coordinate. Five other vibrations correlate to free rotations of the CH_3 radicals, see Problem F. These are the torsion mode of C_2H_6 and the CH_3 rocking modes. Twelve vibrations are "conserved," meaning that they correlate to the vibrations of CH_3.

The frequencies of the conserved vibrations do change as we move along the reaction coordinate from the molecule to the fragments. We know the frequencies at the two ends. From quantum chemistry or from structural considerations let us interpolate the frequencies. We can use the frequencies given as a function of the position along the reaction coordinate in several ways. One is to compute where the flux of dissociating trajectories will be least. This gives a variational transition state formulation, which is particularly important for *flexible transition states*.[47] The more detailed option is to determine the barrier along the reaction coordinate q for each quantum state of the conserved vibrational modes. As we move along q this means adding the (q-dependent) energy of these modes. It is an adiabatic model in the sense that the state of the "conserved" modes is taken to be conserved as we cross the barrier. Otherwise, it is a statistical theory because we must allow energy redistribution in the energy-rich molecule and for the transitory modes. Augmented by an interpolation scheme for the frequencies of the conserved modes, this is the *Statistical Adiabatic Channel Model*[48] (SACM). If the available energy exceeds the height of the barrier, that particular state can dissociate. The end result looks like the RRKM expression, Eq. (6.16),

$$k(E, J) = \frac{N_{0J}(E - E_0)}{h\rho(E, J)}$$

Here however $N_{0J}(E - E_0)$ is the number of states that can energetically correlate to the products. It is the number of open channels. The J is to remind us to conserve the total angular momentum.

6.4.6 Up, up and away

For too long we have stayed on the ground potential energy surface. Beginning with Chapter 7 we explore what can be learned and what can be done by taking advantage of electronically excited states. Experimentally this is made possible by the introduction of lasers. Theoretically this opens up new possibilities for the dynamics, including the option of the control of the collision.

Problems

*A. (a) Detailed balance in transition state theory at a given energy. Given that the forward and reversed reactions proceed via the same transition state, show that the yield function, Eq. (6.4), is the same for the forward and reversed reactions. Next, based on the derivation of Eq. (6.5), show that the rate constant at energy E, computed by transition state theory, satisfies detailed balance. It is easier to derive the condition of detailed balance between the rates $k(E)$ of the forward and reversed reactions starting from the equality of the yield function. (b) Detailed balance in transition state theory at a given temperature. Given that the forward and reversed reactions proceed via the same transition state, show that the rates $k(T)$ of the forward and reversed reactions satisfy detailed balance, meaning that their ratio is the equilibrium constant for the reaction. If you are not familiar with the statistical mechanics result for the equilibrium constant $K(T)$, reverse the question. Derive an expression for $K(T)$ starting with the transition theory expressions for the rates $k(T)$ of the forward and reversed reactions.

*B. Microscopic reversibility is the statement that when i and f are individual quantum states of the reactants and the products then, at a given total energy

E, $k_i^2 \sigma_R(i \to f) = k_f^2 \sigma_R(f \to i)$. The ks are the wave numbers. Show that the yield function as defined by Eq. (6.6) is the same for the forward and reversed reactions.

C. The expression for the thermal rate constant. Equation (3.8) is the thermal rate constant for reactants in a particular state. Equation (A.3.3) shows how to average over the internal states of the reactants. Using the definitions of partition functions from Appendix 6.A, derive Eq. (6.7). This involves a fair amount of changing the variable in integrations.

D. Energy of activation. (a) Compute the energy of activation as given by transition state theory. (b) Show that the energy of activation is not numerically equal to the height of the barrier. (c) Show where the point about the role of the zero-point energy of the reactants and the transition state arises. (d) Compare your result for the energy of activation in TST with the general result of Tolman, Eq. (3.9). (e) Identify the reactive reactants as specified by TST.

E. Entropy of activation. (a) Compute the entropy of activation as given by transition state theory. (b) Show how the qualitative idea of a tight or broad pass at the transition state is reflected in your quantitative result.

F. Steric factor. Make a table with the following format, where n is the number of degrees of freedom:

Species n	Reactant A	Reactant B	Transition state
Translation			
Rotation			
Vibration			

If reactant A has N_A atoms, and reactant B has N_B atoms, then the column sums should be $3N_A$, $3N_B$, and ?. Consider various choices for the species A and B, e.g., atom, diatom, linear triatom, polyatom, and thereby demonstrate how formation of the transition state in a bimolecular A + B reaction is accompanied by loss of freedom of motion and thereby in a steric factor that is below unity.

G. Transition state theory for unimolecular reactions. In the high-pressure limit one can assume that the energy-rich species A* has reached thermal equilibrium. (a) Verify the TST result for the rate of unimolecular dissociation $k(T) = (k_B T/h)(Q^\ddagger/Q) \exp(-\beta E_0)$ where Q is the partition function for A and $Q^\ddagger$ is the partition function for the transition state. (b) This result looks just like the TST expression for the bimolecular thermal reaction rate constant. But this cannot be. A unimolecular reaction rate constant has different dimensions from a bimolecular one. Resolve this dilemma. (c) The thermal dissociation of ethane,

---

* You will need to prove that $S = R \ln Q + \langle E \rangle / T$ and, as you have shown in (A.6.7), $\langle E \rangle = RT^2 \, \partial \ln Q / \partial T$.

$CH_3CH_3 \rightarrow 2CH_3$, has, of course, a very high value for the activation energy. But it also has an unusually high value for the measured Arrhenius A factor, roughly $A = 10^{17.4}\ s^{-1}$. Suggest a possible explanation. Your explanation should take into consideration that many other bond-breaking unimolecular reactions have high A factors. (d) On the other hand, the thermal dissociation $C_2H_5-O-CH=CH_2 \rightarrow C_2H_4 + CH_3CHO$ has a rather low A factor, roughly $A = 10^{11.4}\ s^{-1}$. It is suggested that the reason is that the structure of the transition state is a six-member ring, which enables an H atom to move from one end of the molecule to the other with a more modest barrier. Discuss if this can explain the low A value. (e) Now generalize the TST result for the dissociation rate constant to lower pressures. You may use the steady-state approximation but you must generalize the discussion in Section 6.2.1 because we have shown in Section 6.2.2 that the rate of dissociation is a strongly increasing function of the energy of the molecule.

H. Newton diagram. The $K + HBr$ reaction, exoergic by about $13\ kJ\ mol^{-1}$, has a most probable products' kinetic energy that is about equal to the initial kinetic energy. (a) Draw, to scale, the velocity vector (Newton diagram) for reactant beams that cross at a right angle. Assume that the speed of the reactants is the most probable velocity at 500 K. (b) Show the locus corresponding to the maximal speed of the KBr product and also the locus for the most probable final velocity. (c) In what angular range in the laboratory will you look for the KBr product?

I. The thermal-like prior distribution. The prior distribution for energy disposal into a degree of freedom, say X, contains the factor $(1 - E_X/E_{av})^m$ where E_{av} is the energy available for distribution and the power m depends on how many atoms are involved. The origin of this factor is the density of translational states. What this factor does is describe the shrinking volume in phase space as more and more of the available energy is being put into the particular degree of freedom, X. There are many situations, e.g., the distribution of rotational states of a particular vibrational manifold, where $E_X/E_{av} < 1$. For this range show that $(1 - E_X/E_{av})^m \cong \exp(-E_X/(E_{av}/m))$ is a good approximation. The distribution is then "thermal-like" but note that the "temperature" depends on the available energy. So the distribution of rotational states in different vibrational manifolds will have a different "temperature!"

J. Compute the mean energy in the vibration for a distribution with a linear surprisal, Eq. (6.35), and plot it vs. λ_v, both positive and negative. To do so you will need, as an intermediate step, to compute λ_0 as a function of λ_v. By using a prior distribution from Eq. (6.31) all of this can be done analytically.

K. Quality of fit. Let $P^T(v)$ be a trial form for the distribution of the products' vibration, say of the form Eq. (6.35). Then the quality of the fit to the measured distribution $P(v)$ is $DS^T \equiv \sum_v P(v)\ln(P(v)/P^T(v))$. (a) Using the inequality $\ln(1/x) \geq 1 - (1/x)$, where equality is obtained if and only if $x = 1$, show that DS^T is non-negative and vanishes only for a perfect fit. (b) On the basis of (a) conclude that the best value of λ_v is when DS^T is minimal. Compute DS^T as a function of λ_v, determine the stationary point, and show that it is indeed a minimum. (c) Show that this is the same value of λ_v as determined in the text, that is, by regarding Eq. (6.33) as an implicit equation for λ_v. *(d) Show that DS^T is the difference between the entropy of the distribution $P^T(v)$ and the entropy of the experimental distribution. This is the sense in which the entropy of the distribution $P^T(v)$ is said to be maximal.

L. The table below lists the nascent product vibrational state distributions and the available energy for the three highly exoergic reactions

$$
\begin{array}{ll}
\text{I.} & F + HBr \rightarrow Br + HF(v) \\
\text{II.} & F + DBr \rightarrow Br + DF(v) \\
\text{III.} & CH_3 + CF_3 \rightarrow CH_2{=}CH_2 + HF(v)
\end{array}
$$

Useful input: the vibrational frequency and anharmonicity of HF are 4138.5 cm^{-1} and 90.07 cm^{-1} respectively. The anharmonicity of DF is 45.71 cm^{-1}.

Reaction	$E/\text{kJ mol}^{-1}$	$v = 1$	$v = 2$	$v = 3$	$v = 4$	$v = 5$
I	213.5	0.09	0.22	0.34	0.35	≈ 0
II	215.6	0.06	0.13	0.17	0.22	0.24
III	301.4	0.49	0.32	0.14	0.04	0.01

(a) Reactions I and II should be rather similar yet the results show a significant isotope effect. Suggest how to represent the data so that the effect is largely eliminated. (b) Suggest a way to understand why reaction III, a reaction that is very exoergic, has a qualitatively different vibrational disposal from the first two reactions. If you try you may be able to argue that one can represent the data such that all three reactions are similarly characterized. (c) Can you suggest reasonable entries for the missing $v = 0$ column? And can you extend the results for reaction II to higher vs? (d) The vibrational energy disposal in the H + $CH_2CF_3 \rightarrow CH_2{=}CH_2 + HF(v)$ reaction has a surprisal plot with almost the same slope λ_v as for reaction III (Ben-Shaul *et al.*, 1981). Discuss. (e) Quantum scattering computations have been carried out for the F + H_2 reactive collision. Discuss why such computations for F + HBr will be much more prohibitive. (f) For the Br + HF(v) collision, the vibrational excitation of HF can significantly enhance the reaction cross-section. But in the collision of $CH_2{=}CH_2$ with HF(v) it is not necessarily clear that vibrational excitation of HF will efficiently result in an addition to the double bond. Why?

M. The ion–molecule reaction $CH_2{=}CH_2^+ + CH_2{=}CH_2 \rightarrow C_3H_5^+ + CH_3$ was studied in a crossed-beam arrangement. (a) At low collision energies the product ion was observed to have an angular distribution with a forward–backward symmetry. Why does that make chemical sense? (b) As the collision energy was increased, the angular distribution became increasingly asymmetric and eventually became predominantly forward. Why? And can you suggest at what energy this will happen? (c) Will you use transition state theory to compute the reaction rate at low energies? If not, what will you use? (d) If we use the $CD_2=CD_2^+$ instead of the $CH_2=CH_2^+$ ion, do we expect the ion $CD_2=CH_2^+$ as a possible product? Does the answer depend on the collision energy?

N. Toward biological applications mass spectrometry is increasingly concerned with ions of ever larger mass.* Apart from the technical problem of detection of heavy ions there is the question of a time window. An ion has to dissociate within at most a few microseconds if its fragmentation is to be detected by a conventional mass spectrometer. Assume that ionization is by impact of 70 eV electrons, $e + P \rightarrow e + e + P^+$. (a) Make a reasoned guess for the excess energy of the parent ion and do not forget about the Franck–Condon principle, Section 7.0.1. (b) What is the range of the maximal molecular weight for the parent in order that the fragmentation of the ion can still be detected? Assume that biological molecules are made of atoms of a mean atomic weight of 7.6 and that a mean vibrational frequency is 1000 cm^{-1}. Ionization weakens bonds so take a dissociation energy of 2 eV. (c) Suggest ways for extending the range of application of analytical mass spectrometry toward higher masses. If necessary, make a literature search.

O. Termolecular processes and sticky collisions. Third-atom-assisted processes are often assumed in chemical kinetics. A notable example is in recombination processes, e.g., $O + O_2 + M \rightarrow O_3 + M$, where the function of the third body, M in the example, is to take away energy and thereby stabilize the new ozone molecule. It is reasonable to conjecture that the reaction proceeds by a two-step process where, in the first step, $O + O_2$ (or $O_2 + M$) undergoes a collision where a long-living complex is formed and it lives for long enough to collide with M (or with O). (a) Use RRKM theory to estimate the lifetime of the unrelaxed O_3 complexes formed from thermal $O + O_2$. (Go for, or estimate, the data that you need.) (b) The complex needs to collide with M before it dissociates. Is the estimated lifetime sufficient if the pressure is, say, an atmosphere?

P. Newton diagram. Draw the Newton diagram for the reaction shown in Figure 6.12 and hence explain why certain angular ranges of the products are not easily seen in the experiment.

Notes

1 For the derivation below, the restriction to reactants at equilibrium seems to be essential. It will be interesting to develop a derivation that allows for a generalization.

2 This statement is predicated on the Liouville theorem that allows us to connect from the transition state back to the reactants. Our mental image is that we start with an ensemble of

---

* Such ions need not be formed by electron impact. There are other options. For a very successful method see Fenn and Mann (1989). See Nohmi and Fenn (1992) for an example of molecular weights up to 5 million.

trajectories with different initial conditions, but at the same total energy, such that we mimic a uniform distribution of initial states of the reactants. Each such trajectory will either cross the transition state or will fail to reach it. In the transition-state approximation a trajectory that reaches the point of no return will proceed to products. So the set of states of the reactants that is identified by the trajectories reaching the transition state is the *reactive reactants* of Chapter 3.

3 In addition, special features of the potential can facilitate recrossing. For example, an early release of the reaction exoergicity, as shown in Figure 5.15, drives the trajectory up the repulsive wall and can thereby redirect it toward the reactants' valley. The masses also matter. An exchange of a light atom means that the skewing angle β, Appendix 5.B, is quite small. So there is a sharp bend connecting the reactants' and products' valleys. The result is that the light atom can chatter a few times between the heavy partners. Some, but not all, of these limitations can be overcome by looking for an alternative configuration of no return, one that is not necessarily at the barrier, as discussed in Section 6.4.5.2.

4 Called "the yield" in Levine (1969). See also Miller (1998).

5 Proposed in the form of Eq. (6.11) for the thermal rate constant by Eyring in the 1930s, on the basis of earlier work by Pelzer and Wigner. Transition state theory had a profound influence on the development of chemical kinetics and is discussed in all textbooks. Critical reviews include Pechukas (1976, 1981), Truhlar and Garrett (1984), Truhlar *et al.* (1996).

6 We emphasize again that this is a condition on the system for which the theory is applicable. It is not a trivial condition, and can be violated for a number of reasons. Thus, endoergic reactions will often proceed *preferentially* from reactants that are vibrationally excited. The reaction thus serves to *deplete* such states: only if the population of these states by energy-transferring collisions is faster than the reaction rate will the reagents remain in thermal equilibrium. Similarly, the specific *release* of reaction exoergicity tends to displace the system from thermal equilibrium. Hence, the required condition is obtained only when the mechanisms that restore equilibrium occur on a time scale shorter than that of reaction. The act of crossing of the barrier has to be the rate-determining bottleneck. Because this condition cannot always be insured (particularly for fast reactions) or can be made to "fail" (by keeping the system in thermal disequilibrium), molecular reaction dynamics is relevant to reactions in bulk, in particular in such cases (e.g. combustion reactions) where thermal equilibrium is not maintained.

7 See Problem C. Mathematically, Eq. (6.7) says that $Qk(T)$ is the Laplace transform of $\rho(E)k(E)$. It is natural to want to use an inverse Laplace transform to get $k(E)$ from a measured $k(T)$. Computationally this is not a safe route because taking an inverse Laplace transform is numerically rather unstable and is very sensitive to errors in the measured $k(T)$. What is more satisfactory is to evaluate the Laplace transform by the saddle-point method as introduced in Chapter 3. See Problem P.

8 This idea has been fruitfully generalized, particularly so by the physical organic chemists. One can also apply it to enzymatic reactions (Jencks, 1986; Villa *et al.*, 2000). A partition function can be written in terms of the free energy A as $Q^{\ddagger} = \exp(-A^{\ddagger}/k_BT) = \exp(-E^{\ddagger}/k_BT)\exp(S^{\ddagger}/k_B)$. Here $S^{\ddagger}$ is the entropy of the **other** degrees of freedom at the barrier. If the passage is a narrow one, the structure is highly constrained and its entropy is small. Hence the rate is determined by more than simply the height of the barrier. We also encounter the opposite effect: if the transition state is less constrained than the reactants,

its entropy is large and so is the rate constant. See Problem E in Chapter 12 for a concrete example.

9   The vibrational period is a natural time scale for molecular dynamics. Take a day as a natural unit for a person. By comparison, the molecules in Figure 6.2 live longer than a human lifespan. Take a year as a natural period for the solar system. The molecules in Figure 6.2 live much longer than recorded history. Of course, such long-living energy-rich species can find other modes of relaxation. Sooner or later they will collide with the wall. Given time they can also emit some of the excess energy as IR radiation. It requires special experimental care to keep a molecule isolated for a long time. Under many circumstances, e.g., combustion, energy-rich molecules that can eventually dissociate are formed in the bulk where many collisions can occur. Then we have a similar situation to that considered by Lindemann: the initially energy-rich molecules are deactivated (and also activated) by collisions prior to their dissociation.

10 Bunker (1966), Oref and Rabinovitch (1979), Beynon and Gilbert (1984), Gilbert and Smith (1990), Baer and Hase (1996), Holbrook *et al.* (1996).

11 J. D. Rynbrandt and B. S. Rabinovitch, *J. Phys. Chem.* **75**, 2164 (1971).

12 The original proof by Rice, Ramsperger and Kassel is longer but it had one point that is hidden by our derivation. This is that certain states, $N^\ddagger(E - E_0)$ in number, will dissociate promptly. Physically this is for the reason that these states already have enough energy for crossing the barrier localized in the reaction coordinate. In Section 7.2.2 we will make the point that the RRKM rate, Eq. (6.6), is the mean rate of dissociation and that state-specific rates can differ considerably from the mean. We also suggest that these promptly dissociating states can lead to an ultrafast component in the unimolecular decay.

13 A proof is as follows. The total energy is in the range E to $E + dE$. At equilibrium, all quantum states are equally probable. On the products' side there are $\rho_p(E)dE$ quantum states. At equilibrium the rate of association of reactants to form a complex is $\rho_p(E)\overleftarrow{k}(E)dE$. There are $\rho(E)dE$ quantum states of the complex. At equilibrium the rate of dissociation of the complex is $\rho(E)k(E)dE$.

14 The extreme is when there is a forest of electronically excited states, as is the case when the energy is just below or even above the threshold of ionization (Schlag and Levine, 1997; Campbell and Levine, 2000). Then dissociation and ionization can compete and statistical ideas applied also to the electronic degrees of freedom become useful.

15 An example where such selectivity can be manifest is when the initial state is electronically excited. A *conical intersection*, if it exists, can funnel the system to restricted regions of the ground potential energy surface and dissociation can then be prompt. See Figure 8.8.

16 See Chabinyc *et al.* (1998) for a review of ion–molecule displacement reactions.

17 For ion–molecule collision complexes this has been experimentally implemented by cooling the complex through a supersonic expansion (Levy, 1981, 1984). Even further, one can reactivate these stable complexes either by collisions or by IR multiphoton absorption, as discussed in Chapter 7. The non-covalent bonding in the complex is shown for example by the inequivalence of the two Cl atoms in a $^{37}ClCH_3 \, ^{35}Cl^-$ complex, that dissociates very preferentially to the attacking Cl isotopomer.

18 See Hase (2001) and references therein. Experimental support for these conclusions is provided by studies of the temperature dependence of the rate constant, particularly so if

we can independently vary the internal state of the reactants. See, e.g., A. A. Viggiano *et al.*, *J. Am. Chem. Soc.* **114**, 10477 (1992).

19 Mass spectrometry measures the mass of fragment ions formed in a unimolecular dissociation of an energy-rich parent ion. To apply a transition state theory approach to the determination of the branching into different products requires the assumption that after ionization there is a sufficient delay for the ion to distribute the excess energy over all the available modes. Mass spectrometrists refer to this assumption as the quasi-equilibrium hypothesis, and they are actively concerned about its validity. See Lifshitz (1989), Lorquet (1994, 2000).

20 One can also speculate that the many similar modes are the very reason why biological machines operate so determinedly. This is possible when modes act in a coordinated or cooperative manner. A phase change such as the melting of an atomic solid is the familiar example of *cooperative behavior*, where a single atom of the solid seeking more freedom of motion will be hindered by its neighbors if these remain rigidly localized. But when the neighbors too start moving then the entire system can change its state. In a molecular solid there can be more than one transition because the molecule can remain translationally localized but acquire the freedom to rotate. The contraction of skeletal muscles can be understood as a collective motion where individual fiber proteins move with respect to one another [H. E. Huxley, *Science* **164**, 1355 (1969)].

21 One such possibility, already mentioned in Section 6.2.2.4 and discussed further in Chapter 7, is the preference of modes to resonantly transfer their energy, meaning that any frequency mismatch between modes tends to slow down energy redistribution. Another possibility may be operative in activated chemical reactions in solution, discussed further in Chapter 11. In solution we start with solvent and separated reactants in thermal equilibrium. How then do the reactants, when they come together, collect enough energy to localize enough energy so as to surmount the barrier to a chemical reaction? The internal states of a large molecule form a quasi-continuum. So one possibility, a favorite of many, is to think of the energy as solvent-induced diffusing amongst the quasi-continuum of energy states. Diffusion means taking small steps in every possible direction but doing so often. If we center attention on the localized mode where the energy is needed, diffusion means that this mode goes up and down in its energy, by small amounts but with a net drift upwards. There is however an alternative. This is that a large amount of energy is supplied in one go by a rare event. Mechanical trajectory simulations of reactions in solution do support this alternative as a viable route, [K. R. Wilson and R. D. Levine, *Chem. Phys. Lett.*, **152**, 435 (1988)]. Similarly, a large rare fluctuation can be how the energy is provided in a machine.

22 For the kinetics of energy transduction see Hill (1989).

23 There are several types of myosin. For a detailed discussion of the physics of motor proteins, see Howard (2001) and Nelson (2003). For a review of the mechanochemistry, see Spudich (2001) and Vale (2003).

24 A short introduction by the person who unraveled the mechanism is Boyer (1999). The fuller story is in Boyer (1997). By attaching a fluorescent labeled actin filament to the axis the rotation can be demonstrated as the enzyme cleaves ATP.

25 For an introduction to the probing response of single biomolecules see T. Strick *et al.*, *Phys. Today* **54** (10), 46 (2001).

26 For the control of molecular machines, see Balzani *et al.* (2000) and also Balzani *et al.* (2003). See also the special issue on molecular machines, *Acc. Chem. Res.* **34** (6) (2001).

27 F. M. Raymo, K. N. Houk, and J. F. Stoddart, *JACS* **120**, 9318 (1998), B. Northrup, K. N. Houk, and J. F. Stoddart (private communication).

28 Enzyme catalysis can also be regarded as an example of *reactions in organized media*. Other examples include reactions at interfaces, Section 11.3 and on surfaces, Chapter 12.

29 The problem is to compute long time trajectories for a system with many shallow minima. Special sampling methods are therefore needed, see Elber *et al.* (2003), Bolhuis *et al.* (2002), Dellago *et al.* (2002). See also note 20 in Chapter 5.

30 This section is about the flux of products in the center-of-mass system. It is more concerned with issues of principle rather than the important experimental aspects of specific techniques. At the same time we must recognize that there is a very real problem of how to determine this flux in an experiment. Consider first the simpler case where we use a detector that measures flux in the laboratory system of coordinates. As we discussed in Section *2.2.7 and more below, depending on the mass combinations the scattering in the center of mass may span a wider or narrower range of angles (and velocities) in the laboratory. Even with a flux detector, certain mass combinations are more favorable for experimental study. But not all detectors measure flux. The most common other case, and the one most familiar to chemists, is a detector that measures a number density, i.e., a concentration. Using such a detector requires that the measured number density is converted into flux. Laser detection methods, in particular some of the techniques introduced in Chapter 7, require a careful analysis of what is the observable that is being measured.

31 This is known as DIP, distributed (as) in photodissociation (Herschbach, 1973).

32 Overviews of how it began: Herschbach (1987), Lee (1987), Bernstein (1988). For representative reviews, see Casavecchia *et al.* (1999), Casavecchia (2000), Green and Anderson (2001), Liu (2001). Experimental beam techniques are discussed in detail in Scoles (1988) and Pauly (2000). A survey of different applications is Campargue (2001).

33 Adapted from B. H. Mahan, *Acc. Chem. Res.* **3**, 393 (1970).

34 We reiterate that we count states in the semiclassical limit, treating the final translational energy as a continuous variable, as discussed in Section 6.3. Therefore, the translational energy is not sharply defined but is in the range of E_T to $E_T + dE_T$.

35 These compact results are useful but are not sufficient for an accurate fit to vibrational distributions where there are few levels so that the spacings are too wide for a continuous approximation to be valid. It is then better to integrate only over g_R, while using a discrete vibrational quantum number so that $P^0(v) = B_v^{-1}(1 - f_v)^{3/2} / \sum_{v=0}^{v_{max}} B_v^{-1}(1 - f_v)^{3/2}$.

36 The dynamics literature has reports of a "thermal-like" final state distribution. This is particularly so if one examines a distribution that has only a smaller fraction of the total energy in it, most typically, a rotational distribution or a distribution in one vibration out of many. The essential difference between the prior and the Boltzmann distribution is that the prior has a sharp energy cutoff due to the conservation of the total energy. A Boltzmann distribution has an unbounded tail. Therefore the difference between the prior and a thermal distribution becomes most noticeable for states whose energies are close to the total energy limit.

37 Surprisal is, technically, the amount of information provided by an observation. Wiener's 1948 book *Cybernetics*, MIT Press, is the reference to the point of view that we use. The earlier parts of Wiener's book are much more accessible than you might think from the title. Applications to our subject are reviewed in Levine and Bernstein (1976).

38 Failure to recognize that a given outcome can occur as a consequence of many elementary events arises in many misapplications of the ideas of information theory. In chemical physics, it was only when Planck introduced the notions that led to the result that it is the quantum states of the same energy that are equally probable, that statistical physics gave correct results, e.g., for the specific heat.

39 In some circles the approach that we follow, that of looking for a distribution of maximal entropy, Section 6.4.2.1, is questioned. This occurs because people sometimes overlook the need to include a prior distribution in Eq. (6.32). In this case it follows that in the absence of constraints all groups of outcomes are equally probable. Very often this is manifestly not reasonable and correctly so because different groups can differ in how many mutually exclusive outcomes are members of the group.

40 For applications to molecular collisions see Levine and Bernstein (1976) or Kinsey and Levine (1979). For the general case see Levine and Tribus (1979).

41 Explicitly, $\exp(\lambda_0) = \sum_f \exp(-\lambda_v E_v(f))$. This should remind you of the partition function.

42 This is one possible way to derive the Boltzmann distribution. Given is an isolated large system at equilibrium. It then has a microcanonical distribution, meaning that all quantum states of the large system are equally probable. Now consider a small subsystem. Its energy does not have a sharp value because it is a part of the large system. One can show, e.g., that the subsystem has a Boltzmann (or canonical) distribution of its energy. The prior distribution is not necessarily Boltzmann-like because the group of states of interest need have but a small fraction of the total available energy.

43 Why must it be linear? Physically, because if we increase the number of independent collisions we want the frequency of different outcomes to remain the same. Mathematically, we imposed this condition when we required that the surprisal should be additive.

44 Summing both sides of Eq. (6.38) over f leads to

$$g_i \rho_T(E - E_i)k(i \rightarrow) = \sum_f{}' g_f \rho_T(E - E_f)k(f \rightarrow i) \equiv k_E(\rightarrow i)$$

Therefore

$$\frac{k(i \rightarrow f)/k(i \rightarrow)}{g_f \rho_T(E - E_f)} = \frac{k(f \rightarrow i)}{g_i \rho_T(E - E_i)} \bigg/ \frac{k_E(\rightarrow i)}{g_i \rho_T(E - E_i)} = \frac{k(f \rightarrow i)}{k_E(\rightarrow i)}$$

$$= \frac{k(f \rightarrow i)}{\sum_f{}' g_f \rho_T(E - E_f)k(f \rightarrow i)}$$

The substitution

$$P^0(f) = g_f \rho_T(E - E_f)/\rho(E)$$
$$\rho(E) = \sum_f{}' g_f \rho_T(E - E_f)$$

and the definition of $P(f)$, Eq. (6.39), completes the derivation.

45 For the foundations of phase space theory, see Nikitin (1974), Pechukas (1976), Light (1979).

46 Not so simple to implement in a rigorous fashion. Fortunately, the results are not overly sensitive to the choice made. The cutoffs imposed by the centrifugal barrier, as introduced in Section 3.2.6, are sufficient for many applications but fail to allow for any steric

requirements on the formation of a complex. For example, a collision such as Na + KBr will proceed through the well of the stable salt $(NaK)^+Br^-$. But the formation of the stable cation $(NaK)^+$ requires Na to attack preferentially from the K end. For reactions with an activation energy, delineating the conditions for formation of the complex is even harder. In principle (Pechukas, 1976; Pollak, 1985) one knows how to define a strong coupling regime but the implementation of rigorous criteria is only computationally tractable for collinear collisions.

47 See Wardlaw and Marcus (1988), Baer and Hase (1996).

48 On the SACM model and its applications, see Troe (1992, 1997).

Chapter 7
Photoselective chemistry: access to the transition state region

So far, the energy necessary for a change to take place had to be brought in by the reactants. In this chapter the required energy is provided by light. Often, light allows us to promote the system to an electronically excited state where the potential and hence the dynamics are different from those of the ground electronic state. But light-induced chemistry or "photochemistry" is not only a new way of driving chemical reactions. Photochemistry offers a degree of control – a selectivity – that is unique in its potential and variety. Our ability to tailor light toward specific applications is part of the secret. This control ranges from a simple requirement such as wavelength or polarization selection that allows us to access a particular excited state to more complex tasks such as manipulations that produce short light pulses that can freeze the motion and to *pulse shaping* in frequency and time for the purpose of guiding nuclear motions.* Another important aspect is the ingenuity of scientists in matching up light and matter. In particular we have learned to take advantage of the changing character of the intramolecular dynamics of molecules as the energy increases. At low levels of excitation polyatomic molecules offer us a great individuality through their spectroscopic fingerprints, whereas at higher levels of excitation they have a quasi-continuum of vibrational states and can be made to take up lots of energy like a good heat sink. This saga is nowhere near to a closed chapter. The future seems ever brighter, just like new laser sources.

This chapter is called "access to the transition state region" because we want to either access it directly or to control the motion toward it in a manner designed to allow us to determine the outcome. This is achieved because light allows us to access localized regions of the potential energy surface. Another advantage of this selectivity is that we can avoid the averagings (say over impact parameter, approach angle, etc.) that are inherent when we start two reactants, that are far apart, on their way toward a bimolecular collision. We begin this chapter with

---

* As the wavelength of lasers gets shorter and shorter we can also envisage approaching the limit where the light will be used in a diffraction mode, similar to the current uses of X-rays. Combined with ultrashort pulses, Chapter 8, this will allow tracking the structure of molecules during chemical, biochemical, and material (e.g., fracture) processes.

selectivity achieved by using light of a tightly defined frequency and because of the time–energy uncertainty principle these light pulses need to be long in time. Later we examine the complementary situation of pulses of very short duration. The ultimate is the situation of a "shaped pulse," meaning that the light pulse has a controlled spread in both time and frequency.[1]

7.0.1   The Franck–Condon principle

The Franck–Condon principle plays such a key role in allowing us to localize the excitation that it is necessary to review it. We begin with what is "the sudden limit." A favorite trick is for a magician to astound the audience by quickly whisking away a tablecloth upon which is set dinnerware. If done rapidly enough, the dinnerware appears undisturbed by the process. The trick is to be quick enough. The same thing happens for molecules that absorb light and are excited to a higher electronic state. The electronic rearrangement during the absorption process is so quick that the nuclei make the transition from the ground-state potential energy surface to the excited-state potential energy surface as if they did not move. Loosely speaking, the transition is probable when it is possible for the nuclei to remain in the same position. This limiting behavior is called the Franck–Condon principle in honor of early investigators who first realized the propensity for the nuclei to retain their *momenta* during the fast electron jump caused by the absorption of light. In the terminology of Chapter 1, the nuclei act as spectators.

The above description is classical and depends on the separation of electronic and nuclear motions, which is at the heart of the Born–Oppenheimer approximation. A quantum mechanical description is necessary because we cannot specify at the same time the positions and momenta of the nuclei. The quantum mechanical stationary vibrational state is described by a wave function whose square gives the probability of finding the nuclei at some position. This nuclear wave function is oscillatory in those classically allowed regions of the potential for which the kinetic energy of the nuclei is positive and rapidly decays (exponentially so) for those classically forbidden regions of the potential for which the kinetic energy of the nuclei is negative because the total energy is less than the potential energy.[*] This behavior is shown in Figure 7.1 for several different total energies. Moreover, as first suggested by de Broglie, the wavelength is shortest and so the wave function wiggles most when the nuclei move the fastest but the amplitude of the wave function is greatest for those positions for which the classical motion is slowest (near the turning points of the motion,[2] see Problem A).

The quantum analog of the classical Franck–Condon principle is that the probability of the nuclei making a transition from one potential surface to another is

---

[*] In quantal scattering theory, Chapter 4, we saw that this is equally true for the motion in the unbound part of the potential. Indeed, the shape of the vibrational wave function changes smoothly as we go to higher energies and even as the energy is higher than the threshold for dissociation, see Figure 7.1.

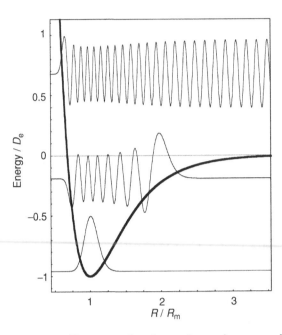

Figure 7.1 Three wave functions at increasing energy for the Morse potential (heavy line, drawn in reduced units). The ground state has no nodes between the two classical turning points. The state with a vibrational quantum number v has v nodes and so it is smaller everywhere except near the classical turning points. Note that the probability is largest to locate the system near the turning point on the right. The wave function uppermost in energy is for a dissociating state. It has a classical turning point only on the left, where it is larger. Otherwise it is highly oscillatory, meaning that it has many nodes. (How else can these different wave functions be orthogonal to one another if they do not differ in the number of nodes?*) Consult Figure 7.2 to conclude that also a dissociating state can be reached by an electronic transition from a bound vibrational state of a lower potential.

greatest for those places in which the square of the overlap between the vibrational wave functions of the initial and final states, $q_{v'v''} \equiv |\langle \psi_{v'}(R)|\psi_{v''}(R)\rangle|^2$, called the Franck–Condon factor, is the greatest.[3] This is the quantal version of the classical description of a vertical transition.** Consequently, if two potential curves are nearly identical in shape and the nuclei in the ground electronic state are also in the ground vibrational state (see Figure 7.2), an electronic transition favors a nuclear transition to the ground vibrational level of the excited state. All the vibrational wave functions (of a given potential) are orthogonal to each other, which means that their overlap is unity among themselves but zero otherwise.

---

* Conclude that the number of nodes of a dissociating state of zero energy is the number + 1 of the bound states supported by the potential.

** Since each one of the two wave functions has the largest amplitude near its classical turning point, the simplest estimate of the Franck–Condon factor is that it is largest when, for the two vibrational states, their classical turning points are vertically above one another.

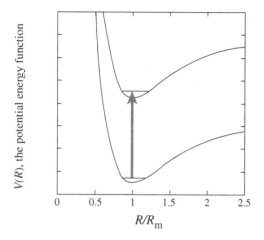
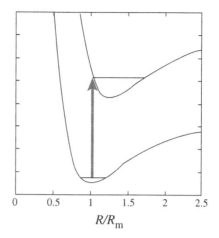

Figure 7.2 Franck–Condon propensities for vibrational change of state upon an electronic transition. Left: the lower and upper potentials are exactly the same. Only $\Delta v = 0$ transitions are allowed and the transition shown is from the ground vibrational state. Right: the upper potential has a significantly higher equilibrium distance (see Figure 7.5 for ICl as an example). The most probable transition is then to a fairly vibrationally excited state, but transitions to nearby vibrational levels are also likely. The absorption will therefore be broad and its peak occurs at a higher frequency than the separation between the two potential minima. Note that the transition from the ground vibrational state is shown not from the classical turning point but from its midpoint. This is because the wave function for the ground state has its largest amplitude at the equilibrium distance R_m as shown in Figure 7.1. For higher vibrational states the amplitude is maximal near the turning points, as expected for a classical-like motion.

Thus, for identical potential curves in the ground and excited states, the vibrational selection rule $\Delta v = 0$ is strictly obeyed for whatever vibrational level v in which the nuclei start. We refer to such a limiting situation as having diagonal Franck–Condon factors.

Suppose that the two potential surfaces are dissimilar. Then the Franck–Condon factors are less than unity and you get different probabilities for making transitions to final v'' vibrational levels depending on the vibrational overlap.[4] We shall make repeated use of the Franck–Condon principle in understanding which vibrational levels are populated in various dynamical processes. In Section 9.2 we will generalize the principle so that it also applies to excitation as a result of a collision (where we need not be in the sudden limit).

7.0.2 Beyond the Born–Oppenheimer approximation

We have just discussed a light-induced transition between two different electronic states. But such transitions can take place also in the absence of light. What we need is that the transition is between isoenergetic states. This is possible even for diatomic molecules in their excited electronic states but is much more common for polyatomic molecules. Why? Because in polyatomics we can have the favorable situation that we already encountered through the Franck–Condon principle: in polyatomic molecules the potential energy surfaces can cross (see the discussion of the conical intersection in Section 5.1), and therefore the electronic state change can take place without a change in the kinetic energy of the nuclei.

A key point is that a change in electronic state is probable only in the localized regions of space about the crossing of the two potentials.* This obviously invites controlling the access to these regions. For example, a very rough rule of thumb is that, in polyatomics, bending and torsion facilitate electronically non-adiabatic** transitions. Hence photoexcitation accompanied by excitation of the bend favors accessing a conical intersection.[5] We will return to the idea that failure of the Born–Oppenheimer approximation occurs in a localized region in configuration space in Section 9.3.

In the context of accessing the transition state it is important to point out why electronically non-adiabatic behavior may be particularly important near a transition state. The transition state is a single point only in a limited description, often used to make maps as in Section 5.1, where all degrees of freedom except two are frozen. But for three independent nuclear displacements the transition state is a two-dimensional object, namely a line. In general, it is a surface in $N-1$ dimensions. The serious failure of the adiabatic behavior is often confined to a manifold of even fewer dimensions,[6] and so it is not all over the place. That is good news because it allows scope for our intervention. But why is the transition state region particularly suspect and when and where is the failure serious?

As shown in Figure 5.9, a chemical reaction involves a change from an electronic state representing the bound reactants to a state representing the bound products.† The potential energy surface corresponding, say, to the electronic state representing the bound reactants looks just like the entrance valley of the potential energy surface of the reaction. But in the region that is physically the products' valley, such a surface is purely repulsive because the old bond is broken and the new bond is not allowed to form. It is the other way round for the potential energy surface corresponding to the electronic state representing the bound products. It behaves correctly in the products' valley but is repulsive in the reactants' region.

---

 * But even improbable transitions are possible if we wait long enough. See Problem M.

 ** Technically, an adiabatic behavior means that quantum numbers are conserved. Grammatically, it means that they are not changing. So if quantum numbers are changing the behavior is, properly speaking, diabatic. Even so, non-adiabatic is also used. Physically, "an adiabatic behavior" can be used in reference to any quantum number. Exoergic reactions, for example, can be electronically adiabatic but vibrationally very much non-adiabatic. Often, which quantum number is meant is implied by context and not spelled out.

 † A point of view known as the Evans–Polanyi or valence bond model, Section 5.1.4. See also Problem G of Chapter 5. The LEP(S) potential energy surface originates from this point of view.

Figure 5.9 showed the energy profile along the reaction coordinate for two such surfaces.

A realistic potential energy surface does not quite look like Figure 5.9. The reactants' and products' valleys do look the same but in the middle, instead of a seam showing a crossing between two surfaces, it looks smooth but has a small barrier where the seam was. Why? Because the electronic state of the system is neither that of H + H—H nor that of H—H + H. Rather, it is a linear superposition of the two.* When the Hamiltonian is diagonalized, Problem G of Chapter 5, you obtain the Born–Oppenheimer description: two (adiabatic) states where each is a superposition of the two states that we started with and two eigenvalues. The lowest of the two eigenvalues is the potential energy surface that describes the exchange reaction and has a (relatively) low barrier. What is the physics of the other eigenvalue? It is an excited electronic state that asymptotically correlates to an antibonding state. But close-in, that upper eigenvalue has a well and it has a well just where the lower state has a barrier. For a collinear configuration of the three atoms, the gap between the top of the barrier and the bottom of the well is large. But for a bent configuration of the three atoms, that well can be deep enough and the barrier can be high enough that the two potentials get quite near to one another. At the equilateral configuration of the three identical atoms, there is, see Figure 5.8, a conical intersection between the two potentials!

Electronically non-adiabatic transitions are significantly more likely when the two potentials are close to one another. This is why there is so much emphasis on the region about a conical intersection and why, in general, such transitions are confined to limited regions in configuration space. We will have several occasions to discuss why proximity of the two potentials is so very important. For the moment we think of it as a kind of Franck–Condon principle.** There are two implications: (i) near the barrier is also where electronically non-adiabatic transitions are likely; (ii) electronically non-adiabatic transitions will take place if the two Born–Oppenheimer surfaces are near enough. But how "near" is "near enough"?

Why do electronically non-adiabatic transitions take place near the barrier? Because above it is the well of the higher electronic state. Is there another way to phrase this? Yes there is: near the barrier the character of the electronic wave

---

* Quantum chemists will correctly point out that, quantitatively, many additional configurations such as ionic ones, like H^-H^+H, etc., will also contribute. Our discussion aims at getting the essence but it is not quantitatively accurate.

** Explicitly: large changes in the kinetic energy of the nuclei are unlikely. The quantitative statement in Section 9.2.2 is that the probability of such changes is exponentially small. The total energy is the same whichever is the electronic state. Change of state means a change in the potential energy and hence a change in the kinetic energy. A cautionary note: if the motion is bound, the system can return again and again to the region where the required change in the kinetic energy is minimal. So it has an improved chance of making the transition as in the predissociation of O_2 or of Na_2 as discussed below. See also Problem M.

function changes from that of the reactants to that of the products. The change can be gradual. This reflects a strong coupling between the two electronic states. It means that there will be a wide gap between the top of the barrier and the bottom of the well. The system will proceed adiabatically across the barrier region. Physically, the gradual change of the electronic wave function means that the electrons have the time to rearrange and adjust while the nuclei are moving. But the change of the electronic wave function can be more abrupt. This indicates a weaker coupling between the two electronic states and a narrow gap between the top of the barrier and the bottom of the well. There is a finite chance that upon approach to the barrier region the system makes a transition to the upper electronic state. Physically, the electrons do not have the time to rearrange and adjust while the nuclei are moving and so their state is maintained as what it was to the left of the barrier. But this means that to the right of the barrier it is the upper state, cf. Figure 5.9.

To conclude: the barrier is one region where electronically non-adiabatic effects are potentially important.* It is actually important if the coupling between the two electronic states is weak enough that the gap between the energies of the two electronic states is not large.

How large is a large gap? It is determined by two factors. One is the velocity with which the nuclei move. Other things being equal, the faster the nuclei transverse a region, the harder it is for the electrons to "instantaneously" adjust. The second consideration is how quickly the electrons need to adjust. This factor is governed by the change of their wave function with the displacement of the nuclei. Problem A of Chapter 5 proves that the smaller the gap between the energies of the two different states, the more adjustment is necessary. Quantitatively

$$\left\langle \psi_f \left| \frac{\partial}{\partial R} \psi_i \right. \right\rangle = \left\langle \psi_f \left| \frac{\partial H}{\partial R} \right| \psi_i \right\rangle \bigg/ (E_i(R) - E_f(R)) \tag{7.1}$$

Here i and f are indices of the two electronic states, R is a nuclear coordinate along which the motion occurs, and the denominator is the gap in the electronic energies at the position R. The strength of the coupling is Eq. (7.1) times the velocity of the nuclei.**

Figure 7.16 is an illustration of such a consideration. The experiment is the competition between C–Cl and C–Br bond fission in the photodissociation of bromoacetylchloride ($BrCH_2COCl$) and bromopropionylchloride

---

* We have seen that a barrier can also separate the reactants from a central well. The change in character of the electronic wave function equally applies to such a barrier because the well signifies a bound species. If there are several inner wells along the reaction coordinate, in going from reactants to products the wave function will change character several times, once at each barrier.

** This product is $\langle \psi_f | (\partial \psi_i / \partial t) \rangle$ where the time dependence of the electronic state is due to its change of character as the nuclei move. If the wave function changes rapidly, the scalar product will be large. The large coupling is because in the Born–Oppenheimer approximation electrons need to track the change in configuration and this fails if the change is fast.

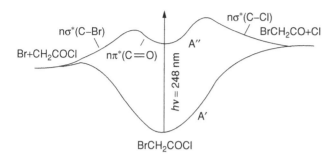

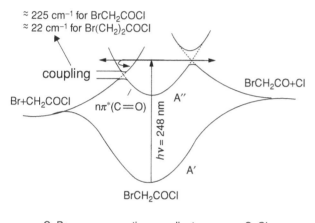

Figure 7.3 Schematics of the energy profiles along the reaction coordinate for C–Cl and C–Br bond fission. The upper panel is the adiabatic point of view. The excitation is to the upper electronic state that correlates to the same products as the ground state. The lower panel shows the avoided crossings at the two barriers to dissociation on the upper electronic state (the coupling is to a bound higher excited state). The hindrance of the dissociation of the C–Br bond is shown schematically as a trajectory that makes a transition to a higher electronic state and is thereby reflected. The gap between the two adiabatic states, whose magnitude is the coupling between the diabatic states, is higher for the C–Cl bond fission [adapted from M. D. Person, P. W. Kash, and L. J. Butler, *J. Chem. Phys.* **97**, 355 (1992); see also Waschewsky *et al.* (1994), Butler and Neumark (1996), Butler (1998), Conroy *et al.* (2001)].

(BrCH$_2$CH$_2$COCl) at 248 nm. On the excited surface accessed by the photon, it is the C=O bond that is weakened by a transfer of a non-bonding electron into an antibonding π orbital located primarily on CO. Dissociation requires an antibonding character of the C—Br or C—Cl bonds. This is achieved by the electronic wave function in the well region changing character on its way out. For the Cl exit it needs to be mixed with a σ antibonding character on C—Cl. For the Br exit it needs to be mixed with a σ antibonding character on C—Br. The top panel in Figure 7.3 shows this electronically adiabatic view of the process. The prediction: the C—Br bond will break preferentially because it has the lower barrier. The experiment:* C—Cl bond fission is much more probable. The possible explanation: above each barrier a well exists owing to a higher electronic state. The computed gap at the barrier configuration for C—Br is not large and is smaller than that at the C—Cl barrier on the other side. (Figure 7.3 is not to scale; if it were, the narrow gap would not be visible.) Upon stretching of the C—Br bond, exit may be frustrated by trapping in the well and reflection from the excited state that is higher up, as schematically sketched in the figure.

7.0.3 Radiationless transitions

If the potential energies of two different electronic states can cross or can get near to one another then the transition between them will be efficient. In this section we discuss a special but important case when the non-adiabatic coupling is weak because we are not near a conical intersection. A weak coupling can still have dramatic implications when we deal with excited but bound states, states that live for far longer than a duration of a direct collision.

An excited electronic state of an isolated atom will fluoresce. It will do so on a nanosecond time scale for an allowed transition. A radiationless transition of an excited molecule means the absence of this fluorescence or its observation with a quantum yield (photons out/photons in) below unity. If the excited molecule eventually dissociates, then it is clear where the missing energy is. It went into bond breaking. But larger polyatomic molecules exhibit low to negligible fluorescence quantum yields even below the threshold energy for dissociation and at very low pressures, so that quenching by collisions is negligible.

Single UV photon transitions from the ground state of a molecule to a highly vibrationally excited level of the same electronic state are very Franck–Condon disfavored. The absorption of light in the near UV is therefore to the lower vibrational levels of the first (singlet) higher electronic state, S_1, a state that for a large molecule is typically bound and whose minimum is below the dissociation threshold of the ground electronic state. We shall refer to the state that is accessed by the light as the *bright state*. We need a special name because at the same total energy as that of the bright state there is a quasi-continuum of vibrational levels on the ground electronic state S_0. This is the first encounter with an idea that will occur again and again in this chapter. Because of the selectivity of the optical excitation, the state that is initially accessed is not necessarily a quantum mechanical stationary state, meaning that there are other, about isoenergetic, states that the bright state can interact with.

---

* M. D. Person, P. W. Kash, and L. J. Butler, *J. Chem. Phys.* **97**, 355 (1992). The C—Br/C—Cl bond dissociation branching ratio is 0.04 in bromoacetylchloride and at most 0.05 for bromopropionylchloride. The computed gap at the C—Br barrier for bromopropionylchloride is only about $20\,\mathrm{cm}^{-1}$.

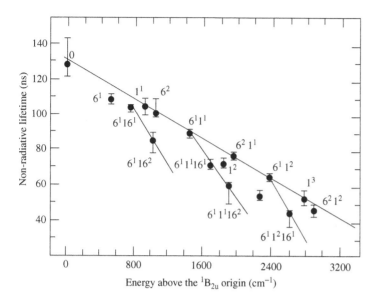

Figure 7.4 The lifetime for non-radiative decay of different low vibrational states in the first excited singlet state of benzene ($S_1 = {}^1B_{2u}$) plotted against the vibrational energy. The vibrational states of S_1 are identified by the number of quanta, shown as superscripts, in the C–C bend, mode 6, symmetric stretch, mode 1, and out-of-plane bend, mode 16. The rates are measured from the rate of disappearance of the bright state (which determines the sum of radiative and non-radiative rates) and the quantum yield of fluorescence [adapted from K. G. Spears and S. A. Rice, *J. Chem. Phys.* **55**, 5561 (1971); for a review see Jortner and Levine (1981)]. Seen in the figure is that not all vibrational states affect the rate only to the extent of their energy. In particular note the selective enhancement of the non-radiative decay upon excitation of the out-of-plane bend, which is therefore referred to as the "promoting mode." Such state-selective effects are observed also for low energies above the barrier to dissociation and even in the selective population of final states, e.g., in the dissociation of the benzene. For the van der Waals cluster following electronic excitation of benzene, see T. A. Stephenson and S. A. Rice, *J. Chem. Phys.* **81**, 1083 (1984).

The coupling between the bright state that is on S_1 and the quasi-continuum of vibrational states of S_0 is small for two reasons. First, it requires a breakdown of the Born–Oppenheimer approximation and furthermore it is disfavored by the Franck–Condon principle because of the poor vibrational overlap between the two manifolds. The overlap of vibrational states is small because for an isoenergetic transition it requires that the vibrational states of S_0 have a rather high vibrational energy, see Problem M. But on the nanosecond time scale of fluorescence the initially excited bright state can vibrate for very many periods. So even if the transition probability per vibration is low, the decay of the bright state in S_1 to the quasi-continuum of S_0 can still compete with the rate of its radiative decay (= fluorescence). As the excitation energy is increased, Figure 7.4, the lifetime for non-radiative decay becomes shorter because of the increasing density of states in the quasi-continuum.

The bright state on S_1 decays into the quasi-continuum of vibrational states of S_0. By microscopic reversibility this process should also be possible in the reversed direction: the molecule makes an isoenergetic transition from S_0 back to S_1. We call this a *revival* of the bright state. If we could keep the molecule isolated for a very long time then the initial bright state will revive and the quantum yield for emission will increase, but such an experiment requires keeping a molecule isolated for a very long time.* We are not in doubt about the revival of the initial state starting with a system in the vibrational quasi-continuum because there is the option of directly observing it starting with vibrationally very hot molecules in the electronic ground state. Such molecules can be prepared, e.g., by the multiphoton IR excitation method discussed in Section 7.2.5. At sufficiently high vibrational energies such molecules should therefore fluoresce from an excited electronic state.[7] In Appendix 7.A we discuss the revival time and show that it is indeed long, of the order of h/D where D is the spacing in energy between states of the quasi-continuum of vibrational states of S_0. This is why we need a polyatomic molecule to observe the radiationless decay: the density of vibrational states of S_0 needs to be high enough that the revival time will be long.

For many molecules pumped to the electronically excited S_1 there is another option because typically below the origin of S_1 is the lowest excited state T_1, a triplet. Except for elements down the periodic table, where coupling of spin and orbit is strong, spin is a relatively good quantum number** and it will be conserved during the optical excitation. But on a much longer time scale the system may cross from S_1 to a higher vibrational state of T_1. This so-called *intersystem crossing* mode of decay will be less Franck–Condon disfavored than an *internal conversion* to S_0 because less energy needs to go into vibration. The first excited triplet state is long-living because its radiative emission

---

* Not an easy experiment because there are other decay channels such as very weak IR emission from the vibrational quasi-continuum. Such radiative emission from very hot molecules in the electronic ground state has been suggested as a cooling mechanism for large molecules excited somewhat above the threshold for fragmentation. The very high number of atoms (e.g., C_{60}) means that just above threshold the rate for unimolecular dissociation is rather low. Yet the hot molecules are monitored to decay. The possible reason: cooling by IR emission. This is a molecular analog of blackbody emission.

** Quantum numbers go with eigenstates and these are stationary. Should not a quantum number be conserved? Yes and no. A quantum number is constant at a level of an approximation that is defined by the neglect of weaker coupling terms. The role of the weak coupling is only felt after a long time. In addition to intramolecular weak-coupling terms, it is almost impossible, and some will say strictly impossible, to completely isolate a molecule. There are always external perturbations and any excited state will, sooner or later, decay by light emission. A quantum number is good for a finite time interval, until processes that have been neglected begin to be important. This time may be long enough for our needs, or not.

down to S_0 is "forbidden," which means in reality that it has an unusually slow rate. It is known as phosphorescence and is often used technically to design paints that reflect incident UV light with a delay and at a lower (hence, visible) frequency.

Above the energy barrier to dissociation of S_0, radiationless decay from an excited electronic state results in photofragments from the ground electronic state. Below this barrier, radiationless decay from S_1 is an effective way to prepare ground-state molecules with chemically significant amounts of vibrational energy, and we will discuss it further in Section 9.1.3.

7.A   Appendix: The picket-fence model

This is a simple model for radiationless transitions,[8] unimolecular dissociation,[9] and vibrational energy redistribution.[10] The starting point for our purpose is the concept of the optically prepared state. This is defined operationally: it is that state that carries transition dipole strength from the ground state. Such a state need not be stationary and it typically is not stationary because of Franck–Condon and other propensities that limit what state can be optically accessed. Nearly isoenergetic with the optically prepared state are other states that are "dark," meaning that they cannot be seen by optical absorption from the ground state. By contrast, the optically prepared state is "bright" and to make life simple we take it that there are no other bright states nearby in energy.

Neither the bright nor the dark states are stationary in the sense that they are not sharp energy eigenstates of the Hamiltonian. To determine the eigenstates we need to allow for the weak residual coupling between the bright and dark states. We take the coupling to be weak because if it were not we would not be able to prepare the optically prepared state as it would couple to other states while our laser pulse was still on.

The coupling assumed in the model is between the bright state and manifold of equispaced (hence, picket fence) dark states. The coupling is weak but it cannot be treated by perturbation theory. The reason is that it is not necessarily weak by the criterion that matters: the dark states have a sufficiently high density that the coupling, while weak, can exceed the mean spacing D between the dark states. This requires an exact diagonalization of the Hamiltonian. Problem G does so analytically. From this analytical solution one can show that if one treats the dark states as a real continuum then the decay rate of the bright state is $(2\pi/\hbar)|\overline{V}|^2\rho$. Here $\overline{V}$ is the mean strength of the coupling and $\rho = 1/D$ is the density of the dark states.

The picket fence model as carried out in Problem G requires that the coupling between the bright and dark states does not vary too much between one dark state and another nearby so that the actual coupling can be replaced by its mean

Figure A7.1 Tiers of increasingly dense manifolds of dark states. The simple picket fence model assumes only one tier of dark states. Only the zero-order bright state, ZOBS, is optically accessible. But it is spread by intramolecular coupling over the dark states. The shading of each dark state indicates how much it contributes to the mixing.

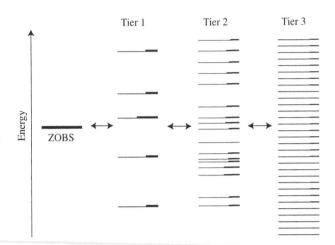

value. Physically one can argue to the contrary, that certain subsets of dark states will be in better resonance with the bright state than others. The refined model, supported by experiment, is to divide the dark states into tiers, Figure A7.1, where the bright state is coupled to the members of the first tier. These dark states are coupled to the (denser) second tier, etc.[11]

Experimental spectra[12] at increasing frequency resolution support the idea of successive tiers of states revealed as the resolution is improved. At relatively low resolution the spectrum is a broad feature centered about the energy of the bright state. At a higher resolution the broad feature splits into a number of narrower features. These correspond to a longer "preparation" time that can therefore discern a set of states that are a superposition of the bright states and states of the first tier. An even higher frequency resolution will show each such narrow feature to be a set of distinct denser lines, Figure A7.2.

The observable spectral tier structure forces on us the recognition that "what" is excited by the optical pulse depends on the experiment. The longer is the pulse, the narrower is the energy width of the optically prepared state. Ultrashort pulses can even fail to excite a vibrational eigenstate in a diatomic molecule if the pulse duration in time is shorter than a vibrational period, recall Figure 1.7. It is, of course, always possible to think of the optically prepared state as a superposition of eigenstates. This is useful because no eigenstate is coupled to the others and it evolves in time in a rather simple manner, namely by acquiring a phase, $\exp(-iEt/\hbar)$, where E is the eigenenergy of that stationary component. But a superposition of eigenstates of somewhat different energies can exhibit all manners of time dependence because it is like a Fourier series (except that the energies are not necessarily equally spaced). If we excite a superposition of just two eigenstates, at a distance D apart in energy, the time evolution will exhibit quantum beats with $\hbar/D$ as the revival time, Problem F and Figure 7.17. The more

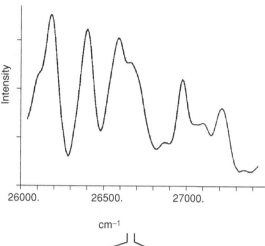

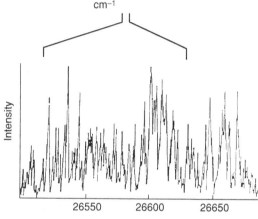

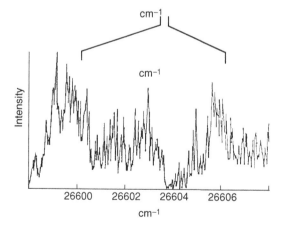

Figure A7.2 The spectrum of acetylene at increasing resolution in a relatively high-energy regime [adapted from J. P. Pique *et al.*, *Phys. Rev. Lett.* **58**, 347 (1987) and J. P. Pique *et al.*, *J. Chem. Phys.* **88**, 5972 (1988)]. The experimental technique used to access such high-energy states is stimulated emission pumping (SEP) (Hamilton *et al.*, 1986; Silva *et al.*, 2001). This technique allows the preparation of states even at energies above the barrier to isomerization and/or dissociation. The initial state is pumped to an electronically excited state whose equilbrium geometry (trans-bent in the case of HCCH) is different from that of the ground state. Rather than being allowed to spontaneously emit, the down fluorescence is stimulated, sometimes described as "dumping" the state down. The dumping allows us to reach vibrational states of the ground electronic state that otherwise have very low Franck–Condon factors. An identification of the nature of the states excited is aided by what one knows about vibrational spacings. The lowest resolution is the excitation of the C–H chromophores.

eigenstates are coherently* excited, the more complex can the time evolution be. In a real molecule at higher levels there can be many eigenstates even within a quite high resolution energy window.

7.1 Laser photoexcitation and photodetection of diatomic molecules

Laser excitation is not only a means for providing the necessary energy for chemistry to take place. We begin the discussion of the selectivity achievable by the excitation and the specificity of the detection for the special case of diatomic molecules. Then the spectrum is simpler in that there is only one vibrational mode as well as only one rotational constant. But once the molecule is electronically excited we will find it necessary to allow for the breakdown of the Born–Oppenheimer separation.

7.1.1 The discrete vibrational energy levels of diatomic molecules

A quantum of vibration of a diatomic molecule is often a sizeable amount of energy on the chemical scale. Vibrational excitation of a reactant diatomic molecule will therefore significantly decrease the activation energy, recall the Tolman interpretation, Section 3.1. As already discussed in Chapter 1, reactant vibrational excitation can have a selective role with the enhancement of the rate being more than when the same additional energy is put into the relative translation. We can do more because the molecular dipole of a heteroatomic molecule is along the bond. Hence, using a polarized light beam we can start with a randomly oriented sample and preferentially excite those molecules whose bond axis is aligned with respect to the polarization vector of the light; as discussed in Section 10.1.2 this will allow us to probe the preferred orientation for reaction.

7.1.2 Electronic excitation

Transitions of diatomic molecules to upper electronic states have been thoroughly studied in spectroscopy. Bound electronically excited states of diatomic molecules can have rather different dynamics from that of the ground state. The propensity of O_2 in its lowest electronically excited state,** shown in Figure 7.7 below, to

---

* This is the operative technical term. It is equivalent to our earlier condition that there is a definite initial state and not an ensemble. The optically prepared state is then a pure state in the quantum mechanical sense, namely, it can be expressed as a superposition of states, each with its own phase, hence "coherent."

** $a^1\Delta_g$, it is a singlet where the two valence electrons, unpaired in the ground state, each being in a different π orbital, are paired and placed in the same orbital.

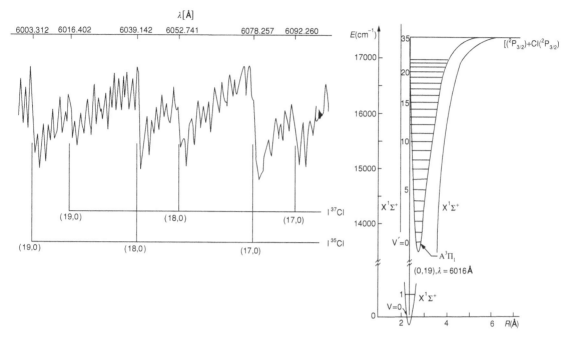

Figure 7.5 A portion of the low-resolution visible electronic spectrum of ICl in the region of maximum absorption. The vibrational quantum numbers of the upper and lower states are shown for the two isotopomers [adapted from D. M. Brenner, S. Datta, and R. N. Zare, *JACS* **99**, 4554 (1977)]. ICl in the $A^3\Pi_1$ electronic state has a significantly longer equilibrium bond distance, as shown in the insert. Therefore the Franck–Condon region for the absorption is to fairly high vibrational states ($v''=17$ to 19 is shown). ICl in the $A^3\Pi_1$ electronic state dissociates to ground state atoms so it has a weaker bond than the ground electronic state. The correlation of bond length–bond strength seen here is a fairly general one.

add to organic double bonds, e.g.,

$$O_2(a^1\Delta_g) + \underset{O}{CH_3\;\;\;\;\;CH_3} \rightarrow CH_3 \underset{O-O}{-O-} CH_3$$

is well known and is of practical importance.* Diatomic halogen molecules also add to double bonds in an electrophylic mechanism. But in the gas phase ICl will not, for example, add to acetylene. It will add when it is excited to its $A^3\Pi_1$ state

$$ICl(A^3\Pi_1) + HC\equiv CH \rightarrow IHC=CHCl$$

We next want to use this reaction to separate the two isotopes of Cl. Despite the heavy masses, the electronic absorption spectrum of ICl, Figure 7.5, exhibits a clear isotopic shift between the peaks due to the two isotopomers $I^{35}Cl$ and $I^{37}Cl$.

* Photodynamic therapy is based on the *in situ* excitation of $O_2(a^1\Delta_g)$ by electronic energy transfer from photoexcited organic molecules.

Even a broad laser can selectively excite one species and not the other. In the presence of a scavenger for the electronically excited ICl, such as acetylene, one can separate the two chlorine isotopes.

7.1.2.1 Angular distribution in photodissociation

Electronic excitation can lead to dissociation of the diatomic molecule. We ask "what is the angular distribution of the products?" The atoms recoil along the direction of the bond but in a sample in the gas phase all orientations of the molecules are possible. We can select the orientation of those molecules that absorbed the photon because the interaction with the photon depends on the direction of the transition dipole with respect to the electric field of the light. What this means is that under the action of light the electrons in the bond can be set in motion either along the bond or perpendicular to it (other options are ruled out by symmetry). Say that excitation is to the repulsive range of the excited state potential and that the two fragments instantly recoil, then the direction of their velocity directly reveals the direction of the transition dipole, as shown graphically in Figure 7.6. If θ is the angle between the electric light vector and the direction of recoil of the fragments, the angular distribution of the photofragments will be proportional to $\cos^2 \theta$ for a transition where the electrons are excited parallel to the bond and to $\sin^2 \theta$ for a perpendicular case. Since both possibilities may arise we show in Chapter 10, Eq. (10.31), that one can write for the angular distribution of the photofragments

$$I(\theta) = [1 + \beta P_2(\cos \theta)]/4\pi \tag{7.2}$$

where $P_2(\cos \theta) = (3 \cos^2 \theta - 1)/2$ is the Legendre polynomial of second order and the parameter β characterizes the anisotropy of the distribution. The cases we discussed are $\beta = 2$ for a parallel transition, $I(\theta) \propto \cos^2 \theta$, and $\beta = -1$ for a perpendicular one, $I(\theta) \propto \sin^2 \theta$. Experimentally, β can fall short of these limiting values because of some rotation of the molecular axis while the products separate from one another.

The direction of the transition dipole with respect to the molecular axis is determined by the symmetry of the lower and upper electronic states. Ω is the projection of the angular momentum of the electrons on the internuclear axis. A parallel transition excites from $\Omega = 0$ of the ground state to an upper state with $\Omega' = 0$.* A perpendicular transition will lead to a state with $\Omega' = 1$. Figures 7.5 and 7.8 show the ground and low excited states of ICl. From the symmetry of the states, transitions in the longer visible wavelengths (say about 560 nm) are

---

* When the transition dipole moment is along the internuclear axis no torque is exercised on that component of the electronic angular momentum that is parallel to the axis. When the transition dipole moment is perpendicular, the angular momentum of the photon (one $\hbar$ unit) adds to the electronic angular momentum.

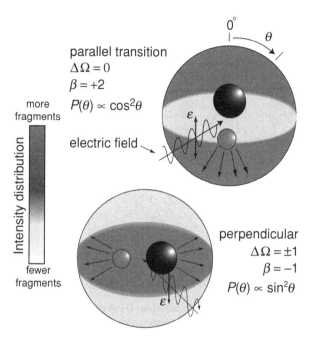

Figure 7.6 Angular distribution of the dissociation products following the absorption of a linearly polarized light by a diatomic molecule. Shown are the two extreme cases of a very fast dissociation for a parallel and a perpendicular transition. ε is the electric field of the light, β is the anisotropy parameter defined by Eq. (7.1), and Ω is the quantum number discussed above and in Chapter 10. The intensity of scattering into a given angular range is indicated by the shading [adapted from A. J. Alexander and R. N. Zare, *Acc. Chem. Res.* **33**, 199 (2000)]. See Section 10.3.1.2 for a more extended discussion. Here we just comment that this is a simple example of a *vector correlation* where we correlate the direction of the products' velocity with the direction $\hat{\varepsilon}$ of the photolysis light.

almost purely perpendicular, $\beta \approx -1$. At shorter wavelengths the transition is to a mixture of symmetry types, $\beta \approx -0$, and it becomes purely parallel, $\beta \approx 2$, at even shorter wavelengths (490 nm). In Section *7.1.2.3 we explore further what really happens at intermediate wavelengths and conclude that the transition there is not to a classical mixture but to a coherent superposition of the two symmetry types. Either of the states leads to the same products. Measuring the energy dependence of the quantum mechanical interference between the two alternatives provides information on the repulsive forces that send the two halogen atoms apart.

7.1.2.2 The photochemistry of molecular oxygen in the atmosphere

Like ICl, molecular oxygen also has more weakly bound excited electronic states that upon dissociation correlate to ground-state atoms. These are shown in

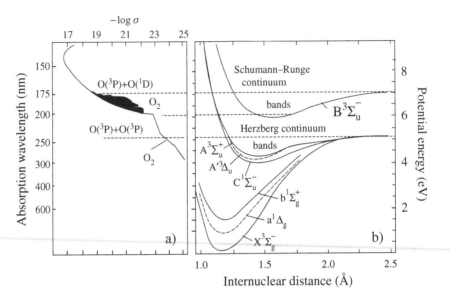

Figure 7.7 (a) Absorption spectrum (logarithmic scale) for O_2 in the near and deep UV. The figure is drawn to correlate the absorption with the potential energy curves. (b) The peak of the absorption ($\lambda \approx 135$ nm, where the cross-section is about 0.1 Å^2) corresponds to the optimal Franck–Condon overlap with the vibrational ground state of the ground $X^3\Sigma$ electronic state. The energy thresholds for dissociation into ground and electronically excited O atoms are shown as dashed lines [adapted from D. H. Parker, *Acc. Chem. Res.* **33**, 563 (2000)].

Figure 7.7 and are seen to have longer equilibrium bond lengths. At the UV region where the absorption begins, say from 300 nm to 250 nm, the Franck–Condon overlap from the ground electronic state is so poor that the absorption cross-section, as shown in the left panel of Figure 7.7, is lower by several orders of magnitude from what it can be for a Franck–Condon favored transition (to the Schumann–Runge continuum, around 150 nm). At wavelengths below about 242 nm the absorption is not structured, but because the energy is above the dissociation threshold, the transition is to the continuum, named after Herzberg. It is an example of a photodisscoiation from an excited electronic state. The very weak UV continuum absorption is the first step in the formation of the ozone layer (at 25 to 50 km above the surface of the Earth):

$$O_2 \xrightarrow{\ h\nu \text{ (from 242 nm down to about 176 nm)}\ } O(^3P) + O(^3P) \qquad \text{Herzberg continuum}$$

The next step in the sequence is a recombination assisted by the presence of a third body that we designate by M

$$O(^3P) + O_2 + M \rightarrow O_3 + M$$

The location of the ozone layer is determined primarily by two competing factors. As we go up from the surface of the Earth, the density of oxygen molecules decreases exponentially while the density of the UV radiation increases exponentially.* At a height of roughly 25 km the rate of the photodissociation of O_2 into the Herzberg continuum is maximal. Even then, due to the low absorption cross-section, the rate is so slow that it takes days to restore a depletion in the equilibrium concentration of ozone.

At shorter wavelengths the photodissociation of O_2 produces electronically excited atoms:

$$O_2 \xrightarrow{\; h\nu \text{ (from 176 nm down)} \;} O(^1D) + O(^3P) \qquad \text{Schumann–Runge continuum}$$

This absorption blocks the short-wavelength UV solar radiation from penetrating the atmosphere below 120 km or so. The longer-wavelength radiation does penetrate further because the absorption in the Herzberg continuum is so weak. It is only blocked when the concentration of O_2 becomes high enough that formation of the ozone layer, as discussed above, is possible. The reactive $O(^1D)$ atom** is quenched by collisions with N_2 or O_2, but given the chance it inserts into H_2 as in Figure 1.6 and it also reacts readily with other hydrogen-containing species, mainly water and methane:[†]

$$O(^1D) + H_2O \rightarrow OH + OH$$

*7.1.2.3 Interference of exit channels

Two open-shell ground-state atoms can combine to form a diatomic molecule in a large number of different electronic states.[13] Quite a number of these states are purely repulsive, but the ground state is typically chemically bound and there can be some excited states with a chemical well. All of these states correlate to atoms in their ground state. We have seen this for both ICl and O_2, and the plurality of states is the rule rather than the exception. What it also means is that optical excitation is typically not to a single state and one purpose of translational spectroscopy is to identify the states that participate in the transition, as a function of wavelength. In this section we point out that it is the wave function of the molecule that is perturbed by the light and the wave function of the excited state can be a superposition of two (or more) different electronic states. For each such excited state we can imagine a trajectory describing the relative motion of the

---

* In the laboratory, lasers can provide a much higher flux of photons. Still, because the cross-section is about seven orders of magnitude below what it can be at maximum, the required laser intensity is such that other, unintended, higher-order processes, as discussed in Section 7.2.5.2, can also contribute.

** Formed also in the photodissociation of ozone in the strongly absorbing Hartley band, $\lambda < 310$ nm. This occurs in the stratosphere. At lower altitudes (the troposphere) it is the weaker absorption of ozone at longer wavelengths, the so-called Higgins band, that is the main filter of the UV radiation.

[†] OH, "nature's atmospheric detergent" (Ravishankara *et al.*, 1998), is the daytime oxidizer that removes pollutants from the atmosphere.

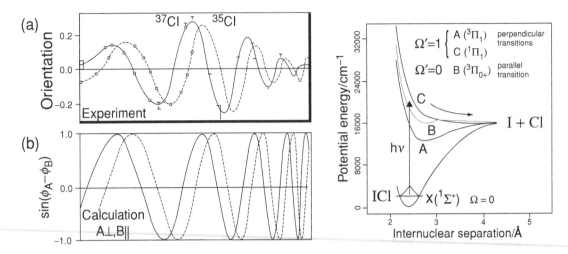

Figure 7.8 (a) Observed orientation of the electronic angular momentum of a Cl atom as determined by ionizing the atom with a circularly polarized light vs. the wavelength of the light used to photolyze ICl. Results are shown for two different isotopomers. (b) Computations that allow for interference between two different histories – excitation to either the A or the B state, either one of which correlates to ground state I and Cl atoms. The relevant states are identified in the panel on the right [adapted from A. J. Alexander and R. N. Zare, *Acc. Chem. Res.* **33**, 199 (2000)].

two atoms out of the Franck–Condon region and into two separated atoms. The two excited electronic states have different potentials and so two such trajectories will not, in general, be identical. Each such trajectory will be associated with a different phase, a phase that is determined by the shape of the potential and the excess energy. Our simple but useful guideline is that when there are two classical ways of reaching a detectable outcome (here, a product atom in its ground electronic state), the corresponding wave functions can interfere and the interference is an oscillatory function of the phase difference.

In the case of ICl, the orientation of the electronic angular momentum of the Cl atom can be determined by ionizing the photofragment with a circularly polarized light. Figure 7.8 shows the measured orientation vs. wavelength as well as the results of computation of the interference term, as determined by the phase difference for motion on the A and B excited states of ICl. As a further check, both experimental and computational results are shown for dissociation of either $I^{37}Cl$ or $I^{35}Cl$. Because the phase shift varies with the reduced mass and, for ICl, this is the mass of Cl, there is a significant isotope effect. Problem D uses our rough estimate for the energy dependence of the phase shift, $-ap/h$, where a is the range of the repulsive potential and p the momentum, to account for the qualitative features of Figure 7.8. The agreement between experiments and computations based on interference between the C and B states of ICl is poor because the computed phase difference is too small. Figure 7.8 thus serves not only to identify the states involved but also to check our quantitative knowledge of the potential energy curves.

7.1.3 Photodetection

Lasers have allowed detection of products in entirely new ways and opened up many new avenues of exploration. This section outlines the essential ideas behind some of the more important methods, but it must be supplemented by reading the more specialized literature for a proper understanding of the experimental implementation.

7.1.3.1 *Imaging and translational spectroscopy*

Imaging is the term used for a direct mapping of the velocity (vector) of the products.[14] It captures the Newton sphere, Section 6.3.1, and projects it for viewing in a number of ways, as shown in Figure 7.9. The figure shows the technique applied to photodissociation but it is equally useful and important for bimolecular collisions and in other instances where an energy-rich species blows up and one wants to monitor the distribution of fragments in velocity space.

To "see" the nascent fragments it is most convenient to ionize them. This can be done by a UV laser and, in common with other spectroscopic techniques, it can be done in a state-selective manner. In Section 7.1.2.3 we adopted this feature by using a (right or left) circularly polarized laser ionization beam to discern the orientation of the electronic angular momentum of the Cl fragment formed in the dissociation of ICl.

Ionization has the important advantage that it hardly perturbs the velocity of the fragment. By applying an electrical field, the just formed ions are collected, if need be, mass selectively, and recorded (say, on a charge-sensitive plate), as shown in Figure 7.9. The practical success of the method depends of course on the abilities of the ion optics used to guide the ions. Resolution comparable or better than vibrational spacings is possible.

We have already seen applications of imaging. When the dissociation of O_2 in the Herzberg continuum is examined, the results are complicated by there being more than one excited state of O_2 in the energy range of interest, see Figure 7.7. Most of the absorption is to the A state but transitions to the A′ and the C states do contribute. The value $\beta \approx 0.6$ extracted from the data is due to a mixture of parallel and perpendicular transitions.

7.1.3.2 *Doppler spectroscopy*

In translational spectroscopy the energy distribution of the photofragments is determined by the distribution of the "times of flight" (TOF) to the detector. An important alternative is *Doppler spectroscopy*,[15] where the velocity (in both magnitude and direction) is determined through the Doppler shift in the absorption. Consider a fragment moving at an angle $\vec{\theta}$ with respect to the electrical field

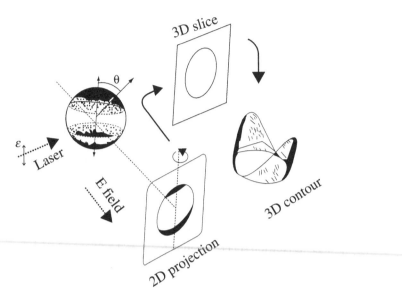

Figure 7.9 Experimental observation of the entire Newton sphere, Section 6.3.1. The sphere of products can be generated by using a linearly polarized light for photodissociation, see Figure 7.6, or from a crossed-beam experiment as in Section 6.3. A second laser ionizes the fragments and this has the advantage that it can be done selectively for different internal states of the products. An electric field allows collection of the ions and produces a 2D image. The cylindrical symmetry of the original sphere allows this collapsed image to be transformed into a 3D contour plot of the intensity of the products [adapted from D. H. Parker, *Acc. Chem. Res.* **33**, 563 (2000)].

of the dissociation laser, the velocity of the fragment along the direction of the propagating light is $v\cos\vec{\theta}$. Therefore the fragment absorbs at the frequency

$$\nu = \nu_0(1 + (v/c)\cos\vec{\theta}) \tag{7.3}$$

where c is the velocity of light, ν_0 is the absorption frequency of the stationary species, and be careful to distinguish frequency ν and velocity v. The detuning from the unshifted frequency ν_0 is not large, often being in the GHz range, but modern lasers can be made stable enough to detect the shift.

In Eq. (7.2) the angle θ is the velocity of the product with respect to the electrical field of the dissociation laser. By using the addition theorem for Legendre polynomials, Eq. (7.2), one can write for the pump–probe Doppler profile of the angular distribution

$$P(\vec{\theta}) = [1 + \beta P_2(\cos\theta)P_2(\cos\vec{\theta})]/4\pi \tag{7.4}$$

We can solve Eq. (7.2) for $\cos\vec{\theta}$ and insert the result in the explicit formula for the second Legendre polynomial. This expresses the angular distribution in terms of the shift, called the *detuning*, between the nominal absorption frequency ν_0

and the actual frequency v:

$$P(\vec{\theta}) = [1 + \beta P_2(\cos\theta)\{3\,[(v - v_0)/v_0]^2(c/v)^2 - 1\}/2]/4\pi \qquad (7.5)$$

The absorption frequency of the photofragment is dependent on its internal state and so a Doppler profile can be determined for each internal state of, say, the diatomic fragment when the dissociating molecule is triatomic. This is not an idle thought because different internal states may well have different angular distributions. For the photodissociation of a diatomic, the absorption of the fragment atom can be simpler and Figure 7.10 shows an early example for the predissociation of Na_2 following a photoselective preparation of an initial level. Despite the long lifetime of the quasi-bound initial state, the fragment angular distribution retains a memory of the preparation because the plane of rotation of the molecule has also been selected.*

7.1.3.3 Laser-induced fluorescence

Spectroscopic detection of the photofragment can usually be done in a manner that is sensitive to its internal state and can therefore determine the state distribution of the product. When the Doppler detuning of the spectroscopic transition can be determined, the angular distribution of products in different states can be extracted. For, say, the diatomic product from the dissociation of a triatomic molecule this can provide a wealth of information. Nor is there excessive redundancy. Different vibrational final states and even different rotational states of the same vibrational ladder can have interestingly different angular distributions.

Laser-induced fluorescence,[16] LIF, is the technique used for generating the data shown in Figure 7.10. The electronically excited $Na(3^2P_{1/2})$ atom formed in the dissociation of $Na_2(B^1\Pi_u)$ is pumped to the higher $Na(4^2D_{5/2})$ state that can then fluoresce. When applied to molecules, the technique requires knowing the relevant Franck–Condon factors so that the intensity of fluorescence from the upper level can be converted to population in the lower level.

7.2 Photodissociation dynamics

Photodissociation,[17] the extension of spectroscopy into the continuum, is a source of reactive species such as H, F, or Cl atoms or OH and other radicals that can be used to initiate bimolecular reactions. Photodissociation processes are important not only in the laboratory but also in the atmosphere** as well as the regions of

---

* It is a Q branch transition with a transition dipole perpendicular to the molecular axis. Therefore the molecule rotates in a plane perpendicular to the transition dipole. The solid curve in the Doppler profile in Figure 7.10 is a fit to $\beta = -1$.

** Nature does not like to only break bonds, it is a waste. But photoisomerization processes, where bonds break and others reform, in concert, are very common in larger molecules, particularly biologically relevant ones, such as rhodopsin, which plays an essential role in vision. More on this in Section 9.3.7.

Figure 7.10 The potential energy curves (simplified) showing the excitation of a quasi-bound (high rotational state, $j = 42$, see Section 4.3.6; W is the energy of the state above the dissociation limit for a state of $j = 0$) level of the $^1\Pi_u$ excited electronic state of Na$_2$ at 457.9 nm. The photoselective detection of the Na(3p) product is by state-selective absorption at 568.82 nm to the 4d state(s) followed by emission at 330 nm. The resulting Doppler profile for the absorption of Na(3p) is shown below, where a perpendicular arrangement of the two lasers is used. The solid line is a fit for a fragment distribution with 400 cm^{-1} excess kinetic energy and $\beta = -1$. The "bumps" on the right shoulder are due to there being two 4d states of Na [adapted from G. Gerber and R. Moller, *Phys. Rev. Lett.* **55**, 814 (1985)].

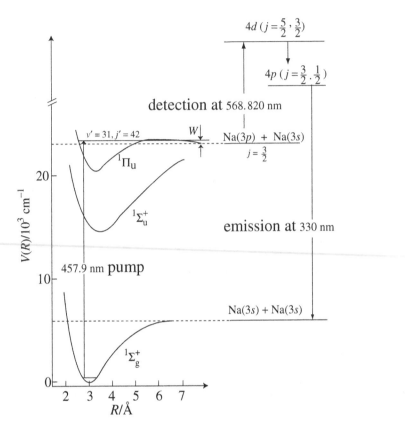

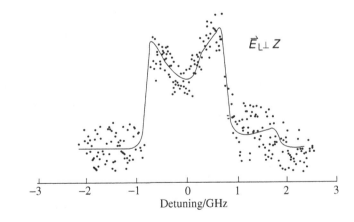

space between stars (the interstellar medium). There are further reasons why we devote a separate section to it. One is conceptual. The initial state for photodissociation is in the transition state region. We refer to it as a "half collision" because we start the process where the fragments that will evolve into the products are close together and interacting. We have no better general methodology for accessing the transition state region and, as we have already seen and we shall further see, different methods sample different regions of configuration space. The other reason is technical: using lasers and supersonic beams we can be very selective in choosing the initial state. A simple but powerful aspect is that by cooling[18] the rotational states of molecules by supersonic expansion, they can mostly populate states of rather low angular momentum. The photon adds (or subtracts) one additional unit. So the optically accessed state has a low angular momentum, a starting condition that is not so easy to probe in collision experiments that tend to favor states of higher angular momentum.* Photodissociation offers an ultimate testing ground for our understanding.

7.2.1 Direct and indirect processes

The excitation may elevate the system to a repulsive region of the potential, pointing toward the exit valley. The products then promptly separate, in a time that is short compared to a rotational period. Much of the available energy will be released in translation, Figure 7.11.

Indirect photodissociation can be the result of more than one type of delay. First, the energy may not be made directly available for the motion along the reaction coordinate. If there is time for energy scrambling then the dissociation will be à la RRKM, as will be discussed further in Sections 7.2.2–7.2.4. There is however another type of delay. It is when electronic energy is made available as vibrational energy of a lower-lying electronic state.

Molecules such as CH_3I or ICN, whose photodissociation dynamics in the gas phase is well studied and analyzed,[19] and which, like H_2O, can form products in more than one electronic state, are being increasingly studied in solution[20] to explore the solvent effects that we discuss in Chapter 11.

7.2.2 Unimolecular dissociation

Energy-rich polyatomic molecules can be photoprepared in a variety of ways. If the barrier to dissociation is not large then even overtone excitation may be

---

* There are other advantages to having a total angular momentum $\mathbf{J}$, $\mathbf{J} = \mathbf{l}' + \mathbf{j}'$, close to zero. In the photodissociation process ABC→A + BC(j'), the rotational angular momentum of BC must then be about equal in magnitude and directed in the opposite direction to the angular momentum l' of the relative motion. So the measured distribution of the states of BC tells us about the impact parameter in the exit valley. If, say, the rotational excitation of BC is high there must be anisotropic exit forces that set the BC product in rotational motion so the geometry of excited ABC is bent, etc.

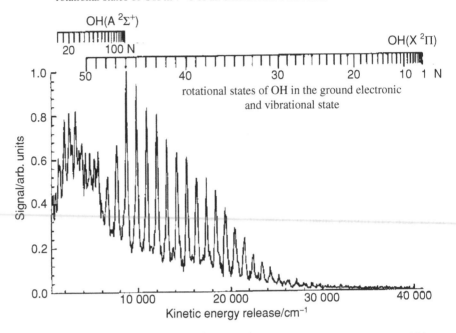

Figure 7.11 Kinetic energy distribution of the fragments that result from the UV photodissociation $H_2O \xrightarrow{\lambda=121.6\,nm} H + OH$ [adapted from D. H. Mordaunt, M. N. R. Ashford, and R. N. Dixon, *J. Chem. Phys.* **100**, 7360 (1994); see also Schnieder *et al.* *Faraday Disc. Chem. Soc.* **91**, 259 (1991), Butler and Neumark (1996), Houston (1996)]. There are two reasons for the very high resolution as shown. One is the high sensitivity for detection of H atoms made possible by immediately exciting them to a long-living Rydberg state. They can therefore be detected after a long flight path,[21] as shown in Figure 1.4. Another facilitating factor is that the light H atom takes the lion's share of the relative energy so that it moves with a high velocity. A third advantage is that Rydberg tagging produces neutral molecules so that space-charge effects can be ignored. Later, the high-lying Rydberg state is converted to an ion and counted. The resolution is sufficient so that, by conservation of energy, Section 6.3.1, the distribution in kinetic energy can be converted to the rotational distribution of the OH, as indicated. There is no evidence for vibrational excitation of OH but, as shown and as is to be expected for a Franck–Condon excitation, high rotational states are produced. The dynamics of this bond breaking is more complicated than a straightforward direct dissociation. There is some OH produced in its electronically excited A state, so the dissociation is not electronically adiabatic. Further analysis shows that there is also some dissociation into three atoms. As a rule of thumb,* when this happens, the two atoms, here both hydrogen,[22] come out with rather different kinetic energies.

---

* This is because the number of accessible final states is much larger for an asymmetric dissociation where first one atom departs and then the other.

sufficient, Section 7.2.4. They can be heated by absorbing infrared radiation and this is so useful that Section 7.2.5 is entirely devoted to this topic. Excitation to the repulsive region of an excited electronic state will cause the molecule to fall apart, but typically it will do so promptly. The reason is that such excitations often involve promotion of an electron from a bonding to an antibonding orbital. It is often the case that the antibonding orbital is fairly localized on a given bond. Upon excitation there is then a sudden repulsion between two particular atoms leading to a selective dissociation of not necessarily the weakest bond. For example, in

$$(CO)_5 Re\text{–}Re(CO)_5 \xrightarrow{h\nu} 2\,Re(CO)_5$$

it is the metal–metal bond that dissociates rather than the weaker Re–CO bond. But even in this case there is enough time for some internal conversion followed by dissociation of the carbonyl bond. In many other molecules internal conversion is the rule rather than the exception. This produces a vibrationally hot molecule in the electronic ground state and if the wavelength is short enough, there is enough energy for bond breaking.

With the potentials shown in Figure 7.12, NO_2 is a well-studied example because it allows comparison with detailed computations for the rate of dissociation as a function of the excess energy, as also shown in Figure 7.12.

The measured temporal decay of the excited NO_2 is not exponential, particularly so for energies just above threshold. This behavior is taken to reflect the variation of the decay rates from one quasi-bound state to the next, as shown in Figure 7.12. The laser (whose temporal width is about 650 fs) excites a superposition of states, each one decaying exponentially but with its individual rate. The observed decay is then a superposition of several exponentials. Why do "states" occur above the threshold? Should the system not be in a continuum? Here we must refer to the nature of the pump–probe experiment as shown in the left panel of Figure 7.12. The short-pulse laser prepares states of NO_2 in the bound region of the $^2B_2$ state. During the short laser pulse these states are stable. These are the bright states of Section 7.1. It is only over a longer time interval that these states couple to the isoenergetic, unbound, continuum states of the lower-lying $^2A_1$ electronic state. In terms of its energy content the molecule is in the continuum yet it has states that are bound but only temporarily so.[23]

The quasi-bound states of NO_2 have the same nature as the states that the Lindemann hypothesis, Section 6.2.1, is all about. The difference is that we refer to the case of NO_2 as *electronic predissociation*, while the typical situation in the Lindemann case is *vibrational predissociation*. To see the similarity (predissociation) and the difference recall that for an energy-rich molecule that can dissociate the total energy is above the threshold for bond breaking but this total energy is spread over more than one mode. There is not enough energy in the reaction coordinate for the molecule to dissociate. In the absence of vibrational energy redistribution, such a state will be stable. In this sense the states are truly bound. But energy does exchange between different motions in the molecule,

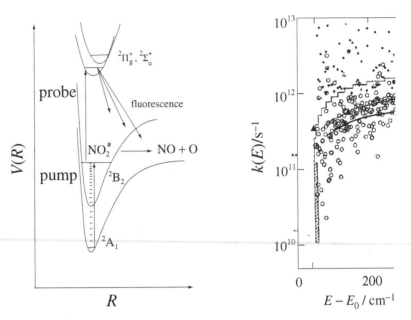

Figure 7.12 Left: a cut through the ground and excited state potentials of NO_2 showing the principle of the time-resolved pump–probe experiment for measuring the rates of dissociation as a function of the excess energy. The pump laser excites molecules to the state $^2B_2$ that is above the dissociation threshold. The excited molecules are probed by excitation to the $^2\Pi_g$ state followed by fluorescence. The graph on the right shows a comparison between the experimental results for the state-resolved rates (triangles and dots) and detailed computations (open circles). By averaging the rates computed for specific initial states over a narrow energy interval we obtain the energy-averaged rate shown as the solid curve. The RRKM-like prediction is the staircase curve, which is computed in the SACM approximation of Section 6.4.5.2. The steps in the RRKM curve are discussed in the text [adapted from S. I. Ionov *et al.*, *J. Chem. Phys.* **99**, 3420 (1993) and B. Kirmse *et al.*, *J. Phys. Chem. A* **104**, 10398 (2000)]. Just above threshold, in the cross-hatched area, one can extract the rates from measurements of the line widths in the frequency domain, see B. Abel, H. H. Hamann, and V. Lange, *Faraday Discuss. Chem. Soc.* **102**, 147 (1995). See Reid and Reisler (1996) for a review.

and eventually it may sufficiently localize in the reaction coordinate so that the molecule will fall apart. In both cases there are states stable for a time sufficient for excitation into them but that will fall apart on a longer time scale. Hence "predissociation." In the case of NO_2 it is coupling between two different electronic states that allows the dissociation. In the traditional Lindemann mechanism it is coupling between different vibrational states.

The dissociation energy of NO_2 is low enough, and it is a small enough molecule, that the density of these quasi-bound states, while high, cf. Figure 7.12, is not overwhelming. Consequently, both experiment and theory can just about see their discrete signature.[24] Higher up in energy or for a bigger molecule, the quasi-bound states are, de facto, a quasi-continuum.

Excited to just above the threshold to dissociation, NO_2 is a very small molecule with just enough excess energy to dissociate. How well can it be described by the statistical RRKM theory? A statistical theory does not seek to resolve state-specific effects because by its very definition all that matters is what is the total energy. The RRKM theory assumes that prior to dissociation the system has "forgotten" how it was prepared. All that we can ask from the statistical theory is to reproduce the energy-averaged behavior, solid line in the right panel of Figure 7.12. Why does the RRKM rate, as a function of the energy, have a step structure? Because the computation is done correctly. Recall that the RRKM rate is $N^{\ddagger}(E - E_0)/h\rho(E)$, where $\rho(E)$ is the density of states of NO_2 at the energy E. It is rather high and hence smoothly increasing. $N^{\ddagger}(E - E_0)$ is the number of states, not counting the reaction coordinate, at the barrier to dissociation. In Figure 7.12 the results shown are for an energy E that is just above the barrier energy E_0. As E increases, more states become accessible, but at threshold there is just one state and so $N^{\ddagger}(E - E_0)$ increases slowly and the count needs to be done discreetly and not by a quasi-classical approximation that produces a smoothly increasing number of states. Each step in the rate, as shown in Figure 7.12, means that one more state at the barrier becomes energetically accessible. Pretty soon, so many states are available that even a discrete count produces a smooth function but, just above threshold, there are distinct increments.

At the statistical level we control the access to the transition state if the steps in the rate of dissociation can be resolved.[25] A triatomic molecule like NO_2 is just about the largest size for which the profile of energy along the reaction coordinate is simple. Figure 7.13 shows such a profile for ketene, CH_2CO, for both the ground state (the lowest singlet state on which there is a high-energy intermediate, oxirene, as shown; in this intermediate the two carbon atoms are equivalent), the triplet state, and an excited singlet. The lowest dissociation threshold is from the triplet surface producing triplet methylene.

The dissociation of ketene has a barrier of about $1300\,cm^{-1}$ above the zero-point energy of the products. Stepwise increases in the dissociation rate constant are seen experimentally, Figure 7.14, and, as for the case of NO_2 in Figure 7.12, these steps can be quantitatively correlated with the quantum chemically computed vibrational energy spacings in the transition state.

Internal conversion allows ketene excited to S_1 to dissociate also to singlet CH_2. For this higher-threshold channel, Figure 7.13, there is no barrier measured with respect to the zero-point energy of the products.* This is a case where the

---

* This is often described by saying that the reverse association reaction has no barrier. In transition state theory terminology, this corresponds to a *loose transition state*. Variational transition state

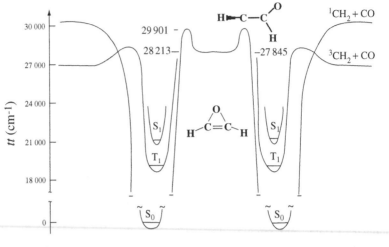

$$CH_2 + {}^{13}CO \longleftarrow CH_2{}^{13}CO \rightleftarrows {}^{13}CH_2CO \longrightarrow {}^{13}CH_2 + CO$$

Figure 7.13 Potential energy profile for the lowest three electronic states of ketene. Note that in addition to the low barrier for dissociation from the triplet state, T_1, there is a higher barrier for formation of a symmetric structure, oxirene, that can rearrange to a ketene molecule in which the two C atoms have exchanged. In the transition state for dissociation the reaction coordinate is a C—C stretch, as shown. The products are detected by laser-induced fluorescence of CO [adapted from A. P. Scott *et al.*, *JACS* **116**, 10159 (1994); see also Moore and Smith (1996)].

transition state is located further out in the products' valley, essentially at the top of the centrifugal barrier. Then the bending vibration of the transition state is essentially a free rotation of the product. Phase-space theory, Section 6.4.5, indeed counts the accessible channels in this way and it matches the observations.*

The other channel in ketene is the possibility of interchanging carbon atoms[26] by going through the symmetric configuration called oxirene, Figure 7.13. Both by direct measurements of the rate of isomerization[27] and by measuring both CO channels in the process

$$
{}^{13}CH_2{}^{12}CO \xrightarrow{\quad h\nu \text{ above } 28\,000 \text{ cm}^{-1} \quad}
\begin{cases}
{}^{13}CH_2 + {}^{12}CO \\
{}^{12}CH_2 + {}^{13}CO
\end{cases}
$$

it is possible to verify that carbon–atom exchange does indeed take place.

---

theory allows for optimizing the location of the transition state, including the option that it will tighten up at higher energies. For a discussion of RRKM theory including such examples see Wardlaw and Marcus (1988).

* The agreement is very good provided the rotational excitation of CH_2 is discriminated against. This is an example of a dynamic constraint as discussed in Section 6.4.

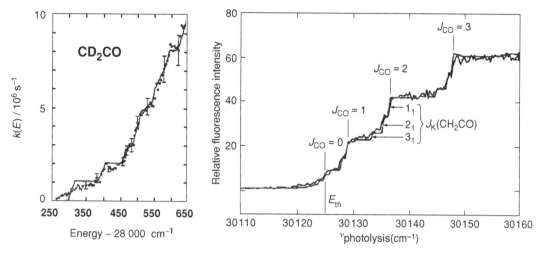

Figure 7.14 Steps in the measured and (RRKM) computed reaction rate for the unimolecular dissociation. Left: the dissociation of CD_2CO to triplet CD_2, as a function of the photolysis energy. This channel has a barrier, see Figure 7.13 [adapted from S. K. Kim, E. R. Lovejoy, and C. B. Moore, *J. Chem. Phys.* **102**, 3202 (1995); see also Lovejoy *et al.* (1992), Marcus (1992)]. Right: a similar plot for the dissociation of CH_2CO to singlet CH_2 in the very threshold region [adapted from Moore and Smith (1996)]. The dissociation is energetically possible already at threshold. Just above the threshold the energies for the steps match the rotational levels of CO. This correspondence is confirmed by a match to the phase-space theory computation shown as a smooth line. The fine structure in the steps can be identified with the low rotational states of the parent CH_2CO.

7.2.3 Access to the transition state region: vibrationally mediated photodissociation

Vibrationally mediated photodissociation[28] starts from selectively vibrationally excited levels on the ground potential energy surface that, upon electronic excitation, fragment differently from photodissociation, at the same total energy but starting from the ground level. This long sentence means that vibrational excitation has a dynamical role beyond the energy that is made available. We have already seen that this is possible in full collisions. Vibrationally mediated photodissociation shows that it is equally possible in half collisions. The point is that, because it is a photoexcitation, the Franck–Condon principle implies that, starting with vibrationally excited reactants, we access a different region of the (upper, dissociative) potential energy surface.

A simplified, one-dimensional, view of vibrationally mediated photodissociation is shown in Figure 7.15. The sequential use of an IR photon and a UV photon accesses the repulsive upper state more efficiently than a single UV photon of their combined energy. The vibrational excitation allows a bond stretch on the

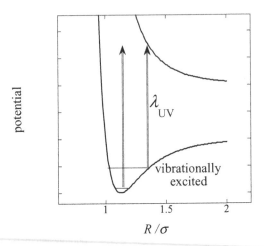

Figure 7.15 Schematic drawing of a one-dimensional cut along the upper state dissociation coordinate, of the ground and electronically excited states potential. If dissociation can be made to start from an excited vibrational motion on the ground potential then a lower-energy UV photon can reach the Franck–Condon region of the repulsive upper state. The energy content in the vibration is exaggerated for clarity. Starting from the bottom of the ground state, a UV photon of the same total energy will have a far smaller dissociation cross-section because of very unfavorable Franck–Condon overlap. Note that IR vibrational excitation of the ground-state polyatomic molecule need not access motion along the dissociation coordinate. It may therefore be advantageous to delay the UV photon until vibrational energy can flow to the required coordinate. When strong lasers are used it may be possible to reverse the order of the two pulses. This counterintuitive route is reviewed by Bergmann *et al.* (1998) and by Vitanov *et al.* (2001).

ground potential such that a lower-energy UV photon can induce a transition to the repulsive upper state.

What is the limitation of a one-dimensional view of vibrationally mediated photodissociation? It is simply that the dissociation coordinate on the upper state is usually not the coordinate that is vibrationally excited on the ground potential. Taking the dissociation of $HONO_2$ as an example,[29] to pump in a significant amount of energy the overtone absorption of O—H is utilized, but it is the O—N bond that breaks:

$$HONO_2 \xrightarrow{\text{IR followed by UV}} HO + NO_2^*$$

The fragmentation is detected by monitoring the fluorescence of the electronically excited NO_2 product. The UV pulse that promotes the system to a higher electronic state is delayed with respect to the overtone excitation. If the delay is just a few picoseconds, the fluorescence is lower because there has not been enough time for energy to flow out of the O—H bond. By measuring the NO_2 fluorescence as a function of the delay time between the two pulses one obtains 12 ps as the rise time for the intramolecular energy transfer. This transfer brings the system upon electronic excitation to a region of the upper potential where the O—N bond promptly breaks.

An example of how a proper positioning on the ground state is important in selective bond cleavage is isocyanic acid, HNCO. It has two chemically distinct photodissociation products

$$\text{HNCO} \xrightarrow{\text{UV}} \begin{cases} \text{H} + \text{NCO} \\ \text{NH}(^1\Delta) + \text{CO} \end{cases}$$

with a minority, spin-forbidden, channel to yield $\text{NH}(X^3\Sigma^-) + \text{CO}$. The main NH + CO channel has a somewhat higher threshold but in a single UV photon dissociation it is the majority channel at energies above threshold. Vibrational excitation of the $v = 3$ overtone of NH followed by a lower-energy UV photon considerably alters the branching ratio, at the same total energy, in favor of the H + NCO channel. The electronically excited state of HNCO is bent while the N—C—O backbone is nearly linear in the ground state. The $v = 3$ overtone of NH is a bright state but it appears to be effectively coupled to a dark state that is a mixed overtone of two quanta of N—H stretch and one quantum of the bend.[30] This dark state allows a much better Franck–Condon access to the electronically excited state.

When reliable potential energy surfaces are available we can explicitly understand how the Franck–Condon limitations determine which product channel is favored. To do so we return to our old workhorse, HOD. In the photodissociation of H_2O, the surviving OH bond is a spectator. For example, vibrationally exciting one O—H to its third overtone (four vibrational quanta) and leaving the other cold produces almost exclusively cold OH fragments. But photodissociating a state with one vibrational quantum in one O—H mode and three in the other produces large amounts of vibrationally excited OH. In vibrationally mediated photodissociation of HOD, the bond that is excited is almost exclusively the bond that breaks, unless we go to shorter UV wavelengths. Then, vibrationally mediated photodissociation produces comparable amounts of OH and OD. Figure 7.16 offers an explanation. It shows the excited state potential and drawn on it are two special contours. One is the total energy for a UV photon of lower energy, when the experimental result is selective bond cleavage. This is because we place the molecule on the excited surface such that the motion carries it preferentially to the desired product. The other contour is the total energy for a more energetic UV photon. The classically allowed regions at these two energies are qualitatively different. At the lower total energy there is no classical passage between the two exit valleys. The Franck–Condon region for an excited O—H stretch is in the exit valley so the bond that is vibrationally excited is the bond that dissociates. At a higher photon energy, the saddle point region on the surface is energetically accessible and the system can be born on one side of the barrier but exit on the other side.

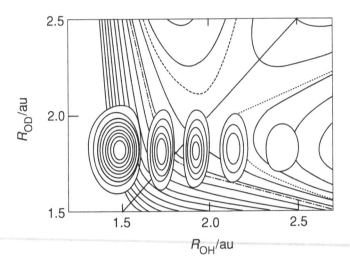

Figure 7.16 Contour plot of the potential energy in the repulsive excited electronic state of HOD. The diagonal line is the bisector between the two exit valleys. Two special energy contours are shown as dashed lines. One energy contour shown as dashed lines corresponds to the total energy made available by exciting the O–H local mode to the fourth state plus a UV photolysis photon, $\lambda = 238.5$ nm, when the experimental result is selective bond cleavage. (The wave function for the stretch is shown as contours. Note the nodes along the O–H coordinate that show that it is the fourth vibrational state.) The Franck–Condon region for the UV photon is near the outer turning point of the O–H stretch, where the total energy is about equal to the potential energy. The other energy contour, dot-dashed line, is for a more energetic UV photon, $\lambda = 218.5$ nm. Now the Franck–Condon region is much closer in and the molecule can exit in either valley [adapted from Crim (1996), see also Schinke (1993)].

7.2.4 Mode-selective chemistry

Vibrational excitation means the stretching of a bond. A chemical change involves a rearrangement of the molecular geometry. Should we not be able to give the molecule vibrational excitation in a mode selective[31] manner such that the energy is made available where we need it at the transition state? Such a goal for many-atom transition states has proved somewhat elusive. In an atom–diatom A + BC collision we can imagine how the modes of the reactants correlate to the motions in the transition state, as discussed in Chapter 6. In a polyatomic system it is not quite so simple to excite vibrations that dynamically enhance the crossing of the barrier.* The motion along the reaction coordinate can be as simple as the

---

* Exciting a vibration adds to the energy of the system and this, by itself, can enhance reactivity, albeit not in a manner that necessarily depends on the initial vibrational mode that was pumped. Even by itself this is worthwhile: how else can we prepare a really hot large molecule? (There are at least two other ways.) The tar on the walls of the heated vessel in an organic chemistry laboratory is one motivation. Collision-free heating, which has no wall effects, is interesting enough and useful

stretching of a bond, but it can also be a mode that is not simply related to modes of vibration of the molecule near its equilibrium.*

The essence of the problem can be seen by thinking of a polyatomic molecule as a number of masses, representing the atoms, connected by springs. If we kick a particular mass then the first thing that happens is that it starts moving. The springs communicate this displacement to the neighboring masses and then these too start to move. Next, atoms further away are set in motion and, after a number of vibrational periods, all the atoms are moving: the initially localized energy is communicated to the molecule as a whole. Our task is to find out why this classical picture, which is not incorrect, is not the whole story either. In particular, why will it possibly fail at lower energies but become increasingly realistic at higher energies? This is the topic of *intramolecular vibrational energy redistribution* or IVR. This is a key idea because it allows us to interpolate between two familiar limiting situations. At quite low energies the molecule is in the spectroscopic regime. It has well-defined vibrational states. These vibrational states are eigenstates and so are stationary with assignable good quantum numbers. At energies below (and above) the dissociation is the RRKM limit: the energy is distributed over all modes. Only the total vibrational energy matters and there are no other good quantum numbers.** In between these two idealized limits is the regime where a molecule can be optically prepared in a bright state but this state need not be stationary. The picket fence model of Appendix 7.A is a good way to think more quantitatively about this regime and we will make use of it below.

For small polyatomic molecules, such as H_2O, NH_3, CH_4, it is possible to excite a vibration and make its energy directly available and we have already seen this, Section 1.2.4, for HOD where the unexcited bond acts as a spectator. For HOD the mechanical picture is clear. The O—H and O—D bonds have rather different frequencies. They are not in mechanical resonance. A vibrational quantum of O—H is sufficiently higher in energy than that of O—D that energy placed in the OH bond remains localized there. Let us carry on with the idea of a mechanical resonance. Such bonds as O—H, C—H, etc. have high vibrational frequencies. Larger amounts of energy can be pumped into them by directly optically accessing

---

enough that it merits its own section, 7.2.5 below. Because it starts with vibrational excitation, it can be made species-selective sufficiently so as to excite one isotopomer and not others.

* Quantum chemical methods can not only locate the transition state configuration but can also evaluate the forces. It is therefore possible to perform a normal mode analysis of the motion. This is just like for an analysis about the equilibrium configuration with one essential difference: the transition state is a saddle point and so the force constant for motion along the reaction coordinate will be negative and the corresponding frequency will be imaginary. Otherwise it is a normal mode of motion and one can, as for any other normal mode, ask what are the displacements of the different atoms that together make up the mode of motion that transverses the barrier. For example, in the transition state of the $Cl + H_2O$ reaction, the reaction coordinate is primarily an O—H stretch. This is consistent with the significant role of OH overtone excitation for bimolecular reactions of water.

** Except for the total angular momentum, the M in RRKM. See also Section 6.4.

their overtone absorption. The key here is the relatively large anharmonicity of these high-frequency vibrations.[32] Say the state corresponding to four quanta of vibration in one O—H bond in water can be reached. Such an initial state has no quanta of vibration in the other O—H bond and let us designate this state as 4,0, meaning four quanta in one OH bond, none in the other. Is it a stationary state? It depends on whether the energy can go to other states. In a harmonic world the 4,0 excited state is isoenergetic with the $v = 3$ excitation of one O—H bond plus one quantum of vibration in the other O—H bond, and with the 2,2 state, etc. But in the real world, anharmonicity means that the initially excited, 4,0, state is lower in energy than the 3,1 state, which in turn is lower in energy than the 2,2 state. These states are nearer in energy to one another than, say, to the 3,0 state. They form a multiplet of states of four vibrational quanta. But the energy differences between the states of the multiplet are such that they are not in mechanical resonance and a laser can distinguish them apart.

For molecules containing a weak bond, say, HOOH, it is possible to provide enough energy by an overtone excitation of an O—H bond so as to dissociate the molecule.[33] For exoergic processes it is only necessary to provide enough energy to surmount the barrier. A well-studied example is the dissociation of the cyclic peroxide tetramethyl dioxane (TMD) that fragments over a barrier of about $110 \, \text{kJ mol}^{-1}$ (about three quanta of C—H stretch) to two acetone molecules

The energy available in the dissociating molecule is high enough that one acetone molecule is electronically excited.

Are there any limitations on the selectivity that we can achieve by vibrational excitation? Unfortunately, yes and these limitations are of two kinds. First, molecules are mechanically not quite so simple. For example, HOOH, that has a trans form at equilibrium, has a barrier to internal rotation about the single O—O bond. The torsional motion has a low frequency. A combined change in both O—H stretch and torsion facilitates the mechanical resonance. In TMD it is the methyl group deformations that provide the bath of low-frequency states. The density of states of an excited molecule need not be high but it is not sparse either. The upshot is that states of molecules with localized excitation do not survive for ever. They may survive long enough for our needs and one can try to be quick, but the states that are optically excited are typically not stationary. The problem is compounded when the energy is higher. At sufficiently high excitation even the vibrational multiplets will begin to overlap in energy and long before that, other modes will offer options for establishing resonances.[34] Typically, at energies well below dissociation one is already in the quasi-continuum of vibrational states, Section 6.2.2.3, where there is a plethora of states. A high density of states acts as a heat sink into which the localized state can dump its energy.* It is possible to be somewhat more quantitative about when a molecule can act as its own heat bath. Say the mean spacing of states is D. Then it takes time of the order of $\hbar/D$ to recognize that we have a discrete manifold of states rather than a real continuum. Problem E puts in some numbers. You will conclude that already at the (fairly low) density of 1000 states cm^{-1} this time is longer than 10^{-9}s, so the molecule in the bulk gas phase will undergo a collision before it has had enough time to discover that it is a quantum mechanical object. In solution, an energy-rich molecule has even less of a chance. The environment helps to very effectively erase any discrete-level structure, but often a large molecule is sufficiently large, Section 6.2.2.3, to practically be in an effective continuum already on its own. We first faced the issue of intramolecular energy redistribution in Section 6.2 and it will remain with us throughout.

To compute the rate of decay we cannot appeal to transition state theory because the rate is Franck–Condon limited. In other words, the system reaches the critical configuration often enough but only rarely does it cross to the other side. In Section 7.1.3.1 we discuss experimental evidence for systems that exhibit such behavior.** In Section 7.1.3.2 we discussed the picket fence model that allows us to write the rate of decay of a state into a smooth quasi-continuum as $(2\pi/\hbar)\left|\overline{V}\right|^2\rho$. Here $\overline{V}$ is the mean strength of the coupling and $\rho = 1/D$ is the density of states of the quasi-continuum to which the initial state is coupled. The perturbation V couples states that differ in the distribution of energy over the modes. It is like the self-anharmonicity of modes except that it is a *cross anharmonicity*. It need not be large to have an observable effect. It is enough if $\overline{V}$ can bridge the energy gap between adjacent states, $\overline{V}/D > 1$, for the decay of the optically prepared state to be faster than the rate of revival.[35]

The density of such states to which the initially excited state can decay can well be much smaller than the total density of vibrational states. But the latter can be so enormous that even a weak coupling will suffice to meet the condition $\overline{V}/D > 1$. But what will happen if we are at a relatively low energy where the bath states are not numerous? Well, the initial state will revive within our observation window. The simplest model of this revival is when there is only one bath (or optically inaccessible) state. Then the initial state periodically disappears and reappears, a phenomenon known as quantum beats and shown in Figure 7.17.

7.2.4.1 Unimolecular dissociation of van der Waals clusters
Dissociation of van der Waals clusters[36] provides a nice illustration of several of the ideas that were just discussed in connection with mode selectivity. Van der

---

* Not all of these vibrational states need to contribute. Quite a number can have too much of the energy in one mode, which disfavors being in resonance with other states.
** We already encountered such behavior in Section 4.3.6, where the rate of decay of tunneling resonances is much smaller than the rate of approach to the barrier that confines these states.

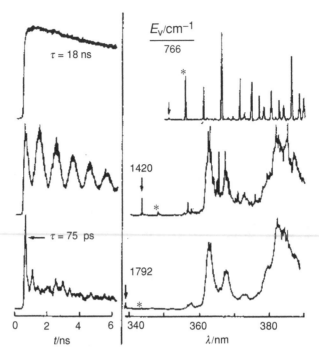

Figure 7.17 Time-resolved decay of the fluorescence of an isolated anthracene molecule in different vibrational levels, of increasing vibrational energy, in the electronically excited first singlet state. Left: time-resolved decay of the population of the excited state. Right: frequency-resolved emission. The energy of the initial vibrational state is given and also indicated by an arrow. By exploring the vibrational energy redistribution in the electronically excited state one can use the fairly prompt fluorescence as a probe for the optically accessible state. At very low vibrational excitation the usual decay of an electronically excited state is observed. The system's temporal evolution is not different from that of an electronically excited atom. The fluorescence spectrum is made up of sharp lines. At a somewhat higher vibrational energy superimposed on the smooth decay are quantum beats due to the revival of the initial state. Further up in the initial vibrational energy, bottom row, there is a fast and essentially irreversible decay. In principle, if one waits long enough, there will be a revival but in practice other perturbations such as collision with a wall or jitter in the detection system or the graduate student having gone home to get some sleep, preclude its observation. Adapted from Felker and Zewail (1988) who have complemented these time-resolved measurements with measurements of the fluorescence spectra in the frequency domain. As expected on the basis of the uncertainty principle, the slow decay results in sharp lines while the spectrum higher up in energy, where the decay is fast, is inherently quite broad. Such natural broadening that is due to the uncertainty principle is often termed *homogeneous broadening* to distinguish it from other causes where the broadening is due to the sample not being a collection of strictly identical absorbers.

Waals clusters,[37] of which we take $I_2 \cdot He$ as a prototype, are species where one or more modes are bound by the far weaker physical-type potential. This shallow potential is still deep enough to support one or a few vibrational states, so in the ground state the He is bound with respect to I_2. However, even if one vibrational quantum of I_2 is transferred to the relative motion of He and I_2, the cluster will promptly fall apart.

A (fairly high) vibrational state of I_2 is prepared when it is electronically excited. This is very similar to the case of ICl discussed earlier and is due to the longer I—I equilibrium bond distance in the electronically excited state. The electronic absorption of $I_2 \cdot He$ is shifted a bit with respect to that of the bare molecule, but is otherwise similar in that one can access $I_2^*(v) \cdot He$ with a vibrational quantum number v in the low twenties. Such an electronically excited cluster can fluoresce and the small shift in wavelength, due to the one-atom "solvation" of $I_2^*(v)$ by He, is sufficient to distinguish it from the fluorescence of bare $I_2^*(v)$. The optically accessed state can dissociate if even one quantum of $I_2^*(v)$ vibration makes its way to the $I_2^*(v)$—He motion. The experimental observation[38] is that some of the fluorescence is from the cluster and some from the free molecule. In other words, dissociation occurs on the same nanosecond time scale as the fluorescence.

On the nanosecond time scale, the I_2^* molecule vibrates some 10^4 times. So only in about one in every 10^4 vibrations does a quantum of energy flow from the I_2 bond to the He—I_2 motion. This is understandable because the resonance between the two modes is of quite a high order.[39] The van der Waals potential is rather shallow so the vibrational frequency is much lower by comparison to the vibrational frequency of I_2^* (which, by itself, is low compared to that of other diatomics). The dissociation of the energy-rich cluster does not conform to the RRK model. It is much more similar to the restricted vibrational energy redistribution of molecules.*

7.2.5 Multiphoton dissociation

Multiphoton dissociation is a way of using infrared photons, from a higher-power laser, to up-pump polyatomic molecules, to an energy sufficient to induce unimolecular reactions.[40] Superficially, it might appear at first thought not to work. But it does, and remarkably well. Monitoring the angular distribution of the fragments by dissociating molecules in a beam shows a forward–backward symmetry, the signature of the sprinkler model, meaning that the energy-rich molecule executed at least a few rotations prior to dissociation. It also verifies that an isolated

---

* What restricts the energy transfer in $I_2^*(v) \cdot He$ is the poor Franck–Condon overlap between $I_2^*(v) \cdot He$ and the dissociation channel, $I_2^*(v-1) + He$. In Chapter 9 we develop the exponential gap principle as a semi-quantitative estimate of how far one can deviate from a naive Franck–Condon view, namely that nuclei are rather reluctant to undergo a change in their momentum. Then the inefficient energy transfer in $I_2^*(v) \cdot He$ will provide quantitative experimental evidence for the principle.

molecule does the absorption so that the process is truly unimolecular. Under-standing why multiphoton dissociation works invokes all the theoretical concepts at our disposal. Proposing new applications for its many capabilities is up to you.

The phenomenon is easy to state. Large molecules can be made to absorb a high number of IR photons, sufficiently so that they acquire enough energy to isomerize or to dissociate. The unexpected aspect is that these photons can come from a laser that has a very well-defined frequency. Say the first photon matches a vibrational absorption line of the molecule.* Then, because of anharmonicity, the second photon will no longer be resonant and the process will stop. This is the conventional wisdom backed by lots of experience. There are three ingredients in the explanation why the process keeps going.

The first point is that we are typically talking here about larger molecules. One of the attributes of "large" is that not all atoms can be near to the center of mass and so large molecules necessarily have large moments of inertia and hence closely spaced rotational levels. This fact helps to explain why the second photon can also be about resonant. Say that the first transition is from $v = 0, j$ to $v = 1, j - 1$; the frequency is $v - 2Bj$. Now go from $v = 1, j - 1$ to $v = 2, j - 1$, a Q-type transition; the frequency is $v(1 - x_e)$, where x_e is the anharmonicity. It should be possible to find a value of j such that the two frequencies are nearly the same. In fact, let's go even higher up. The next transition can be from $v = 2$, $j - 1$ to $v = 3, j$; the frequency is $v(1 - 2x_e) + 2Bj$. It is again about resonant. "Rotational compensation" of the anharmonicity can take us up to $v = 3$.** Taking the typical IR frequency to be that of a CO_2 laser, about $1000 \, cm^{-1}$, rotational compensation can help us deposit about $3000 \, cm^{-1}$ in the molecule.

The second point is that a large molecule with an energy content of $3000 \, cm^{-1}$ is on the threshold of its quasi-continuum, particularly so if you take its initial thermal energy content into consideration. To emphasize this point, Figure 7.18 shows the vibrational levels of SF_6, a molecule whose multiphoton dissociation received much attention. Once a large molecule absorbed three to four photons, it is getting to be black in the sense that it will absorb further at almost any IR-active frequency. The third ingredient is then the quasi-continuum and the mixing of levels within. If so, the energy that is initially selectively put into a particular

---

* Experimentally we know that this is so because the method is species-selective. Indeed it is selective enough to be used for isotopic separation (because of the isotopic shift in the vibrational spectrum) or for dissociating (followed by scavenging of the radicals) trace impurities from chemicals that need to be ultrapure.

** Seemingly this route is not very efficient because only molecules in a particular rotational level can be pumped up. If the rotational spacing $2B$ is small, many rotational states are populated in a thermal sample at room temperature. They will not absorb. But if rotational spacing is large, we show in Chapter 9 that rotational-state-changing collisions are very efficient. So in the bulk, molecules in a given j absorb and are pumped up. This upsets the rotational state equilibrium and collisions restore the population in level j. These molecules absorb, and so on. Hence, in the bulk, many molecules are efficiently heated.

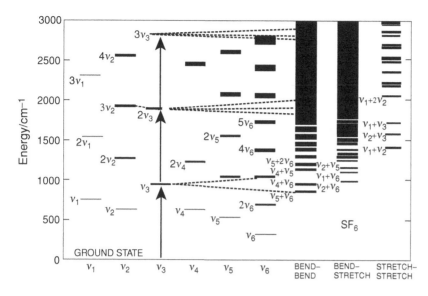

Figure 7.18 A partial listing of vibrational states of SF_6 at energies below 3000 cm^{-1}. The arrows represent excitation of the v_3 mode of SF_6 by CO_2 laser photons. As is evident even for one or two quanta in v_3 there are near-resonant energy transfer pathways to other modes [adapted from M. I. Lester *et al.*, *J. Chem. Phys.* **80**, 1490 (1984)].

mode will eventually spread out over all modes. When we say that multiphoton excitation prepares "hot" molecules we really also mean hot in the technical sense of energy randomly distributed over all available modes.

There is still the problem of fine-tuning the rotational compensation. The quantum number j is discrete so we cannot hope to compensate exactly for the anharmonicity. The third point is that multiphoton excitation was only discovered when high-power lasers became available. These were lasers of $10^6 - 10^8$ W cm^{-2}.* The new aspect at such powers is that power broadening begins to have qualitatively new implications since for a transition dipole of 0.1 debye, the broadening (that scales as the power$^{0.5}$) is 0.1–1 cm^{-1}. From the point of view of the uncertainty principle, power broadening arises from the shortening of the lifetime of the level because, at high densities of photons, light absorption and stimulated emission, that are both proportional to the density of photons, become rather fast. The inherently shorter time that the molecule spends in a state means that its energy acquires a width. We emphasize that this broadening is measurable in the frequency domain as wider lines. The effective onset of the quasi-continuum is therefore not just a property of the molecule but can be brought about at lower energies if the laser power is high enough.

7.2.5.1 *Infrared multiphoton dissociation*
Once the energy pumped into the molecule exceeds the dissociation threshold it is in a real continuum. However, we already recognized that a large molecule

---

* Estimate (order of magnitude) the light power reaching the floor of the room where you are. Your answer is likely to be below 1 W cm^{-2}. On the other hand, nowadays we will not think of 10^6 W cm^{-2} as truly high-power. Table-top lasers with output reaching all the way to 10^{20} W cm^{-2} are available.

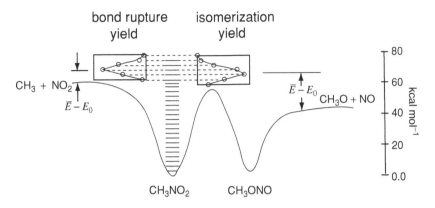

Figure 7.19 The energy profile along the reaction coordinate for the dissociation of nitromethane [adapted from A. M. Wodtke, E. J. Hintsa, and Y. T. Lee, *J. Phys. Chem.* **90**, 3549 (1986)]. Shown are the successive energies made available by the photons of the IR laser. Not all molecules absorbed the same number of photons. See Wodtke *et al.* (1986). $\overline{E}$ is the mean energy in the molecule. In this experiment an attempt was made to fit the potential, including the barrier height for isomerization, to CH_3ONO using the observed translational energy distribution of the products. For this purpose it is necessary to compute the RRKM rates as a function of energy, cf. Figure 7.20. The analysis also makes use of the observed translational energy disposal that can also be computed from RRKM if we assume, as shown, that there is no barrier to the reversed association of the radicals. the transition state is then on the way out in the exit valley and the distribution of energy in the products is therefore accurately reflected by the distribution of energy in the transition state. The extracted potential is as shown in the figure.

will have very many quasi-bound states at energies above the threshold of the continuum. So the molecule will continue to absorb light in competition with its unimolecular dissociation. The multiphoton-pumped molecule has its energy shared among many modes, so the RRKM estimate for the rate constant should be valid. At energies not too high above threshold the rate of dissociation of a large molecule, while rapidly increasing with energy, is not high. Molecules can therefore be pumped by multiphoton excitation to energies well above the lowest threshold for dissociation.

Figure 7.19 shows an energy profile inferred mostly from experimental measurements of the branching ratio between competing channels in the multiphoton dissociation of nitromethane

$$CH_3NO_2 \xrightarrow{\;nh\nu\;} \begin{cases} CH_3 + NO_2 \\ CH_3O + NO \end{cases}$$

and the translational energy disposal.

The methoxy radical, CH_3O, is assumed to form via an isomerization of the hot molecule to CH_3ONO, as shown with an isomerization barrier height of 55.5 kcal mol^{-1}. At lower energies, RRKM rates favor the dissociation over a

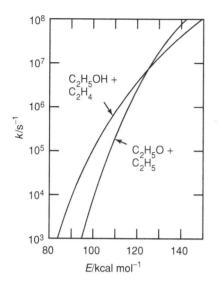

Figure 7.20 RRKM computations of the unimolecular rate constants vs. energy for the two decay channels in the dissociation of diethyl ether, as identified in the insert [adapted from L. J. Butler *et al.*, *J. Phys. Chem.* **87**, 5106 (1983)]. The dissociation into $C_2H_5O + C_2H_5$ has a higher activation energy as compared to the elimination of ethylene. At lower energies it is not the favored channel, but its transition state is looser and so has softer vibrations and less hindered rotations. Entropy therefore favors this higher channel as summarized by its having a higher A factor ($10^{17.3}$ vs. $10^{13.9}$ s^{-1} for $C_2H_5OH + C_2H_4$). At high energies the entropic factors are more dominant and the dissociation into $C_2H_5O + C_2H_5$ has a higher rate. IR multiphoton pumping can heat molecules to well above the lowest barrier to dissociation and reach higher-energy channels.

lower barrier as shown in Figure 7.19. It is also shown that the importance of the two dissociation channels is reversed at higher energies. This is an example of the energy vs. entropy competition that we first discussed in Chapter 6. At higher energies a channel with a higher threshold can become dominant if the corresponding passage through the transition state is broader. If we write the RRK rate as $A((E - E_0)/E)^{s-1}$ then, as follows from Section 6.2.2.1, A is the Arrhenius A factor. An entropically favored channel is one with a higher value of A. This competition is shown for the two decay channels of diethyl ether in Figure 7.20. The dissociation into $C_2H_5O + C_2H_5$ has a higher A factor and a higher threshold energy. It therefore becomes dominant at higher energies.

The observed rate of dissociation (and the branching between different decay channels) reflects the distribution of (total) energy because not all molecules have absorbed the same number of photons. Above threshold, the distribution is determined by a competition between dissociation, which depletes energy-rich molecules, and further up-pumping by the laser. Is it possible to pump molecules quickly enough so that the sideways siphoning off of energy-rich molecules is too slow to compete? If so, we could thereby reach higher up in energy and maybe even beat the branching ratios as governed by RRKM. The alternative scenario for what happens at fast up-pumping is known as *ladder switching*. The parent molecule dissociates statistically but the (hot) fragments, which are typically already in their quasi-continuum, can further absorb light and themselves dissociate, and so on. Either mechanism leads to extensive fragmentation as shown experimentally by the appearance of high-energy end products.

7.2.5.2 Multiphoton ionization/dissociation

Ions can be collected very efficiently and detected mass-selectively. The method of choice for the production of ions for analytical purposes is the species-selective *multiphoton ionization*,[41] with lasers in the visible or the UV. When used for identifying the absorbing species the method is implemented via a "resonance-enhanced" route as discussed below. But using very high laser powers, it is possible to ionize in an indiscriminate (or not resonant) fashion any species that is present in the path of the laser beam. In dynamics, multiphoton ionization is used to probe reaction products as noted in Sections 7.1.2.1 and 7.1.2.4 and in specific examples.

Resonance-enhanced multiphoton ionization (REMPI) means that the first transition is to a bound excited electronic state of the molecule, typically the first excited singlet, S_1, or to another state with a strongly allowed absorption. Because it is a resonance transition it is species-selective. The second transition, using one or more photons, is from the intermediate level to the ionization continuum. This transition can be carried out, if so required, with a second laser.

Power broadening of allowed electronic transitions (those with a large transition dipole) means that when the laser is high-powered enough a non-resonant absorption is also possible. The scaling of power broadening means that a 1 cm^{-1} broadening at 10^8 W cm^{-2} (for a transition dipole of 0.1 debye) scales to[42] several hundred wavenumbers at powers of 10^{13} W cm^{-2}. So this high power is more or less the point where all discrete transitions are broadened beyond recognition. The molecule becomes "black," meaning that it absorbs at any frequency.

The molecular ion formed upon ionization can be stable or it can fragment if, due to Franck–Condon factors, it is formed in a non-equilibrium configuration and hence it is born energy-rich. A fragment can also continue to absorb photons in a mechanism that we called *ladder switching*.[43] The secondary fragments can continue to absorb photons. At high laser power we can break a molecule all the way down to atomic ions and electrons,[44] Figure 7.21.

Stable polyatomic molecules live in a relatively deep well on a potential energy surface. When there are known isomers, say toluene ($C_6H_5CH_3$) and cycloheptatetraene (C_7H_8), they are separated by a high barrier,* as shown in Figure 7.19

---

* For less rigid species such as clusters or linear molecules there are many more isomers, some of which differ only in conformation ("conformers") with a low barrier between them.

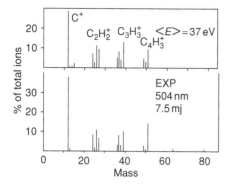
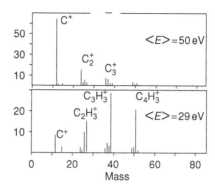

Figure 7.21 Left: observed, labeled EXP, and computed fragmentation pattern of benzene by multiphoton ionization. The experimental results, adapted from D. A. Lichtin, R. B. Bernstein, and K. R. Newton, *J. Chem. Phys.* **75**, 5728 (1981), are for a laser wavelength and intensity such that the same pattern can be obtained at a different wavelength and intensity. This supports the idea that the pattern depends only on the mean energy absorbed from the laser. The computations [J. Silberstein and R. D. Levine, *J. Chem. Phys.* **75**, 5735 (1981)] assume that this is so and the mean energy is shown. Right: computed fragmentation pattern at a lower energy, where the fragmentation is less extreme and at a higher mean energy, where the C^+ ion is the main product.

for the simpler case of CH_3NO_2 and CH_3ONO. At the high energies made available by multiphoton pumping, such barriers can be transversed. The multiphoton ionization–fragmentation spectra of isomers can therefore be essentially indistinguishable.

7.2.6   On to quantum control and the time domain

In this chapter we are using light to make the system access the transition state region. The selectivity that is possible in the frequency domain has often invoked the limitations provided by the Franck–Condon principle to place a molecule where we want it on the upper potential energy surface. To restrict the range of accessible configurations on the ground state potential we can go down from an excited state. The dumping down onto the ground state can be light-induced (as in stimulated emission pumping) or spontaneous (as in internal conversion). Accessing the ground state from an excited state via a conical intersection is a particularly promising way of reaching a localized region in configurations, and we will return to it in Chapter 8. An alternative option is to start from a stable transition state precursor such as an anion and detach the extra electron. But if an initially localized state is what we are after,[45] can we not directly impose it? In Chapter 8 we shall examine how far this is possible by working in the time domain. In Section 7.4 we continue with this theme, achieving selectivity by optical means where the new point is that we specifically rely on the quantal aspects of the motion

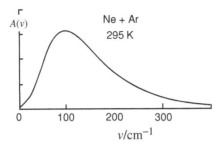

Figure 7.22 The bimolecular absorption coefficient for a mixture of Ne and Ar in the infrared. This collision-induced absorption arises from the transient dipole moment that arises during the flyby collisions of Ne with Ar atoms. In other words, it is the heteronuclear NeAr transitory species that is absorbing light. Adapted from D. R. Bosomworth and H. P. Gush, *Canad. J. Phys.* **43**, 751 (1965).

of the nuclei. So far however we concentrated on unimolecular processes. Before going further we need to discuss the effect of light on bimolecular collisions.

7.3   Bimolecular spectroscopy

Only at very low densities do spectroscopic observations refer to isolated atoms or molecules. At higher densities collisions can perturb the act of light absorption or emission. At the high densities characteristic of liquids and solids the molecular interactions can be so severe that the spectrum is changed beyond recognition. In this section we begin first with collisions as perturbations but then we turn the point of view around and speak of the spectroscopy of the collision act and thereby provide yet another means to access the transition state.

7.3.1   Collision-induced light absorption

An example of a bimolecular spectroscopic phenomenon is collision-induced absorption. In its essential form it is exhibited as a broad-band absorption (usually in the far infrared) that is observed in a gaseous sample containing different rare gases. The absorption of photons at the frequency v is found to depend upon the product of the densities of both gases, and, as usual, on the path length l,

$$\log[I(0)/I(l)] = ln_A n_B \varepsilon(v) \tag{7.6}$$

The *binary absorption coefficient* $\varepsilon(v)$ is independent of the density, but does depend upon the species present. This absorption is *not* found in the pure rare gases and obviously disappears in the very low density limit. It is a bimolecular phenomenon and requires the participation of dissimilar atoms.

   The spectrum shown in Figure 7.22 can be considered as the spectrum of the transient heteronuclear AB dimer that fleetingly exists during the A + B collision. AB is a very short-lived species and this is reflected in the spectrum as we discuss

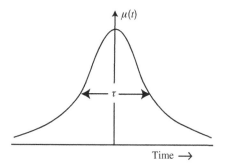

Figure 7.23 A schematic representation of the time-dependence of the (transient) dipole moment associated with a flyby collision of unlike atoms, at a given b and E. The "width" in time, τ, of the dipole-moment function can be related via the uncertainty principle to the frequency of the most intense absorption of radiation.

below. The far IR absorption is not found in the pure gases for the same reason that stable homonuclear, A_2, diatomics do not have an infrared spectrum.* In classical language, light is absorbed or emitted when a dipole is oscillating or, in general, changing with time. A heteronuclear diatom has a dipole moment and will have an infared vibrational spectrum due to the oscillations of the AB distance. Similarly, a transient AB heteronuclear pair has a dipole moment $\mu(R)$ that is a function of the AB distance. During the collision the AB separation varies with time and so does the dipole moment, i.e., $\mu(R) = \mu[R(t)]$, as shown in Figure 7.23. This variation with time is how the transient dimer can absorb light, when the frequency of light is within the range of frequencies characteristic of the time dependence of $\mu[R(t)]$. Technically speaking, what we need to do is to expand the time dependence of $\mu[R(t)]$ in a Fourier series and check which frequency components have a large contribution. We will simplify this task, see below, by appeal to the uncertainty principle: the range of important frequencies is inverse to the range in time over which $\mu[R(t)]$ varies.

It is important to stress that we are considering truly transient dimers that exist during a flyby collision. These transient dimers have an absorption spectrum with a width $\Delta \nu$ of about 200 cm^{-1}, Figure 7.22. Hence the average duration of their existence is

$$\tau \cong h/hc\Delta\nu = (c\Delta\nu)^{-1} \cong 2 \cdot 10^{-13} \text{ s} \tag{7.7}$$

which is indeed comparable to a typical duration for a collision between two rare gas atoms at ordinary temperatures, where a is the range of the force between the atoms

$$\tau \cong a/\nu \cong 10^{-8}/5 \cdot 10^4 = 2 \cdot 10^{-13} \text{ s} \tag{7.8}$$

---

* With a sensitive enough detection one can observe a Raman spectrum from a single component rare gas mixture due to the time-dependent polarizability that characterizes an A–A collision.

In Section 7.3.4 we discuss collision-induced UV absorption where the AB dimer is pumped to an electronically excited state.

As we already discussed, broad absorption spectra can also be seen for half collisions, that is, for an AB species that has a finite lifetime because it can dissociate. Examples include the electronic predissociation of O_2 due to crossing from a stable to a repulsive electronic state, Section 7.1.1, vibrational predissociation when a state with energy partitioned over many vibrational modes is isoenergetic with the dissociation continuum, Section 7.2.2, and the orbiting states, trapped behind the centrifugal barrier, Section 4.3.6. Typically such quasi-bound states have narrower absorption lines than the broad spectrum shown in Figure 7.22. The reason is clear. The width of the spectrum in Figure 7.22 is inverse to the time the system spends over the (shallow) rare gas–rare gas potential. Hence the width is the vibrational frequency ν. For other quasi-bound states the width is typically of the form $\nu p, p \ll 1$, where the probability p is the likelihood of dissociation per vibrational period. For example, for orbiting states, p is the probability of tunneling through the barrier.

7.3.2　Pressure broadening of spectral lines

An important spectroscopic manifestation of state-changing collisions is the phenomenon of pressure broadening of molecular absorption lines.[46] Consider an excited vibrational state of the molecule, which in a very low-density gas (i.e., in the absence of collisions) would lose its excitation by radiative decay. Collisions can change the state or reorient the molecule and hence serve as a mechanism to quench the emission of that excited state.* The lifetime of the molecule in the absence of collisions, the radiative lifetime, is long compared to the duration of a collision.** The quenching and "interrupting" collisions shorten the lifetime of the excited state so that the spectral line will then be broadened. To argue that this broadening is proportional to the pressure we proceed as follows. An excited state can decay either through a unimolecular fluorescence

$$X^* \rightarrow X + h\nu$$

or through a bimolecular quenching, that is, by a change of state induced by collision with some partner M,

$$X^* + M \rightarrow X' + M'$$

---

* In addition, the perturbation of the molecular levels by collision modifies the emission process owing to changes in the time evolution of the transition dipole. We consider such processes as "interrupting" collisions.

** Typically 1–10^{-3} s for emission in the infrared and decreasing with increasing frequency as ν^{-3} so that it is in the nanosecond range for allowed transitions in the UV. We say "allowed transitions" because the radiative lifetime also scales inversely with the squared transition dipole moment.

where the prime denotes new states. Hence, the kinetic equation for the concentration of the excited species is of the form

$$\frac{d[X^*]}{dt} = -[X^*](k_r + k_q[M]) \tag{7.9}$$

$\tau_r = 1/k_r$ is the radiative lifetime and k_q is the bimolecular rate constant for the quenching process. The rate for the decay of the concentration of the excited state is given by

$$\tau^{-1} \equiv -[X^*]^{-1} \frac{d[X^*]}{dt} = k_r + k_q[M] \tag{7.10}$$

meaning that the lifetime τ of the state will be shorter than the radiative lifetime τ_r. The width $\Delta\nu$ of the line is therefore larger than the natural (or radiative) width $\Delta\nu_r$

$$\Delta\nu = (c\tau)^{-1} = k_r/c + nk_q/c = \Delta\nu_r + nk_q/c \tag{7.11}$$

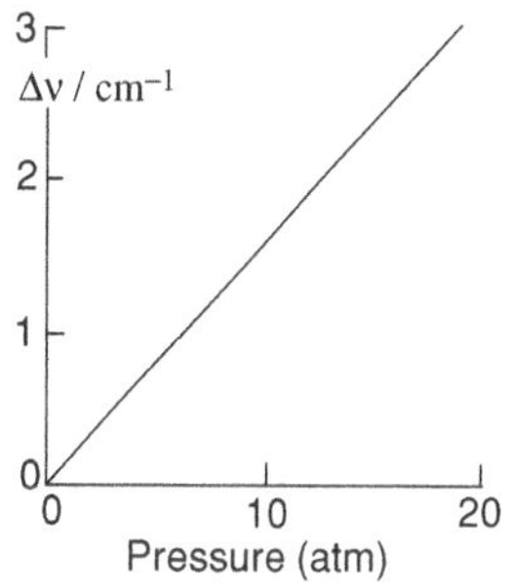

Here k_q is the bimolecular rate constant for quenching the excited level and n is the number density of the quenching collision partner. As shown in Figure 7.24, the observed spectral line width increases linearly with pressure and the slope allows the determination of the quenching rate constant.

The infrared spectra of molecules contain many adjacent lines that correspond to different rotational–vibrational transitions. As the pressure increases, these lines broaden and begin to overlap. Ultimately, at high pressures, they merge together and the detailed structure is lost. This collapse due to pressure broadening is typical of spectra at high pressures and of molecules in the liquid phase.

Emission from electronically exited states is also pressure broadened (Myers, 1998). Application of the Franck–Condon principle allows us to provide an intuitive picture of how collisions perturb the emission. We use the familiar* yellow sodium D line, Na(3^2P) → Na(3^2S), as an example. Figure 7.25 shows the interaction potentials of the ground and excited states of Na with an Ar atom. Because the excited Na atom has a significantly higher polarizability, the upper potential has a stronger long-range attraction,[47] leading to a deeper well. When the rare-gas atom is far away, the spacing between the two potential curves is the energy corresponding to emission on the D line. But if emission takes place when the two atoms are nearer to one another, and the emission occurs without a change in the momenta of the two atoms, then it must occur vertically and the energy of the photon is now somewhat lower (vertical arrow in Figure 7.25). This emission

Figure 7.24 Width of an absorption line of HCl in the presence of Kr at increasing pressures. The slope of the line provides a determination (cf. Eq. (7.11)) of the rate constant k_q for state-changing collisions of HCl with Kr. The very high rates of rotational state changes, cf. Section 9.2, too fast to be measured by conventional time-resolved kinetics, were first determined by such a frequency domain measurement [adapted from D. H. Rank *et al.*, *J. Mol. Spec.* **10**, 34 (1963)].

---

* The Na D emission can be efficiently generated by an electrical discharge through sodium vapor and so such lamps are used for high-power illumination. But some people find the intense yellow color uncomfortable. The solution: use a high-pressure lamp so as to shift the emission toward the red, as we discuss below.

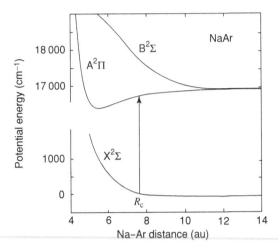

Figure 7.25 Potential energies (cm^{-1}) for the interaction of a ground- and an excited-state Na atom with an Ar atom vs. the internuclear distance. The ground state correlates to Na(3^2S) + Ar while the excited state asymptotically corresponds to Na(3^2P) + Ar. On the energy scale needed to plot both curves, the well in the ground-state potential is barely visible. In the Franck–Condon vertical transition approximation the energy of the photon emitted when the two atoms are at a distance R apart is the difference between the two potentials at that distance R, $hV = V_{upper}(R) - V_{lower}(R)$. Such a transition is shown at a particular distance R_c. Note that over most of the accessible range at thermal velocities, the photon will be red-shifted compared with the emission of an isolated Na* atom, as seen in the experimental results in Figure 7.26. The arrow shows a vertical transition at a frequency shifted to the red. The repulsive B$^2\Sigma$ excited state plays a role in the observation of absorption during the collision, Section 7.3.4. Here, because the upper state is rather steeply repulsive, the Franck–Condon approximation indicates that the photon needs to be detuned to the blue.

is shifted to the red of the D line because if the transition is vertical, the energy $h\nu$ of the photon emitted when the atoms are a distance R apart is

$$h\nu = V_{\text{excited}}(R) - V_{\text{ground}}(R) \tag{7.12}$$

The extent of the shift is sometimes referred to as the detuning. If the emission takes place when the two atoms are rather close to one another, the emission will be shifted to the blue. This requires a close-in approach (meaning a low impact parameter collision) and preferably an above thermal collision energy (kT at room temperature is about 200 cm^{-1}). Typical experimental results are shown in Figure 7.26 and confirm that there is a broad shoulder of red-shifted emission.

There are two immediate implications of our discussion. If electronically excited atoms can emit to the ground state during a bimolecular collision, why can the ground-state atoms not absorb light during their collision? Well, they can, and we take up this theme in Section 7.3.4. There we will also return to our theme

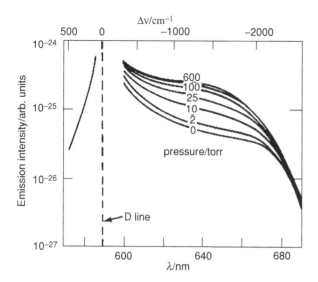

Figure 7.26 Measured line profile (logarithmic scale) for Na D line emission in the presence of Ar, at a pressure of below an atmosphere and 418 K. Most of the emission is at the wavelength of the unperturbed D line (the collisionally unperturbed line is inherently broad due to the spontaneous decay). Shifts primarily to the red but also to the blue are evident. Below an atmosphere the blue wing is not pressure dependent. The upper scale gives the shift (= detuning) with respect to the position of the D line. Some of the emission is from bound states of the upper electronic state [adapted from G. York, R. Scheps, and A. Gallagher, *J. Chem. Phys.* **63**, 1052 (1975)].

of interference and argue that there can be interference between absorption on the way in or absorption on the way out. Our next topic is emission of molecules in the process of falling apart, picking up this theme from Section 5.1.2.1.

7.3.3 Emission in half collisions

If an electronically excited sodium atom can emit during its bimolecular collision with an Ar atom, why can it not emit during a half collision when it recedes, say, from an I atom? Why an I atom? Because NaI can be excited in the UV to an electronically excited state that dissociates to a $Na(3^2P)$ state, the state that emits the D line, and an I atom, Figure 7.27.

The ground electronic state of NaI is much more strongly bound than the electronically excited state so that, in this case, the D line emission is shifted to the blue. To estimate the fraction of molecules that emit we take a radiative lifetime of several nanoseconds and the time of about 10^{-13} s for Na and I to recede from one another, cf. Eq. (7.8). So only fewer than one in 10^4 molecules

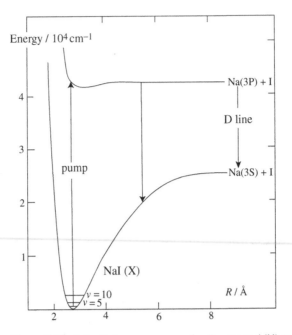

Figure 7.27 Potential energy curves for the ground (X) and the electronically excited state of NaI that can be reached by a UV excitation ($\lambda \sim 235$ nm). This state dissociates to an electronically excited Na(3^2P) atom and a ground-state I atom. The vertical excitation from the ground state, shown as an arrow, accesses the inner region of the upper state with enough excess energy to allow dissociation. In the course of falling apart the emission from the excited state, vertical arrow, will be spread over a wide frequency range and be very blue-shifted. This emission leads to bound (but possibly highly vibrationally excited) NaI in its ground electronic state. The vibrational structure is hard to resolve for a molecule with low vibrational spacing such as NaI, but can be seen in the case of photodissociation of CH_3I as discussed in Figure 5.7. The earlier on its way out the NaI molecule emits, the larger is the shift to the blue [adapted from H.-J. Foth, J. C. Polanyi, and H. H. Telle, *J. Chem. Phys.* **86**, 5027 (1982)].

will emit on their way out. Moreover this emission will be spread over a wide frequency range so that the wing of the D line will be of low intensity.*

As is clear from Figure 7.27, the earlier on its way out the NaI molecule emits, the larger is the shift to the blue. This is quite interesting because by a measurement in the frequency domain we find out about where the system was. This is just what we are missing so far. The absorption spectrum tells us how quickly the molecule departs from the Franck–Condon region it was excited to, but it does not tell us where the system went to. This Raman-type experiment

---

* This is unlike pressure broadening where the intensity of the wings can be made to increase by increasing the pressure.

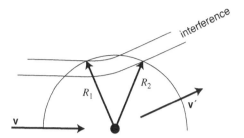

Figure 7.28 A construction showing the two possible locations of the interatomic distance, solid arrows, at which absorption of a photon of a given frequency is possible in a collision of a ground-state Na atom with Ar [adapted from Grosser *et al.* (1999)]. For a transition from the $X^2\Sigma$ ground state to the $B^2\Sigma$ excited state, the transition dipole lies along the interatomic distance R. The circle is that value of R that is the implicit solution of Eq. (7.9) for a photon that is blue-shifted by a given detuning from the D line. For this detuning the absorption occurs when the trajectory crosses the circle. Two such circles correspond to two trajectories that traveled for different lengths on the ground- and excited-state potentials. One made the transition at R_1 and the other at R_2. The deflections of these two trajectories are such that they result in the same angle of scattering of the excited Na atom.* Two such trajectories can exhibit quantum mechanical interference and the experimental manifestation is shown in Figure 7.29.

(light in and light out) provides just what we wanted to know. Problem L outlines a time-dependent point of view that allows us to further develop this idea.

7.3.4 Spectroscopy of elastic collisions

We already have all the machinery that is needed to think about the UV absorption of molecules in the process of their collision. In this subsection we consider the simpler case where the collision is elastic so that, in the absence of radiation, the only result is the deflection of the atoms. Even for this simple case we will go all the way up to control. Reactive collisions are even more of a challenge and are discussed in the next subsection. As always in this chapter the final goal is to access the transition state.

We return to a Na + Ar collision. For an incident photon with a frequency to the blue of the Na D line, absorption will lead preferentially to the $B^2\Sigma$ excited state, as shown in Figure 7.25. Here the detuning of the photon selects, via Eq. (7.12), the interatomic distance at which the absorption takes place. Figure 7.28 shows two trajectories for a Na + Ar collision. Superimposed is

---

* It is the condition that both trajectories lead to the same scattering angle that excludes the absorption of the photon taking place at the other point where the trajectory crosses the circle. For example, if the trajectory that absorbed at R_2 had instead absorbed earlier it would have traveled more on the more attractive upper potential and have collected a different deflection.

Figure 7.29 The
observed (points) and
computed (solid curve)
laboratory differential
cross-section of excited
Na atoms from a Na + Ne
collision with a photon
detuned by 240 cm^{-1}
from the sodium D line.
The collision velocity is
1125 m s^{-1} [adapted from
Grosser *et al.* (1999)].

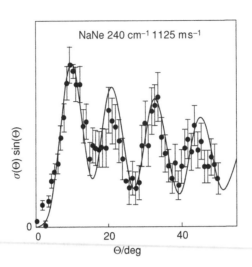

a circle at that particular value of R, as determined by Eq. (7.12), for which a vertical transition is possible. In the absence of absorption the scattering results in a deflection. It is seen that absorption of the photon is possible either as the two atoms are on their way in or on their way out.

The excited Na atom can be efficiently detected by exciting it further to a high Rydberg state, cf. Figure 7.11, and its angular distribution can be determined at high resolution as shown in Figure 7.29.

As shown in Figure 7.28 there can be two classical trajectories that lead to the same deflection angle. As in Section 2.2.6, the differential cross-section can therefore be written as a superposition of two scattering amplitudes

$$I(\theta) = |f_1(\theta) + f_2(\theta)|^2 \qquad (7.13)$$

The resulting interference pattern is very clearly seen in Figure 7.29.

When the excited state is attractive and can support bound states, light absorption during the collision of ground-state atoms results in their photoassociation.* Figure B3.1 showed the potentials for a Hg + Hg collision leading to a bound Hg$_2^*$ excimer.

The ability to use light at a frequency detuned from the absorption of the asymptotically separated reactants has the advantage that only partners in the very process of a collision will absorb the light. So it is potentially very selective. But the required high-intensity light source has the disadvantage that it may induce unwelcome multiphoton processes.

---

* When the collision partners are not atoms, the excited-state potential can conically intersect the potential of the ground state and absorption during the collision then results in quenching and the exit of the products in the ground electronic state. M + H$_2$ collisions, M alkali atom, have been extensively studied.

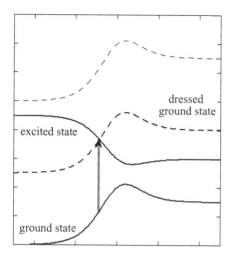

Reaction coordinate

Figure 7.30 Potential energy along the reaction coordinate of the ground electronic state and the same potential dressed by a UV photon, cf. Figure B3.1 and Appendix 3.B, dashed curve. The dressed ground state crosses an excited state potential that has a well in the region of the transition state of the ground surface. The dressing photon is shown as an arrow. The system can exit as excited products or on the way out the excited state can emit and thereby bring the system to the other side of the barrier on the ground state. The lighter dashed curve is one cautionary note. If the laser is intense enough to well dress the ground state by a single photon it is already able to dress it by two photons, bringing it up to the light dashed curve. There is usually a significant density of higher excited electronic states for this doubly dressed state to cross into.

7.3.5   Spectroscopy of the transition state and laser-assisted collisions

We have discussed a variety of optical methods that can place reactants in the transition state region. What we specifically aim for in this section is to have reactants absorb light during their approach motion so as to spectroscopically probe the transition state region and/or to facilitate a reaction taking place. At the moment, theory is ahead of experiment. There have been many valiant efforts, but currently there are more theoretical arguments and detailed computations showing that this should be possible than clear-cut experimental examples.

In principle, the situation is clear-cut. We want to use an intense laser to dress the ground-state potential energy surface, a surface that often has a barrier, Section 5.1, so that it crosses an excited state surface, a surface that typically has no barrier,[48] Figure 7.30.

Laser-assisted collisions are most readily demonstrated experimentally when reaction is only possible on the electronically excited state. The harpoon reactions of excited rare gas atoms,[49] as discussed in Section 3.2.4.1, e.g., $Xe^* + Cl_2 \rightarrow$ $XeCl^* + Cl$, is an example where pumping of the rare-gas atom during the collision is possible.

7.4 Quantum control

Organic chemists are masters of control. They not only optimize solvent, temperature, catalyst, etc. to get the desired chemical product, they are even able to direct the process toward one particular stereoisomer out of very many. In this section we too want to be able to direct the reaction to one out of several alternatives. Of course, we want to do so on the microscopic, molecular level. Starting with an initial state we want to prescribe a particular final state. But much of this book and almost all of this chapter is about doing just that. For example, in Section 7.2.3 we showed how at a given energy it is possible to vary the branching ratio for HNCO photodissociation by providing the energy in different ways. So why do we need a special section about control? What is new in this section is that we want to be able to take advantage of the unique opportunities that are only available courtesy of quantum mechanics.

In this section we discuss quantum control in the frequency domain.[50] In Section 8.3 we discuss control in the time domain.

From quantum mechanics we need an idea that has already played a central role: superposition of states and the resulting interference.

First, by our definition of control we aim to favor one particular outcome. So obviously there needs to be more than one possible outcome. Say that there are two. To make it very simple, say two final quantum states, each one being a separate outcome. In Section 4.3.2 we showed how the wave function of the collision is specified in terms of the initial state and the Hamiltonian of the system. On the molecular level time runs just as well forward as backward. Therefore,[51] a wave function can just as well be specified in terms of the final state it will evolve into! For our problem we have two wave functions, at the same energy, each evolving exclusively into a different final state. An arbitrary wave function will be a superposition of these two "target" states. If we can experimentally alter the contribution of these two states we can vary the branching ratio.

A more technical but still simple analysis suggests that we have two control parameters. Say that experimentally we can prepare two different wave functions ψ_1 and ψ_3.* We know that we can, in principle, do that because there are two independent wave functions defined by the two different outcomes.** We denote

---

* Why the indices "1" and "3?" Because a simple way to prepare these two states is by a one- or a three-photon process, see below.

** Note the correspondence between physical reality and theory. At the same energy there are two distinct final states. Therefore there are two independent solutions of the Schrödinger equation. We denote the two solutions (that each correspond to a definite final state) as ψ_a and ψ_c. These two solutions have the same energy and so any superposition of them, say $\psi = \alpha\psi_a + \chi\psi_c$, is

the two target states as ψ_a and ψ_c. Say that we prepare a state ψ that is a super-position of ψ_1 and ψ_3, $\psi = c_1\psi_1 + c_3\psi_3$. Here the c's are complex amplitudes whose amplitude and phase are determined by the laser pulse and are therefore under our control.

The probability of getting products in state a is the overlap of the wave function ψ that we prepared with the target state ψ_a:

$$|\langle\psi_a|\,\psi\rangle|^2 = |\langle\psi_a|\,c_1\psi_1 + c_3\psi_3\rangle|^2 = |c_1|^2\,|\langle\psi_a|\,\psi_1\rangle|^2 + |c_3|^2\,|\langle\psi_a|\,\psi_3\rangle|^2$$
$$+ c_1c_3^*\,\langle\psi_a|\,\psi_1\rangle\,\langle\psi_a|\,\psi_3\rangle^* + c_1^*c_3\,\langle\psi_a|\,\psi_1\rangle^*\langle\psi_a|\,\psi_3\rangle \qquad (7.14)$$

The first two terms are the usual classical "OR" addition of probabilities: we can get to state a either through route 1 **or** through route 3. The second set of terms in Eq. (7.14) is the, by now, familiar interference between the two classical alterna-tives. There are two control parameters because the magnitude of the interference is governed by the absolute values of the two "experimental" coefficients and, sep-arately, by their relative phase. We could write that $c_1^*c_3 = |c_1c_3|\exp(i(\vartheta_3 - \vartheta_1))$ and so conclude that it is the phase difference $\vartheta_3 - \vartheta_1$ between the two paths that determines the interference. This is almost, but not quite, correct. The properties of the molecule also enter in: wave functions in the continuum are complex num-bers and so have a phase. Scalar products such as $\langle\psi_a|\,\psi_1\rangle$ are therefore complex numbers. If we want to emphasize interference we should rewrite Eq. (7.14) as

$$|\langle\psi_a|\,\psi\rangle|^2 = |\langle\psi_a|\,c_1\psi_1 + c_3\psi_3\rangle|^2 = |c_1|^2\,|\langle\psi_a|\,\psi_1\rangle|^2 + |c_3|^2\,|\langle\psi_a|\,\psi_3\rangle|^2$$
$$+ 2|c_1c_3\,\langle\psi_a|\,\psi_1\rangle\,\langle\psi_a|\,\psi_3\rangle|\cos(\vartheta_1 - \vartheta_3 - \delta) \qquad (7.15)$$

where δ is the phase of the scalar products of the molecular wave functions; δ is due to molecular properties. The two control parameters are the amplitude and the phase angle of the interference term.

How are we going to implement this in the laboratory? We need a process with two (or more) distinct outcomes. Experimentally we need two (or more) distinct ways of preparing the system with an energy E. For photodissociation, for example, one way is by absorption of one photon of frequency $\hbar\omega = E$. Another way is by absorption of three* photons of frequency $\hbar\omega = E/3$. Indeed, our notation above, where we called the two experimental routes "1" and "3," was with this choice in mind. An early experimental result is shown in Figure 7.31. The control is to alter the branching between dissociation and ion-ization of HI by using lasers such that one- and three-photon excitation leads to the same total energy

$$\text{HI} \xrightarrow{\;10.6\,\text{eV}\;} \begin{cases} \text{H} + \text{I} \\ \text{HI}^+ + e \end{cases}$$

also a solution of the Schrödinger equation. In general, an experiment prepares a superposition of ψ_a and ψ_c so that both possible final states are produced. Control means being able to change the coefficients α and χ in the superposition prepared by the experiment.

* Two photons will not do the trick because the selection rules for one- and two-photon absorption are different. But three photons can access the same state as one photon.

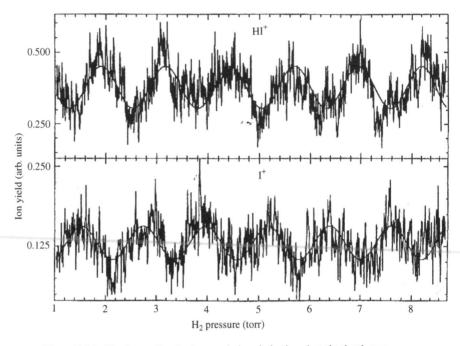

Figure 7.31 The (smoothed) observed signals in the photoionization vs. photodissociation of HI vs. H_2 pressure. The dissociation products are detected as ions. The phase difference $\vartheta_1 - \vartheta_3$ is proportional to the pressure of H_2. Note that the signals of the two channels are out of phase and they are shifted by about 150°, showing an (energy-dependent) molecular contribution δ in addition to the phase difference between the two light beams [adapted from L. Zhu *et al.*, *Science* **270**, 77 (1995); see also Gordon, Zhu *et al.* (1999, 2001)].

The phase difference between the two preparations is imposed by passing both beams through a cell of H_2. The index of refraction for the two frequencies is different and depends on the pressure in the cell, and by varying the pressure the phase difference, $\vartheta_1 - \vartheta_3$ above, is varied.

7.4.1 Strong field control

The control scheme that we have discussed does not require high-power lasers. The electric field of the light is weak enough that the levels of the system are not shifted by it. The laser induces transitions but these can be described as transitions between states of the free molecule. This is no longer the case when the laser is strong. The very potential on which the nuclei move can be altered by the external field because, if it is strong enough, its effect on the electrons can become comparable to the electrical field due to the nuclei or that of the other electrons. Once that happens a variety of other processes, such as non-resonant ionization,

(A)

NO YAG

(B)

(C)

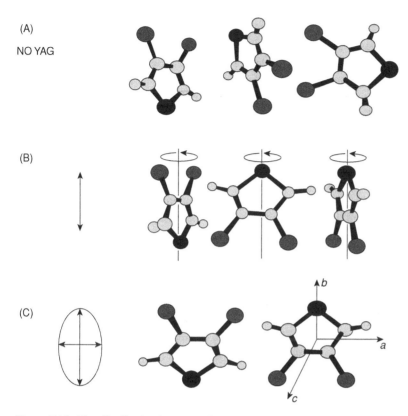

Figure 7.32   The distribution in space of a 3,4 dibromothophene molecule (black circle: S; dark gray circles: Br) when: (A) no field is present and the rotation is unhindered; (B) a linearly polarized laser aligns the symmetry axis of the molecule but it is still free to rotate about this axis; (C) an elliptically polarized laser forces all three axes into alignment. That the alignment is achieved is probed by coulomb exploding the molecule in a second intense laser pulse [adapted from J. M. Larsen *et al.*, *Phys. Rev. Lett.* **85**, 2470 (2000)].

Section 7.2.5.2, also become important. Strong laser fields can be used to induce mixing between different electronic states and thereby tailor the molecule in a desired direction. We are not quite yet as good as the organic chemists who can tune the solvent to perturb the system in just the right way. But we are getting better. Here we discuss a different aspect: the control of the alignment of the molecule.[52] We give a technical meaning to the term "alignment" in Chapter 10 where we discuss stereodynamics. For now let us think of it as inhibiting the free rotation of the molecule, which one can do by a strong laser field. The effect is due to the interaction of the electric field of the light with the (permanent or induced) dipole moment of the molecule. Figure 7.32 shows the schematic idea for aligning all three rotation axes of a polyatomic molecule.

Problems

A. The quasi-classical vibrational motion and the Franck–Condon region. We consider a bound diatomic molecule with a given vibrational energy. We want to mimic the quantal distribution of bond distances in a stationary vibrational state. Conclude from Chapter 5 or otherwise that the probability of finding the atoms in the range R, $R + dR$ apart is proportional to $dR/v(R)$ where $v(R)$ is the classical velocity at R, $\mu(v(R))^2/2 = E - V(R)$ and μ is the reduced mass. It is therefore more likely to find the system where the motion is slowest. At the turning point itself the quasi-classical probability diverges. This point-wise divergence is not serious because the integrated probability is still finite. The quantum mechanical probability is maximal near the turning points, Figure 7.1, and decreases exponentially in the classically forbidden regime. For the ground vibrational state, the quantum mechanical probability is maximal in the middle and so is rather different in character from the quasi-classical distribution.

B. IR multiphoton dissociation. Attempts to dissociate O_3 or other triatomics under the same conditions that work for, say, SF_6, fail. There are at least two reasons why this is consistent with the mechanism proposed in the text. Discuss.

C. Delayed ionization of vibrationally hot molecules. Attempts to dissociate C_{60} by multiphoton absorption reveal electron emission and the formation of C_{60}^+ ions. (a) Why is the dissociation of energy-rich C_{60} slow?[53] Of course, the threshold energy for breaking the cage structure is high, but all the evidence is that the molecule has energy above the threshold. (b) Are there any other examples where a hot body emits electrons? (c) Size and one other criterion should be relevant to when delayed ionization will be an important decay channel. Discuss.

D. The reaction $Br + HCl(v = 0) \rightarrow HBr + Cl$ is endoergic by $65 \, kJ \, mol^{-1}$ and the reaction rate constant at room temperature is estimated to be $1 \, cm^3 \, mol^{-1} \, s^{-1}$. (a) Refer to the potential energy surface that you have drawn earlier, or make one now, and discuss whether it is consistent with this result. (b) Two surfaces of rather similar barrier heights were proposed for this reaction. The difference was that one surface led to a far higher frequency for the Br–H–Cl bending vibration. For which surface does TST predict a higher reaction rate constant? (c) A vessel containing HCl was pumped with light from a HCl chemical laser. A vibrational quantum of HCl is $34 \, kJ \, mol^{-1}$. The reaction rate constant at room temperature for $Br + HCl(v = 2) \rightarrow HBr + Cl$ was measured as $10^{12} \, cm^3 \, mol^{-1} \, s^{-1}$. Propose a *simple* argument explaining why the increase in the rate constant by 12 orders of magnitude is to be expected. The rate of the reaction of Cl with HBr was measured at room temperature to be $4.5 \cdot 10^{12} \, cm^3 \, mol^{-1} \, s^{-1}$. Is this consistent with your explanation? (d) Naturally occurring Cl atoms are a mixture of two stable isotopes, ^{35}Cl and ^{37}Cl in a ratio of 3:1. The Cl atoms produced by the $Br + HCl(v = 2) \rightarrow HBr + Cl$ reaction were efficiently scavenged by the reaction

$Cl + Br_2 \rightarrow ClBr + Br$. It was found that about 90% of the ClBr molecules contain the lighter Cl isotope. Provide a simple explanation for the isotopic enrichment.[54]

E. Interference between dissociation channels. As discussed in connection with Figure 7.8, provide a simple estimate for the difference in phase shifts for the two different isotopes and show that the trends you expect when the mass is heavier are consistent with the observations.

F. A quasi-continuum. It takes of the order of $\hbar/D$ to discern that a set of quantum energy levels that have a mean energy spacing D apart are discrete and not a continuum. Plot this time on a log–log scale against D. Make your plot practical by expressing D as, say, the number of states cm^{-1} or another convenient measure of your choice. Conclude that even fairly small polyatomic molecules with an energy excess of, say, one vibrational quantum per mode, are, for practical purposes, in a quasi-continuum.

G. The picket fence model. This is one of the oldest workhorses. It is necessary to diagonalize the Hamiltonian. To build the Hamiltonian matrix we take the bright state to have the index 1. Its energy is the matrix element H_{11}. The dark states have indices from 2 on and are not coupled to one another. They are equally spaced in energy, D units apart, but this is only needed when you compute the time evolution. Otherwise, all that we need is their energies $E_n = H_{nn}$. The only other non-vanishing matrix elements of $\mathbf{H}$ are along the first row and first column and, apart from the diagonal element, they all have a common value $\overline{V}$. Diagonalizing the Hamiltonian means solving the secular equation $\mathbf{HC} = E\mathbf{C}$ where $\mathbf{C}$ is a column vector whose components are the amplitudes of the wave function on the different states. (a) Show that the amplitude on dark state n is largest when E is near H_{nn}. (b) By formally solving for C_1 obtain an equation for the energy. (c) Use the fact that the wave function needs to be normalized to get an explicit solution for C_1. (d) Now make the dark states equally spaced and try to sum $\overline{V}^2 \Sigma_n (E - H_{nn})^{-m}$, $m = 1,2$. It can be done, see O. K. Rice, *Phys. Rev.* **35**, 1551 (1930) and *J. Phys. Chem.* **65**, 1588 (1961), but it is not trivial so, as an alternative, (e) devise a graphical way to solve for the energy (there is more than one solution). (f) Use the solution for $\mathbf{C}$, replace summation over a quasi-continuum by an integration, and derive the decay rate as given in the text. (g) Show that the bright state revives after the time h/D.

H. The picket fence model criterion. When a bright state is coupled by a mean strength $\overline{V}$ to a quasi-continuum of dark states of density ρ, the picket fence model predicts that it will decay irreversibly when $\overline{V}\rho > 1$. (a) IVR. The coupling is due to cross anharmonicity. Anharmonicities of increasing order scale as $h\nu/N^m$ where ν is the vibrational frequency, N is the number of bound states of that (anharmonic) mode, and m, $m > 1$ is the order. By what size molecule and in what energy range will you predict facile IVR? (b) Radiationless transitions. The coupling is due to the breakdown of the Born–Oppenheimer approximation. Use, say, input from the data of Figure 7.3 for the energy range and the coupling. By what size molecule will you predict an onset of radiationless transitions?

I. Competitive unimolecular decay. Consider an excited species A^* that can decay in two different ways: $A^* \xrightarrow{k_1} P$ and $A^* \xrightarrow{k_2} Q$. Write kinetic rate equations and thereby show that the rate constant for the decay of the concentration of A^* is $k = k_1 + k_2$, $[A^*](t) = [A^*](0) \exp(-kt)$. As discussed, in quantum mechanics we can reproduce a unimolecular decay by endowing the energy of the state with an imaginary part that we call the width. Hence, in the absence of interference effects, the widths add up.

J. The threshold for coulomb explosion. A strong laser field can strip many electrons off an atom, molecule, or cluster on a time scale comparable to or shorter than that for nuclear motion. A multicharged molecule or cluster will then fragment owing to the coulomb repulsion between the like charges. Consider the removal of the first electron from a one-electron H atom. The electron is bound in a coulomb potential around the nucleus. Draw this potential as a function of the distance along the z axis of the electron from the nucleus. Now add an electric field F along the z axis. This adds a term ezF to the potential. (a) Draw the new potential. The barrier for an escape of the electron is now lowered. Examine the drawing and convince yourself that except for just under the barrier top, the barrier is quite wide so that tunneling through it is not probable. (b) What is the field (in $V \, cm^{-1}$) for which an electron formerly bound in a $1s$ level can escape? (c) Translate this electrical field to the intensity of the laser beam.[55]

*K. Electronic and spatial control. Construct molecular orbitals for the one-electron molecule H_2^+ as a symmetric and an antisymmetric linear combination of $1s$ wave functions on each one of the H atoms, as in Section 5.1.5. The anti-symmetric state is unbound. In HD^+ the dissociation can be to either $H + D^+$ or to $H^+ + D$. Show that these two channels can be written as a linear combination of the molecular orbitals and further that they have a dipole moment opposite in direction. Consider a linearly polarized electric field composed of two frequencies ω and 2ω, with amplitudes E_1 and E_2 and a relative phase ϕ, $E = E_1 \cos(\omega t) + E_2 \cos(2\omega t + \phi)$. Show that this field has a maximal amplitude in the forward or backward direction depending on the magnitude of the relative phase. Using such a field one can control the direction in space for the dissociation products. See the experimental results in B. Sheehy, B. Walker, and L. F. DiMauro, *Phys. Rev. Lett.* **74**, 4799 (1995).

*L. The time-dependent view of spectroscopy (Heller, 1981). Suppose that at time $t = 0$ we make a Franck–Condon transition to an upper electronic state. This takes the initial vibrational wave function up in energy and onto a different potential for the motion of the nuclei. This initial state is no longer stationary and it starts evolving in time. The first thing that will happen is that the wave function will depart from the Franck–Condon region. What we want to know is how quickly it will do so. If the molecule is a diatomic, the initial state will periodically revive. But for a polyatomic, IVR will, over time, reduce the revival. Hence for a polyatomic we also want to know where the wave-packet goes to. This problem is to show that the first question is answered by the frequency dependence of the envelope of the absorption spectrum while for the second problem we need a Raman excitation spectrum.

M. The energy gap law for radiationless transitions. To understand the role of the Franck–Condon principle for radiationless transitions we consider a simple model of one, harmonic, nuclear coordinate as shown in the figure below. We further simplify by assuming that the force constant for this vibration is the same in the ground and the electronically excited state and that the only difference is that the equilibrium distance is shifted to a slightly higher value. As seen in the figure, excitation to the first excited singlet state, S_1, at a low energy E, say just above the ground vibrational level, results in a very poor Franck–Condon overlap with vibrational levels of the ground singlet state, S_0, at the same total energy. (a) Using harmonic oscillator wave functions show that the overlap integral from the $v = 0$ level of S_1 to the level v of S_0 is given by

$$|\langle 0|v\rangle|^2 = (\lambda^v/v!)\exp(-\lambda)$$

where

$$\lambda = \omega(\Delta Q)^2/2\hbar$$

ω is the harmonic frequency and ΔQ is the difference in the equilibrium geometries of the two states. In the harmonic approximation the final vibrational quantum number, v, is determined by E, $E = \hbar\omega(v + 1/2)$.

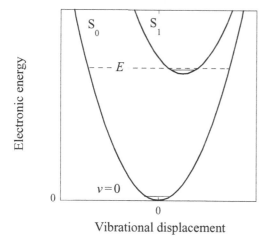

Vibrational displacement

(b) Plot the dependence of the FC factor on the excess energy E remembering that for typical values of E, $v > 1$. By Chapter 9 we will come to call this dependence the exponential gap principle. (c) Conclude that, as observed, replacement of C—H vibrational modes by C—D ones will further reduce the FC factor. (d) The rather low FC factor means that radiationless transitions are relatively slow. Can you suggest changes in the assumptions of the model above that will allow a facile transition from S_1 to S_0?

Notes

1 The reason we want to shape the pulse is that ours is a world of high-amplitude motion where bonds stretch and molecular geometries deform. Shaping (or tailoring) light pulses is a way to adjust to the dynamics of anharmonic motion. The motivation for the shaping is that the frequency at which a molecule absorbs light is a constant characteristic of the molecule but this is so only at, or very near, equilibrium. Consider a diatomic molecule that we want to dissociate by exciting its vibrational motion with a laser pulse. Because of anharmonicity, the vibrational frequency decreases monotonically the more energy is put into the molecule. As the laser pulse sets the nuclei in motion, it makes sense to adjust the frequency to the changing geometry of the molecule, if we want the absorption of the same molecule to continue. We typically do not do it but use a fixed frequency. But to drive the molecule to the transition state region it may be necessary to shape the pulse by adjusting its frequency on the time scale of the molecular vibration so as to track the deformation of the molecular geometry. This is known as a chirped pulse where the intensity contours are the ellipses and their tilt is meant to show that the main frequency components of the shaped pulse go down with time, as would be necessary for pumping an anharmonic mode. For more on pulse shaping see Kawashima *et al.* (1995), Kohler *et al.* (1995). Learning algorithms can be used to advantage to improve the pulse shape so as to approach a desired outcome, see Rabitz and Zhu (2000), Rabitz *et al.* (2000), Brixner and Gerber (2003).

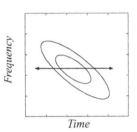

Frequency / Time

2 In the semiclassical limit the amplitude squared of the wave function goes as $1/\text{velocity}$, $|\psi(R)|^2 \propto 1/v(R)$, where $v(R)$ is the classical, local value of the velocity, $(\mu/2)(v(R))^2 + V_{\text{eff}}(R) = E$. ($V_{\text{eff}}(R)$ is the effective potential, Section 2.2.2). This estimate for where the system can be found is equally true when the energy is in the continuum and the motion is unbounded.

3 Strictly speaking, the transition probability is determined by $|\langle\psi_{v'}(R)|\mu_{\text{ge}}(R)|\psi_{v''}(R)\rangle|^2$ where $\mu_{\text{ge}}(R)$ is the transition dipole moment, the expectation value of the dipole operator between the ground and excited electronic states. The *Condon approximation* takes the transition dipole moment to be a constant and therefore it can be factored out. The success of the Condon approximation is because, as we argue, only a very limited region in R values contributes to the integral over R. Strictly speaking, the transition dipole moment can vary significantly with R.

4 When the two potentials are quite different the Franck–Condon factors can vary non-monotonically with the final vibrational state. This behavior is observed in the fluorescence of the I_2 molecule after it has been prepared in one particular high-lying vibrational level. Condon called these oscillations, a function of v'' for a given v', "internal diffraction" and suggested that they were just as real a proof for the intrinsic wave nature of matter as external diffraction experiments. The oscillatory nature of the wave function is even more important when the energy is so high that it is in the continuum. For the general problem of the Franck–Condon principle in bound–free transitions see J. Tellinghuisen, *Adv. Chem. Phys.* **60**, 299 (1985).

5 See Michl (1972), Bernardi *et al.* (1996), Yarkony (1998), Haas *et al.* (2000), Worth and Cederbaum (2004).

6 See Yarkony (1998) for a discussion of the dimension of the manifold and for coordinates that are especially suited for describing a conical intersection.

7 Such an "inverse electronic relaxation" is indeed observable [A. M. Ronn, *Adv. Chem. Phys.* **47**, 661 (1981); H. Reisler and C. Wittig, *Adv. Chem. Phys.* **47**, 679 (1981)].

8 There is a rich literature dealing with radiationless transitions. Selective references include Henry and Kasha (1968), Jortner *et al.* (1969), Schlag *et al.* (1971), Parmenter (1972), Jortner and Mukamel (1975), Avouris *et al.* (1977), Freed (1978), Jortner and Levine (1981).

9 Pioneering studies of dissociation dynamics are by O. K. Rice, *Phys. Rev.* **35**, 1551 (1930) and later, *J. Phys. Chem.* **65**, 1588 (1961).

10 Jortner and Levine (1981), Felker and Zewail (1988), Crim (1996), Moore and Smith (1996), Nesbitt and Field (1996).

11 Should we not have a zeroth tier consisting of bright states? Yes, this will be an even better model. But at some point, making schematic models should give way to introducing a Hamiltonian and doing a more fundamental job. For physical insight it may be useful to sort states into sequences of coupling, but a real Hamiltonian has so many states that the sorting is better done by a systematic procedure that can be implemented by the computer rather than by hand, see Wyatt *et al.* (1995).

12 To bear on this question the spectra need to be as free as possible from inhomogenous broadening. This is the obvious requirement that there is a definite initial state so that what is measured is not a mix of different spectra originating from different states. It is easy to state but not quite so easy to implement because polyatomic molecules have many (mostly rotational, some vibrational) states close in energy to the ground states and these are thermally populated. The measured spectrum can therefore contain many "hot bands." For the present purpose, these need to be avoided.

13 The states that can be formed from two atoms in a given state are enumerated by the Wigner–Witmer rules (Herzberg, 1950; Zare, 1988). See also P. Pechukas and R. N. Zare, *Am. J. Phys.* **40**, 1687 (1972).

14 For experimental aspects see Eppink and Parker (1997), Chandler and Parker (1999), Yonekura *et al.* (1999), Whitaker (2003).

15 Zare and Herschbach (1963) Houston (1989), Gordon and Hall (1996).

16 Zare and Dagdigian (1974), Kinsey (1977).

17 Simons (1984), Crim (1993), Dixon (1994), Butler and Neumark (1996), Ashfold and Baggott (1997), Sato (2001).

18 In a suitable setup, e.g., Amirav *et al.* (1983), cooling is routinely achieved to rotational temperatures below 10 K (Scoles, 1988; Pauly, 2000). The limiting aspect is the heat of vaporization, the energy that is being released upon condensation. The higher it is, the higher is the limiting temperature that can be reached. With effort and using He as the carrier gas one can go down to 1 K or even below (Even *et al.*, 2000).

19 For $CH_3I \rightarrow CH_3 + I$ and $CH_3 + I^*$ See Y. Amatatsu, S. Yabushita, and K. Morokuma, *J. Chem. Phys.* **94**, 4858 (1991). For ICN see Y. Amatatsu, S. Yabushita, and K. Morokuma, *J. Chem. Phys.* **100**, 4894 (1994).

20 For ICN in different solvents see A. C. Moskun and S. E. Bradforth, *J. Chem. Phys.* **119**, 4500 (2003).

21 See Schnieder *et al.* (1995) for application of this method toward a high-resolution study of the H + D$_2$ → HD + D bimolecular reaction.

22 So should we expect quantal interference between trajectories dissociating as H···O–H vs. H–O···H? See Y. Dixon *et al.* (1999).

23 The special nature of these quasi-bound states is usually expressed by writing their energy as a complex number, $E_j - i\Gamma_j/2$. The imaginary part, $\Gamma_j/2$, indicates the finite lifetime of the state as is most easily seen by writing the time-dependent part of the wave function of state j: $\exp(-i(E_j - i\Gamma_j/2)t/\hbar) = \exp(-iE_j t/\hbar)\exp(-(\Gamma_j/2)\,t/\hbar)$. The probability of finding an isolated state decreases exponentially with time as $\exp(-\Gamma_j\,t/\hbar)$. So the decay rate is $\Gamma_j/\hbar$. The complex energy is a convenient device for a quantitative expression of the implication of the time–energy uncertainty principle that an unstable state cannot have a sharp value of its energy. A hermitian Hamiltonian with the usual boundary conditions on the wave function cannot have complex eigenvalues. A state with complex energy is equivalent to a coherent superposition of states of sharp real energies where the states are weighted (by a Lorentzian function) such that the spread in energy is Γ. So a well-resolved spectroscopic experiment will see a broadened line. Of course, such an experiment will not be able to probe a time decay of the state because it takes an energy resolution better than Γ to see the energy dependence of the line profile. To probe the decay you need to prepare the state in a time short compared to $\hbar/\Gamma$. Such a preparation must have an energy spread wider than Γ. Note that the pump used in the experiment shown in Figure 7.17, that has a time duration of 650 fs, just about satisfies this requirement. Actually, the discussion above was simplified in one respect that bears on the experiment. We took the quasi-bound state to be isolated; its neighbors are more than Γ away in energy. In a real system isolation is not necessarily valid. The widths can overlap and this is why the pump used in the NO$_2$ experiment prepares a superposition of quasi-bound states and the temporal decay is not exponential.

　　To compute states with a complex energy an effective Hamiltonian is often introduced that treats the states as bound at the price of adding a non-hermitian term representing the coupling to the continuum (Levine, 1969). A few of the eigenvalues will have a rather large energy width. These are the states that have enough energy localized near or at the reaction coordinate. Such states dissociate promptly, cf. Section 5.3.3.2.

24 The fluctuations in the decay rates of states nearby in energy illustrate that such states can differ considerably in how the total energy is partitioned and therefore in the strength of coupling to other states. The variations can be far more extreme than shown for NO$_2$ in Figure 7.17, because the coupling in NO$_2$ is strong, as can also be judged by other indicators. The fluctuations themselves can be treated statistically [W. Pollik, C. B. Moore, and W. H. Miller, *J. Chem. Phys.* **89**, 3584 (1988); Levine, (1988)].

25 Nature need not fully cooperate. There are steps only if you either have enough energy to cross the barrier or you do not. But a real molecule, with energy just short of the barrier height, has the option of quantum mechanically tunneling through the barrier [W. H. Miller, *Chem. Rev.* **87**, 19 (1987)]. This will smooth the sharp step structure. There is no question that tunneling is an option, but one must also be careful not to invoke it without proper examination, particularly so in many-atom systems that sometimes have an option of going around rather than going through a barrier.

26 For a discussion of isotopic scrambling in general, see J. I. Brauman, R. N. Zare, and R. D. Levine, *Chem. Phys. Lett.* **172**, 231 (1990).

27 E. R. Lovejoy and C. B. Moore, *J. Chem. Phys.* **98**, 7846 (1993).

28 Crim (1996, 1999), Bar and Rosenwaks (2001). See also Schinke (1993), Zare (1998).

29 D. Bingemann *et al.*, *J. Chem. Phys.* **107**, 661 (1997), A. Sinha, R. L. Vanderwal, and F. F. Crim, *J. Chem. Phys.* **91**, 2929 (1989).

30 E. Woods, C. M. Cheatum, and F. F. Crim, *J. Chem. Phys.* (2000). See also Crim (1999).

31 Whitehead (1988), Manz and Parmenter (1989), Jortner *et al.* (1991).

32 Henry (1977), Child and Halonen (1984), Lehmann *et al.* (1995).

33 Crim (1984, 1996, 1999).

34 The notion of a resonance is when n_a quanta of mode a are about equienergetic with n_b quanta of mode b, $n_a\omega_a \approx n_b\omega_b$. We then speak of the $n_a{:}n_b$ order resonance. Resonances are most effective when they are of low order. The earliest example of a resonance is the 1:2 resonance between the symmetric stretch and bend in CO_2, as analyzed by Fermi. For this reason all resonances are sometimes referred to as "Fermi resonances." At a more accurate level note that, because of anharmonicity, the frequencies of the modes will depend on their energy content. As the energy varies, modes can get tuned into and also out of resonance. Further, a resonance may involve more than two modes. It is customary to represent this case as $\Sigma_i n_i \omega_i \approx 0$, where now the integers n are not all positive. Finally, unlike what it might seem, this notion of a resonance is very well defined in classical mechanics. Indeed, the onset of classically chaotic motion, Section 5.2.2.1, is when the resonances are no longer isolated but their ranges of influence start to overlap (Rice, 1981; Sagdeev *et al.*, 1988).

35 It is useful to think of a *dilution factor*, namely to how many other states is the optically prepared state coupled. If the state can decay, it has an uncertainty in its energy equal to its rate of decay. N, the number of coupled states, is then the number of states that lie within the energy width, $N = (2\pi/\hbar)\,|\overline{V}|^2\,\rho/D = 2\pi\,|\overline{V}|^2\,\rho^2$ or the decay rate is $N/\hbar\rho$. Note that unlike the RRKM rate, here N is the effective number of states and it is defined in terms of the rate. If $N > 1$ we have an irreversible intramolecular decay.

36 See, for example, Wittig *et al.* (1988), Bacic and Miller (1996), Lester (1996), Rohrbacher *et al.* (2000), Oudejans and Miller (2001).

37 Hutson (1990), Bacic and Miller (1996), Schlag *et al.* (1998), Hutson (2001).

38 Levy (1981). One can tell from the fluorescence that the major product of the dissociation is $I_2^*(v-1)$.

39 Meaning that one quantum of vibration of $I_2^*(v)$ is equivalent in energy to many quanta of the He–$I_2^*(v)$ vibration. Jortner and Levine (1990) discuss selective chemistry from this point of view.

40 See the extensive review by Lupo and Quack (1987).

41 See Bernstein (1982) for systematics of multiphoton ionization and the reviews of Johnson (1980), Schlag and Neusser (1983), Gobeli *et al.* (1985).

42 The broadening can be even more for stronger transition dipoles, which are quite possible for an electronic transition. We can think of this broadening in terms of the dressed states picture, Appendix 3.B. We can also think of it as a kind of Stark effect brought about by a strong electric field. Say first that the field is static. The figure below shows the potential seen by a hydrogenic electron when, in addition to a proton, there is also a strong DC field. The barrier to ionization is seen to be lowered but, in addition, levels below the barrier are broadened because the electron can tunnel out. The figure also shows a diatomic molecule where the motion of the electron is very dependent on the interatomic distance.

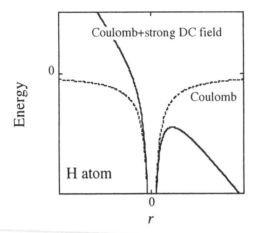
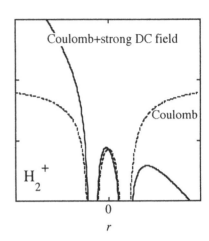

The electric field of a laser is oscillating. Keldysh (Landau and Lifschitz, 1971) provided an estimate for the role of an oscillating field that is not resonant. The Keldysh estimate for the onset of non-resonant ionization of an atom, for a field frequency lower than any natural frequency of the system, is consistent with our value of 10^{13} W cm^{-2}. Molecules are not like a hydrogen atom and, in particular, the gap between electronic states is very dependent on the internuclear distances (Dietrich and Corkum, 1992; Chelkowski and Bandrauk, 1995). This has interesting potential applications, e.g., S. Lochbrunner *et al.*, *J. Phys. Chem. A***102**, 9334 (1998). For diagnostic non-resonant ionization of molecules at high laser powers, see Levis and DeWitt (1999).

43 "Above threshold ionization" (ATI) is another manifestation of ladder switching. The outgoing electron, which after all is also a product, can absorb photons and be accelerated. This is detectable by secondary peaks in its kinetic energy distribution. A particularly neat application of ATI is that in the intense electric field of a pulsed ultrafast laser, the electron may scatter from the nuclei and give a diffraction pattern that provides information on the nuclear geometry. Ultimately, the molecule will be blown apart, that is, reduced to a rubble of highly charged ions and electrons, but if you can work quickly enough it might be possible to obtain structural information. See Corkum *et al.* (1997).

44 A "coulomb explosion" occurs if a molecule (or a cluster) is at least doubly ionized before any fragmentation takes place. The strong electrostatic repulsion sends the fragments rapidly apart. With high-power lasers one can reach very high charge states. In clusters, coulomb explosions are a fertile field (Last and Jortner, 1998; Zhong and Castleman, 2000; Castleman, 2001). For time-resolved coulomb explosion imaging see Constant *et al.* (1999).

45 Localization of the excitation is only one way to direct chemistry. Placing a charge on an atom typically weakens its bonding. Can we localize the charge on molecular ions (Williams and Beynon, 1976) and can we direct this localization? Can we localize the odd electron in a radical?

46 See, for example, Rabitz (1974), Birnbaum (1976).

47 See Duren *et al.* (1982) for the alkali atoms–rare gas potentials.

48 George *et al.* (1977), Brooks (1988), Polanyi and Zewail (1995). We can become more fancy and have the system return a photon to you, that is, have the excited state as shown

in Figure 7.30 emit on its way out. This is sometimes called laser catalysis (Lau, 1982; Vardi and Shapiro, 2001).

49 See T. O. Nelson, D. W. Setser, and J. Qin, *J. Phys. Chem.* **97**, 2585 (1993); J. Kohel and J. W. Keto, *J. Chem. Phys.* **113**, 10551 (2000).

50 The subject of quantum control is extensively reviewed. Special reference to the frequency domain will be found in Brumer and Shapiro (1992, 1995), Shapiro and Brumer (1994).

51 For a detailed technical discussion of this point see Levine (1969).

52 Lasers can also be used to guide the translational motion of molecules, e.g., H. Sakai *et al.*, *Phys. Rev. A* **57**, 2794 (1998). Laser cooling of the translational motion is also possible, e.g., C. Cohen-Tannoudji, *Rev. Mod. Phys.* **70**, 707 (1998), and is also applied to cooling of internal degrees of freedom of molecules.

53 In mass spectrometry, the rather slow dissociation of large ions is known as a kinetic shift. See C. Lifshitz, *Eur. J. Mass Spec.* **8**, 85 (2002).

54 The actual experimental situation [D. Arnoldi and J. Wolfrum, *Ber. Bunsenges. Phys. Chem.* **80**, 892 (1976)] is a shade more complex. The HCl laser pumps the HCl molecules in the sample to $v = 1$. HCl molecules in $v = 2$ are produced by the energy pooling process $HCl(v = 1) + HCl(v = 1) \rightarrow HCl(v = 2) + HCl(v = 0)$, a process that we discuss in Chapter 9. The straightforward argument predicts a somewhat higher isotopic enrichment than is actually observed. It is necessary to appeal to the exponential gap principle, Chapter 9, to identify the process that acts to reduce the isotopic selectivity.

55 Intensity of light is the flux of energy, meaning energy per unit time through a unit area. Since power is energy per unit time, intensity is expressed as power per unit area, typically quoted in $W\,cm^{-2}$. The SI units are $W\,m^{-2}$. Intensities of up to, say, $10^{20}\ W\,cm^{-2}$ are available from table-top lasers.

Chapter 8
Chemistry in real time

In this chapter we follow the chemical change as it unfolds in time. We have three primary motivations. The first is that the time-dependent view matches the way most of us think about a reaction. We have an image of atoms moving, changing partners, etc., rather like a movie on a molecular scale. Time-resolved experiments provide insights that are intuitively appealing.[1] On a more technical level, working in the time domain allows us not only to access the transition state region but also to probe the system as it exits from the transition state. We can see how things evolve rather than just the integrated effect of the dynamics as revealed by a post-collision analysis. Thirdly, time-resolved experiments reveal what happens at short times and this kind of information is otherwise hard to come by experimentally.*

In earlier chapters we discussed experiments at a well-defined energy; we know where we start before the event and we can probe what happens well after it. But we have to infer what happens in the middle. In this chapter we discuss time-resolved experiments, experiments that are able to probe what the nuclei are doing throughout. This then paves the way for studies in the condensed phase where time-resolved experiments are a main tool.

To implement our program we have first to address two key issues, one of principle and one of practice. The uncertainty principle inherently imposes a loss of energy resolution when the time resolution becomes better. We have to show that for the time intervals of interest to us, this loss is acceptable on the chemical scale. We further have to discuss the practical problem of how we are going to measure time on the scale of intramolecular motions when the electronic circuits at our disposal have much longer response times.

What is the time resolution that we need? Chemical forces are short-ranged so the transition state region around a barrier is typically quite localized, at most a few Ångströms wide. The motion slows down as we cross the barrier so if the

---

* In the gas phase some of that information can be inferred from spectroscopic experiments as discussed in Section 7.3. Such experiments, that use long time pulses with a well-defined frequency, are referred to as "working in the frequency domain." The time–energy uncertainty principle forces us to make a choice between working at a better frequency resolution or working at a better time resolution, so the two types of experiment are complementary.

velocity is in the thermal range, say $300 \text{ m s}^{-1} = 3 \cdot 10^{12} \text{ Å s}^{-1}$, and we want a spatial resolution of, say, 0.2 Å, the required time resolution is $0.2/3 \cdot 10^{12} \text{ s} \approx 70 \text{ fs}$. Such ultrashort pulses are available nowadays from commercial lasers with the Ti–Sapphire laser being the more common choice. Typically we need at least two such pulses, one, the pump pulse, prepares the system. The pump is followed by the probe pulse that interrogates the system. The two pulses are delayed in time with respect to one another and it is this delay that is varied in order to reveal the motion.*

The time resolution required for preparing a system in a localized vibrational state is a shade more stringent.** The vibrational period is the time required to span the available range of the vibrational motion, from the right to the left turning points. To (at least partially) freeze the vibrational motion we need a pulse that is short compared with the period. We can more easily freeze rotational motion, because its period is so much longer, but it is not quite so easy to localize the vibrational motion because some vibrational periods can be as short as 10 fs.[2]

In classical mechanics there is no limit to how well one can localize the motion in time. Indeed we launch a classical trajectory with well-defined values for all the coordinates and momenta. As already discussed in Section 5.2.2, this is inherently not possible in quantum mechanics: if the coordinates are sharply defined there is complete uncertainty about the momentum. This is not quite as detrimental as it might seem because the value of Planck's constant is moderate on the scale of interest to us. With a finite but small uncertainty in position we still can have a small uncertainty in momentum. Let us start with the preparation. The uncertainty in time needs to be shorter than the vibrational period. So the uncertainty in energy is larger than a vibrational spacing: with an ultrashort excitation that can localize the vibration we must give up knowing a sharp value of the vibrational quantum number, as first seen in Figure 1.7. The ultrashort pulse prepares not a stationary vibrational state but a localized state. Such a state is not stationary and will move in time just as a classical trajectory will, if we launch it with some vibrational energy. We can think of the non-stationary state as a superposition of stationary vibrational states and for some purposes this view is particularly useful. But it is not essential to think of the non-stationary localized state as a

---

* The delay can be introduced by splitting a pulse and making one beam take a slightly longer route to the sample (light travels 30 cm in 1 ns). That what we measure is the probe signal vs. the delay overcomes the technical problem that our detectors are too slow to clock the motion of the nuclei.

** A vibrational quantum of energy is high compared with $k_B T$ so the velocity is higher than for thermal motion. An ultrafast photoexcitation, on a time scale short compared with a vibrational period, is often referred to as "impulsive." This is a terminology that we do justice to in Section 9.2.2. What it means is that we kick the oscillator on a time scale shorter than its response time. An atom colliding with a molecule also applies a force and then the time scale is the duration over which the force acts between the atom and the molecule. An ultrafast photoexcitation has a shorter duration than a molecular collision at thermal velocities and so the photoexcitation can be in the impulsive regime but the vibrational excitation by collisions is typically not.

superposition. It is a state in its own right. In that it is not stationary it is similar to other non-stationary states that we encountered in Chapter 7. How localized in space is a state prepared in a time shorter than a vibrational period? If we start from the time–energy uncertainty principle for a pulse width $\Delta \tau$, $\Delta E \Delta \tau > \hbar$ we can go over to the position–momentum principle as follows: $\Delta E = (\overline{p}/\mu)\Delta p$ where $\overline{p}$ is the mean momentum. Hence the uncertainty in position is $\Delta x = \hbar/\Delta p = \hbar \overline{v}/\Delta E = \overline{v}\Delta \tau$. But this is the same estimate as we used above to go from the required resolution in position to the time span of the pulse, $\Delta \tau = \Delta x/\overline{v}$. The shorter is $\Delta \tau$ than a vibrational period, the smaller is Δx compared with the span of the vibrational motion between the two classical turning points, and hence the more adequate is the preparation for our purpose of creating a localized state.

*8.0.1   The coherent state – a wave-packet

The state prepared by an ultrafast excitation from an initial state is often called *coherent*, meaning that it can be written as a superposition of wave functions of stationary states. The term coherent is used because we are adding wave functions (= amplitudes) rather than probabilities so that the components of the non-stationary state have a definite phase with respect to one another. The different stationary states have different energies and so, under the Schrödinger time evolution, each component in the superposition evolves independently and differently: as time moves on, the phase of each term in the superposition, $\exp(-iEt/\hbar)$, changes. In physics one often uses the term *coherent state* in a more restrictive sense. The most common restriction is that a coherent state evolves in time without changing its shape. Such a coherent state[3] is rather near to what a classical motion does. Indeed the center of such a coherent state travels along a classical trajectory. A simple and often used such coherent state is a wave function localized in a Gaussian shape (Heller, 1981),

$$\psi(x; \overline{x}(t), \overline{p}(t)) = (1/2\pi\sigma^2)^{1/4} \exp(-(x - \overline{x}(t))^2/4\sigma^2)$$
$$\times \exp(i(x\overline{p}(t)/\hbar + \delta)) \qquad (8.1)$$

Here $\overline{x}(t)$ and $\overline{p}(t)$ are time-dependent parameters. This wave function is centered at $\overline{x}(t)$ and so, as time progresses, the wave function moves with its center along $\overline{x}(t)$, as shown in Figure 8.1. But the shape remains Gaussian.* If the potential is harmonic, this wave function is an exact solution of the Schrödinger time-dependent equation and, and this is the point, the parameters $\overline{x}(t)$ and $p(t)$ are the coordinate and momentum of a classical trajectory at the same energy as the (mean) energy of the quantum system.[4]

For potentials that are not harmonic the Gaussian shape, Eq. (8.1), is not an exact solution: if we take the initial state as Gaussian and proceed to numerically solve the time-dependent Schrödinger equation, the solution first broadens and

---

* The way we have written it, it remains not only Gaussian but Gaussian with constant width.

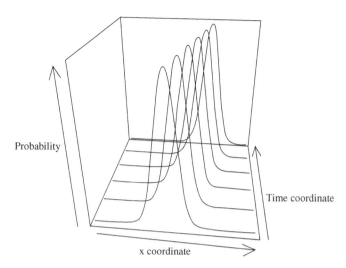

Figure 8.1 The probability density $|\psi(x(t), \overline{p}(t))|^2$ vs. x for different values of time. Computed for a wave-packet, Gaussian in the position x, Eq. (8.1), and centered at the mean position $\overline{x}(t)$, moving in time along the classical trajectory $\overline{x}(t)$. Not shown is the phase of the wave function that is also varying with time.
The plot assumes, as does Eq. (8.1), that the width σ is not changing as the packet moves. For anharmonic potentials this is not quite accurate, see Figure 8.2.

then bifurcates into several localized components. On the other hand, as a practical matter, even for anharmonic motion a Gaussian retains its shape for times of the order of a few vibrational periods and this is often sufficient for our needs.[5]

A localized non-stationary state is also referred to as a "wave-packet." A wave-packet is a chunk of a wave. But a strict wave, meaning one with a definite frequency, needs to extend forever in time and space. To create a localized entity we need to superpose waves of different frequencies. Over most of space these waves interfere destructively and therefore cancel out, but in one region they add up constructively.* Figure 8.9 below illustrates that this interference is an experimental reality.

8.1 Watching the breaking and making of chemical bonds

8.1.1 Photoinitiated bond-breaking

The object of the most elementary experiment is to watch a direct bond-breaking, i.e., when we rapidly excite the molecule, what is the time scale for the

---

* Using the technique of Fourier transformation one can show that the Gaussian in position as given by Eq. (8.1) can be written as a superposition of waves of different frequencies, where the amplitude of the different waves is a Gaussian in the momentum variable. The width of the Gaussian in the momentum variable is inverse to the width of the Gaussian in position. So a state described by a Gaussian wave function has a finite uncertainty in both its position and its momentum. By making such a state more localized in its position, we make it more spread in the momentum, or vice versa.

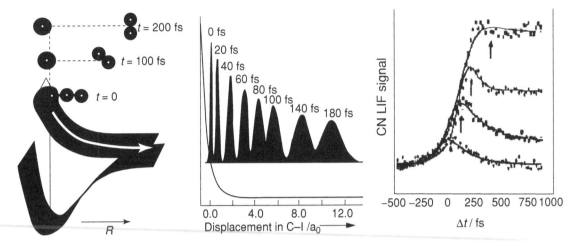

Figure 8.2 Three views of the photodissociation of ICN. Left: the ground and excited state potential in the quasi-diatomic view where CN is regarded as an atom, plotted vs. the I–CN distance R. Middle: the time evolution of the excited wave-packet, shown as probability $|\psi (R,\Delta t)|^2$ vs. the I–C distance for different values of the delay time Δt. The initial wave function $\psi (R,\Delta t = 0)$ is quite localized and hence has a large spread in momentum. As the delay before probing increases, the wave-packet is moving out at a mean speed of about $2 \cdot 10^5$ cm s^{-1} and broadens. Right: experimental LIF signal profiles of CN vs. the delay time between the pump and probe pulses, for a series of frequencies. The topmost curve is at the resonance CN adsorption. Lower curves are for probing as the products are closer and closer in, so the peaks occur at earlier times [adapted from M. J. Rosker, M. Dantus, and A. H. Zewail, *Science* **241**, 1200 (1988) and G. Roberts and A. H. Zewail, *J. Phys. Chem.* **95**, 7973 (1991); see also Bernstein and Zewail (1988), Zewail (2000)].

appearance of products? To do so, an ultrafast pulse promotes the system to a repulsive state, and the probe pulse is used to determine for how long the fragments remain near one another. The pump pulse must be shorter than the time it takes the products to separate.

The photodissociation of ICN in the UV has been studied extensively; experiments using polarized lasers, cf. Section 7.1.2, and quantum chemical computations have characterized the relevant potentials. For our present simplistic discussion, we use a "quasi-diatomic" model[6] where the CN fragment is treated as an atom.[7] There is then only one coordinate, R, the relative separation of I and the center of mass of CN. The potentials relevant to this picture are shown in Figure 8.2. It is seen that the excited state rises very steeply in the Franck–Condon region, so that even a pulse of duration comparable to the vibrational period of the ground state will create a localized dissociative state. As this state propagates outward it quantum mechanically broadens, which is what we mean by *dephasing*. However, as shown in Figure 8.2, the broadening is tolerable.

Next to probing the outcome: CN has an electronically excited state that is very convenient for laser-induced fluorescence (LIF, Section 7.1.3.3) detection. The probing is carried out both at the resonance absorption frequency of CN and at a second frequency, detuned to the red. Both are measured as a function of the delay between the pump and probe pulses, Figure 8.2. When the probe is at the resonance CN absorption frequency, its rise time, about 200 fs, is the time for appearance of the separated dissociation products. The absorption of the CN resonance line probes the products while they are still near to one another. This absorption depends on the extent of detuning, Figure 8.2, and the data can be well accounted for if the system spends some 20–50 fs in the strong interaction region. This time is determined primarily by the steepness of the repulsive potential and the available excess energy.

The direct dissociation as discussed for ICN is also observed in other reactions. However, once we go beyond a quasi-diatomic model, the motion of the other degrees of freedom needs to be addressed.

8.1.1.1 *Bond-breaking occurs along more than one dimension*

Chapter 5 emphasized the polyatomic nature of reaction dynamics. There we saw that it is not enough to ask about the reaction coordinate. During the motion on the potential energy surface the system can and does take large excursions from that path. Working in real time one can directly probe and demonstrate these more intricate dynamics. Here we discuss a simple example, but the very same coherent behavior is seen in more complex systems, including those of photobiological interest as discussed in Section 9.3.7.

HgI_2 dissociates in the UV. At 310 nm there is sufficient energy to break both bonds but the two-body dissociation is still an important channel

$$HgI_2 \xrightarrow{\lambda = 310\,nm} \begin{cases} HgI + I\left(^2P_{3/2}\right) \\ HgI + I\left(^2P_{1/2}\right) \\ Hg + I + I \end{cases}$$

HgI is formed in its ground electronic state and can be probed by excitation to its well-characterized fluorescing $B^2\Sigma$ state. HgI can be probed on its way out and also after the HgI and I products have separated. So far this is as shown for CN from ICN in a quasi-diatomic view in Figure 8.2.

However, now we want more than probing the motion along the reaction coordinate. We also want to characterize the motion in the modes perpendicular to the reaction path, the modes that after the dissociation are the bound Hg–I vibration and rotation.

In a two-dimensional view, Figure 8.3, the non-stationary wave-packet begins at the Franck–Condon region and descends along the symmetric coordinate. For those events where only one bond breaks, the symmetrically placed wave-packet needs to bifurcate. If it bifurcates one way, one bond breaks and if it goes the other

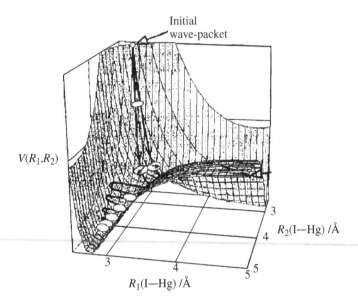

Figure 8.3 A schematic two-dimensional view of the potential energy surface and wave-packet dynamics in the ultrafast photodissociation of HgI_2 [adapted from Voth and Hochstrasser (1996), Zewail (1996)]. The transition state for the I + HgI reaction is along the bisector, dashed line, with the lowest barrier at the bottom of the potential along that line. The UV excitation creates a localized wave-packet along the bisector. The center of the packet is displaced from the saddle point to a compressed configuration along the symmetric stretch. During the dissociation the wave-packet bifurcates, as shown, and each component is followed in the figure. It shows the coherent vibrational motion in the Hg–I well.*

way, it is the other bond that will break.** During this descent from the barrier the motion changes character, from being along the symmetric stretch to some energy going into the asymmetric stretch. This energy will appear as the relative translation of the HgI and I products. The energy remaining in the symmetric stretch will appear as vibrational excitation of the bound Hg–I.

8.1.1.2 Coherence

The two-dimensional view, as shown schematically in Figure 8.3, gives us more than just the realization that HgI will be born vibrationally excited. The initial wave-packet is a coherent state and propagates out in the exit valley just as a

---

* Note that what is plotted is a contour within which most of the wave-packet is located. As time evolves, the packet moves along the line shown and the contour is shown only at a number of points in time. As expected from the Franck–Condon principle, the packet is initiated with most of its energy as potential energy. As time evolves, much of this is converted to kinetic energy. The total energy remains constant but the plot shows the potential energy along the center of the packet.

** Recall a similar discussion for photodissociation of H_2O in Figure 7.16.

classical trajectory will. What this means is that it undergoes oscillatory motion along the symmetric stretch coordinate. Asymptotically, it oscillates along the Hg—I distance: the nascent products are not simply vibrationally excited but are coherently excited.* We must recognize the implications and the limitations of this conclusion. The motion is coherent, but the potential along which it occurs is anharmonic. So the vibrational motion may be initially localized, but on the picosecond time scale it will delocalize and spread as motions tend to do in anharmonic potentials.[8] The time for exit is measured in the experiment to be about 300 fs for the I-atom-producing channel so the detection signal, similar to Figure 8.4, shows oscillations reflecting not only the coherent vibrational motion of Hg—I but also the rotational coherence for which the period is much longer.

Coherence in the nascent products is one manifestation of the limitation of a one-dimensional point of view because it refers to motion not along the reaction coordinate. Will the coherence survive for more complex systems or for situations such as reactions in solution where the coupling to the environment is a key? The answer to both questions is yes. In solution, the coherence of Hg—I motion following ultrafast photodissociation of HgI_2 is observable, Figure 8.4.

Another well-studied case is I_2^- coherence in the photodissociation of I_3^-.[9] Coherent motion also survives in molecules solvated in clusters.[10] In Section 9.3.7 we discuss photobiological processes and there too we will find a need for going beyond a one-dimensional view. Also in this class of many-atom reactions there is clear experimental evidence for coherence in the response to ultrafast excitation.[11]

8.1.2 Bimolecular collisions

We want to access the transition state and we want to probe the motion in a time-resolved fashion, so we need to establish an origin of time from which the motion starts. Initiating a bimolecular collision at a definite time is an experimental challenge. One way is to start with a precursor that is photopumped to start the reaction, just as in Section 1.2.5, except that the excitation is done with an ultrashort pulse and the products are to be probed on their way out. The HBr·I_2 van der Waals adduct is an example of a suitable precursor for the study of the Br + I_2 → BrI + I reaction, Figure 2.12, a member of a family that has been thoroughly studied in collision experiments and is known to proceed over a shallow well.[12] The reaction is initiated by photolyzing HBr in the UV, offering two advantages. First, the light H atom moves away at a far higher velocity than the Br atom. So the H atom is out of the scene of chemical action before much happens.[13] Second, the

---

* Meaning that we have a vibrational wave-packet, made up as a superposition of several stationary vibrational states, where the different components have definite phase relations.

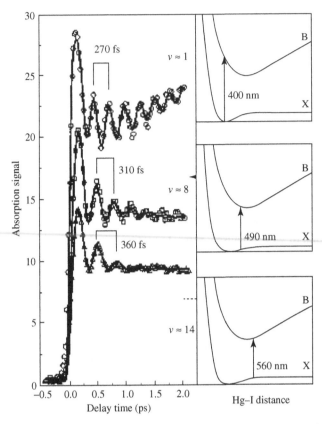

Figure 8.4 Visible absorption of HgI to the upper B state, at a short delay after an impulsive photodissociation of HgI$_2$ in ethanol [adapted from N. Pugliano *et al.*, *J. Chem. Phys.* **99**, 7273 (1995); see also Voth and Hochstrasser (1996)]. Shown are results at three wavelengths. The upper trace corresponds to a Franck–Condon region near the well of the ground state of HgI, labeled as $v \approx 1$. As shown in the side panel this pulse samples molecules at low vibration and the period of the oscillation is that which is expected. The middle plot is for a pulse that probes the ground state of HgI at higher vibrational excitation. The lower trace is for a Franck–Condon region near the outer turning point in the well of the ground state of HgI. It samples vibrationally excited molecules ($v \approx 14$) so the vibrational period is longer and the dephasing is faster.

slow-moving Br atom does not have enough energy to cross any large barrier. So one is studying the Br + I$_2$ reaction near its threshold. To increase the available energy one can start with DBr·I$_2$, which (about) doubles the velocity of the Br atom.

The experimental detection of IBr by LIF, Figure 8.5, shows that there is essentially an exponential increase with time of the fluorescence signal. The rise time constant, 53 ± 4 ps, confirms that the collision is sticky, with a duration of the order of several rotational periods, suggesting that it proceeds through a

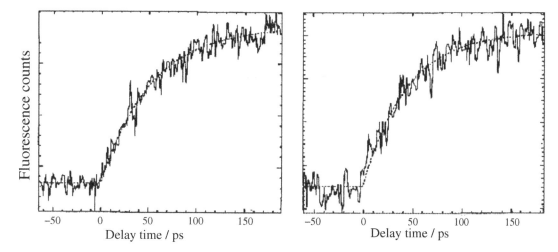

Figure 8.5 The rise of the LIF signal of IBr vs. time (fitted to an exponential function $[1-\exp(-\text{delay time}/\tau)]$) as a function of the time delay between the pump and probe. The fit is shown for both HBr·I$_2$, available energy 145 cm^{-1}, $\tau = 53$ ps, left panel and DBr·I$_2$, available energy 286 cm^{-1}, $\tau = 44$ ps, right panel precursors [adapted from I. R. Sims *et al.*, *J. Chem. Phys.* **97**, 4127 (1992); see also S. A. Wright, M. F. Tuchler, and J. D. McDonald, *Chem. Phys. Lett.* **226**, 570 (1994), who used multiphoton ionization to detect the products].

complex that lives over a shallow well. The rise of the products is faster when DBr is used, as shown.

The results shown in Figure 8.5 indicate that reaction occurs even when the available energy is low. If there is any barrier to reaction it is therefore late, that is, in the exit valley. The simple picture, see Section 5.1.5.1, is that of an occupied p orbital on Br oriented along the I–I axis, sharing its electron with the lowest unoccupied orbital of I$_2$. This I$_2$ orbital is antibonding and has a node between the two I atoms. So the Br–I bond is formed at the expense of the I–I bond. But a more detailed picture is required to fully analyze the reaction. For example, there are three p orbitals on Br, two of which are perpendicular to the I–I axis. So there has to be more than one potential energy surface correlating to the reactants. Somewhat higher up there are the surfaces that correlate to the spin-orbit excited state of Br, and since the photodissociation of HBr also produces Br* atoms, other features of the dynamics are there to be studied.

8.2 Chemical transformations

As for scattering of crossed molecular beams, real-time experiments have greatly profited from the introduction of probing via mass spectrometry. The experiment is made possible by having the probe be an ionization laser, as in many examples in Chapter 7. This addresses issues of chemical interest because we can determine

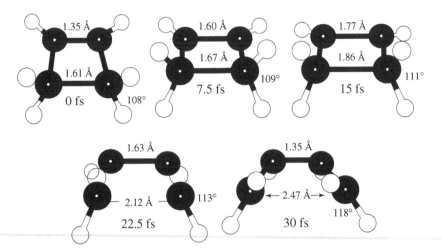

Figure 8.6 Snapshots of the computed geometry of cyclobutene for ultrashort times following excitation of a π electron to an antibonding π^* orbital at time zero [adapted from M. Ben-Nun and T. J. Martinez, *JACS* **122**, 6299 (2000)]. Shown are the bond distances for the initially double bond and the C—C bond that is broken and the HCH hybridization angle. By about 20 fs the hybridization angle has increased and the motion is well toward a disrotatory ring opening, as expected from the Woodward–Hoffmann rules for a photochemical ring opening in a ring with $4n + 2\,\pi$ electrons, see Problem A. For the well-studied case of ring opening in cyclohexadiene the motion is conrotatory, as expected for $4n\,\pi$ electrons [see M. O. Trulson, G. D. Dollinger, and R. A. Mathies, *J. Chem. Phys.* **90**, 4274 (1989)].

the entire range of species that are present after one laser shot of a given delay after the pump. Varying this delay provides snapshots of the **chemical** dynamics vs. time. Of course, we can also lock the detection on a particular mass and watch how its number density evolves in time.

8.2.1   Concerted vs. sequential bond forming

Pericyclic reactions in general and electrocyclic reactions* in particular, for example the ring opening of cyclobutene to 1,3 butadiene as shown in Figure 8.6, have been central to physical organic chemistry and its interpretation by orbital theories of electronic structure.[14]

---

* In an electrocyclic reaction (Woodward and Hoffmann, 1970) a molecule with a conjugated system of $m\,\pi$ electrons cyclizes to form a ring of $m - 2\,\pi$ electrons and one σ bond. For this transformation to happen, the two ends of the relevant π system must approach each other in such a way as to enable the end p orbitals to overlap constructively. To form a σ bond the two terminal p orbitals need to rotate. When the molecule is substituted the rotations can be in the same or in opposite directions. This leads to different stereoisomers and is of much interest to physical organic chemists. Therefore, such reactions are extensively studied. They are a special case of a pericyclic isomerization where the transition state is cyclic.

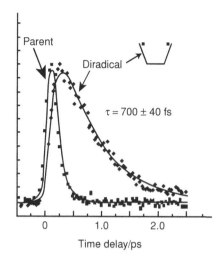

Figure 8.7 Dynamics following ultrashort UV excitation of cyclopentanone. Shown is the (time-of-flight) mass spectrum at different delays between the pump and ionization laser. When there is no delay, only the mass of the parent (84 amu) is detected. The parent signal decays exponentially, with a lifetime of about 120 fs. There is a corresponding buildup of an intermediate of mass 56, the mass of the tetramethylene diradical that results from elimination of CO (this is also the mass for elimination of ethylene and formation of CH_2CH_2CO). The peak of the intermediate (mass 56) decays exponentially in about 700 fs. The decay of the intermediate becomes even longer if the parent ketone is substituted so that there are more vibrational modes and the decay time does become shorter when the excess energy brought in by the UV photon is higher. This is as expected if the energy is randomized. The decay time is rather short, but not short enough and, even at significant excess energies, the lifetime for the disappearance of the intermediate is not as short as expected for a direct descent from a barrier region [adapted from Pedersen *et al.* (1994)]. The high power of the ultrashort pump laser means that it is necessary to properly resolve one- and two-photon processes. [The nature and fragmentation of the ions created by the ionization laser also needs to be elucidated, see T. I. Solling *et al.*, *Chem. Phys. Chem.* **3**, 79 (2002)].

Another Woodward–Hoffmann allowed process is the ring opening of electronically excited cyclobutane to two ethylene molecules. This is unlike the thermal process that has a high barrier. Upon crossing the transition state region, do the two bonds break in concert or sequentially? It is quite conceivable that both paths are possible on one and the same multidimensional potential energy surface (Hoffmann, 1999). It is further reasonable that the relative importance of the two paths will depend on the total energy and on the detailed manner in which this energy is supplied. So the essential question to ask is whether a sequential mechanism for ring opening is possible. To access the transition state region where the intermediate in a sequential mechanism is located we need a suitable precursor. This can be a cyclic ketone. Figure 8.7 illustrates the clocking of the mass spectral pattern.

A different aspect of concerted vs. sequential routes is illustrated by the high-energy dissociation of linear ketones

$$R-CO-R' \xrightarrow{h\nu} (RCOR')^* \rightarrow \begin{cases} RCO + R' \rightarrow R + CO + R' \\ (RCOR')^\dagger \end{cases}$$

that are found experimentally to dissociate in a stepwise manner, with one radical departing well before the other. The alternative route is to form vibrationally hot parent molecules through radiationless transitions. The first bond break is a stretch of the C—C bond accompanied by a C—C—O bend to form the acetyl radical. The time constant for the first bond break increases somewhat as the molecule size increases, from about 60 fs in acetone to 290 fs in $(CH_3)_2CHCOCH(CH_3)_2$ with comparable values for cyclic ketones. This increase is rather small compared with what is expected from the statistical RRKM theory of Section 6.2 when the number of vibrational modes roughly doubles. It strongly suggests that not all degrees of freedom participate in the bond-breaking process. This is further confirmed by the effect of deuterium atom substitution. If D is substituted on the near C atom (the α position in the terminology of organic chemistry), the decay slows significantly to 230 fs for $CH_3CD_2COCH_3$ compared with 100 fs for $C_2H_5COCH_3$. Substitution on the next C atom, the β position, brings the time down to 110 fs for $CD_3CH_2COCH_3$. This is particularly telling because, apart from differences due to zero-point energy, the potential energy landscape should not change upon isotopic substitution.

The non-statistical character of the first bond breakage does, of course, depend on the nature of the excitation. The UV excitation promotes an electron from a lone pair on the O atom. The unpaired electron that is left behind can then form a new π-type bond when the C—C bond is weakened. The lowest excited singlet state $(\tilde{A}^1A_2)$ has the unpaired electron placed in a π^* orbital. Computations suggest that the motion out of the Franck–Condon region is along vibrations (CO stretch and out-of-plane bend, CH_3 torsion) that are not part of the reaction coordinate for the C—C bond break but that favor the bonding. The wave-packet moves rapidly downhill in potential energy. There is a barrier along the reaction coordinate and if the energy is sufficient, a combination of C—C bond stretch and an increase in the C—C—O angle will take the motion over the barrier and toward a conical intersection with the ground surface. Crossing of the intersection produces either a hot parent molecule or the first C—C bond break. The nascent acetyl radical will be nearly linear but it can deform on the ground state or the trajectory can bend on the upper state prior to crossing. Figure 8.8 is a two-dimensional view, which indicates that a correlation between the C—C stretch and the C—C—O bend is necessary for dissociation.

It is possible to access the first excited state of acetone at energies that are not sufficient for the motion to surmount the barrier and cross to the ground state in its

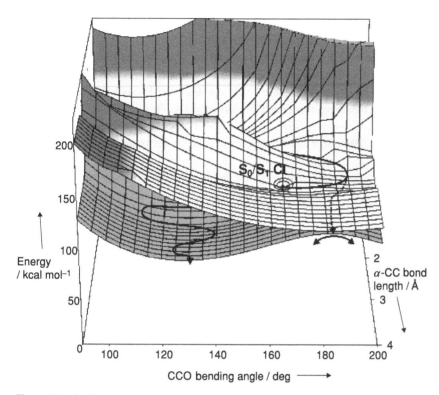

Figure 8.8 A 3D view of the potential energy surfaces for the first excited state and the ground state of acetone. The plot is as a function of the two most important coordinates for dissociation: the C–C bond distance and the C–C–O bending angle. Crossing from S_1 to the ground state is possible either at a linear C–C–O configuration, identified in the figure as CI, or at a bent one. The solid/dashed line traces the motion of the center of the wave-packet on the excited/ground state potential. The concerted motion required for bond-breaking makes the dissociation non-statistical [adapted from E. W. G. Diau, C. Kotting, and A. H. Zewail, *Chem. Phys. Chem.* **2**, 273 (2001)].

exit valley. The excited molecule will then live much longer. It will (intersystem) cross to the lowest triplet state and dissociate on the nanosecond time scale. On an even longer scale, it can internally convert to the ground electronic state.

At shorter wavelengths and/or for a two-photon process the excitation can reach higher up excited states. These correspond to the electron on the O atom reaching a Rydberg orbital. These higher states have several conical intersections with the first excited singlet state. Here too the crossing is non-statistical and the first bond-breaking is quite prompt.

In cyclic ketones, the first bond break leads to the formation of the $\cdot CH_2(CH_2)_{n-2}CO\cdot$ diradical. As for the acyclic ketones, the second bond break in these energy-rich species is delayed but the delay does not increase with size in the manner expected from a statistical theory.

In Section 7.2.4 we argued that excitation along the reaction coordinate can propel a system across a barrier even when there are many other degrees of freedom. The dissociation of energy-rich ketones provides such an example since there are two motions that control the dissociation, the C—C stretch and the CCO bend. When the species is nearly linear and energy is placed in the stretch, direct dissociation can occur in a vibrational period. On the other hand, if some of the energy is placed in the bend motion, the molecule will miss the narrow bottleneck and rattle around. Thus, at the same energy, the molecule can exhibit both direct and complex dynamics depending on the finer details of the preparation process.

8.3   Control of chemical reactions with ultrashort pulses

Ultrashort pulses can help us place a wave-packet in a well-localized region on an excited state potential.[16] Moreover, the details of the optical excitation, specifically the momentum in the different vibrations, can have a decisive influence on the outcome. In Section 7.4 we discussed control in the frequency domain and emphasized interference between different pathways as an essential element in making the process go one way or the other. Here we consider how we might control chemical reactions[15] in the time domain.

A very vivid experiment that forges a link between the two points of view is to let two wave-packets interfere. By being able to adjust the relative phase of two, otherwise identical, laser pulses we can proceed as follows. The first ultrashort pulse prepares a localized state at the left (inner) turning point of the bound excited electronic state of I_2. This packet begins to propagate to the right. The excited state potential is fairly shallow so it takes a while (300 fs) for the packet to reach the outer turning point and come back to where it started. At that time the second laser pulse, with a π phase difference, is applied. It too creates a wave-packet at the inner turning point and this packet destructively interferes with the first packet so there should not be any fluorescence from the electronically excited I_2 molecule. On the other hand, if the phase difference between the two pulses is 0, the interference should be constructive and the fluorescence will be enhanced. This shows the imprint of the coherence of the light on the motion of matter. In addition the wave-packets acquire a phase during their motion on the upper potential. So a further refinement is to examine the fluorescence also as a function of the delay time between the two pulses, Figure 8.9.

To generalize this approach consider a pulse that is tailor-made to have prescribed phase relations between its frequency components.[17] Such a "shaped" pulse is what was used in the example mentioned already in Section 1.1. A *learning feedback approach* is to start from an arbitrary pulse shape and feed the output to a (computer) controller that then seeks to change the shape of the

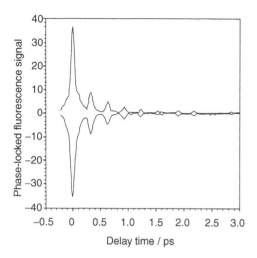

Figure 8.9 The interference contribution* to the fluorescence (heterodyne detection) signal from I_2^* excited by two ultrashort identical pulses as a function of the delay between them. Upper trace: no phase difference between the two laser pulses. The wave-packet promoted by the first pulse and the newly promoted wave-packet can interfere constructively whenever the time delay between the pulses is a multiple of 300 fs so that the first wave-packet is back in the Franck–Condon window. Lower trace: a π phase difference between the two laser pulses. At intervals of 300 fs the returning wave-packet interferes destructively with the new wave-packet [adapted from N. F. Scherer *et al.*, *J. Chem. Phys.* **95**, 1487 (1991). For other experiments showing interference between wave-packets excited by two pulses see Baumert *et al.* (1997). Even an electron can be so controlled during electron transfer (Barthel *et al.*, 2001; Martini *et al.*, 2001; Bardeen, 2001)].

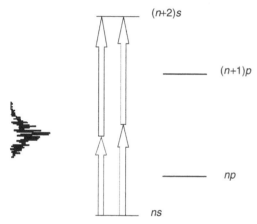

Figure 8.10 Energy level for a two-photon excitation of a Cs atom by a shaped pulse [adapted from D. Meshulach and Y. Silberberg, *Nature* **396**, 239 (1998)]. The frequency spectrum of the pulse is shown schematically so as to indicate why the pulse will contain two or more pairs of photons whose frequencies add up to the required energy.

pulse so as to enhance the desired outcome. The process is iterated. A simple example, where one can understand why this approach should work, is shown in Figure 8.10.

The $ns \rightarrow (n+2)s$ transition is two-photon allowed. Detection of the upper level is through its one-photon fluorescence to the np state. The object of the control is either to populate or not to populate the upper state using the interference between different two-photon transitions. The principle is very much the same as in Section 7.4 but the implementation is different. Here only one pulse is used but this pulse is ultrashort so it is a superposition of many pulses each with its own well-defined frequency. It is the interference between these components that allows us to direct the outcome. The shape of the pulse, meaning the amplitude

---

* The interference term between the two amplitudes, see, e.g., Eq. (2.37), is determined by heterodyne detection.

Figure 8.11 Potential energy curves for Na_2 and Na_2^+ for the two-photon coherent excitation of a packet on the $2^1\Pi_g$ electronic state. By probing this packet at the left, arrow 1, and the right, arrow 2, turning points respectively one accesses two different higher excited states. The probing at the left produces the ground state of Na_2^+. The probing at the right produces a doubly excited state, Na_2^{**}, that dissociates to a ground-state Na atom and a Na^+ ion. The ground state can be probed by transitions to a repulsive excited state, arrow 3.

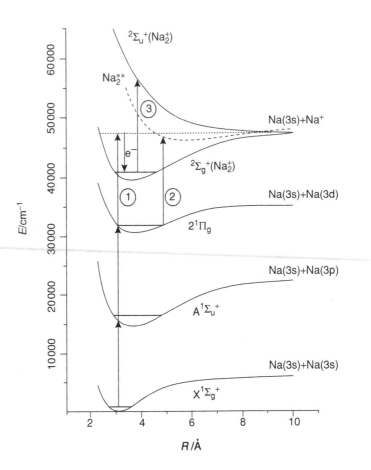

and phase of its different frequency components, has been sequentially adapted, using a feedback loop, to optimize the output.

8.3.1 Control by pump and probe

The pump–probe experiments discussed earlier in this chapter lend much scope for control. The idea is to localize the initial state by the pump pulse and to time the probe so as to catch the system as it is following one path rather than another. The pumping sequence for the Na_2 molecule for which this has been realized is described in Figure 8.11, with the results shown in Figure 8.12.

Examination of the signals of the ions as shown in Figure 8.12 shows that the Na_2^+ production oscillates with the 320 fs vibrational period of the $A^1\Sigma_u$ potential. This is because the $2^1\Pi_g \rightarrow {}^2\Sigma_g^+$ transition to form Na_2^+ is possible at all interatomic distances because the light departing electron takes up any excess kinetic energy. On the one hand, the Franck–Condon region for pumping the $2^1\Pi_g$ state through the $A^1\Sigma_u^+ \rightarrow 2^1\Pi_g$ transition only occurs near

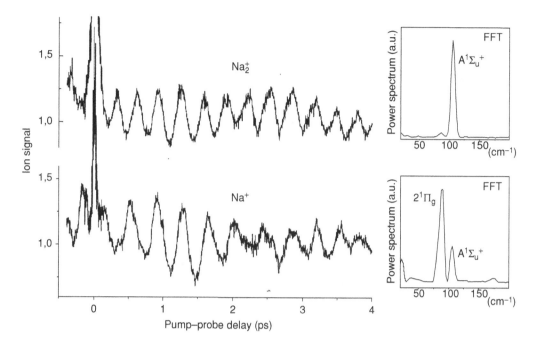

Figure 8.12 Left: the out-of-phase Na_2^+ and Na^+ signals as a function of the delay between the pump and probe pulses (each 80 fs in width at 618 nm). The potential energy curves are shown in Figure 8.11. By taking a fast Fourier transform (FFT) of the signals, right panels, the main frequency components can be identified and thereby the electronic state on which the motion took place [adapted from Baumert *et al.* (1997)].

the inner turning point. On the other hand, the Na^+ production oscillates primarily with the 380 fs vibrational period of the $2^1\Pi_g$ potential. This is because the formation of Na_2^{**} is possible only near the outer turning point of the $2^1\Pi_g$ potential.

The probe laser can also be used to dump the system back to the ground state. Stimulated emission pumping[18] (SEP) shows that dumping can reach higher-energy regions on the ground potential energy surface that are not otherwise accessible. By using a localized initial wave-packet, its motion on the upper surface can bring it to where it can be effectively dumped down into an exit valley. Note that here, too, one can think of the selectivity being achieved as an interference between alternative paths being made possible by the many components of the localized state.

Control of molecular motion need not refer only to species selectivity. For a number of purposes, including molecular-scale computing, one may want to send the system to one or another internal state. Examination of the options of doing so have also drawn attention to the advantages that sometimes accrue of reversing the order of the probe and pump pulses. A counterintuitive sequence can dress the intermediate and final states and thereby allow a very selective pumping (Bergmann *et al.*, 1998).

Dephasing limits our ability to control. In larger systems, energy redistribution and coupling to the environment are important additional sources of dephasing. Learning algorithms, where the pulse is optimized by monitoring the performance, have allowed us to make progress (Brixner and Gerber, 2003). Quantum computing is a particular form of control and the issue of dephasing is paramount for such applications.

Problems

A. The Woodward–Hoffmann correlation diagrams (Woodward and Hoffmann, 1970). The text states that the thermal ring opening of cyclobutane is forbidden but that the photophysical process is allowed. (a) Draw the four molecular orbitals for a ring of four C atoms where only one p_z orbital per C atom is considered. The z axis is taken as normal to the plane of the ring. Also draw the orbitals for two ethylene molecules, two π orbitals per molecule. (b) Using the symmtery of the orbitals with respect to the plane of the molecule, correlate the four orbitals. (c) Why in Figure 8.6 is the ring opening in cyclobutene called "disrotatory" and how is this related to the number of π electrons?

B. How long does it take to break a bond? Consider a direct dissociation along a repulsive potential. The optical excitation places the system high up on the potential, essentially at the classical turning point, R_0. (a) Take the repulsive potential to be exponential as in Eq. (2.13). Say that as in ICN the potential correlates with the ground state of the fragments. For a given (mean) frequency of the light pulse, calculate the available energy and hence relate the classical turning point to the parameters of the potential and to the photon energy. (b) As shown in Figure 8.2, the products are detected when they have separated to a larger distance, say R_d. Compute analytically how long it takes the products to move from initiation to detection. Hint: if the velocity v is constant, the time it takes to transverse a distance ΔR is $\Delta R / v$. But the velocity increases as the products separate. (Why?) So it is necessary to divide the distance $R_0 - R_d$ into short intervals and add together the contributions from the separate intervals. (c) Verify the experimental value for ICN, about 200 fs, if the range parameter for the exponential repulsion is 0.8 Å.

C. The revival time. Taking the ground vibrational state to be at zero energy, the higher vibrational states have energies that are not quite multiples of the harmonic frequency. Rather, $E_v = \hbar(\omega v - \omega x v^2)$. There is an uppermost bound state. Its quantum number N is given as Nx where we approximate $1/x$ as an integer. Therefore the energy levels are $E_v = (\hbar\omega/N)v(N - v)$. Say we build an arbitrary wave-packet of bound vibrational motion. Its generic form is $\psi(t) = \sum_{v=0}^{N} A_v \exp(-iE_v t/\hbar)\psi_v$. The amplitudes A_v determine the shape of the packet. (a) Show that in the harmonic limit, $E_v = \hbar\omega v$, the wave function at times $t + 2\pi/\omega$ recovers the same shape it had at time t. (b) Show that in the anharmonic case the revival time is N times longer. (c) Back to the harmonic limit. Can you specify such amplitudes A_v that the wave function retains its shape at all times? Try: $A_v = (\alpha_v/\sqrt{v!})A_0$. This is, in fact, the wave function Eq. (8.1). See C. Cohen-Tannoudji, B. Diu, and F. Laloë, *Quantum Mechanics*, New York, John Wiley, 1977, complement Gv. (d) The anharmonic limit. Can you do the same as you did in (c) for the anharmonic case?

D. Coherent rotational motion. Build a wave-packet of (rigid rotor) rotational states and show that it will revive.

E. Directed molecular states. Can you build a wave-packet of rotational states, giving a different amplitude to different m states, such that the wave-packet will be initially localized in a given direction and will then rotate as the hand of a clock?

F. Spreading of wave-packets. In an anharmonic potential, a localized wave-packet will, in general, spread. Show this for the simplest case, a constant potential. Take an initial state that is a Gaussian like Eq. (8.1), but allow its width σ to depend on time. By substituting this function in the Schrödinger time-dependent equation show that it is a solution. Thereby show how σ increases with time. See C. Cohen-Tannoudji, B. Diu, and F. Laloë, *Quantum Mechanics*, New York, John Wiley, 1977, complement GI.

Notes

1 Theorists like it too because you do not need to propagate the numerical solution of the quantum dynamics for a long time. A short propagation interval is often enough: you have computed just as much detail as is really needed. Working at a sharp energy is, in some sense, wasteful. You generate answers that are more precise than they need to be. There are other pleasing aspects. One is that one can explore large systems using only few coordinates because intermode energy exchange is initially limited to a small subset of modes. A more technical way of saying this follows from our considerations in Chapter 7: you do not have to include weaker coupling terms until such propagation times are of the order of $\hbar$/coupling strength. Because $\hbar$ is small, a strength of 0.1 cm^{-1} corresponds to 50 ps, which is long on the scale of vibrational periods.

2 Would it not be interesting to localize electrons? Seemingly it is not going to be practical because by the Born–Oppenheimer approximation, electrons move faster than nuclei. True but not always true. Say we go up in electronic energy. The spacings between excited states get smaller so the period of the electronic motion, which is inverse to the spacings, gets longer. Indeed, if you use the Bohr model of the atom, the orbital period increases as n^3 where n is the principal quantum number. So we should be able to localize electrons with high n's, known as Rydberg electrons. Once localized they will revolve about the nucleus almost just as Bohr and Sommerfeld imagined them to do. See, for example, Nauenberg *et al.* (1994), Verlet and Fielding (2001).

3 The first example, our Eq. (8.1), was reported by Schrödinger himself in 1926.

4 The general case was already known to Ehrenfest who showed in 1927 that the mean position and momentum of a wave function strictly follow what look like classical equations of motion, $d\bar{x}(t)/dt = \bar{p}(t)$, $d\bar{p}(t)/dt = -\overline{dV(x)/dx}$. The quantal nature of the dynamics enters in the equation of motion for the momentum. For a classical trajectory $\bar{x}(t)$, $\bar{p}(t)$ the expected result is $d\bar{p}(t)/dt = -dV(\bar{x})/d\bar{x}$. For a harmonic potential, where $V(x) = kx^2$, the two results are the same but, in general, the average (over the wave function) of the force, $-\overline{dV(x)/dx}$, is not equal to the force at the average position. The Ehrenfest theorem is powerful but can be potentially misleading: it is valid, say, if the wave function bifurcates into two components. Under such circumstances, the system is very much not at its average position.

5 The Gaussian is also an exact solution if there is no potential, that is, for a free motion. In this case the width increases with time. This is useful for setting a time scale for broadening. Note that this broadening has a classical interpretation: for a Guassian wave function there is a distribution in the values of the momentum and, classically, the components of higher velocity move out faster. Explicitly, if Δv is the spread in velocity as determined by the initial width σ, the additional spread in x is $\Delta x = t\Delta v = t\Delta p/\mu = \hbar t / \mu\sigma$. Given any initial state that is localized, the time-dependent Schrödinger equation enables us to propagate it in time. Numerical methods for doing so are available, e.g., Kosloff (1994), Jackson (1995). An important advantage of such methods is that the complete wave function can be simultaneously propagated on several electronic states. For each state there can be a different wave-packet for the nuclear motion, so that the motion can be bound on one state and dissociative on another. See Figure 9.12 for an example.

6 The pioneering study of ICN photofragmentation is K. E. Holdy, L. C. Klotz, and K. R. Wilson, *J. Chem. Phys.* **52**, 4588 (1970).

7 This means that we overlook that the equilibrium configuration in the excited state(s) is probably bent so the Franck–Condon region for excitation from the linear ground state will create an I—CN configuration with a force that will set CN rotating (this is detected as fairly rotationally excited CN fragments). We also neglect the presence of several electronically excited states and the possibility of non-adiabatic coupling between them. Furthermore, there are also electronic states that correlate asymptotically to the spin-orbit excited state of the iodine atom that is about 1 eV above the ground state. For the accurate description of the potentials from *ab initio* quantum chemistry, see Y. Amatatsu, S. Yabushita, and K. Morokuma, *J. Chem. Phys.* **100**, 4894 (1994).

8 In principle and in the absence of environmental perturbations, the coherence will revive. The revival time is expected to be much longer than a vibrational period. Problem C addresses the not trivial task of estimating that the revival time is about $2N$ times longer than a vibrational period where N is the (finite) number of bound states of a Morse-type anharmonic potential. $N (\approx 60)$ is quite high for HgI. The (harmonic) vibrational period is 267 fs.

9 On the coherence in the ultrafast dissociation of I_3^-, see Banin *et al.* (1994); T. Kuehne, R. Kuester, and P. Voehringer, *Chem. Phys.* **233**, 161 (1998). See also Ruhman and Scherer (1998) for a collection of papers on coherence effects.

10 J. K. Wang, Q. Liu, and A. H. Zewail, *J. Phys. Chem.* **99**, 11309 (1995).

11 Wang *et al.* (1994), Zhu *et al.* (1994), Bardeen *et al.* (1998), T. Ye *et al.*, *Chem. Phys. Lett.* **314**, 429 (1999). See also El-Sayed *et al.* (1995). Of course, coherent motion can be seen

in larger inorganic molecules. One can make the case for a parallelism between the decay of coherence and the phenomenon of IVR. By analogy to the terminology of nuclear magnetic resonance (NMR), one can call them a T_2 and a T_1 process. Such analogies have been beneficial to coherent optical spectroscopy and have suggested many useful extensions, particularly so in relation to 2D NMR (Mukamel, 1995, 2000; Dantus, 2001; Jonas, 2003).

12 Migratory dynamics is discussed by J. J. Wang, D. J. Smith and R. Grice, *Chem. Phys.* **207**, 203 (1996).

13 The structure of the adduct favors this because in similar molecules it is known that they are "anti" hydrogen bonded, meaning that the hydrogen atoms point away from the halogens. The prompt departure of the H atom also eliminates the possibility that in reactions initiated, e.g., by photolysis of $BrH \cdot CO_2$, which is hydrogen bonded, the heavy Br atom stays around and may play an unintended guest role.

14 For more on electrocyclic reactions, see Borden *et al.* (1988), Dolbier *et al.* (1996), Fukui (1982, 1971), Pross (1995), Rauk (1994), Woodward and Hoffmann (1968, 1970).

15 For control in the time domain, see deVivie-Riedle *et al.* (2001), Gaspard and Burghardt (1997), Gordon and Rice (1997), Rabitz and Zhu (2000), Rice and Zhao (2000), Tannor and Rice (1988).

16 In principle, using IR and the coordinate dependence of the (permanent) dipole moment one could also do this on the ground state. One can also dump the packet onto the ground state from higher up as in SEP, e.g., Pausch *et al.* (2000).

17 Wave-packets moving in an anharmonic potential dephase see problem F. We can think of the wave-packet as a coherent superposition of stationary vibrational (and rotational) states, each having its own phase. These components are born with a phase imprinted by the pump but as time goes on each acquires an additional phase and if the potential is anharmonic the phase factors get out of 'sync' because, unlike the case of a harmonic potential, the period of the vibrational motion depends on the energy. This dephasing can be relatively slow because its time scale is determined by the anharmonicity. But it can also be turned to advantage. Using chirped (that is, phase-modulated) pumping one can compensate for the dispersion in the phases by exciting each component with a proper phase. The proof of principle is that one can set the chirp in the opposite direction so as to enhance the spreading.

18 Stimulated Emission Pumping is reviewed in Hamilton *et al.* (1986), Silva *et al.* (2001).

Chapter 9
State-changing collisions: molecular energy transfer

The transfer of molecular energy by collisions is the slow step in the return of a macroscopic system to thermal equilibrium.* These state-changing collisions are called inelastic because conservation of energy requires that the change in the internal energy of the molecules is balanced by a change in the (relative) kinetic energy of the two partners. When they gain internal energy, the molecules exit from the collision slower than they approached and vice versa. You already know that the heavy atoms are resilient to changes in their kinetic energy. So our main theme in this chapter is an elementary discussion of what determines the propensities of molecular energy transfer. This will provide a more quantitative version of the Franck–Condon principle, a formulation that is not limited to the sudden regime. We will examine a variety of processes of interest that are made possible or are limited by these propensities. Our last section deals with how and when it is possible to transfer electronic energy, a process that nominally results in a large change in the internal energy.

9.0.1   Equilibrium and disequilibrium

Thermal equilibrium in the gas phase can be characterized by a constant (time-independent) fraction of molecules in any given energy level. At a temperature T the relative population in the ith energy level is given by the Boltzmann distribution

$$P_i = \frac{N_i}{N} = g_i \exp(-E_i/k_B T)/Q_1 \tag{9.1}$$

Here N_i is the number of molecules in level i and N is the total number of molecules. As usual, the *degeneracy* g_i is the number of possible quantum states

---

* In principle, the system can also relax to equilibrium by light emission. Unless the circumstances are unusual, emission of IR photons is too slow a process to significantly compete with collisional relaxation. An extreme case is that of molecules in outer space where the density is very low so the time between collisions is exceedingly long. Another unusual situation is a chemical laser, Section 9.0.3, where the presence of a high density of photons stimulates the emission of other photons.

of the molecule corresponding to the energy level E_i. Q_I is the internal partition function, Eq. (A.6.4), that insures the conservation of matter, $\sum_i N_i = N$.

Perturbing the system can lead to a transient population distribution that deviates from Eq. (9.1). The system is then in a state of disequilibrium. Such a state can be achieved, for example, by very rapid heating as in a shock tube. Alternatively, and more selectively, a preferential population of excited states can be created either by physical activation (e.g., absorption of light) or by chemical reactions that are selective in their energy disposal (compare Figures 1.2 and 1.3).

Following a transient disturbance the gas "relaxes" into an equilibrium distribution. One can follow this relaxation process in terms of the changes in the population of the different energy levels. Such changes are the manifestations of the energy-transferring molecular collisions. For example, a vibrationally excited HCl molecule can lose its excess energy in a collision with a vibrationally cold DCl molecule

$$HCl(v = 1) + DCl(v = 0) \rightarrow HCl(v = 0) + DCl(v = 1), \qquad Q = 775\,cm^{-1}$$

Such an inelastic collision is not reactive yet it is not elastic either since the final translational energy is not equal to the initial translational energy. The initial vibrational energy of HCl is converted partly to vibrational excitation of the DCl partner and the balance goes into the translational energy of the receding molecules, which therefore move faster after the collision. The energy balance is shown in Figure 9.1, which also shows an experiment that monitors the rate of energy transfer. Q is the translational energy release for the process indicated. Below we shall loosely call Q the *gap* and set up a scale for assessing when the gap is large or small for a given experimental condition.

In the experiment shown in Figure 9.1 a mixture of HCl and DCl in an excess* of Ar buffer gas is irradiated by a brief pulse of infrared light (say, from an HCl chemical laser, Section 9.0.3). HCl molecules are thereby selectively excited from the $v = 0$ to the $v = 1$ level. Thereafter, the population of the HCl($v = 1$) level is depleted both by transfer of its entire vibrational energy to rotation and by translation, mainly owing to collisions with the excess Ar buffer gas, HCl($v = 1$) + Ar $\rightarrow$ HCl($v = 0$) + Ar, and by vibrational energy transfer to DCl($v = 1$). The $v = 1$ population of DCl is depleted mainly by transfer of its energy to rotation and translation by collisions with an Ar atom, DCl($v = 1$) + Ar $\rightarrow$ DCl($v = 0$) + Ar. This process has a large gap but not as large as the corresponding transfer in the case of HCl. The smallest gap, $Q = 775$ cm^{-1}, is for the vibrational energy transfer from HCl to DCl.

The experiment monitors the IR fluorescence** of HCl($v = 1$) and DCl ($v = 1$). For DCl($v = 1$) it shows, Figure 9.1, both a fast-rising exponential

---

* The Ar acts as a buffer gas, that is, as a heat bath for the released translational energy.
** The loss of molecules due to fluorescence is negligible compared with the loss due to energy transfer collisions. But the sensitivity of detectors in the near IR is sufficient to detect the photons that are emitted and thereby to follow the changes in the concentrations.

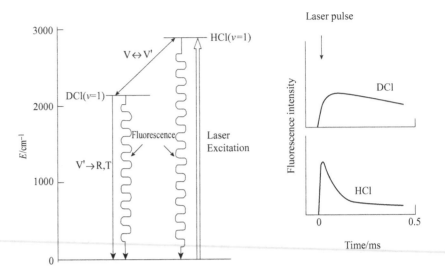

Figure 9.1 Time-resolved IR fluorescence intensity from DCl(v = 1) and HCl(v = 1) following a laser excitation pulse that excites HCl to v = 1 [adapted from H.-L. Chen and C. B. Moore, *J. Chem. Phys.* **54**, 4072 (1971)]. The energetics of the different processes are shown in the left panel. Three different values* of Q, for three different inelastic collisions, are indicated: the HCl/DCl transfer, the V–R, T transfer in HCl + Ar collision, and the V–R, T transfer in DCl + Ar collision. The time dependence of the fluorescence signal from vibrationally excited HCl and DCl is shown on the right. HCl decays faster by the lower Q process, the V–V′ transfer to DCl. The decay of DCl is by V–R, T transfer and is slower.

due to the rapid energy transfer from HCl(v = 1) and a slower decaying exponential due to V–R,T, transfer due to collisions of DCl(v = 1) with the buffer gas. The time evolution of the populations measures the role of collisions because the slow loss of molecules by IR emission is only a small fraction on a millisecond time scale. On the other hand, on that scale at a moderate pressure, a molecule undergoes millions of collisions with the buffer gas.

9.0.2 A hierarchy of relaxation rates

The rate of change of state due to inelastic collisions depends on the pressure. To factor out this dependence we compare the rate of energy transfer to the number ω of all collisions that the molecule undergoes per unit time, Eq. (2.8). The ratio is defined as the (dimensionless) collision number Z_r for a particular relaxation process r. An equivalent definition of the collision number is as the ratio of the relaxation time τ to the time interval ω^{-1} between collisions:

$$Z_r = \omega\tau \qquad\qquad (9.2)$$

* At a total energy E, $E = E_I + E_T = E'_I + E'_T$ or $Q \equiv E'_T - E_T = E_I - E'_I$.

Thus the expected transition will occur, on average, once in every Z_r collisions. Alternatively we can regard Z_r^{-1} as the probability of inducing the particular inelastic transition r in a single collision. The transfer between rotation and translation, R–T, that has a rather low gap, is very fast. Measurements of pressure broadening of the line shape, see Figure 7.24, show that R–T transfer occurs on nearly every collision, or that Z_{RT} is typically between 1 and 10. Studies of V–T transfer as shown in Figure 9.1, or of many other examples, say O_2 in Ar at room temperature, yield $Z_{VT} \sim 10^6$, meaning that a vibrationally excited O_2 molecule will survive a long time before it is deactivated by transfer of the vibrational energy to translation. An early example of the importance of these differences is the lasing pattern of chemical lasers, as discussed next. Another practical application is that workhorse, the CO_2 laser, as discussed in Section 9.1.2.1.

Studies of relaxation in the bulk have yielded valuable information, summarized in Figure 9.2, on the rates of energy transfer processes. It is clear from the summary and the HCl/DCl example that low translational exoergicity is favored. But why? We will argue that this requires an extension of the Franck–Condon principle, Section 7.0.1, beyond the sudden limit. We will call the resulting rule of thumb the *exponential gap principle* and show that the collision number Z_r is exponentially large in the "gap" (= the change, Q, of the translational energy), $Z_r \approx \exp(? \cdot Q)$. As indicated by the question mark, Q cannot be the only relevant factor because Q carries dimensions and the exponent must be dimensionless. For the moment we continue to examine the implications of an exponential decrease of the probability of the process with increasing Q.

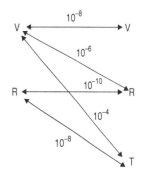

Figure 9.2 Schematic diagram showing the energy transfer processes occurring in thermal molecular collisions. V, R, and T refer to vibrational, rotational, and translational energy, respectively. The numbers are typical relaxation times, in seconds, characterizing the particular mode of energy transfer for a gas at atmospheric pressure. The slowest process is the V–T transfer and so it is rate-determining for the return to thermal equilibrium [adapted from W. H. Flygare, *Acc. Chem. Res.* **1**, 121 (1968)].

9.0.3 The HF chemical laser

The HF chemical laser operates by a flash photodissociation of a suitable precursor for F atoms in the presence of H_2 and an excess buffer gas. The F atoms react with H_2 and produce rovibrationally excited HF molecules. A few microseconds after the flash the system begins to lase in the IR, Figure 9.3. It does so when an excited HF molecule spontaneously emits an IR photon that then stimulates another excited HF molecule to emit a clone of that photon. These two photons, which are of the same frequency and direction of propagation, next stimulate two excited HF molecules to emit two more photons. As long as there are HF molecules in the excited level and photons of the right frequency this stimulated emission process can continue to cascade. Opposing this is the (stimulated) absorption process. As long as there are photons of the right frequency, HF molecules in the final state can absorb these photons and go back to the emitting level. To have a net gain of identical photons the stimulated emission process must be faster than the absorption of photons (and of other losses due to photons going out, etc.). In 1917 Einstein argued that there should be not only spontaneous but also stimulated emission of light. He further showed that the rate constant for stimulated

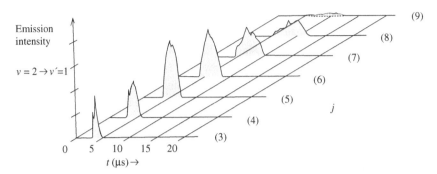

Figure 9.3 Time-resolved laser emission from HF($v = 2$) molecules produced by the F + H$_2$ reaction initiated at time zero by a photodissociation of an F atom-producing precursor. The different axes are labeled by the rotational quantum number, j, of the emitting molecule [adapted from M. J. Berry, *J. Chem. Phys.* **59**, 6229 (1973)]. Note the microsecond time scale in reference to the relaxation times shown in Figure 9.2.

emission, the so-called Einstein *B* coefficient, equals the rate constant for absorption of light. Say we consider an emission process where the HF molecule is going down from state v, j to v', j' (where $v' = v - 1$ and $j' = j \pm 1$). It follows from net gain, meaning that the system operates as a laser,* producing a net increase in cloned photons, that there is a "population inversion" defined quantitatively by

$$\frac{[\mathrm{HF}(v, j)]}{g_j} > \frac{[\mathrm{HF}(v', j')]}{g_{j'}} \tag{9.3}$$

Here the square brackets denote the concentrations of molecules in the particular rovibrational level and g is the degeneracy of the level, $g_j = 2j + 1$ for a rovibrational level. Note that it is the concentration of the initial, emitting, species that is on the left-hand side.

The laser emission occurs several microseconds after the reaction is initiated. During this interval a nascent HF molecule has had time to undergo hundreds of collisions. Such collisions could bring the molecules to thermal equilibrium. Obviously they do not because the condition of population inversion, Eq. (9.3), is the very antithesis of equilibrium. If there was equilibrium

$$\frac{[\mathrm{HF}(v, j)]}{g_j} \bigg/ \frac{[\mathrm{HF}(v - 1, j')]}{g_{j'}} = \exp\left(-(E(v, j)\right)$$

$$-E(v - 1, j')/k_\mathrm{B}T) \ll 1, \qquad \text{thermal equilibrium}$$

---

* Laser: light amplification by stimulated emission of radiation. The rate of stimulated emission for the transition from state v, j to v', j' is $[\mathrm{HF}(v, j)] \, \rho B(v, j \to v', j')/g_j$ where ρ is the density of photons at the frequency of the transition. The rate of transition from v', j' to v, j by absorption of photons is $[\mathrm{HF}(v', j)] \, \rho \, B(v', j' \to v, j)/g_{j'}$. For lasing, the rate of stimulated emission needs to be larger. Why do we divide by the degeneracy g? Because the concentration divided by g is the number of molecules in any particular quantum **state**, per unit volume.

whereas the experiment shows that the left-hand side is larger than 1! How can that be? Understanding the mechanism was one of the early triumphs of the principle of exponential gap.

It follows from the time scale axis of Figure 9.3 that prior to the onset of lasing there is enough time for hundreds of collisions. Therefore, for any given vibrational state of HF there is complete exchange of energy between the rotations and the translation. (Recall that there is an excess of a buffer gas.) But the HF quantum of vibration is rather large. There is a large gap when such a large energy needs to be dumped into translation. Therefore, on the microsecond time scale of interest, the nascent vibrational population of HF is hardly perturbed. It takes a far longer time for vibrational relaxation of HF, see Figure 9.1 where the vibrational relaxation of DCl occurs on the millisecond time scale (and the vibrational quantum of HF is larger than that of DCl). In other words, it is an implication of the principle of exponential gap, as summarized in Figure 9.2, that on the microsecond time scale the HF sample acts like a mixture of distinct species. Each species is an HF molecule in a given vibrational state. We can surely tell apart an HF molecule in $v = 2$ from that in $v = 1$, so there is no problem. They really can be regarded as distinct chemical entities. Each species is in thermal equilibrium because the R–T energy exchange is fast. But the species are not in "chemical" equilibrium. The "reactions" that bring the system to "chemical" equilibrium are of the type $HF(v) + M \rightarrow HF(v') + M$ where M is an atom of the buffer gas and the process interconverts the different species. These "reactions" are too slow to occur on the time scale of interest.

The recognition that a gas can be in equilibrium in one sense yet fully in disequilibrium in another is actually familiar once you pause to think about it. An extreme case* is a gaseous mixture of O_2 and H_2 that is in thermal equilibrium. It can be stable for a very long time but it is not in chemical equilibrium, as one verifies by accelerating the rate of reaction by some external means. The reason we care about the principle of exponential gap is that it allows us to understand that, on the shorter time scale, some internal modes can be in equilibrium while others will not be. In other words, what we usually call thermal equilibrium can be undone in a variety of ways and we will see several examples.

*9.0.3.1 The funnel mechanism

The funnel mechanism shows why the fast rotational relaxation and the far slower vibrational relaxation allow an efficient operation of an IR chemical laser. To illustrate the mechanism Figure 9.4 shows, as a function of the rotational quantum number j, the population in two vibrational manifolds, v and $v - 1$. An IR transition occurs not quite vertically on such a plot. A state v, j emits to $v - 1, j + 1$ or

---

* A very extreme case is a gaseous sample that is in both thermal and chemical equilibrium. It is typically not in equilibrium with respect to nuclear reactions, meaning that the distribution of elements is not the one that would be obtained if nuclear reactions took place at a measurable rate.

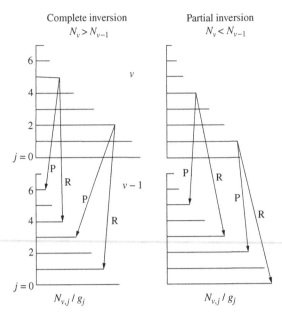

Figure 9.4 The funnel mechanism for chemical lasers. Plotted, with the energy on the vertical scale, are the populations (per quantum state), $[HF(v, j)]/g_j$, for the upper and lower vibrational states of the laser transition. The rotations are taken to be at thermal equilibrium (with $B/k_B T = 0.05$, which is realistic for HF at room temperature). Left plot: an excess population in the upper vibrational level. Right plot: $[HF(v)]/[HF(v-1)] < 1$. Shown are two $j \to j+1$ transitions and these show that at higher js one can have laser emission even when there are fewer HF molecules in the level v than in the level $v-1$. Problem A determines the range of j values for which this is possible.

$j - 1$. The plot is made under the assumption that rotational relaxation is fast enough so that the rotational states of a given j are in thermal equilibrium. But if so,

$$\frac{[HF(v, j)]}{g_j} \bigg/ \frac{[HF(v - 1, j + 1)]}{g_{j+1}} = \frac{[HF(v)]}{[HF(v - 1)]}$$
$$\times \exp\left(-B(j(j+1) - (j+1)(j+2))/k_B T\right) \tag{9.4}$$

where, for a $j \to j + 1$ transition, as written, the Boltzmann factor is larger than unity (B is the rotational constant). So positive gain (= population inversion, left-hand side of (9.4) > 1) is possible even when there are fewer HF molecules in the level v than in the level $v - 1$, $1 > [HF(v)]/[HF(v - 1)]$. One can therefore recover a lot of the chemical energy as laser radiation. Lasing is not limited to the case where there is an excess of HF molecules in the vibrational level v.

The funnel begins to operate once the system begins to lase. The laser output means that the concentration of molecules in the level v, j decreases while the concentration of molecules in the level $v-1$, $j+1$ increases. So even if we had gain it would shortly be exhausted. To keep the laser going we need to drain the final level $v-1$, $j+1$ and/or to provide a fresh supply of molecules into the emitting v, j level. Both needs are satisfied by a fast rotational energy transfer. The shortage of molecules in level v, j due to their (stimulated) emission is made good by transfer from other rotational levels of the same v, thereby restoring thermal equilibrium. The excess of molecules in level $v-1$, $j+1$ is depleted by transfer to other rotational levels of $v-1$. The funnel mechanism explains why, once lasing begins, the system acts collectively to lock onto that transition: the cascading of cloned photons means that the stimulated emission (whose rate is proportional to the density of photons) favors this transition and the funnel insures that there is a continuous fresh supply of molecules to the upper level and a draining of molecules from the final level. Indeed, it is a characteristic of chemical lasers (e.g., the excimer lasers of Section 3.2.4.1) that they operate not only by preferential pumping into the emitting levels but also by the efficient draining of the final level.

While not of much practical use, it is an interesting implication of the principle of exponential gap that lasing is also possible on pure rotational transitions, v, $j \rightarrow v$, $j-1$. Seemingly we just argued that this is not possible because of efficient rotational energy transfer. True, but recall that the spacing between rotational levels increases with increasing j. If we could access states of such very high j that the gap between two adjacent rotational levels, $2Bj$, is large enough then such high j states will not efficiently relax! How would we go about it? One option is to note that the process

$$\text{HF}'(v, j) + \text{Ar} \rightarrow \text{HF}(v-1, j') + \text{Ar}$$

can have a small gap and therefore be efficient, **if** the internal energy of the final state is about comparable to that of the initial state. This requires that almost an entire vibrational quantum has been converted to rotational energy. For HF a V–R transfer with a low gap requires dumping over $4100\ \text{cm}^{-1}$ into rotation. As a hydride, the rotational constant of HF is large, $B \approx 21\ \text{cm}^{-1}$, and so for a low initial j it requires j' of about 14 or 15 for the gap to be small.[1]

9.1 Vibrational energy transfer

We discuss first collisions that induce vibrational–vibrational energy transfer, first in diatomics and then in larger molecules. Such processes can have a low gap and therefore occur fairly efficiently. Then we return to a topic that played a key role throughout the history of chemical kinetics: the collisional acquisition (or loss) of vibrational energy of larger molecules. We know from the Lindemann mechanism that such processes can drive a molecule all the way to isomerization or dissociation. But they are also important in any chemical system in disequilibrium.

Figure 9.5 The relative population of vibrational states of CO in an O_2–CS_2–He flame plotted vs. the vibrational quantum number v for different times after ignition. The $CO(v)$ molecules are mainly produced by the very exoergic, see Figure 5.17, $O + CS$ reaction. The Treanor distribution, that neglects the V–T relaxation at high vs, is shown as a dashed line [adapted from S. Tsuchiya, N. Nielsen, and S. H. Bauer, *J. Phys. Chem.* **77**, 2455 (1973)].

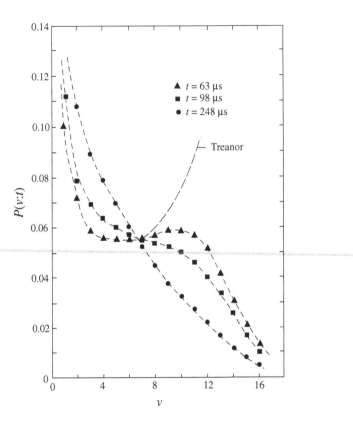

The rate of energy transfer collisions increases with the pressure, and in Section 11.1.4 we examine vibrational relaxation in liquids.

The experimental techniques that can be brought to bear toward understanding inelastic collisions are the same as for reactive collisions. In particular, the techniques that were discussed in Chapter 7 for photoselective chemistry have been useful in preparing initial states and for probing the products. The theoretical machinery is also similar.[2]

9.1.1 V–V′ processes in diatomics

Spectroscopy of a flame rich in CO molecules reveals an unexpected deviation from thermal equilibrium, Figure 9.5. For the very lowest vibrational states the distribution is Boltzmann-like, exhibiting a monotonic decrease in concentration of $CO(v)$ with increasing vibrational quantum number. The only feature worth remarking on is that the temperature of the distribution is high, but this is to be expected for a flame. The unexpected feature is the maximum in the distribution for intermediate values of v. Theoretical considerations[3] have suggested that the

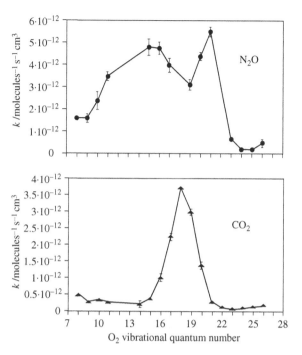

Figure 9.6 The rate constants for loss of $O_2(v)$ in collisions with N_2O and CO_2 plotted against the initial vibrational state of O_2. For CO_2 the gap for the two-quantum transition $O_2(v) + CO_2(000) \rightarrow O_2(v-2) + CO_2(001)$ is minimal at $v = 18$. [Data from J. A. Mack, K. Mikulecki, and J. Wodtke, *J. Chem. Phys.* (1996) and M. Klatt *et al.*, *J. Chem. Soc. Faraday Trans.* (1996); adapted from Flynn *et al.* (1996); see also Silva *et al.* (2001).]

excess of population in higher v states is pumped by the vibration to vibration, V–V, processes

$$CO(v=1) + CO(v=1) \rightarrow CO(v=2) + CO(v=0)$$
$$CO(v=2) + CO(v=1) \rightarrow CO(v=3) + CO(v=0)$$
$$\cdots$$

Such processes have no gap if the molecule is harmonic, but the real molecule is not, and so the processes as written are a shade exothermic. Problem J shows that the process is faster in the exothermic than in the reversed endothermic direction. At very high vs the vibrational levels become close enough that other relaxation processes such as V–T begin to compete and the up-pumping ceases.

The same V–V up-pumping is observed in other diatomics and in mixtures of gases. When the two molecules are not the same one can have either a one- or a two-quantum transfer and the relaxation of O_2 by N_2O, as shown in Figure 9.6, is a particularly nice example because it shows the role of anharmonicity. The efficient transfer from O_2 at $v \approx 15$ is due to the one-quantum loss that goes into the symmetric stretch of N_2O:

$$O_2(v) + N_2O(000) \rightarrow O_2(v-1) + N_2O(100)$$

where the notation for the vibrational quantum numbers of N_2O is discussed in Section 9.1.2. The facile rate of transitions from O_2 at $v = 21$ is due to the

Figure 9.7 Low-lying
vibrational levels of
CO_2 (simplified).
Their spectroscopic
designations are also
shown [adapted from
C. B. Moore, *Acc. Chem.
Res.* **2**, 103 (1969)].

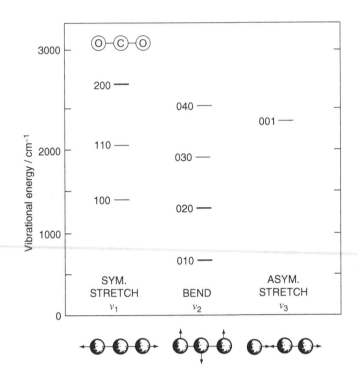

two-quantum loss to the (higher-frequency) asymmetric stretch of N_2O:

$$O_2(v) + N_2O(000) \rightarrow O_2(v-2) + N_2O(001)$$

Computing the gap for a one- and a two-quantum loss explains why the second process is more selective for the initial vibrational state of O_2.

Vibrationally excited O_2 is formed in atmospheric photochemical processes and its relaxation competes with its removal by reactions, particularly with another O_2 to form O_3. Satellites monitoring the IR emission of the atmosphere have also uncovered the formation and relaxation of rovibrationally excited NO.

9.1.2 V–V′ processes in polyatomics

In polyatomic molecules, new types of V–V′ transfer processes are possible. To discuss these we begin with Figure 9.7, which shows the low-lying vibrational states of CO_2 together with their spectroscopic designation. The triplet of numbers m, n, p indicates the numbers of vibrational quanta in the symmetric stretching (m), bending (n), and asymmetric stretching (p) modes, respectively.

The new feature for polyatomics is illustrated by the *intra*molecular collision-induced V–V transfer

$$CO_2(020) + M \rightarrow CO_2(100) + M \qquad Q = -103 \text{ cm}^{-1}$$

that is very efficient* ($Z_r \sim 10$). Next are the two *inter*molecular V—V "sharing" processes

$$CO_2(100) + CO_2(000) \rightarrow CO_2(010) + CO_2(010) \qquad Q = 54\,cm^{-1}$$

and a V—V process familiar to us from diatomics

$$CO_2(010) + CO_2(010) \rightarrow CO_2(020) + CO_2(000) \qquad Q = 49\,cm^{-1}$$

for which $Z_r \sim 50$. In CO_2, intermolecular sharing "equilibrates" among the levels of the $0n0$ manifold, while the intra- and intermolecular transfer processes couple the $m00$ and $0n0$ levels. Thus, the whole $mn0$ manifold of vibrationally excited states is rapidly equilibrated and is then depleted slowly by the inefficient V—T transfer out of the lowest vibrationally excited state

$$CO_2(010) + M \rightarrow CO_2(000) + M \qquad Q = 667\,cm^{-1}$$

to the vibrational ground state. In pure CO_2 this process is rather inefficient: $Z_r \sim 5 \cdot 10^4$. This, then, is the slowest step in the bulk relaxation of CO_2.**

On the other hand, the $00p$ manifold of states reaches equilibrium within the manifold by intermolecular sharing, but it is not efficiently coupled to the $mn0$ manifold, as there are no states near enough in energy to the 001 state, cf. Figure 9.7. Thus V—V transfer processes out of the 001 state are comparatively inefficient, with $Z_r \sim 2 \cdot 10^4$. We thus see a somewhat uncommon case of a slow V—V step, which is nearly rate-determining.† Hence, vibrationally excited CO_2 can be viewed as a mixture of two species, each species being in thermal equilibrium. Molecules of one species are CO_2 molecules in vibrational states of the $00p$ manifold and those of the other species in the $mn0$ manifold. Collisional transfer between the two manifolds is inefficient.

---

* The near degeneracy of the two zero-order states 100 and 020 means that they are already mixed by anharmonicities. This was the first example of a Fermi resonance, here a 1:2 resonance. The gap is however wide enough that it requires a perturbation by a collision to fully mix the two states. When the energy is higher so that there are many nearby states, as in Chapter 7, intramolecular coupling can be sufficient to overcome the (small) gap.

** Indeed, the proliferation of nearby excited vibrational states of polyatomics, see Figure 7.22 for example, suggests that this would be a general trend: the rate-determining step in the vibrational relaxation is the V—T transfer out of the lowest vibrational level. The principle of exponential gap therefore implies that the lower is the vibrational quantum of that mode, the faster is the relaxation. This is a long-known observation, as the Lambert–Salter rule, Eq. (9.16) below, and it holds over several orders of magnitude. Note that its empirical validity tells us more than just that the principle of exponential gap is useful. It also implies a facile V—V transfer among the excited vibrational states.

† Transfer out of the 001 state can be made to become the rate-determining step when the inefficient V—T relaxation of $CO_2(010)$ is facilitated by adding other molecules to which it can relax by V—V transfer. Finding molecules that effectively drain $CO_2(010)$ is a key to increasing the efficiency of the CO_2 laser.

9.1.2.1   The CO₂ laser

The seemingly esoteric considerations of V—V′ transfer form the basis for the workhorse* of all lasers, the CO_2 laser. In the first stage nitrogen molecules are efficiently vibrationally excited by collisions with electrons** during an electrical discharge,

$$N_2(v = 0) + e \rightarrow N_2(v = 1) + e$$

The vibrationally excited N_2 molecules then undergo a rather facile V—V transfer with CO_2

$$N_2(v = 1) + CO_2(000) \Leftrightarrow CO_2(001) + N_2(v = 0) \qquad Q = -18\,\text{cm}^{-1}$$

and promote a significant fraction of CO_2 molecules into the 001 level. The basis for lasing is the stimulated emission from the 001 level, Figure 9.8,

$$CO_2(001) \rightarrow \begin{cases} CO_2(100) + h\nu(10.6\mu) \\ CO_2(020) + h\nu(9.6\mu) \end{cases}$$

provided that (a) collisional loss of $CO_2(001)$ is kept to a minimum and (b) collisional loss of $CO_2(100)$ and $CO_2(020)$ can be enhanced so that the upper-level population can be maintained higher than that of the lower levels. The first condition is obtained owing to the inefficient V—V′ transfer out of the 001 state. The second condition is fulfilled by depletion of $CO_2(010)$ and (020) due to the efficient sharing and V—V processes as discussed above.

*9.1.2.1.1   Chemical pumping for the CO₂ laser

The CO_2 need not be pumped by vibrationally excited N_2. Chemical means can also be used. As an example, vibrationally excited DF molecules produced in the fast exoergic reaction $F + D_2 \rightarrow D + DF(v')$ can undergo efficient V—V′ transfer with CO_2:

$$DF(v') + CO_2(000) \rightarrow CO_2(001) + DF(v' - 1)$$

From here on, the CO_2 fate is the same as before. The necessary F atoms can be produced by the precursor reaction

$$NO + F_2 \rightarrow NOF + F$$

---

* For everyday applications this laser is the universal "knife." It can cut through most materials because it is so efficiently absorbed. But if one uses CO_2 with the heavier carbon isotope, ^{13}C, the laser emission is in an atmospheric "window," meaning that it is not highly absorbed by the atmosphere. This has not escaped the attention of the (former Soviet) military. The isotopic enrichment was achieved by IR multiphoton dissociation of organic molecules, as discussed in Section 7.2.5.

** Why would that be? Surely an electron is too light to impart much momentum to the far heavier N atoms. Quite right, but the mechanism is the temporary attachment of an electron to form N_2^-. The extra electron goes into an antibonding orbital and so the equilibrium position of N_2^- is larger than that of N_2. The anion is therefore formed in a vibrationally excited state and when it autodetaches the extra electron it retains some of that vibrational excitation.

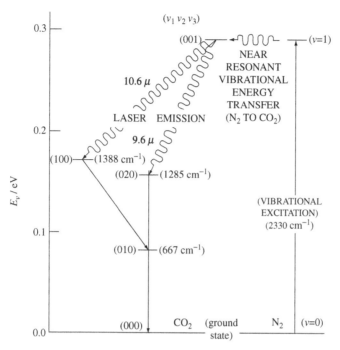

Figure 9.8 Level diagram appropriate for the N_2–CO_2 chemical laser. Vibrationally excited N_2 molecules are the source of the excitation of $CO_2(001)$. The final states of the laser emission, $CO_2(100)$ and $CO_2(020)$, are depleted by V–V' transfer. See Figure 9.7 for the relevant levels of CO_2.

In this way, by mixing tank gases (NO, F_2, D_2, CO_2) it is possible to effect a direct conversion of chemical to coherent radiation energy.

At the center of our discussion in this section has been the concept of a state-to-state rate constant for energy transfer, and the wide variations possible in such rates. In Section 9.2 we begin to explore the basic dynamical factors that determine such rates. Before that we turn to energy-rich polyatomic molecules.

9.1.3 Energy-rich polyatomics

Molecules with energy sufficient for dissociation or isomerization deserve a special mention. Even for diatomics, the high amplitude of the vibrational motion and the small energy spacings for high v states make for more efficient collisional deactivation.* This is even more so for polyatomic molecules because of the exceptionally high density of vibrational states, Section 6.2.2.3. The Lindemann mechanism of unimolecular reactions, Section 6.2.1, assumes that energy-rich molecules are produced by collisions and, in the high-pressure regime, they are mostly depleted by deactivating collisions rather than reacting. It is therefore of interest to directly observe such energy-loss processes.[4]

Methods of photoselective chemistry can be used to pump molecules up and to monitor the collisional processes. It can even be done in the bulk as long as the pressure is low enough that the pump process is over in a time that is short

---

* Detailed balance then implies that the collisional activation into such states is also more efficient.

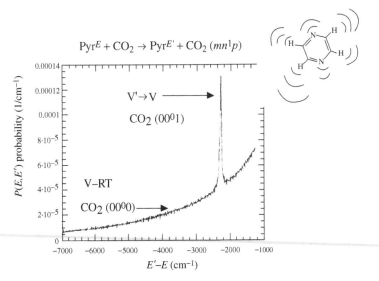

Figure 9.9 Probability of a given energy transfer for pyrazine ($C_4N_2H_6$, insert), excited at 248 nm, colliding with CO_2 at a thermal initial velocity. The final translational and internal states of CO_2 are probed by (Doppler) spectroscopy [adapted from Flynn (2001); see also J. Park *et al.*, *J. Chem. Phys.* **117**, 5221 (2002)]. There is a spike corresponding to a resonant transfer to the 001 level of CO_2, but there is also a wide range of energy transfers where CO_2 is left in its ground vibrational state. The initial vibrational energy of pyrazine is about 41 000 cm^{-1} and an energy transfer of $\approx$6000 cm^{-1} corresponds to a loss of two C–H stretch quanta.

compared to the interval between collisions. A particularly used pump process is internal conversion, Section 7.0.3. In a relatively short time, of the order of nanoseconds, a molecule raised by a UV laser to its first electronically excited state will convert the excess energy (say, of the order of 5 eV $\approx$ 40 000 cm^{-1} or more[5]) to vibrational excitation on the ground electronic state. When such a vibrationally excited molecule collides with another polyatomic, the most probable energy transfer is a low-gap V—V process with not much energy dumped into translation. So far, this is as expected. The transfer can be monitored both by inferring the energy content remaining in the initially rich molecule, say, by how much energy is required to ionize it. The initially cold partner can be probed spectroscopically and Doppler measurements will further determine that not much energy went into translation.

It is also observed that energy-rich polyatomic molecules can lose large amounts of vibrational energy into the translation and these are high-gap processes. The energy loss can be so extreme that one sometimes speaks of *supercollision*s.[6] One must not be misled; these are relatively rare as shown in Figure 9.9, but they do occur.

There are at least two mechanisms that can induce large energy transfers. One is a sticky collision where the relative motion of the partners undergoes several vibrations before they recede from one another. The other is to recognize that at, say, a vibrational energy content of 41 000 cm^{-1}, as in Figure 9.9, the molecule is quite a bit contorting. There are therefore atoms that execute large-amplitude motions and are able to deliver a large kick to the collision partner. Of course, for such a close encounter there are steric requirements on the approaching partner so the supercollisions are low-impact-parameter, low-probability processes. But they do take place and thereby drain large amounts of energy out of the hot molecule. Do note that there must be a corresponding time-reversed process that is able to pump large energies into the vibrations, in a single collision. An interesting implication is that, in the Lindemann mechanism, it need not be the case that the polyatomic molecule works its way up the ladder of increasing energy in many small steps. Rather, warm molecules could get all the way up to a chemically interesting region in one or a few steps each involving a large energy transfer. We shall return to such a possibility when we discuss activated barrier crossing in solution.

9.2 Understanding the essentials of energy transfer

The reluctance of nuclei to undergo a change in their momenta has so far been discussed in terms of the Franck–Condon principle. In its simplest form it says that a photon-induced electronic excitation is not accompanied by a change in the kinetic energy of the nuclei. But an inelastic collision is necessarily characterized by such a change, that we called "the gap." We have seen ample experimental evidence that the transition is more likely when the gap is small. How small is "small" is the subject of this section.

9.2.1 Two extremes of vibrational energy transfer

The simplest inelastic event, conceptually, is the collinear collision between an atom A and a "harmonic oscillator molecule" BC. In such a collision all three atoms are confined to a line, and the atom A hits the near atom of the oscillator, say atom B. Two limiting situations can be considered. One extreme is that the oscillator spring is extremely stiff, meaning that BC is practically a rigid body. When A collides with the oscillator, any kinetic energy lost by A is gained by the "united atom" BC as kinetic energy of its center of mass. In other words, after A hits B, the atoms B and C recoil together without any change in their relative velocity. In the limit of a stiff BC bond the collision is purely elastic, as the total kinetic energy has not changed.

The opposite extreme is that of a very loose BC oscillator spring. Here the atom C is so very weakly connected to B that it initially does not respond to a change in the velocity of atom B. In the limit of a vanishingly small oscillator force constant, atom C is a spectator to the A—B collision. Any change in the velocity of B is thus a change in the relative velocity of B and C and hence a change in the vibrational energy of BC.

We next need to decide what is the scale against which we regard the oscillator as stiff or loose.

9.2.2 The adiabaticity parameter

In physical terms, the limit of a loose BC oscillator spring is the limit where the duration, τ_c, of the A—B collision is short compared to the time required by atom C to "know" that atom B has changed its momentum. B communicates with C via their mutual force. So the time required for C to respond is the vibrational period,* t_v, of the oscillator. The limit of a loose spring is therefore when

$$\tau_c < t_v \quad \text{or} \quad \nu\tau_c < 1 \qquad \text{sudden limit} \tag{9.5}$$

where ν is the oscillator frequency. Since the duration of the collision decreases with increasing relative velocity, the short collision time is a high-velocity or a *sudden* limit. The opposite extreme

$$\tau_c > t_v \quad \text{or} \quad \nu\tau_c > 1 \qquad \text{adiabatic limit} \tag{9.6}$$

is a low-velocity or an *adiabatic*[7] limit. Energy transfer is very inefficient in the adiabatic limit. In this limit the duration of the collision is long compared with the period of oscillation, so that the perturbation by the collision is "slow." The oscillator can accommodate itself to the perturbation or (in our earlier language) it presents a rigid, unyielding front.[8]

To determine the range of velocities that correspond to the different limits we estimate for the duration of the collision

$$\tau_c = a/v \tag{9.7}$$

Here a is the "range" of the intermolecular force and v is the relative velocity during the collision. In the adiabatic regime $v \leq a/t_v$ or, in terms of the vibrational energy spacing ΔE, $h\nu = h/t_v = \Delta E, v \leq a\Delta E/\hbar = av$. For many diatomic molecules $v \cong 10^{13}$ s^{-1} and taking as an estimate $a = 2$ Å, we obtain $v < 2 \cdot 10^5$ cm s^{-1}. At ordinary temperatures, where velocities in the gas phase are typically of the order of 10^4 cm s^{-1}, collisions are in the adiabatic range for vibrational energy transfer.

We can summarize our qualitative considerations by the introduction of an adiabaticity parameter ξ,

$$\xi = \tau_c/t_v = \frac{a|\Delta E|}{h\nu} \tag{9.8}$$

---

* The period (or the inverse of the frequency) is a measure of the oscillator force constant, $k : t_v = v^{-1} = 2\pi(\mu_{BC}/k)^{1/2}$. A weakly bound (= loose) oscillator has a long vibrational period.

where $|\Delta E|$ is a measure of a typical amount of energy transferred into (or out of) the translation, so that ΔE is the translational exoergicity. Large values of ξ correspond to adiabatic collisions when energy transfer is inefficient. Below $\xi \cong 1$ we enter the region of efficient transfer and for $\xi < 1$ we reach the sudden or spectator regime.

9.2.2.1 Rotational energy transfer

The spacings between rotational levels (with the exception of hydrides) are some three orders of magnitude smaller than vibrational energy spacings.* Hence, for T—R transfer, the adiabatic velocity range is more typically $v < 10^2 - 10^4 \mathrm{cm\,s^{-1}}$. At ordinary temperatures R—T transfer is in the sudden range and hence should occur quite readily. The one special feature is that rotational spacings increase with increasing rotational excitation so that, for higher js, the transition may not be in the sudden limit.

Similar considerations apply to rotational energy transfer where the adiabaticity parameter is defined by

$$\xi = \tau_c / t_r \tag{9.9}$$

Here t_r is the rotational period. Since typically $t_r \gg t_v$, ξ for rotation is usually much smaller than ξ for vibration. Rotational energy transfer tends to be in the sudden regime.

9.2.2.2 Adiabatic behavior

The notion of an adiabatic process is a very general one and conforms also to our everyday experience. The adiabaticity parameter ξ has been introduced as the ratio of the time scale for the perturbation to the time scale within which the system can adjust:

$$\xi = \tau_{\mathrm{perturbation}} / \tau_{\mathrm{response}} \tag{9.10}$$

If you perturb a system slowly enough on its own time scale, it can adjust and therefore it will be found in its original state when the perturbation is over. This does not mean that the system is not changing as the perturbation is acting. It can and generally will change, but, because the system changes quickly as the perturbation is acting, its instantaneous state is adjusted to the perturbation. So the system will return to the initial state when the perturbation is over. The way electrons adiabatically adjust to the motion of the nuclei is a good example. But a perturbation can also be sudden-like, $\xi < 1$, and its action can leave the system in a different state.

---

* The classical period of the motion of a mode at a given level of excitation corresponds to the difference between the spacings of adjacent levels. As discussed in Chapter 7, we need at least two stationary quantum states to define a state that mimics a classical motion, Problem O.

9.2.3 The exponential gap

A more quantitative approach is to define a resonance function, $R(\xi)$, a measure of the efficiency of the energy transfer at a given value of ξ. Thus if $\langle \Delta E \rangle$ is the average energy transfer in the collision, we write

$$\frac{\langle \Delta E \rangle}{E_T} = \frac{\langle \Delta E \rangle_{\xi=0}}{E_T} R(\xi) \tag{9.11}$$

where $\langle \Delta E \rangle_{\xi=0}$ is the energy transfer in the sudden limit. This is the part that is mass-dependent, Problems H and I.

The computation of the resonance (or "energy mismatch") function requires a solution of the collision dynamics. We then find that, at least for $\xi > 1$, we can approximate the expected decrease with increasing ξ as exponential

$$R(\xi) \cong \exp(-\xi) \tag{9.12}$$

Qualitatively similar results apply for R–T transfer. The exponential decline of the efficiency of energy transfer with increasing ξ is a useful rule in the field of energy transfer. This is not to say that deviations from this rule are unknown (see, for example, Section 9.3.3). Rather, deviations are taken as a diagnostic indication that some special features are present.

Some guidance to the physical significance of the resonance function is provided by our model of a collinear A + BC collision. The energy transfer to the vibration was very small when it was difficult to change the relative momentum of B with respect to C (a stiff bond), while it was large in the opposite extreme. The combined role of the internal motion and the perturbation is characterized by the value of the adiabaticity parameter ξ. For a given value of ξ, $R(\xi)$ is a measure of the readiness of the BC bond to change its momentum. In other words, $R(\xi)$ is a quantitative measure for the more qualitative guidance given by the Franck–Condon principle, see also Section 9.2.4. Rather than saying that there is a propensity for the momentum not to change, we say that the larger the change in the momentum of the nuclei, the larger is ξ and the (exponentially) smaller is $R(\xi)$. Quantitatively, if p and p' are the initial and final momenta we can write

$$\xi = (p - p')a/h \tag{9.13}$$

with a being the range of the force that is applied. To show the equivalence to Eq. (9.8) we put $p - p' = [p^2 - (p')^2]/(p + p')$ or in terms of the energy gap $p - p' = \Delta E / v$ where $v = (p + p')/2\mu$ is the mean velocity.

9.2.4 Light as a bridge of the exponential gap

An appealing way to see the connection of our considerations to the Franck–Condon principle is to think of the light as dressing the ground state as discussed in Appendix 3.B. The potential for the dressed ground state is raised up by the energy of the photon as shown, for example, in Figure B3.1. This potential crosses the potential of the excited state and the point of crossing is the Franck–Condon

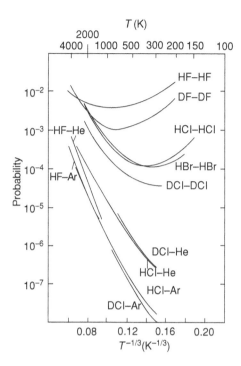

Figure 9.10 Landau–Teller plot of the temperature dependence of the vibrational relaxation efficiency vs. temperature for hydrogen halides. A linear dependence on $T^{-1/3}$ is observed at higher temperatures. The low-temperature results are discussed below as evidence for the formation of bimers [adapted from Ben-Shaul *et al.* (1981)].

window. Why? Because at a given total energy, when the potential energies are equal the kinetic energies are also equal. The transition can then occur without any gap. The gap has been bridged by the photon. What the exponential gap estimate shows is that the Franck–Condon window is not a point but a finite interval in the internuclear distance. The width of the window is that range where there can be finite changes in the momenta provided that they satisfy $\xi = a(p - p')/\hbar < 1$. In Section 9.3.3 we will have more to say on the finite window of opportunity when two potential curves cross. We will there estimate the range parameter a whose value is needed to compute the magnitude of ξ.

9.2.5 Propensity rules for energy transfer

We can thus summarize the necessary conditions for an efficient energy transfer (i.e., a resonance function near unity). At a given energy gap, the resonance function $R(\xi)$ will be exponentially small at low velocities, and will increase with increasing velocities. This version is the form most useful for collision theory. An indirect illustration of this exponential gap principle is the strong positive temperature dependence of the V–T relaxation rate. For most diatomic molecules the temperature dependence is best fitted by a so-called Landau–Teller equation

$$\ln k = A - BT^{-1/3} \qquad \text{V–T transfer} \qquad (9.14)$$

which is quite different from the Arrhenius T^{-1} temperature dependence of bimolecular reaction rate constants. See Figure 9.10.

Saddle point integration, Problem F, shows that Eq. (9.14) is the expected result for an inelastic cross-section that increases near exponentially with increasing collision velocity

$$\sigma_{VT} \propto \exp(-v_0/v) \qquad (9.15)$$

The parameter v_0, with dimensions of velocity, is determined by the saddle point procedure discussed in Problem F. It is found to be a nearly linear function of the vibrational frequency, as expected from Eq. (9.8). Similarly, the collision number Z_{VT} for V—T transfer in polyatomics is found to be an approximate exponential function of the lowest normal mode vibrational frequency

$$\log Z_{VT} \propto v_{min} \qquad (9.16)$$

Other fairly general statements can also be made, e.g., collisions at higher impact parameters will tend to be more adiabatic, because only the longer range of the force will be sampled. High-impact-parameter, grazing collisions, which lead to nearly forward scattering, will therefore be essentially elastic. The inelasticity increases at larger scattering angles. It is the "head-on," backward scattering collisions that are the most inelastic. A corollary is that collisions resulting in a large change in the internal energy (and hence a large change in the vibrational quantum number) will occur mostly at low impact parameters. The cross-section for energy transfer being an integral over the impact parameters that contribute would then be rather small.*

A less obvious but valid qualitative generalization is that it is easier to transfer energy into (or out of) an excited oscillator than into a ground-state one. The physical reason is that the amplitude of oscillation of the bond is greater. An experimental test is to determine the efficiency of a V—T transfer process, as a function of the initial vibrational energy. This has been done for a number of V—T processes. A rough linear relation between the efficiency of the transfer (as measured, say, by the reciprocal of the collision number) and the initial vibrational energy is obtained.

At low temperatures bulk vibrational relaxation shows evidence of the long-range attractive intermolecular forces. This is particularly so between molecules with polar bonds that also strongly absorb radiation in the near infrared. Such strong absorption indicates that the molecule can easily be distorted from its equilibrium position by the electrical field of the light. In a collision experiment such a time-dependent field can be provided by the motion of the other polar

---

* A caveat is that events with exponentially small probabilities can still be chemically important. Experimental evidence for large changes in the vibrational quantum number in a single collision comes from molecular and ion beam measurements. In bulk systems such rare events can compete with gaining energy by a succession of collisions each involving a smaller change in the internal energy. This is because climbing up the energy ladder by small steps is like a random walk as the molecule can also efficiently lose and not only gain energy by collisions. The event is rare but in one collision it accesses a chemically interesting range.

molecule. When the temperature is low enough, dimers can be formed. Then, as in Section 7.2.4.1, although the transition probability per collision is low, the two molecules of the dimer collide again and again, at a frequency much higher than the bulk collision frequency, until a V—T process causes dissociation of the dimer. The anomalous temperature dependence shown in Figure 9.10 is therefore not an indication of the failure of the exponential gap principle.

9.2.6 Detailed balance

Detailed balance relates the rates of a particular activation and deactivation energy transfer process. Detailed balance thus provides a quantitative exact relation between rate constants that correspond to the same gap. This is unlike the principle of exponential gap that provides an estimate of how the rate constants vary when the gap changes. The qualitative implication of detailed balance is that on a quantum state-to-quantum state basis, the rate constant for the activation process is always smaller than the rate constant for the reverse deactivation process. Take as an example the V—T process that we started this section with, $A + BC(v = 0) \rightarrow A + BC(v' = 1)$ and the reverse deactivation process, $A + BC(v = 1) \rightarrow A + BC(v' = 0)$. Detailed balance states that at equilibrium the rates of these two detailed ways of transferring populations between $BC(v = 1)$ and $BC(v = 0)$ must be equal. This is to be so even though there may be other processes that can transfer populations, such as transitions in the IR. Therefore, using the subscript eq to designate concentrations at equilibrium,

$$k\,(v = 0 \rightarrow v = 1)\ [A]_{eq}\,[BC(v = 0)]_{eq}$$
$$= k\,(v = 1 \rightarrow v = 0)\,[A]_{eq}\,[BC(v = 1)]_{eq} \qquad (9.17)$$

Detailed balance therefore determines the exact ratio of the two rate constants:

$$\frac{k\,(v = 0 \rightarrow v = 1)}{k\,(v = 1 \rightarrow v = 0)} = \frac{[BC(v = 1)]_{eq}}{[BC(v = 0)]_{eq}}\ (= \exp(-(E_{v=1} - E_{v=0})/k_B T) < 1) \qquad (9.18)$$

The principle of exponential gap tells us that either one of the two rate constants in Eq. (9.18) is small if the gap in a V—T process is large. It does not however imply that the two rates are equal, meaning that they depend only on the magnitude, $|\Delta E|$ and not on the sign of the translational release, see Problem J.

9.3 Electronic energy transfer

There is considerable interest in the collision dynamics of electronically excited species not only due to their practical importance but also because such processes raise novel theoretical points. Since the electronic excitation energy is usually quite large, the adiabatic criterion would seem to rule out the efficient conversion of electronic to vibrational (or translational) energy upon collision. Yet, as we have already seen, such processes *do* occur, often with reasonable efficiency. Thus an electronically excited mercury atom will induce E–V transfer upon collision, e.g.

$$Hg^* + CO(v = 0) \rightarrow Hg + CO(v')$$

where all CO states with $v' \leq 9$ have been observed via their IR emission. Often, enough energy is transferred into the vibration to dissociate the bond.[9] (The excitation energy of Hg^* is 112 kcal mol^{-1}.) Highly endoergic chemical reactions can proceed quite readily using electronically excited reactants. The reactions of excited oxygen or nitrogen atoms and molecules are particularly interesting because of their contribution to the chemistry of the atmosphere.

Our purpose in this section is to suggest what **special** features in the electronic structure of the reagents allow electronically non-adiabatic processes, with their nominally very high gap, to occur efficiently. The required feature is not invariably present. Many processes are electronically adiabatic. The exponential gap principle is valid. We will sketch a mechanism that, if present, allows the gap to be smaller than what you would expect.

9.3.1   Non-adiabatic processes

At this point we already know that the occurrence of non-adiabatic processes[10] requires that the adiabaticity parameter ξ, Section 9.2.2, does not have too high a value. In this section we ask when and why is an electronically non-adiabatic process efficient even though the gap is seemingly high. In Section 7.0.2 we have already sketched a possible answer: the actual gap that the nuclei need to bridge may be quite a lot smaller than the nominal one.[11] In this section we return to this point: what are the special circumstances for which we can expect the gap to be small, and small compared to what? We point out ahead of time that a key ingredient in the answer is that the gap will be small in a localized region in the configuration space of the nuclei. This is not only important conceptually, it also allows you the option of control. You may want to access or, alternatively, to avoid this localized domain.

The simplest case for discussing a localized non-adiabatic transition with a small gap is the model of curve crossing, Section 3.2.4, when there is just one nuclear coordinate. We turn now to a more quantitative analysis of this model.

9.3.2   Curve crossing

An example of an electronic energy transfer collision that is more efficient than the high nominal translational exoergicity would suggest on the basis of the adiabaticity criterion is the process of *charge neutralization*, e.g., $Na^+ + I^- \rightarrow$ Na + I or the reverse process of "collisional ionization":

$$Na + I \rightarrow Na^+ + I^-$$

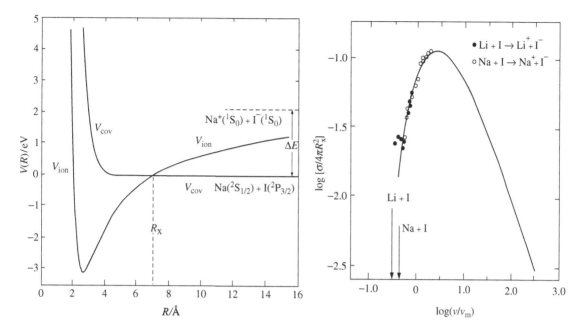

Figure 9.11 (Left) Potential energy curves for the covalent (NaI) and ionic (Na$^+$I$^-$) forms of sodium iodide. ΔE is the nominal (i.e., asymptotic) energy gap between the two states, i.e., the difference between the ionization potential of Na and the electron affinity of I. Thus, from ΔE_0 (eV) $\cong 14.35/R_x$ (Å) (Eq. (3.18)), $R_x \cong 7$ Å. It is known that for this system the actual electronic energy gap as discussed below is the much smaller value of $\cong 0.025$ eV. Note that the curves shown are "diabatic" ones. The electronic state is frozen and is not allowed to adjust to the position of the nuclei. The Born–Oppenheimer or adiabatic states do **not** cross. (Right) Cross-section for collisional ionization, in reduced units, logarithmic scale vs. the velocity in reduced units. The velocity v_m is defined in Eq. (9.24). The thresholds are indicated. Points: experimental. Adapted from Moutinho *et al.*, *Physica*, **53**, 471 (1971) and Baede (1975).

Figure 9.11 shows experimental results as well as the interaction potentials for the NaI system. The nominal energy release in the neutralization process is

$$\Delta E_0 = \text{IP(Na)} - \text{EA(I)} = 5.14 - 3.06 = 2.08 \text{ eV}$$

It is evident from the figure that the gap is ΔE_0, the energy required for the electronic state change when the atoms are far apart. At closer separation there exists a distance, R_x, where the electronic energy of Na$^+$I$^-$ equals that of NaI. Near the separation R_x the system can change its electronic state with only a very slight change in the kinetic energy of the nuclei.* Such a "switch" in the diabatic

---

* The kinetic energy of the relative motion of the nuclei at the separation R is $E - V_{\text{eff}}(R)$. Near R_x, $V_{\text{eff}}(R)$ is the same for both electronic states because the angular momentum is unchanged.

electronic state, accompanied by only a small change in the kinetic energy of the nuclei, is precisely what is needed to make the process efficient!

For E $\rightarrow$ V transfer such as Na* + N$_2$($v = 0$) $\rightarrow$ Na + N$_2$($v' > 0$) the quenching probably occurs through a sequence of two such changes. The first switch is a "harpoon"-type change Na* + N$_2$ $\rightarrow$ Na$^+$ + N$_2^-$, where N$_2^-$ has a higher equilibrium separation than N$_2$ so that it is born vibrationally excited, followed by a "neutralization" process Na$^+$ + N$_2^-$ $\rightarrow$ Na + N$_2$. If this is the case, the quenching cross-sections for Na* + M collisions should depend on the nature of the quencher M. For such diatomics as NO, CO, or N$_2$ the effective cross-section can be up to 50 Å^2, while it is less than 1 Å^2 for quenching of Na* by the rare gases. This is as expected on the basis of a harpoon mechanism where the crossing distance R_x (R_x (Å) $\cong 14.35/\Delta E_0$ (eV)) would increase with increasing electron affinity of M.

The convergence of two potential energy curves so as to allow an efficient switch of the nuclei between them is known as *curve crossing*.* How can we know if curve crossing is possible? The only guidance is either physical reasoning or an appeal to a proper quantum chemical computation or, *post factum*, from the nature of the observed results.

9.3.2.1 *Diabatic vs. adiabatic*

The diabatic and adiabatic descriptions of the electronic states are two complementary points of view, each of which has its merits and neither one of which offers a perfect approximation. We have first used these two alternatives in Chapter 7. Here we need to return. The adiabatic description is easier to define. At each configuration of the atoms we allow the electrons all the time that is needed to adjust. Technically, we diagonalize the Hamiltonian while keeping the nuclei in a fixed position. When the nuclei are moving at a finite velocity the electrons no longer have all the time to adjust. Hence the adiabatic approximation, meaning that the electrons maintain the same adiabatic state, may fail and the higher the collision velocity the more likely it is to fail. In the adiabatic description the electronic states are stationary states as long as the nuclei are stationary. They are no longer exactly stationary when the nuclei are allowed to move.

In the diabatic description we do not allow the electronic states to adjust. Therefore the diabatic states do not change as the nuclei move. It is a seemingly more natural description for a collision. There is however a price to pay. The diabatic states are not stationary states for the electrons. We first encountered that in the harpoon mechanism where, owing to the attraction to the halogen atom, the electron "jumps" over, leading to the covalent (diabatic) state changing

---

* Strictly speaking this is a slightly misleading term. The Born–Oppenheimer or adiabatic curves do not cross, they only converge and diverge again beyond R_x. Figure 9.11 uses a diabatic picture where the electronic state is "frozen" as either covalent or ionic. The adiabatic state adjusts to the change in internuclear distance. For a collision of Na + I, the ground adiabatic state is covalent at large distances but is primarily ionic for $R < R_x$.

to an ionic (diabatic) state. In the diabatic description the electronic states are not stationary because of electronic coupling terms.

The electrons do typically move faster than the nuclei; vibrational periods are two or three orders of magnitude longer than electronic periods. So for slow collisions the adiabatic approximation makes a better starting point in that it does allow the electrons to fully adjust. But it will fail and this failure is the point of this section.

*9.3.3 The adiabaticity parameter for curve crossing

When two diabatic potential curves cross there is a finite energy gap $\Delta E(R)$ between the corresponding adiabatic states; at the "crossing point" R_x the gap is smallest and we denote it $\Delta E(R_x)$. If the system spends a time τ traversing the crossing region its energy is indefinite by an amount h/τ. If this indefiniteness is greater than $\Delta E(R_x)$ the system can efficiently emerge on the curve that is otherwise excluded. If the uncertainty is less than $\Delta E(R_x)$ the system maintains its adiabatic state. Thus the condition for adiabatic behavior is that passage through the crossing zone must be sufficiently slow that $\tau > h/\Delta E(R_x)$. The adiabaticity parameter is therefore

$$\xi = \tau \, \Delta E(R_x) \big/ h \tag{9.19}$$

From a complementary point of view, $h/\Delta E(R_x)$ is the time for the system to adjust its electronic state and the adiabaticity parameter can be understood, as before, see Eq. (9.10), as the ratio of the duration of the perturbation to the response time of the system.

Equation (9.19) generalizes our earlier result for ξ in two ways. First, the gap $\Delta E(R_x)$ can be much smaller than the asymptotic (or "nominal") value of ΔE and the relevant time interval τ is not the duration of the entire collision but the (shorter) time during which the nuclei are in the localized region where the gap is small. Both factors act so as to reduce the value of the adiabaticity parameter from what it might otherwise be.

Electronic energy transfer can thus be quite efficient, even for a large nominal ΔE, **if** curve crossing can take place so that there is a range of R values for which the gap between the adiabatic states, $\Delta E(R)$, is small.

To estimate τ, the duration of the passage through the localized curve crossing, we examine the concrete example of Na + I collision where the switch is from a covalent NaI potential for $R > R_x$ to an ionic Na^+I^- form for $R < R_x$, Figure 9.11. It is evident that the region where the two potential curves converge and where ξ, Eq. (9.19), is a minimum, is a very localized region in R, about R_x. If at all, transitions between the two curves will occur only during the time spent by the system in the vicinity of R_x. As usual, we can set $\tau = a/v$ where a is the range in R along which the gap between the two potential curves is a narrow one, v is the radial velocity at R_x (and hence is impact-parameter-dependent).

An estimate of a is provided by regarding the two diabatic potential curves as linear functions of R, in the vicinity of R_x. The two straight lines diverge because of their different slopes, intersecting at R_x. Hence if we expand each potential as a Taylor series about R_x

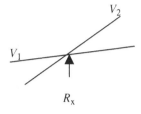

$$\Delta E(R) = V_1(R) - V_2(R)$$
$$= V_1(R_x) + (R - R_x)\frac{dV_1}{dR}\bigg|_{R_x} - V_2(R_x) - (R - R_x)\frac{dV_2}{dR}\bigg|_{R_x}$$

we can estimate the range a as the distance on either side of R_x over which the gap roughly doubles

$$\Delta E(R_x) = (a/2)\left|\frac{dV_1}{dR} - \frac{dV_2}{dR}\right|_{R_x} \equiv (a/2)\,|\Delta F_x| \qquad (9.20)$$

$|\Delta F_x|$ is the absolute value of the difference in slopes at R_x. Combining Eqs. (9.19) and (9.20),

$$\xi = 2(\Delta E(R_x))^2 / h v |\Delta F_x| \qquad (9.21)$$

As before, the adiabaticity parameter scales inversely with the collision velocity. But the other inputs, namely the nominal gap and the forces at the crossing, require an appeal to quantum chemistry.

Solving the dynamics for the probability of state changing during a passage across R_x yields the expected exponential dependence on the adiabaticity parameter

$$P = \exp(-\pi^2 \xi) \qquad (9.22)$$

We emphasize that P is the probability of a non-adiabatic transition, which means entering the crossing range on one adiabatic state and exiting on the other adiabatic state. $1 - P$ is the probability of a switchover of the nature of the adiabatic state.* For a very slow collision, when ξ is very large, a change in the diabatic electronic state will occur with a high probability upon passage via R_x. In other words, the passage will be adiabatic. We have already encountered this phenomenon in the harpoon mechanism (Section 3.2.4), where at about R_x the charge transfer occurs with essentially unit probability for reactants with thermal velocities.

*9.3.4 The Landau–Zener transition probability

From an experimental viewpoint the important quantity is not the adiabaticity parameter ξ itself but the probability of an overall change in electronic state induced by the collision, e.g., $Na + I \rightarrow Na^+ + I^-$. If we could determine this

---

* The adiabatic electronic state changes its character as R varies across R_x. Therefore an adiabatic passage means a switchover from one diabatic state to another, or a non-diabatic transition. On the other hand, a non-adiabatic transition means a diabatic behavior.

probability (as a function of the impact parameter) we could obtain the cross-section for this collisional ionization process. Of course, our discussion assumes that the total energy exceeds the endoergicity ΔE_0, Figure 9.11, of the collision. As the Na and I atoms approach, a curve crossing can occur in the region near R_x with the probability $1 - P$. The two newly formed ions will continue to approach, governed by the attractive ionic potential, until they reach the turning point of their motion. They will then start receding from each other. If the two particles are to separate as ions, then, as they recede across R_x to larger separations, a change of the adiabatic state must not occur. The probability of forming $Na^+ + I^-$ in this way is $P(1 - P)$. Alternatively, if the change did not occur on the way in (probability P) then, if ions are to be formed, a non-adiabatic transition must occur on the way out. The overall probability of collisional ionization is then $2P(1 - P)$.*

The result for P, Eq. (9.22), can be written as

$$P = \exp(-v_m/v) \tag{9.23}$$

where we have an explicit form for the characteristic velocity v_m. In the particular case of the Na + I collision where the covalent potential is only weakly R-dependent and $V_2 = -e^2/R$ (at large R),

$$|\Delta F_x| = e^2/R_x^2 \quad \text{and} \quad v_m = \pi R_x^2 \Delta E_x^2/2\hbar e^2 \tag{9.24}$$

The probability of producing ions, $2P(1 - P)$, cannot exceed $1/2$ regardless of the magnitude of v. When v is small (large ξ, adiabatic collision), P is small and the behavior is adiabatic. When v is large (low ξ), $1 - P$ is small. Thus, there will be a maximum in the energy dependence of the cross-section for collisional ionization, as shown in Figure 9.11. From the position of the maximum one can estimate v_m.

*9.3.5 The localized crossing range

The ability of fs laser excitation to prepare localized initial states has been discussed in Chapter 8. Here we describe a remarkable early experiment that shows the localization of the region where non-adiabatic transitions are important. The system is again NaI where, cf. Figure 9.11, the upper adiabatic electronic state, that near equilibrium is essentially covalent in character and has its shallow minimum much to the right compared with the ground state. A 50 fs UV pulse prepares a localized state with a fair amount of vibrational energy, at the left turning point of the upper adiabatic electronic state. This non-stationary localized initial state

---

* Similarly, if the collision does not lead to ions, then either curve crossing failed to occur both on the way in and on the way out (probability P^2) or curve crossing occurs both on the way in and on the way out, probability $(1 - P)^2$. Checking: $(P^2 + (1 - P)^2) + (2P(1 - P)) = 1$.

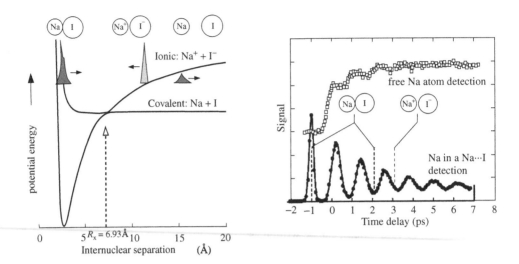

Figure 9.12 The ground and excited potentials for NaI with the localized state created by fs UV excitation shown both initially and as it moves. Right panel: the probe signals for Na at a finite separation from I and for a Na atom far away as a function of the delay between the pump and probe pulses [adapted from T. S. Rose, M. J. Rosker, and A. H. Zewail, *J. Chem. Phys.* **88**, 6672 (1988); see also Zewail (1996) and C. Jouvet *et al.*, *J. Phys. Chem.* **101**, 2555 (1997)]. This and similar experiments provide a direct experimental demonstration of the localized region of the non-adiabatic transition. It took over 50 years for the theoretical ideas of Landau and Zener to have this direct experimental test.

starts to oscillate in the upper well, much as a classical motion will. The vibrational period is relatively long because the potential is shallow and the reduced mass is not low.

By a suitable choice of the probe frequency the experiment can probe the Na atom while it is in the vicinity of I or when the two atoms have separated. Based on the curve-crossing model we expect the following: on the picosecond time scale the upper state will be stable except when it transverses the curve-crossing range as it is moving out. Then it can make a non-adiabatic transition to the ground state. If such a transition occurs the Na and I atoms are in a covalent state (the ground adiabatic state for $R > R_x$) and receding from one another. If the transition is localized one should detect a covalent Na–I pair first after half a vibrational period and then every vibrational period as is seen experimentally, Figure 9.12. The probe detuned from the free Na absorption detects the covalent Na–I pair at some Franck–Condon window and the receding motion of the atoms removes them promptly from that window so the probe spectrum is made up of spikes. A probe for Na shows a staircase accumulation of Na atoms.

The results of this experiment can be validated both by changing the vibrational energy content (and hence the period) of the localized initial state by going to shorter wavelengths and by changing the system, e.g., to NaBr.

*9.3.6   Donor–acceptor systems and photoinduced charge separation

Electronic energy transfer is particularly efficient between donors and acceptors.[12] These are pairs of molecules that can be apart or can be held together by a rigid or a flexible bridge. The donor is optically excited. Instead of fluorescing, it transfers the excitation to the acceptor that is thereby excited. The signature of the transfer is the fluorescence characteristic of the acceptor. It is useful to think of the transfer as the (down transition) dipole of the donor coupling to the (up transition) dipole of the acceptor, and to judge its efficiency by the fluorescence spectrum of the donor overlapping the absorption spectrum of the acceptor. Restricting the freedom of the molecules to reorient with respect to one another can therefore enhance the transfer.

An alternative outcome is that the excited donor transfers the excited electron to the acceptor

$$D + A \rightarrow D^* + A \rightarrow D^+ + A^-$$

Such a photoinduced charge separation is very thoroughly studied[13] because it is a primary step in photosynthesis where a pair of adjacent chlorophyl molecules acts as the donor. There are many subsequent steps that can prevent the initially separated charges from being used as a source of energy by keeping the charges apart. These include back electron transfer, formation of triplet states, etc. It is therefore of interest to monitor the process in real time. A system exhibiting a fast and rather high efficiency of charge separation is a carotenoid pigment (Car for short) covalently linked to the C_{60} fullerene [H. Imahori *et al.*, *Photochem. Photobiol.* **62**, 1009 (1995)]. Ultrafast pumping was in a short wavelength where Car does not absorb and probing was at an absorption of the Car^+ species. The rise time of the absorption was with a time constant of 0.8 ps and the decay occurred on a much longer (0.5 ns) time scale. With improvements in laser technology it is becoming possible to probe for finer details, on even shorter times, as we now discuss.[14]

*9.3.7   From photons to perception

Understanding vision is a very cross-disciplinary activity.[15] But the very first stage is in our domain. It is a fast light-induced isomerization of a polyene (an isomer of retinal) linked (by a Schiff base, see Figures 9.13 and 9.14) to an amino acid inside a bundled protein. Other photobiological processes also occur in conjugated organic systems. Therefore, both model compounds and biophysically interesting molecules (in their natural environment or otherwise) have received much attention.[16] The general trend has been from a more kinetic description toward an understanding of the dynamics as it unfolds in time. The realization that many such photochemical processes proceed via conical intersections has provided an additional motivation for such studies.

9.3.7.1 Ultrafast dynamics in bacteriorhodopsin

We consider the ultrafast dynamics in bacteriorhodopsin (bR), which is a light-harvesting system. A related class of reactions is electrocyclic[17] ring opening or closure as discussed in Section 8.2.1. Then we turn briefly to quantum chemistry to recognize that even the simple process of photoassisted rotation about a double bond is not quite so simple.

Light absorption in retinal proteins results in structural transformations in both the retinal chromophore (a cis $\rightarrow$ trans or a trans $\rightarrow$ cis isomerization) and in the surrounding protein. It was expected that the primary stage is the transient rotation about the double bond made possible by light absorption, as shown for bacteriorhodopsin in Figure 9.13. The initial exit from the Franck–Condon region to the already partly twisted fluorescent state was taken to proceed in about 100 fs. Rotation about the bond between carbon atoms 13 and 14, which brings the molecule to the 13-cis ground state (denoted J in Figure 9.13) is completed in about 500 fs. On a much longer time scale there is isomerization[18] back to the trans form. This one-dimensional view of the photocycle, as shown in Figure 9.13, turns out to be only a part of the story. The protein seems to respond to the light absorption on a scale faster than 100 fs. Model compounds in which either steric hindrance or even outright rigid bridges preclude a facile rotation about the bond between carbon atoms 13 and 14 upon photoexcitation still show a similar response of the protein. The primary stage is thus the ultrafast charge redistribution in the excited state due to the bond stretch. Only somewhat later does the slower rotation begin the photocycle that finally returns the retinal to its starting state.

Dynamical computations are not limited by the number of atoms in the problem but by the plurality of electronic states. Valiant attempts are however being made and a snapshot of a trajectory of bR linked to the protein is shown in Figure 9.14.

A simpler system that undergoes a similar cycle is photochemical concerted ring opening in 1,3 cyclohexadiene. Here too one can summarize the essentials in a one-dimensional view (Figure 9.15), and here too a one-dimensional view is not the complete story.[19]

Problems

A. The funnel mechanism for chemical lasers. (a) Assuming fast rotational relaxation determine the initial rotational levels for which we can lase in HF on

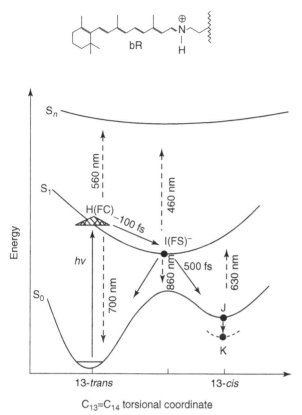

Figure 9.13   A one-dimensional view of a photocycle in bacteriorhodopsin (bR) where the initial ground state is all-trans. The retinal is shown, bound covalently, as a Schiff base, to the amino acid lysine at position 216 of the protein, at the top of the figure. The energy of the different electronic states is shown vs. the torsional angle of the bond between carbon atoms 13 and 14. In the ground electronic state it is a double bond so rotation about it is strongly hindered. The high barrier to rotation is shown both for the native state, all-trans and for the particular isomer, denoted J, where the configuration about the bond between carbon atoms 13 and 14 is cis. Photoexcitation leads to a Franck–Condon region that can be probed either by its fluorescence (that can be stimulated) or by further absorption. In the excited state the bond is weakened and is much less "double" in character so the barrier to rotation around the bond is lower. In about 100 fs the molecule twists into a strongly fluorescing state, denoted I. This state can be detected by its intense absorption. Isomerization is completed when state I makes the transition back to the ground electronic state in about 500 fs. The return to the ground state bifurcates to the two isomers. By using model compounds in which the rotation about the double bond is mechanically hindered or prevented, the photocycle can be stopped [adapted from Gai *et al.* (1998), Aharoni *et al.* (2001)].

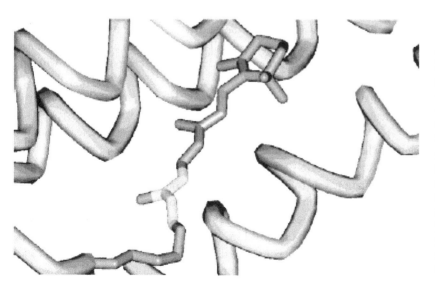

Figure 9.14   The instantaneous positions of the atoms of bR, see Figure 9.13, and the surrounding seven-helix bundle protein. By allowing the atoms to move and computing the instantaneous value of the gap between the upper and lower electronic states, see Figure 9.13, it is possible to determine an optimal time for the trajectory hopping from the upper to the lower state [adapted from Warshel (2002)].

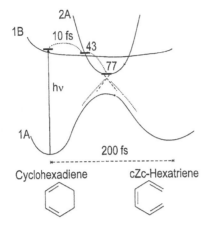

Figure 9.15 A one-dimensional view of the electronic states and their potentials for the electrocyclic ring opening of cyclohexadiene. Shown is an optical absorption to a bright electronic state, 1B, with another dark electronic state, 2A, close by. A conical intersection allows a facile conversion from the bright to the dark state and a second conical intersection allows a return to the ground electronic state on which the cyclic and open ring are two isomers, as shown (estimates for the time of passage via the intersections are also shown, in femtoseconds). A delayed process leads to the linear trans isomer. Experimental and computational studies show that the one-dimensional view, while providing a summary of the essential features, misses some key details, for example the role of the twist in reaching the second conical intersection. Such a view, even when another coordinate is included, is complementary to the description using states that we employed in the picket fence model in Chapter 7.

the v, j to $v - 1$, $j + 1$ levels at 300 K. The rotational constant of HF is 21 cm^{-1}. (b) How does the answer depend on the ratio of the concentration of HF molecules in the two vibrational states and how will this ratio change as the laser operates? (c) Based on your considerations, suggest why the laser emission shifts with time to higher js, as shown in Figure 9.3.

B. Grade the following three processes in order of their efficiency at room temperature: I. $H + D_2(v = 1) \rightarrow H + D_2(v = 0)$, II. $D + H_2(v = 1) \rightarrow D + H_2(v = 0)$, III. $D + HD(v = 1) \rightarrow D + HD(v = 0)$. If the temperature is higher, other channels will contribute. What are they?

C. In a mixture of N_2 and CO_2 heated by a shock wave, the final relaxation toward vibrational thermal equilibrium is the same as that in pure CO_2 and is faster than the similar relaxation in shock-heated air. Explain.

D. The vibrational frequencies of the symmetric stretch, bend, and asymmetric stretch of SO_2 are, approximately, 1152, 517, and 1362 cm^{-1}, respectively. Shock-heated SO_2 is found to relax to toward thermal equilibrium with two relaxation times rather than the single relaxation time as found for CO_2 and many other molecules. (a) Suggest an explanation. *(b) Suggest an experiment that will test your explanation.

E. Compare the magnitude of the adiabaticity parameter for vibrational and rotational relaxation of HF at room temperature. The harmonic vibrational frequency of HF is about 3140 cm^{-1}. (It is only a single bond. Why is the frequency in HF so much higher than the stretch frequencies for SO_2 or CO_2 or even N_2? The rotational constant of HF is an order of magnitude higher than that of, say, N_2. Why? And how does it affect your considerations about the efficiency of rotational relaxation)?

*F. The Landau–Teller temperature dependence, Figure 9.10. (a) Infer from Section 9.2.2 that the cross-section for vibration to translation energy transfer in an atom–diatom collision (roughly) scales with the collision velocity as $\exp(-v_0/v)$. (We have seen in Section 9.3.3 another case of a non-adiabatic collision with a similar scaling.) (b) Determine the dependence of the scale factor v_0 on the frequency of the diatomic molecule. (c) Use saddle point integration, see Problem D of Chapter 3, to derive the Landau–Teller temparature dependence.

*G. Deviations from the Landau–Teller temperature dependence, Figure 9.10. Estimate the lifetime of, say, an $HCl(v = 0) \cdot HCl(v = 1)$ van der Waals dimer, using RRK theory, and argue that at low temperatures it will be long enough for a sufficient number of "internal collisions" to take place and relax $HCl(v = 1)$.

H. A hard-sphere model. (a) Develop the hard-sphere model of Problem D in Chapter 5 for a non-reactive $A + BC$ collinear collision. In this model the interaction time is infinitesimally short so that energy transfer is efficient. Hence the model is used to determine the pre-exponential factor for the efficiency of the collision. (b) The role of the masses. Is it more efficient for A to come from the direction of the B atom or from the direction of the C atom? (c) Draw a trajectory for a BC molecule that is initially vibrationally excited. Does the outcome depend on the phase, cf. Section 5.2.2, of the initial vibration? (d) Under what conditions will A collide more than once with the B atom? (e) If you are geometrically minded, show that you can draw a construction that, for a given mass combination, allows you to determine the collision trajectory as a straight line superimposed on your drawing. There can be more than one line. What is the physics of the different lines? (f) Hint for part (e). Show that certain mass combinations, with a BC molecule that is initially vibrationally cold, will not result in any energy transfer.

*I. Super collisions, which, for smaller molecules, are also known as ballistic collisions. Fisk and Crim (1977), Flynn *et al.* (1996). See the $Ar + CsI$ example of H.-J. Loesch and D. R. Herschbach, *J. Chem. Phys.* **87**, 2038 (1972) or the $Ar + KBr$ results of Fisk *et al.* Devise a hard-sphere model that will predict a high conversion of the initial vibrational excitation to translation or vice versa for $Ar + CsF$ collisions.

J. Detailed balance for energy transfer collisions. It is sometimes loosely stated that detailed balance implies that the rate constant for a deactivating collision is larger than the rate constant for the reversed, activating collisions. Even more loosely it is stated that a reaction is faster in the exothermic direction. (a) Explain why this is not unreasonable, particularly so when the gap is much higher than $k_B T$. There are however important exceptions and you should identify at least two counter examples. (b) Compare reaction rate constants for pure rotational energy transfer, $A + BC(v, j) \rightarrow A + BC(v, j')$. (c) Vibrational energy transfer to or from a vibrationally hot polyatomic molecule P, $A+P(E_v) \rightarrow A + P(E_v')$. The second exception shows what is meant when we say that a large polyatomic can act as a heat bath. Suggest other examples.

K. Collisions of the second kind. Electronically excited (2.1 eV) Na* atoms are quenched efficiently by O_2, $Na^* + O_2 \rightarrow Na + O_2$ and much less so by the rare gases. (a) How would you demonstrate this difference in efficiency by a frequency domain experiment. (b) Sketch the relevant potentials for the quenching of Na* by O_2. The electron affinity of O_2 is 0.44 eV and the ionization potential of Na is 5.8 eV.

*L. Collisions of the second kind. The one-dimensional potentials that you sketched in Problem K overlook the role of the vibration of O_2. Examine this point in at least one of two ways. (a) Draw potential energy surfaces that are functions of both the Na—O_2 distance R and the O—O distance r. (b) Draw one-dimensional potential energy curves where for each curve both the electronic state of Na and the vibrational state of O_2 are fixed (such potentials are diabatic in two senses). Discuss the dynamics of the quenching and show why we expect that O_2 will be vibrationally excited after a quenching collision. The final vibrational state distribution in processes in which an electronic state change occurs are often quite close to the prior limit. Can your considerations above rationalize why this will be so?

M. The *harp model*. Highly exoergic reactions of alkali metal dimers with molecules of high electron affinity (Grice, 1975) produce alkali atoms in many excited electronic states all the way up to the conservation of energy limit, e.g., $K_2 + Cl \rightarrow KCl + K^{**}$ [W. S. Struve *et al.*, *J. Chem. Phys.* **62**, 404 (1975)]. The first step is presumably a harpoon process, $K_2 + Cl \rightarrow K_2^+ + Cl^-$. Propose a model for the second step.

N. The Landau–Zener model. Take 0.025 eV as the value of the smallest gap, $\Delta E(R_x)$, between two adiabatic potentials in the curve-crossing model. When we discussed harpoon reactions in Section 3.2.4 we took it for granted that for thermal reactants the system behaved adiabatically, that is, at R_x it jumped with unit probability from the covalent to the ionic diabatic states. (a) Is this an accurate assumption? If you need more input, assume that the ionic potential is purely coulomb and that at long range the covalent potential can be neglected compared with the ionic. (b) Using data from Figure 9.11 compute the threshold energy for collisional ionization. In the post-threshold region is collisional ionization a probable process?

O. The quantum mechanical period of a vibrational motion of a diatomic molecule. Write the initial (time = 0) wave function as a linear superposition of two adjacent vibrational states. (a) Solve the Schrödinger time-dependent equation by showing that the linear combination $\psi(R, t) = c_v \exp(-iE_v t/\hbar)$ $\psi_v(R) + c_{v+1} \exp(-iE_{v+1}t/\hbar)\psi_{v+1}(R)$ satisfies the equation. (b) Compute an expectation value of, say, the displacement from equilibrium, R, and show that it changes with time with a period that equals $(E_{v+1} - E_v)/\hbar$.

Notes

1 Lasing, which involves a change only in the rotational quantum number, has indeed been observed [E. Cuellar and G. C. Pimentel, *J. Chem. Phys.* **71**, 1385 (1977)].

2 Quantum effects are important at two levels. First, the obvious one that inelastic collisions deal with transitions between the discrete internal quantum states of the colliders. This is clear but even that is not always easy to implement in a computation. Take, for example, the method of classical trajectories. The products of the collision emerge with a continuous distribution of the final kinetic energy. This presents two problems. (i) Where do you put the boundary between inelastic and elastic collisions? Not a trivial issue because the results will show a large fraction of collisions with very small gaps. So which gaps are to be defined as effectively equal to zero can significantly change the output. This problem does not occur for reactive collisions. But for inelastic collisions it is a worry. A secondary aspect is that one needs to assign the final states to discrete bins. Most practical ways of doing this carry the penalty that they violate detailed balance because initial and final states are treated on an unequal footing. (ii) Molecules will exit with less than their zero point of internal energy. For a polyatomic molecule this energy can be quite large. Then there are quantum effects in the dynamics.

3 The interpretation of the V–V up-pumping distribution as an equilibrium subject to conservation of the number of vibrational quanta is due to C. E. Treanor, J. W. Rich, and R. G. Rehm, *J. Chem. Phys.* **48**, 1798 (1968).

4 For energy-rich polyatomics, see Hippler and Troe (1989), Troe (1992), Weston and Flynn (1992), Flynn *et al.* (1996), Flynn (2001).

5 By going somewhat higher up in excitation one can provide enough energy for the isomerization or dissociation. See, for example, V. A. Lobastov *et al.*, *J. Phys. Chem.* A **105**, 11159 (2001).

6 Supercollisions are discussed by Oref and Tardy (1990), Clary *et al.* (1995), Oref (1995), Mullin and Schatz (1997).

7 The term "adiabatic" is used in a consistent sense throughout the text. It means no change in the occupation of the quantum state of the system. If you are good in Stat. Mech. you may want to show that this is the same sense as used in thermodynamics. Hint: write the mean energy of the system as $\langle E \rangle \equiv \sum_i E_i p_i$ where i is an index of the possible states. So a change in the energy is made up of two parts, $\delta \langle E \rangle \equiv \sum_i \delta E_i \, p_i + \sum_i E_i \, \delta p_i$. The first term we are familiar with from mechanics. It is the change in the energies of the states and it is what we call "work." The second part is possible if the occupations of the different states change. It is the part we can "heat." Note that, as you know, neither "work" nor "heat" are the differentials of some function. Only their sum is. But in the adiabatic limit "work" is an exact differential because there is no change in the occupation of the different states.

8 Since $\tau_c > t_v$ the B–C bond oscillations are very rapid and the colliding atom A, which measures time on a τ_c scale, sees only the average position of the atom B and cannot probe the oscillations about the average.

9 The field began in the 1920s when Cario and Franck discovered how these so-called *collisions of the second kind* could be put to practical use, e.g., the Hg*-sensitized activation of paraffins. This is due to initiation of chain reactions by radical formation, Hg* + RH → HgH + R or Hg + H + R. Franck was led to look for such processes by arguments based on detailed balance. For how far the study of this problem has advanced, see Dagdigian (1997).

10 For the one-dimensional Landau–Zener curve-crossing problem that we discuss, see Nikitin (1975, 1999), Child (1991), Zhu *et al.* (2001), Nakamura (2002). The more general case of crossing of potential energy surfaces is discussed by many authors. Koppel *et al.* (1984), Whetten *et al.* (1985), Lorquet (1996), Yarkony (1996), Ben-Nun *et al.* (2000), Baer (2002), Worth and Cederbaum (2004).

11 As we have already seen in Chapter 7, in special cases there can be another option. The adiabaticity parameter is not small but there are many repeated attempts to make the change. Then, even if per attempt the probability is low, eventually it may take place. Radiationless transitions of a polyatomic molecule, Section 7.0.3, provide examples of this slow route. Recall that in a radiationless transition a molecule dumps a large amount, ΔE, of electronic energy into the vibrational manifold of a lower electronic state. The perturbations are the vibrations of the nuclei so the adiabaticity parameter is $\xi = \tau \Delta E / \hbar$ where τ is a vibrational period or $\hbar/\tau$ is a vibrational spacing. Since such spacings are small compared with an electronic excitation energy, ΔE, the adiabaticity parameter is large. But the nuclei do not cease to vibrate and eventually the transition can take place. If we are correct then there is a clear prediction: it is those vibrations with the shortest period that are most effective. This immediately puts the blame on the C–H stretch motions and suggests (as is observed) that deuteration will significantly slow down the rate of radiationless transitions. There are therefore examples where the actual gap is the nominal one. It need not be the case that the gap has to be small for a breakdown of the Born–Oppenheimer to be observable. At the same time we emphasize that radiationless transitions can also occur via conical intersections and then they are quite facile.

12 Speiser (1996), May (2000).

13 For photo-electron transfer, see Wasielewski (1992), Gust *et al.* (1993), Bixon and Jortner (1999), Lin *et al.* (2002), Ritz *et al.* (2002).

14 For carotenoids as donors and the role of higher excited states see P. J. Walla *et al.*, *J. Phys. Chem. A* **106**, 1909 (2002).

15 See, for example, the reports from the colloquium *Proc. Natl. Acad. Sci. USA* **93** (1996).

16 Mathies *et al.* (1988), Pollard and Mathies (1992), Lawless *et al.* (1994), Wang *et al.* (1994), Fleming and Cho (1996), Fleming *et al.* (1997), Bardeen *et al.* (1998), Ben-Nun *et al.* (2000), Haas *et al.* (2000); Walla *et al.*, *J. Phys. Chem. A* **106**, 1909 (2002). For more general accounts: Bonacic-Koutecky *et al.* (1987), Michl and Bonacic-Koutecky (1990), Turro (1991), Klessinger and Michl (1995). There is, of course, a thriving theoretical effort (Karplus, 2002) carried out both within the experimental groups and by dynamicists interested in uncovering details that cannot be easily experimentally probed and in designing control schemes.

17 In an electrocyclic reaction (Woodward and Hoffmann, 1970) a molecule with a conjugated system of m π electrons cyclizes to form a ring of $m-2$ π electrons and one σ bond. For this transformation to happen, the two ends of the relevant π system must approach each other in such a way as to enable the end p orbitals to overlap constructively.

To form a σ bond the two terminal p orbitals need to rotate. When the molecule is substituted the rotations can be in the same or in opposite directions. This leads to different stereoisomers and is of much interest to physical organic chemists. Therefore, such reactions are extensively studied. They are a special case of a pericyclic isomerization where the transition state is cyclic.

18 In some retinal systems this stage is enzyme-catalysed and in others it is thermally induced.

19 For a long time, the cis–trans photoisomerization of stilbene through a torsion about the central C–C double bond has served as a model for photoisomerization. Even for this simple process the interplay of the different electronic states is not uncomplicated and here too the one-dimensional energy level scheme, which is qualitatively similar to that shown in Figure 9.15, is not the entire story. As is often the case, the problem begins with the electronic affairs near the barrier for isomerization on the ground electronic states. The two p orbitals that make for the ethylenic double bond are rotated away from one another and the character of the ground state is essentially that of a diradical. It is then about isoenergetic with the triplet state. The first excited singlet state, S_1, increases in energy toward the twisted configuration and it is a conical intersection with a higher, second excited state S_2 that allows a return of the photoexcited molecule to the ground state where it bifurcates to the cis and trans isomers. The evidence is that there is more than one state that can be called the "second" higher singlet state. Which one of the possible states is lower in energy depends on the region of configuration space. See W.-G. Han et $al.$ $Chem.$ $Phys.$ $Chem.$ **3**, 167 (2002) for a view of this problem.

Chapter 10
Stereodynamics

This chapter explores how the directed nature of chemical bonding affects molecular collisions and chemical reactions. Whereas the concept of size or cross-section leads to numbers (scalar quantities), the concept of chemical shape leads to vectors (numbers tied to directions). Consequently, this topic is more mathematically challenging, but its study yields important insights into the nature of chemical transformations. As expected, just as the size of a molecule depends on the probe selected to measure this quantity, the chemical shape of a molecule also depends sensitively on the probe chosen for the measurement. It is important to stress that the physical shape of a molecule is most commonly determined from molecular spectroscopy; it is usually expressed in terms of bond angles and bond lengths. The chemical shape of a molecule refers to the apparent size and shape of a molecule as experienced by another atom or molecule that collides with it.

10.0.1 The steric factor and early history of stereodynamics

The notion of spatial requirements for a chemical reaction dates to the introduction of a steric factor p in simple versions of the collision theory of reactions, Section 3.2.5. This was forced upon the theory by the smaller than expected, or even much smaller, overall magnitude of the reaction cross-section. Unfortunately, the early theory gave no clue for determining the value of p, which has largely been treated as an adjustable parameter to make simple collision theory agree with observation. It was transition state theory that first accounted for the magnitude of p and, most importantly, showed that p is systematically smaller when the reactants are structurally more complex, Section 6.1.4.1. Collision theory with a barrier to reaction that depends on the attack angle can provide a detailed understanding of the dependence of the steric factor on both the translational energy and internal excitation of the reactants. For a bimolecular $A + BC$ abstraction reaction the theoretical result is conveniently represented as, Eq. (3.36),

$$p = \tfrac{1}{2}(1 - \cos \gamma_{\max}) \tag{10.1}$$

where γ_{max} is the maximum angle for which reaction is possible. Thus, p corresponds to a solid angle within a cone of opening half-angle γ_{max}. This treatment leads naturally to the concept of a *cone of acceptance*[1] for the reaction to occur. Attack angles between the BC bond and the atom A within the cone of acceptance result in reaction, whereas attack angles greater than γ_{max} do not.

For generations of chemists steric considerations have been an important aspect in the choice of a synthetic procedure. More recently, organic chemists have developed computing programs using semi-empirical inter- and intramolecular force fields for the determination of minimum-energy structures, minimum-energy paths, and transition-state geometries. This so-called "molecular mechanics" method has been applied to predict regio- and stereoselectivity in reactions of chiral reagents and to guide the stereospecific synthesis of chiral products. When we say that chemistry is local, the local steric requirements for reactivity are foremost in our minds. The short-range atom–atom repulsive forces play a dominant role in such considerations. In addition, long-range attractive forces tend to steer the approaching reagents, particularly if they are highly anisotropic. Because the rotational motion of molecules is usually slow compared to the time of a collision, it is possible to influence the outcome of a collision, often dramatically, by orienting one of the reagents with respect to the approach direction of the other. This idea gave birth to dynamical stereochemistry (Figure 10.1) and we first need to discuss how an orientation of reagents can be achieved.

10.1 Controlling reagent approach geometry

10.1.1 Preparing oriented molecules in electric fields

Bernstein (1988) conceived the first idea of how to control reagent approach geometry in a chemical reaction. He proposed to use an electric field to align the dipole moment of a polar molecule in the laboratory frame. The idea is easy to grasp. The interaction of the electric dipole moment of the molecule with an electric field results in a lowest-energy orientation with the dipole moment of the molecule pointing antiparallel to the electric field direction. But molecules rotate and further consideration shows that the most favorable case is for the molecule to precess with its dipole moment opposed to the electric field direction. This treatment requires that the molecule has a permanent dipole moment (so that it undergoes a first-order Stark splitting), that is, it should be a symmetric top. Specifically, Bernstein proposed orienting beams of symmetric-top molecules in hexapole electric fields. This is how dynamical stereochemistry started.[2]

For a symmetric-top molecule in the state $|JKM\rangle$ where J is the total rotational angular momentum quantum number of the molecule, K is the projection of $\mathbf{J}$ on the top axis, and M is the projection of $\mathbf{J}$ on the electric field direction, the $|JKM\rangle$ state experiences a radial force within a hexapole electric field that is proportional to the product $KM/[J(J+1)]$. Molecules in states for which the

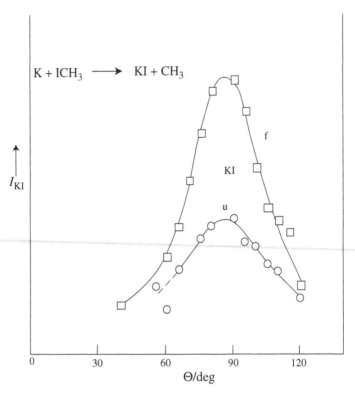

Figure 10.1 Early days of dynamic stereochemistry. An inhomogeneous electrical field is used to orient the CH₃I molecule so that the I-atom end points in the direction of the initial relative velocity of an incoming K atom, the f(avorable) case or with the CH₃ end pointing toward the incoming K atom, the u(nfavorable) case. The difference in reactivity, plotted as the KI signal vs. the (laboratory) scattering angle, is considerable [adapted from G. Marcelin and P. R. Brooks, *Faraday Disc. Chem. Soc.* **55**, 318 (1973)]. More detailed experiments can provide a complete mapping of the reactivity vs. the angle of approach, as shown in Figure 1.5.

product KM is negative are defocused and pushed away from the beam direction, whereas molecules for which KM is positive are focused. Moreover, the focal point depends on the value of $KM/[J(J + 1)]$, allowing different $|JKM\rangle$ states to be individually studied (Parker and Bernstein, 1989).

In these experiments we call orientation any situation in which the two ends of a molecule can be distinguished, and we call alignment the situation in which we distinguish the sides of a molecule from its ends, but in which the two ends cannot be told apart. Again, the idea of a "cone of acceptance" through which reactive collision trajectories must pass in order to access the transition state is a key early concept in interpreting these experiments with oriented reagents. Reaction is possible only when the angles of approach and the impact parameter lie within the angular range of the cone of acceptance. Those that fall outside this cone do not cause reaction. Thus, the "chemical shape" of the reagent molecules is conferred in part by the anisotropy of the interaction potential. But the cone of acceptance is not simply geometrical but depends also on the collision energy and the internal motions of the reagents. As might be expected, the cone of acceptance usually widens as the collision energy increases from the threshold sufficient to permit reaction.

Orientation of polar molecules that are not symmetric-top molecules, such as the molecule HCl, is possible with homogeneous electric fields via the second-order Stark effect. But for commonly populated rotational states, the sizes of the electric fields required to make an appreciable orientation are prohibitive. As a polar molecule with a second-order Stark splitting rotates in an applied electric field, it slows down when the dipole moment opposes the electric field direction but speeds up when the dipole moment points along the electric field direction. At extremely high electric fields, the polar molecules line up parallel to the field, and they execute frustrated (incomplete) rotations, called librations, about the field direction.[3] For significant orientation to be achieved it is necessary for the second-order Stark splitting energy (between the electric dipole moment of the molecule and the applied electric field) to be larger than the rotational energy of the molecule. Under such conditions, the molecules are said to be in pendular states, which are not characterized by one rotational quantum number j but by a mixture of rotational states. Because thermal molecular rotation resists orientation by the applied electric field, the idea of orienting molecules by brute force application of a large electric field was not pursued until experiments[4] showed that supersonic cooling of polar molecules in molecular beams, which populates only the lowest rotational levels, makes this approach feasible.

A particularly illuminating example of what can be learned by brute force orientation is the study of the reaction of potassium atoms with oriented iodine monochloride.[5] The direction of the orienting field was reversed so that the reactivity of K attack on the I-end could be compared to attack on the Cl-end of ICl. The potassium halide reaction product is forward-scattered for both orientations, that is, in the same direction as the incoming K atom, but attack on the I-end of ICl yields KCl whereas attack on the Cl-end of ICl yields KI. The velocity distribution of the KCl product, however, shows both fast and slow components, but the KI product has only one velocity component. It is thought that the reaction proceeds by a harpoon mechanism in which the K atom transfers an electron to ICl, which dissociates into a neutral halogen atom and a negatively charged halogen anion. The K^+ then picks up the X^- to yield the KX product. ICl dissociates to $I + Cl^-$ on its ground-state potential energy surface with little release of translational energy into these fragments. ICl dissociates from an excited state to yield $I + Cl^-$ with significant translational energy release into these fragments. ICl also dissociates from an excited state to yield $I^- + Cl$. Thus, the reaction of K with oriented ICl reveals a rich chemistry involving reactions on more than one potential energy surface. The attack of K on the Cl^- end of ICl to form KI has a cone of acceptance with an estimated apex half-angle of $40°$, whereas attack of

Figure 10.2 Relation between the transition dipole moment μ, the internuclear axis **r**, and the rotational angular momentum j of a diatomic molecule, in the high-j limit. Note that μ is perpendicular to j for P- or R-branch excitation, whereas μ is parallel to j for Q-branch excitation.

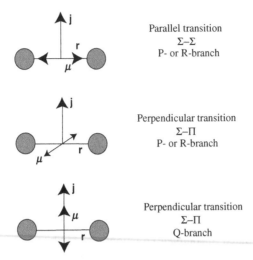

Parallel transition
Σ–Σ
P- or R-branch

Perpendicular transition
Σ–Π
P- or R-branch

Perpendicular transition
Σ–Π
Q-branch

K on the much larger I-end of ICl to form KCl has an apex half-angle of about 100°.

10.1.2 Preparing aligned molecules with polarized radiation

Just as an applied electric field can be used to control the reagent approach geometry by its interaction with the molecule's dipole moment, it is also possible to use the interaction of polarized radiation with the molecule's transition dipole moment when the molecule absorbs radiation (Zare, 1972, 1982). In this case, however, the molecule becomes aligned with respect to the electric field of the beam of radiation. The origin of the alignment effect is the transition probability as a function of the rotationally averaged position of the molecule with respect to the electric vector of the light beam. By symmetry, the transition dipole moment points in only certain directions with respect to the rotational angular momentum vector **j**. For a diatomic molecule, transitions may be characterized as parallel or perpendicular, depending on whether the angular momentum projection upon the internuclear axis remains unchanged as in a Σ–Σ transition or changes by one unit as in a Π–Σ or a Σ–Π transition. In the high-j limit, which is rapidly approached with increasing J, the internuclear axis **r** is at right angles to the rotational angular momentum vector **j**. An electric dipole allowed transition causes the rotational quantum number j to change at most by one unit, that is, $\Delta j = 0$, or ± 1. A transition in which $\Delta j = -1$ is called a P-branch member, $\Delta j = 0$ a Q-branch member, and $\Delta j = +1$ an R-branch member. Moreover, the transition dipole moment must lie either along the internuclear axis **r**, perpendicular to **j** for a parallel transition, or must lie perpendicular to the internuclear axis **r**, either along **j** for a Q-branch transition or perpendicular to **j** for an R- or P-branch transition. These three different possibilities are pictured in Figure 10.2.

The transition probability is proportional to $|\mu \cdot \mathbf{E}|^2 = |\mu|^2 |\mathbf{E}|^2 \cos^2\theta$, where $\mathbf{E}$ is the electric vector of the light beam, and θ is the angle between μ and $\mathbf{E}$. Figure 10.2 shows how the location of the dipole moment μ is related to the location of the rotational angular momentum vector $\mathbf{j}$. The excitation process favors molecules with their transition dipole moments μ along $\mathbf{E}$. Therefore, the absorption of light selects molecules with $\mathbf{j}$ vectors pointing preferentially in some direction. In the high-j limit $\mathbf{j}$ is perpendicular to the internuclear axis $\mathbf{r}$, and thus the absorption of light chooses preferential directions for the plane of rotation of the diatomic molecule. For electromagnetic radiation the electric field $\mathbf{E}$ oscillates in time so the direction of the internuclear axis $\mathbf{r}$ is selected but not the signed sense of the internuclear axis, that is, the light causes alignment but not orientation. Consequently, by preparing vibrationally or electronically excited molecules by the absorption of a beam of linearly polarized light (Zare, 1982), collisions can be studied in which the reagent target is struck primarily either face-on or edge-on with respect to the plane of rotation,[6] as shown in Figure 10.3.

10.1.3 Electronic orbital control

Not only is it possible to control the nuclear framework in a reactive encounter by the absorption of a beam of plane polarized radiation, it is also possible to control the direction of the electron charge cloud for open-shell reagents. For example,[7] Rettner and Zare prepared excited calcium atoms [Ca $3s3p$ ^{1}P] in which the p orbital pointed either along the approach direction or perpendicular to the approach direction of the Ca atom to various halogen-containing reactants. In a beam-gas scattering geometry they found that the total chemiluminescence cross-section for the Ca(^{1}P) + HCl reaction was insensitive to Ca(^{1}P) alignment, but that the branching ratio into the electronic excited states CaCl ($A^2\Pi$) and CaCl ($B^2\Sigma^+$) depended markedly on alignment, with parallel approach of the p orbital favoring production of the CaCl $B^2\Sigma^+$ state whereas perpendicular approach of the p orbital favored the CaCl $A^2\Pi$ state. These findings are readily interpreted in terms of an electron jump model (harpoon mechanism) in which the symmetry of the reagents is preserved during the electron transfer. These results led to the notion of orbital following in a chemical reaction, in which the direction of the p orbital becomes locked onto the target for large impact parameters so that the initial symmetry is maintained. Figure 10.4 illustrates this behavior. The experimental results shown imply that the Ca(^{1}P) + HCl reaction is predominantly adiabatic in nature.

It is also possible to apply the same idea of using radiation to control the electron charge distribution in open-shell molecules. For example, the lithium dimer (Li$_2$) has a closed-shell electronic structure in its $X^1\Sigma_g^+$ ground state in which both valence electrons are spin paired in the same σ molecular orbital. With visible radiation, Li$_2$ is readily excited to its $A^1\Pi$ state in which one electron

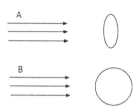

Figure 10.3 Collision with an aligned diatomic molecule in which the most likely alignment is (A) face-on and (B) edge-on to the plane of rotation of the diatomic molecule. As an example, for Sr + HF, face-on attack was favored at low collision energies but edge-on attack at higher collision energies [H. J. Loesch and F. Stienkemeier, *J. Chem. Phys.* **100**, 740, 4308 (1994)]. At low collision energies the transition-state bend angle is estimated to be about 70°, and clearly reflects the divalent character of the Sr atom. On the other hand, the reaction of K + HF($v = 1, j$) showed a clear preference for edge-on attack, as would be expected for a favored collinear approach geometry. Thus the chemical shape of HF($v = 1, j$) varies markedly depending on whether it is probed by collisions with Sr or K atoms.

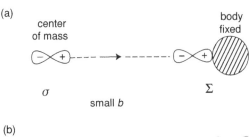

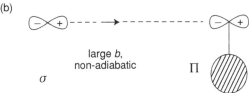

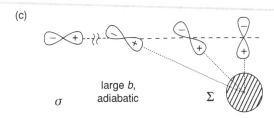

Figure 10.4 Orbital following. (a) At small impact parameters, the parallel approach of the p orbital to the spherical target, denoted as σ, transforms smoothly to parallel body-fixed alignment of the p orbital, resulting in the formation of a Σ state. (b) At large impact parameters, the parallel alignment in the center of mass can give rise to perpendicular body-fixed alignment of the p orbital, yielding a Π state for non-adiabatic behavior, or (c) can yield a Σ state for adiabatic behavior if orbital following occurs. [Adapted from C. T. Rettner and R. N. Zare, *J. Chem. Phys.* **77**, 2416 (1982); for orbital following in atomic collisions see Hertel *et al.* (1985), Campbell *et al.* (1988).]

remains in the σ molecular orbital and the other is promoted to a π molecular orbital. According to symmetry,[8] the electronic charge cloud in the high-j limit either lies perpendicular to the internuclear axis $\mathbf{r}$ and along the rotational angular momentum vector $\mathbf{j}$ or perpendicular to $\mathbf{j}$ as well. The former is called an A″ Λ-doublet level and is antisymmetric to reflection in the plane of rotation, whereas the latter is called an A′ Λ-doublet level and is symmetric to reflection in the plane of rotation. Despite promising beginnings,[9] the collisions of electronically aligned molecules remains a largely unexplored topic.

An experiment that similarly shows the differential reactivity of an aligned electronic orbital is the photodissociation of the T-shaped van der Waals dimer $Hg^* \cdot H_2$. The Hg $6^1S_0 \rightarrow 6^1P_1$ transition, where a single valence electron promoted to a p orbital, is split by the close presence of H_2 because now the

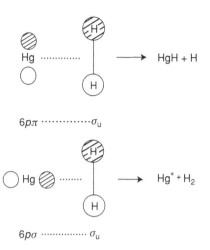

Figure 10.5 Nodal structure of the two possible T-shaped van der Waals dimers $Hg^* \cdot H_2$. The different shading indicates opposite phases of the orbital. The lower-energy $6s6p$ Hg orbital favors reactive dissociation because the p orbital has a favorable overlap with the empty antibonding σ orbital of H_2 [adapted from Jouvet *et al.* (1987), Simons (1987)]. It is also possible to use the promoted p electron as a harpoon starting with $Hg \cdot X_2$ van der Waals bound dimers, X halogen.

atom is not in a spherical environment. Excitation to the red of the atomic transition makes for a p orbital oriented along the H_2 bond, a $6s6p\pi$ configuration, Figure 10.5. This p electron effectively interacts with the σ antibonding orbital of H_2 as seen from the nodal structure exhibited in the figure and the complex dissociates reactively. Excitation to the blue of the atomic transition makes for a p orbital oriented perpendicular to the H_2 bond, a $6s6p\sigma$ configuration. This orbital interacts repulsively with either the σ bonding or antibonding orbital of H_2 and so its energy is higher. The complex then dissociates non-reactively as detected by observation of the Hg fluorescence.

Although the different alignment studies have shown that reaction rates can be changed less than an order of magnitude, they do provide an impressive probe of the three-dimensional nature (architecture) of the reaction pathway.

The most general form of optical preparation of the reagents is to use a radiation field whose spatial and temporal characteristics are continuously tuned (chirped) through the reactive encounter. A note of caution is in order, however. Although optical methods can be highly selective, the degree of orientation or alignment initially achieved may not always be well conserved during the time between the preparation step and the collisional encounter. The reason for this lack of fidelity is the presence of nuclear spins of the atoms in the selected reagent. These act as randomly oriented flywheels whose weak coupling to the rest of the molecule (through hyperfine interactions) acts to reduce whatever initial orientation and alignment was achieved. If the j level of the molecule is large compared to the nuclear spin I, the effect is small.[10]

10.2   Analyzing product polarization

If collisions are studied in a bulb so that the distribution of directions for collisions is uniform, then a measurement of the products cannot be expected to yield directional information about individual collisions. If the distribution of directions for collision is restricted, however, as in beam–gas scattering experiments, or well defined as in crossed-beam scattering experiments, or in some photoinitiated studies (see below), then the products of the reaction are, in general, polarized. We call a distribution polarized whenever the distribution is described as containing orientation or alignment or both. Polarized products can be measured in much the same way that polarized reagents are prepared, either by using applied electric fields to cause deflection of the polarized products or using polarized light to probe the products. In what follows, we highlight a few of these experiments and results.

Herschbach and co-workers were the first to observe product alignment in various chemical reactions. They used an inhomogeneous electric field as an analyzer. One system of particular interest is the reaction K + HBr → KBr + H. The resulting KBr product was strongly rotationally aligned with its rotational angular momentum $\mathbf{j}$ pointing perpendicular to the relative velocity vector $\mathbf{v}$ of the collision. This finding is a consequence of the choice of masses for reactants and products and is referred to as a kinematic effect or a *kinematic constraint*. This behavior happens in any reaction in which a heavy atom H strikes a heavy–light reactant H$'$L to form a heavy–heavy product HH$'$ and a light collision partner L. To understand why, we need to consider briefly what factors control collision outcome.

10.2.1   Conservation of angular momentum

The possible sates of the products may be limited not just by the conservation of energy but also by the conservation of total angular momentum. In the simplest case of an atom–diatom collision the total angular momentum $\mathbf{J}$ is the vector sum of the orbital angular momentum $\mathbf{L}$ (Section 4.1.1) and the rotor's angular momentum $\mathbf{j}$ (assuming that the atom is a closed shell or its angular momentum can be ignored):

$$\mathbf{J} = \mathbf{L} + \mathbf{j} \tag{10.2a}$$

The two components of $\mathbf{J}$ need not point in the same direction and so its magnitude J satisfies the triangular inequality, $|L - j| \leq J \leq L + j$. How high J can reach in a specific reactive collision depends on the system. For the F + H$_2$ system the reduced mass μ is atypically low and the reaction cross-section in the post-threshold regime is small ($\approx 2\,\text{Å}^2$) so the range of impact parameters that contribute to the reactive collisions is small. Hence the range of $L = \mu v b$ is unusually constricted. The H_2 molecules have large rotational spacings. Hence, at thermal equilibrium, the most probable j values for H_2 are also atypically small. It follows that the possible range of J values is quite limited. Estimates suggest that $J \leq 30h$, as compared to a range of hundreds for usual heavy-particle collisions with large reaction cross-sections (for example, $K + Br_2$, where $J \leq 700h$).

The total angular momentum can also be written in terms of the products' angular momenta

$$\mathbf{J} = \mathbf{L}' + \mathbf{j}' \tag{10.2b}$$

where $L' = \mu'v'b'$. For the $F + H_2$ reaction the reduced mass of the recoiling products μ' is even smaller than the reduced mass of the incoming reactants μ so the range of L' is also limited. This reduction in the magnitude of the product orbital angular momentum is particularly the case when the reaction releases most of the exoergicity into vibration so that v' is small. For the $F + H_2$ reaction we therefore expect that the range of HF rotational angular momentum will be limited.

An opposite example, where the product rotational excitation is high, is the family of reactions where the departing atom is significantly less heavy than the other two. The reaction

$$K + HBr \rightarrow KBr + H$$

is an extreme but not unusual example. Here the initial reduced mass is quite high so the L values for which reaction is possible can be large and $j \ll L$ so that to good approximation $J \approx L$. The relative motion of the reagents controls almost completely the total angular momentum. The reaction is not very exoergic and the magnitude of the product orbital angular momentum L' is small, especially when it is realized that $\mu' \ll \mu$ so that $L' \ll L$. By conservation of angular momentum, the angular momentum of the reactants must equal the angular momentum of the products of a reactive collision. It thus follows that to good approximation $\mathbf{j}' = \mathbf{L}$. Essentially all the orbital angular momentum from the approaching reagents becomes rotational angular momentum of the diatomic product. This result also implies that the rotational angular momentum of the HBr product points along the orbital angular momentum of the reactant. Consequently, $\mathbf{j}'$ is perpendicular to the relative velocity vector $\mathbf{v}$. This alignment effect is also seen optically (Orr-Ewing and Zare, 1994), either in the polarized character of the emission if the reaction products are excited or by studying the dependence of the product fluorescence as a function of the polarization of the excitation source.

An even more extreme example of a kinematically constrained heavy plus heavy–light reaction system is that of $Ba + HI \rightarrow BaI + H$. Recall that for this mass combination $j' \approx L$ and $L = \mu v b$. It follows that the rotational state

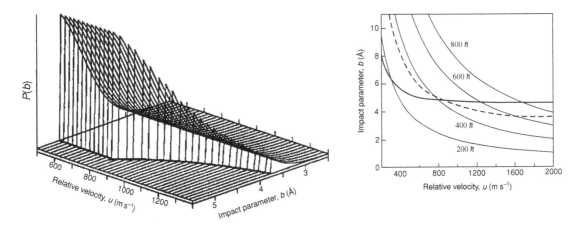

Figure 10.6 Left panel: opacity function, $P(b)$, for the Ba + HI $\rightarrow$ BaI($v = 0$) + H reaction inferred from the measured rotational distribution of BaI in its ground vibrational state, at different collision energies, as shown. The reaction is exoergic and populates BaI up to quite high vibrational states [adapted from K. Kalogerakis and R. N. Zare, *J. Chem. Phys.* **104**, 7947 (1996)]. The result requires comment on both the low and high ends of the b scale. As is the case for other reactions, the peak of the rotational distribution for a given v shifts to lower energies as v increases. The kinematic correlation $L \rightarrow j'$ means that opacity functions for formation of BaI(v) for higher v values will peak at lower impact parameters. The upper cutoff on the b value that leads to reaction is expected to be due to the centrifugal barrier, Section 3.2.6, and this is indeed the case at lower collision energies, right panel. At higher energies, j'_{max} does increase but $b_{max} \propto L_{max}/\sqrt{E_T}$ decreases, as shown in the right panel. It shows the cutoff on b values that lead to reaction owing to the centrifugal barrier, solid line, and the conservation of energy limit on j'_{max}, dashed line. Another kinematically dominated example is the H + HI $\rightarrow$ HI + H reaction (Aker and Valentini, 1993), where the exchanged atom is heaviest. Equation (10.3) below enables us to examine the role of the masses in the partitioning of the total angular momentum.

distribution of BaI reflects the distribution of impact parameters that contribute to the reactive collisions. The products' j' distribution should thus mimic the incoming distribution of impact parameters in the reactive collision, $2\pi b P(b)/\sigma_R$, which is likely to peak toward high b values. The choice of impact parameters cannot be controlled. Nevertheless, measurement of the product rotational state j' distribution for this reaction maps directly onto the impact parameter distribution that caused reaction. By this means the opacity function $P(b)$ for this reaction was derived, Figure 10.6.

The correlation $\mathbf{j}' \cong \mathbf{L}$ just discussed for the Ba + HI reaction is valid irrespective of the precise details of the forces that operate during the collision. It is an example of a *kinematic constraint*. The very description of the motion imposes the result. In Section*10.2.2 we show that for any mass combination the transformation from reactants to products in an atom–diatom $A + BC \rightarrow AB + C$ reaction is of the form[11]

$$\begin{aligned}
\mathbf{j}' &= \mathbf{L} \sin^2 \beta + \mathbf{j} \cos^2 \beta + \mathbf{d} \cos^2 \beta \\
\mathbf{L}' &= \mathbf{L} \cos^2 \beta + \mathbf{j} \sin^2 \beta - \mathbf{d} \cos^2 \beta
\end{aligned} \qquad (10.3)$$

Here $\mathbf{d}$ is a vector that does depend on the dynamics and β is the mass-dependent angle, $\cos^2 \beta = m_A m_C /(m_A + m_B)(m_C + m_B)$ of Appendix 5.B. Equation (10.3) is readily seen to conserve the total angular momentum $\mathbf{j}' + \mathbf{L}' = \mathbf{j} + \mathbf{L}$. The kinematic constraints discussed earlier are for $\cos \beta \rightarrow 0$, which is the case for a light ejected atom C or for a heavy exchanged atom B. The behavior in the opposite extreme, $\cos \beta \rightarrow 1$, is also simple, as we already saw in Chapter 1 for a light exchanged atom. In Section* 10.2.2 we shall refer to $\cos \beta \rightarrow 1$ as the *stripping limit*. For every type of collision, kinematic considerations govern, at least in part, the detailed outcome.

*10.2.2 Kinematic models

Models seek to bypass the need for solving the classical equations of motion. By making plausible assumptions (for example, in Section 3.2) we are to provide useful guidelines that can be tested. The spectator limit (Section 1.2) is an example of a simple model that correlates the states of the products with those of the reactants. We seek here to examine it in some further detail and to incorporate in the model the possibility of forces operating during the collision. To retain the simplicity of the model, we shall assume that the switching of partners is very rapid, i.e., that the forces are impulsive. As an example, consider the $H + Cl_2 \rightarrow HCl + Cl$ reaction. In the c.m. system, the light H atom approaches Cl_2 very rapidly. As soon as the H atom is within the range of chemical forces, the HCl bond is formed, while the Cl atoms repel one another because the electron from the H atom goes into an orbital that is antibonding between the Cl atoms, Section 5.1.5.1. It is this "instant" *product repulsion* that must be incorporated in the model. The resulting, so-called DIPR (Direct Interaction Product Repulsion) model[12] is just one of several that can be developed on the basis of our derivation below.

The starting point is the set of equations (B.5.1) for the coordinate of relative motion $\mathbf{R}$ and the internal coordinate $\mathbf{r}$ that we scale in terms of the masses and transform to the coordinates for the products. The transformation is very simple if the exchanged atom is rather light or if it is rather heavy. In the general case, using Problem A, the transformation is

$$\begin{aligned}
\mathbf{Q}'_1 &= \cos \beta \mathbf{Q}_1 - \sin \beta \mathbf{Q}_2 \\
\mathbf{Q}'_2 &= \sin \beta \mathbf{Q}_1 + \cos \beta \mathbf{Q}_2
\end{aligned} \qquad (10.4)$$

The result is exact and can be generalized to more than three atoms, Problem B. The transformation, Eq. (10.4), remains exact if we differentiate the coordinates with respect to time. The velocities at a given time t are therefore related by the same transformation. We now make the key approximation that the reaction is **sudden.** The atoms come in with their original velocities and promptly exit as products with their own final velocities. We completely neglect any forces that might have acted so that, using a dot to denote the time derivative, we relate the **initial** and **final** velocities by

$$\dot{\mathbf{Q}}_1' = \cos\beta\,\dot{\mathbf{Q}}_1 - \sin\beta\,\dot{\mathbf{Q}}_2$$
$$\dot{\mathbf{Q}}_2' = \sin\beta\,\dot{\mathbf{Q}}_1 + \cos\beta\,\dot{\mathbf{Q}}_2 \tag{10.5}$$

This result, while only valid in the *sudden approximation*, is rather neat. Knowing the final velocities we can compute every property of the products and Eqs. (10.3) are derived from this result. All we need for application of Eqs. (10.5) are the initial velocities and the masses.

There is one snag. The result (Eq. (10.5)) cannot possibly be the whole story, if only for the simple reason that it does not necessarily conserve the total energy. In Problem A you are asked to show that what (Eq. (10.5)) conserves is the kinetic energy. This failure is to be expected; we completely neglected the role of the forces so of course it is the kinetic energy that is conserved. Exoergic reactions, where the potential energy changes by a large amount, require a more refined treatment. Before we do so, let us see the angular momentum correlations. With the definition $\mathbf{L} = \mu\mathbf{R} \times \dot{\mathbf{R}}$, where $\mathbf{Q}_1 = a\mathbf{R}$, $a = \sqrt{\mu}$, $\mathbf{L} = \mathbf{Q} \times \dot{\mathbf{Q}}$, and similarly for $\mathbf{j}$, $\mathbf{L}'$, and $\mathbf{j}'$, Eq. (10.3) is readily derived with $\mathbf{d} = \tan\beta(\mathbf{Q}_1 \times \dot{\mathbf{Q}}_2 + \mathbf{Q}_2 \times \dot{\mathbf{Q}}_1)$. Checking, we see that the approximation (Eq. (10.3)) does conserve the total angular momentum!

A special case of our consideration is the *spectator limit*. Here the old bond has no energy and so $\dot{\mathbf{Q}}_2 = 0$. Because $E_T = \dot{\mathbf{Q}}_2/2$ (and similarly for E_T'): $E_T' = \cos^2\beta\,E_T$. In the DIPR model, one modifies the basic approximation by assuming that bond switching is accompanied by adding a sudden repulsion. This repulsion imparts a large extra velocity, say $\mathbf{q}$, along the old bond $\mathbf{Q}_2$, so that

$$\dot{\mathbf{Q}}_1' = \cos\beta\,\dot{\mathbf{Q}}_1 - \sin\beta\,\mathbf{q} \tag{10.6}$$

Hence the products' relative velocity (i.e., $\dot{\mathbf{Q}}_1$) has a component along the direction of the old bond. For a light atom attack ($\sin\beta \to 1$), that component can dominate so the products' angular distribution reflects the distribution of the orientation of the old bond (with respect to the initial velocity) at the instant of reaction.

Problem B derives the transformation between the coordinates suitable for the reactants and products in the four-center $AB + CD \to AC + BD$ reaction and hence shows that reactant vibration is the primary energy needed in the kinematic limit.

Kinematic considerations are also useful for inelastic collisions.[13] For example, when large changes in the rotational angular momentum occur in an $A + BC(j)$ collision, these changes are often caused by off-center collisions with the inner repulsive core of the BC molecule that is ellipsoidal in shape. Assuming such a localized transition, the change in angular momentum is $\Delta \mathbf{L} = \mu \mathbf{R} \times \Delta \dot{\mathbf{R}}$. Conservation of total angular momentum requires that $\Delta \mathbf{j} = -\Delta \mathbf{L}$. This relation has a clear observable implication about $\Delta \mathbf{L}$ and $\Delta \dot{\mathbf{R}}$, where $\Delta \dot{\mathbf{R}}$ is the change in the relative velocity, which is a vector in the plane defined by the initial and final velocity vectors. The change in BC angular momentum is to be oriented perpendicular to that plane. Is it? We return to this question later.

*10.2.3 The degree of orientation and alignment

Before proceeding, we take time out for a rather mathematical interlude. Stereodynamics deals with vectors, and these have both a magnitude and a direction. Ideally, we want to know the complete distribution of the possible directions of the vector as, for example, in an angular distribution after a collision, where we want to know how the final velocity is distributed with respect to the initial direction of approach. Orientation and alignment are the technical terms that are used to describe deviations from a purely random direction in space. In this section we want to express more precisely how to characterize orientation and alignment. This treatment leads naturally to multipole moments, which serve as a basis in which the distribution of oriented and aligned molecules can be expanded.

Consider a molecular distribution having the total angular momentum $\mathbf{J}$. Recall that, according to quantum mechanics, $\mathbf{J}$ can only take on integral (or half-integral) values, and the possible orientations of $\mathbf{J}$ are quantized. Specifically only $(2J+1)$ different $\mathbf{J}$ orientations can exist, denoted by the magnetic quantum number M, which ranges from $M = -J$ to $M = +J$ in unit steps, where J is the rotational quantum number. The wave functions are denoted by $|JM\rangle$ and are characterized by the total angular momentum quantum number J and its projection M on the axis of quantization, which we choose to be the Z axis. In the limit of high J values, the $\mathbf{J}$ vector of length $\sqrt{J(J+1)}$ can be regarded to precess about the Z axis making a constant projection M, as shown in Figure 10.7 where the cosine of the included angle θ satisfies the relation $\cos\theta = M/[J(J+1)]^{1/2}$. Thus, the operators for the square of the total angular momentum $\mathbf{J}^2$ and the projection $\mathbf{J}_Z$ (of $\mathbf{J}$ on the Z axis) have the expectation values

$$\langle JM|\mathbf{J}^2|JM\rangle = \langle \mathbf{J}^2\rangle = J(J+1) \tag{10.7}$$

and

$$\langle JM|\mathbf{J}_Z|JM\rangle = \langle \mathbf{J}_Z\rangle = M \tag{10.8}$$

We are interested in describing the distribution of $\mathbf{J}$ vectors. For simplicity and in agreement with many common situations, we will assume that the $\mathbf{J}$ vector distribution has cylindrical symmetry about the Z axis. For a system with such axial symmetry, the $\mathbf{J}$ distribution is simply specified by the occupation of states with the fixed possible projections of $\mathbf{J}$ on the Z axis. To make this discussion

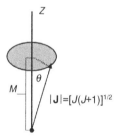

Figure 10.7 Vector model for the state $|JM\rangle$. The angular momentum vector $\mathbf{J}$ precesses about the Z axis at a fixed angle $\cos\theta = M/[J(J+1)]^{1/2}$ making a projection M on the Z axis.

less abstract, let us consider a molecule in a particular rotational level $J = 2$. Then a five-dimensional vector

$$\mathbf{N} = \begin{pmatrix} N_2 \\ N_1 \\ N_0 \\ N_{-1} \\ N_{-2} \end{pmatrix} \tag{10.9}$$

represents the states of such a system. N_M is the number of molecules in the state $|JM\rangle$. This vector can of course be expanded as a linear combination of five "Cartesian" basis vectors, $\mathbf{C}_i$, $i = -2, -1, \ldots, 2$:

$$\mathbf{C}_2 = \begin{pmatrix} 1 \\ 0 \\ 0 \\ 0 \\ 0 \end{pmatrix}, \quad \mathbf{C}_1 = \begin{pmatrix} 0 \\ 1 \\ 0 \\ 0 \\ 0 \end{pmatrix}, \quad \mathbf{C}_0 = \begin{pmatrix} 0 \\ 0 \\ 1 \\ 0 \\ 0 \end{pmatrix}, \quad \mathbf{C}_{-1} = \begin{pmatrix} 0 \\ 0 \\ 0 \\ 1 \\ 0 \end{pmatrix}, \quad \mathbf{C}_{-2} = \begin{pmatrix} 0 \\ 0 \\ 0 \\ 0 \\ 1 \end{pmatrix} \tag{10.10}$$

Explicitly,

$$\mathbf{N} = N_2 \begin{pmatrix} 1 \\ 0 \\ 0 \\ 0 \\ 0 \end{pmatrix} + N_1 \begin{pmatrix} 0 \\ 1 \\ 0 \\ 0 \\ 0 \end{pmatrix} + N_0 \begin{pmatrix} 0 \\ 0 \\ 1 \\ 0 \\ 0 \end{pmatrix} + N_{-1} \begin{pmatrix} 0 \\ 0 \\ 0 \\ 1 \\ 0 \end{pmatrix} + N_{-2} \begin{pmatrix} 0 \\ 0 \\ 0 \\ 0 \\ 1 \end{pmatrix} \tag{10.11}$$

The Cartesian basis vectors $\mathbf{C}_i$ have the property of satisfying the orthonormality condition, namely, $\mathbf{C}_i \cdot \mathbf{C}_j = \delta_{ij}$, that is, the dot product of any Cartesian vector with another vanishes whereas the dot product of any Cartesian vector with itself is unity.

To bring out the symmetry character, however, it is much more convenient to introduce spherical basis vectors that transform under rotation like multipole moments, that is monopole, dipole, quadupole, etc. The spherical basis vectors are defined by

$$\mathbf{T}_0 = \frac{1}{\sqrt{5}} \begin{pmatrix} 1 \\ 1 \\ 1 \\ 1 \\ 1 \end{pmatrix}, \quad \mathbf{T}_1 = \frac{1}{\sqrt{10}} \begin{pmatrix} 2 \\ 1 \\ 0 \\ -1 \\ -2 \end{pmatrix}, \quad \mathbf{T}_2 = \frac{1}{\sqrt{14}} \begin{pmatrix} 2 \\ -1 \\ -2 \\ -1 \\ 2 \end{pmatrix}$$

$$\mathbf{T}_3 = \frac{1}{\sqrt{10}} \begin{pmatrix} 1 \\ -2 \\ 0 \\ 2 \\ -1 \end{pmatrix}, \quad \mathbf{T}_4 = \frac{1}{\sqrt{70}} \begin{pmatrix} 1 \\ -4 \\ 6 \\ -4 \\ 1 \end{pmatrix} \tag{10.12}$$

where the numerical factors in front insure that they are normalized. The spherical basis vectors behave under rotation like spherical harmonics $Y_{LM}(\theta, \phi)$ but with

$M = 0$ because there is no ϕ-dependence owing to the assumed axial symmetry of the problem with respect to the Z axis. Thus, the spherical basis vectors transform under rotation as the Legendre polynomials $P_L(\cos\theta)$, see Eq. (10.23) below, where classically θ is the angle between $\mathbf{J}$ and the Z axis, see Figure 10.7. The spherical basis vectors also satisfy the orthonormality condition, $\mathbf{T}_i \cdot \mathbf{T}_j = \delta_{ij}$. Then $\mathbf{N}$ may be expanded in terms of the spherical basis set as

$$\mathbf{N} = \sum_{L=0}^{4} n_L \mathbf{T}_L \tag{10.13}$$

where the coefficients $n_L = \mathbf{T}_L \cdot \mathbf{N}$ represent multipole moments of order 2^L of the distribution $\mathbf{N}$.

Specifically, for $L = 0$, the monopole moment is given by

$$n_0 = \frac{1}{\sqrt{5}}(N_2 + N_1 + N_0 + N_{-1} + N_{-2}) \tag{10.14}$$

This quantity is proportional to the total number of molecules with $J = 2$, that is, the population of the $J = 2$ system summed over all different M values. All systems must possess some overall population; hence every distribution has a monopole moment. If the system has only a monopole moment, then the $\mathbf{J}$ distribution is uniformly distributed in space and the system is said to be unpolarized. If higher-order multipole moments exist, however, the system is polarized.

The dipole moment, $L = 1$, is given by

$$n_1 = \frac{1}{\sqrt{10}}(2N_2 + N_1 - N_{-1} - 2N_{-2}) \tag{10.15}$$

which in terms of the expectation value of the $\mathbf{J}_Z$ operator can be written

$$n_1 = \frac{1}{\sqrt{2}}n_0\langle\mathbf{J}_Z\rangle \tag{10.16}$$

This term is known as the *orientation*. Note that the orientation of a system requires an imbalance in the number of molecules in the state $|JM\rangle$ compared to those in the state $|J - M\rangle$. In other words, the distribution of $\mathbf{J}$ vectors does not have reflection symmetry in a plane perpendicular to the Z axis. Thus, the system has a head and a tail with respect to the Z axis. Orientation is often reported as the value of

$$O^{(1)} = \langle\mathbf{J}_Z\rangle/J = P_1(\hat{\mathbf{J}} \cdot \hat{\mathbf{Z}}) \tag{10.17}$$

The range of variations of $O^{(1)}$ is from -1 to 1 and $O^{(1)} = 0$ for a randomly oriented molecule.

For $L = 2$, the quadrupole moment

$$n_2 = \frac{1}{\sqrt{14}}(2N_2 - N_1 - 2N_0 - N_{-1} + 2N_{-2}) \tag{10.18}$$

is called the *alignment*. An examination of its form shows that the alignment is positive if $\mathbf{J}$ points more along the quantization axis Z than perpendicular to it,

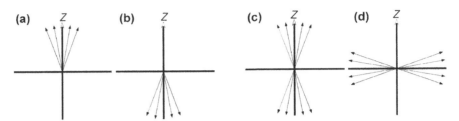

Figure 10.8 Orientation and alignment: (a) and (b) examples of pure orientation of positive and negative sign; (c) and (d) examples of pure alignment of positive and negative sign.

whereas the alignment is negative for the reverse situation. The alignment is often reported using the parameter

$$A^{(2)} = 2P_2(\hat{\mathbf{J}} \cdot \hat{\mathbf{Z}}) = \left(3\left\langle J_Z^2\right\rangle - J(J+1)\right)\Big/ J(J+1) \tag{10.19}$$

where P_2 is the second Legendre moment. The range in variations of $A^{(2)}$ is from $+2$ to -1 where $A^{(2)} = 0$ for a randomly oriented molecule. A positive value of $A^{(2)}$ shows a preference for polar regions whereas a negative value of $A^{(2)}$ shows a preference for equatorial regions. The n_3 term represents the octopole moment, and the n_4 term represents the hexadecapole term. The odd-order multipoles contribute to defining the orientation and the even-order multipoles for $L > 0$ to the alignment. It is possible for the **J** distribution in a system to have both orientation and alignment at the same time. If the system is axially symmetric and has only orientation, then its **J** distribution is defined by a distribution of single-headed arrows that point either preferentially parallel to the Z axis (positive orientation) or antiparallel to the Z axis (negative orientation). If the system is axially symmetric and has only alignment, then its **J** distribution is defined by a distribution of double-headed arrows that cluster either close to the Z axis (positive alignment) or cluster in a direction perpendicular to the Z axis (negative alignment). Figure 10.8 pictures these four situations. If a system is isotropic, then it is unpolarized, and all M states are equally populated. If a system is anisotropic, then it is polarized, and the M states are not equally populated.

As the rotational quantum number J becomes large, we approach the classical correspondence limit, and each n_L expansion coefficient becomes the average value of the Legendre polynomial $\langle P_L(\hat{\mathbf{J}} \cdot \hat{\mathbf{Z}})\rangle$, that is, the Lth classical multipole moment for the angular momentum distribution.

So far, we have restricted our attention to systems having axial symmetry. For example, in a crossed-beam experiment, the collision system has axial symmetry about the relative velocity vector **v**. If we observe the products scattered into a particular solid angle element, however, we break the axial symmetry and the system no longer possesses a cylindrically symmetric **J** distribution. Then the description of the polarization of the system is more complex. It depends on the choice of the X and Y axes. As a consequence, we have additional polarization

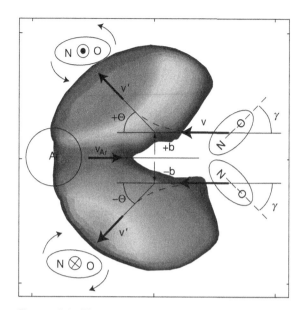

Figure 10.9 The essence of an experiment detecting the sense (clockwise or anticlockwise) of NO rotation in a given internal quantum state and scattering angle. The shading shows an NO state-specific experimental result at different scattering directions, $\otimes$ for clockwise, $\odot$ for anticlockwise [adapted from Lorenz *et al.* (2001); see also Crim (2001)]. Note the cylindrical symmetry about the initial velocity and the variation of the orientation with scattering angle.

parameters, which, in the high-J limit, become the average values of spherical harmonics.

10.2.4 Inelastic collisions

One particularly powerful means of observing products of a collision is by ion imaging in which each product is converted to an ion by some carefully chosen multiphoton ionization process, Section 7.1.3.1. By varying the degree of polarization of the ionizing radiation we learn about the degree of polarization of the product. A striking example is shown in Figure 10.9.

At the high level of final state resolution provided by such experiments we can discern quantal interference effects. The more prominent feature for inelastic excitation is a rotational rainbow that arises by a mechanism similar to the intense scattering of the final velocity into certain directions (Section 2.2.5). Here too, the rainbow arises from different trajectories scattered into the same final state except that the state is specified not only by the direction of $\mathbf{v}'$ but also by the rotational state of the molecule, NO in the case of Figure 10.9. This is a stereodynamic effect because the final state is determined not only by the impact parameter but also by the angle of approach, as shown for scattering by a hard ellipsoid in Figure 10.10.

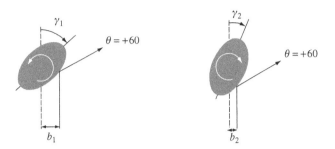

Figure 10.10 Scattering of an atom off a hard ellipsoid for a given impact parameter and angle of approach, γ. The scattering angle is θ. The recoil momentum $\Delta\mathbf{p}$ corresponds to a change $\Delta\mathbf{j} = \mathbf{R} \times \Delta\mathbf{p}$ in the rotational angular momentum of the ellipsoid. The two panels correspond to two different impact parameters that lead to scattering into the same final state.

For a given θ, as γ is varied $\Delta\mathbf{j}$ will pass through a maximum. This behavior is the origin of the stereodynamic rainbow: classical trajectories of different initial b and γ accumulate into the same angular range.[14] When we add an attractive part to the potential, it is also possible for the scattering angles to go through an extremum.

10.2.5 Surface scattering

The scattering of N_2 from the surface of a silver crystal is a particularly interesting example of a rotational rainbow, which becomes more pronounced with increasing collision energy. The broad excess population at high j is very apparent at the highest collision energy, Figure 10.11.

The scattered N_2 from the nominally flat 111 surface of Ag displays both alignment and orientation. The alignment increases with rotational excitation and approaches the limiting value for J being at right angles to the surface normal. It closely follows the expectations for an ellipsoid (American football) bouncing off a hard wall. The degree of orientation is large and the sense of orientation changes sign for the same J state, depending, in general, on whether the molecule is scattered at angles less than or greater than the specular scattering angle in which the angle of incidence equals the angle of scattering, both measured with respect to the surface normal. Understanding the observed orientation requires a modification of the simple model that assumes a flat surface. Rather we must recognize that the face of a crystal is not flat but is corrugated, reflecting the atomic structure of the solid. The major experimental features can be reproduced qualitatively by an ellipsoid scattering from a flat surface to which a tangential frictional force is added to represent the laterally averaged in-plane surface forces. The striking changes of sign in the orientation result from a splitting of the rotational rainbow by these in-plane forces. In other words, to an incoming N_2 molecule, the Ag(111) surface close-up is not perfectly flat. We will have more to say on this in Chapter 12.

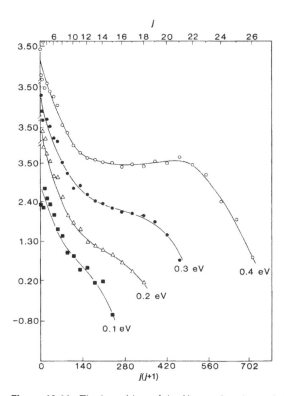

Figure 10.11 The logarithm of the N_2 rotational population (corrected for nuclear spin statistics) is plotted against $j(j + 1)$, which is proportional to rotational energy, for inelastic scattering of N_2 from the 111 surface of a silver crystal as a function of incident beam energy for an incident scattering angle of $15°$ and a final scattering angle of $20°$, close to the specular direction. The N_2 beam is supersonically cooled and has a rotational distribution that peaks at $j = 0$ and does not extend beyond $j = 3$. The surface temperature is 90 K [adapted from G. O. Sitz, A. C. Kummel, and R. N. Zare, *J. Chem. Phys.* **89**, 2558 (1988)].

There we shall also discuss the anisotropic *steering forces* between the molecule and the surface. Experimental evidence shows that the escaping NO molecules with large j from a Pt surface are found to exhibit a clear preference for alignment with $A^{(2)} > 0$, that is, for helicopter motion in which the rotational angular momentum **j** points along the escape velocity. This shows that just before they escape, the NO molecules have a preference for surface in-plane rotation. We shall come to interpret this as showing a preference for NO molecules adsorbed on the surface to be oriented with the N end attached to the surface. An out-of-plane rotation would bring the repulsive O end of NO closer to the surface.

10.2.6 Bimolecular reactions

Let us return now to elementary molecular reactions. For example, consider a beam-target gas arrangement with a chemiluminescent exchange reaction where the light emitted is monitored at right angles to the beam

$$A + BC \rightarrow AB^* + C$$
$$\downarrow h\nu$$
$$AB$$

Initially, $\mathbf{j}$ (the angular momentum of BC) is randomly (i.e., spherically symmetrically) distributed, whereas the orbital angular momentum of the reagent approach, $\mathbf{L}$, is randomly distributed in a plane perpendicular to the initial relative velocity vector $\mathbf{v}$. For simplicity, consider a reaction where many partial waves contribute, so that $j \ll L$, and that, e.g., because C is a light atom, AB^* is formed at high rotational excitation, so that $j' \gg L'$. Conservation of total angular momentum implies that for our assumed conditions, $\mathbf{j}' \approx \mathbf{L}$; $\mathbf{j}'$ will then be nearly parallel to $\mathbf{L}$ and hence confined in a plane perpendicular to the initial relative velocity. The AB product is fully aligned, with $\langle (\hat{\mathbf{j}}' \cdot \hat{\mathbf{k}})^2 \rangle \approx 0$, that is $A^{(2)} \approx -1$. The rotational alignment of AB is measured by the polarization of the emitted light with results showing the approach to a pure spectator limit at higher collision energies (Figure 10.12).

10.2.6.1 PHOTOLOC

Reactions producing ground electronic state products can equally well lead to product alignment, and observation of this alignment is most readily accomplished by optical means, such as laser-induced fluorescence or resonance-enhanced multiphoton ionization. This is illustrated by the reaction of ground-state chlorine atoms $(^2P_{3/2})$ with vibrationally excited CH_4 having one quantum in the C—H asymmetric stretch ν_3 and C—H stretch excited CD_3H. The experiment is carried out using the so-called *PHOTOLOC* technique (Alexander *et al.*, 1998; Simons, 1999), which stands for photoinitiated bimolecular reaction analyzed using the law of cosines. In this experiment, Cl_2 and methane are co-expanded into a vacuum chamber and ground-state Cl atoms are generated by photolysis of Cl_2. Hence, to good approximation, the collision energy is completely controlled by the choice of photolysis wavelength. Either the HCl or the CH_3 products are detected in a quantum-state-specific manner by resonance-enhanced multiphoton ionization. The experiment fixes the speed of the collision in the center of mass and the speed at which a reactant departs in the center of mass in a specific quantum state. A measurement is then performed to determine from time of flight the speed distribution in the laboratory frame of the reaction product. Knowledge of all three speeds is equivalent to knowing three sides of a triangle from which the angles of the triangle may be found using the law of cosines. Specifically, the speed distribution in the laboratory is used to determine the distribution in scattering angles in the center of mass, that is, the differential cross-section. The reactants are also selected in that an infrared laser prepares vibrationally excited methane. The infrared wavelength is chosen to cause the methane to have one quantum of vibration in its asymmetric stretch, denoted as

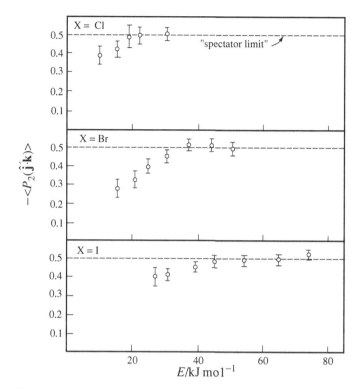

Figure 10.12 Rotational alignment coefficient of the XeX* product vs. the collision energy for the reaction of a fast Xe* ($^3P_2$) beam with HX, X = Cl, Br, I. The dashed horizontal lines represent the maximum theoretical limit on the alignment [adapted from Simons (1987)]. The deviations of the experimental results from the spectator limit can be accounted for if there is a repulsive release of the exoergicity. Such an impulse can contribute significantly when the ejected atom is light and if the transition state is not collinear. With the available understanding of kinematic effects, experimental conditions can be chosen such that dynamical features arising from forces operating during the collision process can dominate.

$CH_4(\nu_3 = 1)$. By observing how the scattering signal changes with the infrared laser on and off, it is possible to extract only those reactive collisions of the form

$$Cl\left(^2P_{3/2}\right) + CH_4(\nu_3 = 1) \rightarrow HCl(\nu', j') + CH_3$$

Shown in Figure 10.13 is the differential cross-section obtained for various HCl ($\nu' = 1, j'$) product states.

In contrast, the reaction of ground-state Cl with vibrationally unexcited CH_4 shows primarily backward-scattered HCl products, suggesting a preference for a linear Cl–H–C arrangement in the transition state region. The reaction with vibrationally unexcited CH_4 is believed to result from small impact parameters and for approach geometries in which the Cl atom lines up with a C–H bond. The effect of vibrational excitation of the methane is to allow reaction for a

Figure 10.13 Angular distribution of HCl($v' = 1$, j') products from the reaction of ground-state Cl atoms with methane (CH$_4$) excited with one quantum in the asymmetric stretch vibration. With increasing rotational excitation of the product, the angular distribution moves from predominantly forward (cos $\theta = +$ 1) to backward (cos $\theta = -1$) scattering.

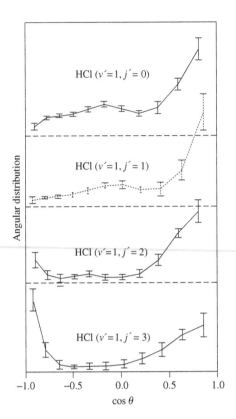

wider range of impact parameters, and hence to increase the cone of acceptance for reaction. Experiments show a factor of 30 or more increase in the reaction cross-section. This is consistent with the fact that many more collisions occur at larger impact parameters and suggests that the barrier to reaction is later along the reaction coordinate, Section 5.1. In particular, the production of HCl($v' = 1$, low j') seems to involve a *peripheral reaction mechanism*, resulting in forward scattering.

10.2.6.2 Peripheral dynamics

Peripheral dynamics is the pictorial term for describing steric hindrance by the central atom (or group). The impact parameter is the miss-distance with respect to the center of mass. Hence, unlike what is often the case, here low-impact-parameter collisions are unfavorable because they tend to direct the attacking atom toward the wrong target. Reaction is thus preferred for higher impact parameters that sample the periphery of the central atom. Figure 10.14 shows the case of the H + HI → H$_2$ + I collision as an example.

Peripheral dynamics, as a higher-impact-parameter collision, is characterized by more forward scattering of the product. This tendency is often augmented by a repulsion of the products along the old bond. There are other examples where obvious steric requirements indicate peripheral dynamics. The origin of the

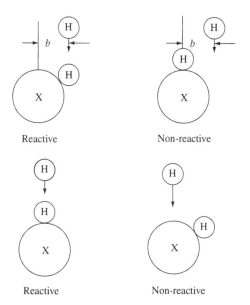

Reactive   Non-reactive

Reactive   Non-reactive

Figure 10.14 H + HI collisions at a large impact parameter b (top row) and a low b (bottom row) but for different orientations of HI. Left: reactive. The H–H approach occurs with a low impact parameter so that the chemical interaction between these two atoms can be expressed. Right: non-reactive. The H–H approach occurs with a higher impact parameter, almost as high as b itself [adapted from Aker and Valentini (1993)]. Because the impact parameter for the H–H approach can span the range from 0 to b, reaction is also possible at low b values. However, there are more collisions at higher bs so it is reaction at the largest bs that determines the reaction cross-section.

requirements can however be dynamical in origin. Consider an exoergic reaction leading to a highly internally excited state of the product, e.g., $F + H_2 \rightarrow H + HF(v' = 3)$. There is little translational energy for the products to separate. Now imagine the reaction trajectory running in the opposite direction. Unless it is a peripheral approach of H–H, reaction will not be favored.

Returning to the $Cl + CH_4(v_3 = 1)$ reaction, the forward-scattered $HCl(v' = 1$, low $j')$ products arise from collisions with large impact parameters and go through a transition state that has an approach geometry in which the Cl atom moves nearly perpendicular to the C–H bond under attack. Varying the polarization direction of the infrared excitation laser so that it prepares the vibrationally excited methane in different collision geometries, we can test this proposed mechanism. To make the analysis more clear cut, experiments were performed for the reaction of vibrationally excited CH_3D with Cl with the vibrationally excited C–H bond alternately parallel and perpendicular to the direction of motion of the Cl atom. The production of a forward-scattered $HCl(v' = 1, j')$ product is substantially larger for side-on than end-on attack of the C–H bond. The experiments thus confirm that the forward-scattered $HCl(v' = 1, j')$ product originates from a peripheral reaction mechanism.

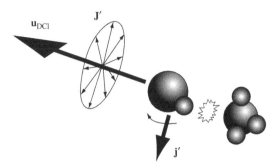

Figure 10.15 A pictorial representation of the preferential alignment of the DCl angular momentum in the Cl + CD$_4$ reaction [adapted from T. P. Rakitzis *et al.*, *J. Chem. Phys.* **107**, 9392 (1997)]. The strong repulsion of the products from one another is one of the reasons that allow us to approximate the angular distribution of the products as that of hard spheres.

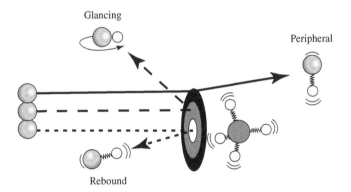

Figure 10.16 A pictorial interpretation of the results in terms of the initial impact parameter for the collision of Cl + CH$_4$(v_3 = 1). [The results, adapted from A. J. Orr-Ewing *et al.*, *J. Chem. Phys.* **106**, 5961 (1997), are based on the hard-sphere relation between final scattering angle and initial impact parameter. See also Section 4.4.3 and Alexander and Zare (1998).]

The HCl(v' = 1, j' = 1) product, which is predominantly forward scattered, could be studied in full detail because the rotational angular momentum has such a small value. Its rotational angular momentum was found to lie preferentially perpendicular to the products' relative velocity $\mathbf{v}'$, as is consistent with a late barrier meaning that the rotation of the products is determined as they separate, Figure 10.15.

For a hard-sphere collision, the scattering angle θ varies with the impact parameter b according to $\cos\theta = (2b^2/d^2) - 1$, where d is the distance between the hard-sphere centers at the point of contact. With this relationship the reactivity can be mapped as a function of impact parameter using the measured differential cross-section data. The results and their interpretation are displayed in Figure 10.16.

This example is remarkable in terms of how much detailed information is revealed about the transition state region from a study of the reaction stereo-dynamics.

10.3 Vector correlations

As we have seen, the vector properties of molecular collisions offer much richer information than that provided by scalar properties, such as the total cross-section of a reaction or the energy content of the reaction products. To illustrate this point, consider a simple atom-transfer reaction, which will be abstractly written as A + BC → AB + C. For this process, we can readily identify four vectors. These are the initial relative velocity $\mathbf{v}$ of the reagents (A, BC), the final relative velocity $\mathbf{v}'$ of the products (AB, C), the initial rotational angular momentum of the reagent molecule BC, denoted by $\mathbf{j}$, and the final rotational angular momentum of the product molecule AB, denoted by $\mathbf{j}'$. Here we have assumed, for simplicity, that no photons are emitted or absorbed in the collision process, and that electronic or nuclear spin angular momenta are non-existent or are randomly oriented and do not couple to other angular momenta present. A simple example of such a case would be the atom-transfer reaction O + CS → CO + S.

Of these four vectors, there are six different combinations, two at a time. Each combination describes a vector correlation. Specifically, the correlation of $\mathbf{v}$ with $\mathbf{v}'$ is the differential cross-section (angular distribution of the products) and $\mathbf{j}$ with $\mathbf{j}'$ the rotational tilt of the reagent molecule compared with that of the product molecule. In addition, $\mathbf{v}$ with $\mathbf{j}'$ and $\mathbf{v}'$ with $\mathbf{j}'$ tells us about the product polarization with respect to the initial and final scattering directions. Similarly, $\mathbf{v}$ with $\mathbf{j}$ and $\mathbf{v}'$ with $\mathbf{j}$ tells us about the reagent polarization with respect to the initial and final scattering directions. We can also go on to define triple vector correlations, such as $\mathbf{v}, \mathbf{v}', \mathbf{j}$ and $\mathbf{v}, \mathbf{v}', \mathbf{j}'$, which give information on how the differential cross-section depends on the polarization of the reagent or the polarization of the product, respectively. Of course, there is only one four-vector correlation, which contains a full description of the directional character of the scattering dynamics. The experimental measurement of these quantities requires that more than one quantity be measured at the same time, that is, that some type of coincidence measurement be made. As might be expected, such measurements of two or more quantities are more demanding than the observation of scalar properties. Consequently, vector correlations represent a largely untapped source of information about the scattering dynamics.

10.3.1 $\mathbf{v}, \mathbf{v}'$ correlation

Let us consider in more detail just one of these vector correlations, namely, that between $\mathbf{v}$ and $\mathbf{v}'$. We shall consider here only a classical treatment in which no distinctions are made between quantized and unquantized vectors and in which we regard the angular momentum to be large. In the absence of external fields,

Figure 10.17 The first three Legendre polynomials vs. $\cos\theta$ in the range -1, 1. Note that l is the number of nodes in the interval and that the polynomials are even or odd upon reflection so that the polynomials for odd l must vanish at the origin.

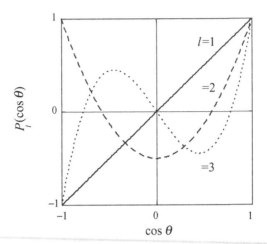

vector properties of reactions are independent of the coordinate frame chosen to describe them. For two unit vectors $\hat{\mathbf{a}}$ and $\hat{\mathbf{b}}$, only the angle θ between the vectors is invariant to a change of coordinates. Consequently, the vector correlation may be expressed solely in terms of this angle. In particular, the vector correlation between $\mathbf{v}$ and $\mathbf{v}'$, which is the differential cross-section, may be expressed in the complete set of Legendre polynomials:

$$I(\theta) = \frac{1}{4\pi}\sum_{l}(2l+1)a_l P_l(\cos\theta) \tag{10.20}$$

where

$$a_l = \int_0^{2\pi}d\phi\int_0^{\pi}P_l(\cos\theta)I(\theta)\sin\theta\,d\theta \Bigg/ \int_0^{2\pi}d\phi\int_0^{\pi}I(\theta)\sin\theta\,d\theta \tag{10.21}$$

and the integration is over the solid angle element $\sin\theta\,d\theta\,d\phi$. We see that $a_0 = 1$ and in general, because of the orthogonality of the Legendre polynomials

$$a_l = \langle P_l(\cos\theta)\rangle = \langle P_l(\hat{\mathbf{a}}\cdot\hat{\mathbf{b}})\rangle \tag{10.22}$$

Here the brackets $\langle\ldots\rangle$ mean an average over the quantity inside the brackets weighted by the differential cross-section $I(\theta)$, where the differential cross-section is normalized so that its average over all solid angles equals unity.

The Legendre polynomials are so important to this development that it is worth reminding ourselves of some of their properties. The first few Legendre polynomials, as plotted in Figure 10.17, are

$$P_0(\cos\theta) = 1, \quad P_1(\cos\theta) = \cos\theta, \quad P_2(\cos\theta) = (3\cos^2\theta - 1)/2$$
$$P_3(\cos\theta) = (5\cos^3\theta - 3\cos\theta)/2, \quad \text{etc.} \tag{10.23}$$

Note that the Legendre polynomial $P_l(\cos\theta)$ is a polynomial in even or odd powers of the argument $\cos\theta$ depending on whether l is even or odd, see Figure 10.4.

In discussing vector correlations it is often useful to invoke the spherical harmonic addition theorem:

$$P_l(\hat{a} \cdot \hat{b}) = \frac{4\pi}{2l+1} \sum_m Y^*_{lm}(\theta_a, \phi_a) Y_{lm}(\theta_b, \phi_b)$$

$$= P_l(\hat{a} \cdot \hat{z}) P_l(\hat{b} \cdot \hat{z}) + 2 \sum_{m=1}^{l} \frac{(l-m)!}{(l+m)!} P_l^m(\hat{a} \cdot \hat{z})$$

$$\times P_l^m(\hat{b} \cdot \hat{z}) \cos(m(\phi_a - \phi_b)) \tag{10.24}$$

Here ϕ_a and ϕ_b are the azimuthal angles of the unit vectors $\hat{a}$ and $\hat{b}$. Consequently, the quantity $(\phi_a - \phi_b)$ is the dihedral angle between the plane containing the vectors $\hat{a}$ and $\hat{z}$ and the plane containing $\hat{b}$ and $\hat{z}$. It is easy to see that an average over the azimuthal angles will remove the second term. Hence, from Eq. (10.22)

$$a_l = \langle P_l(\hat{v} \cdot \hat{z}) \rangle \langle P_l(\hat{v}' \cdot \hat{z}) \rangle \tag{10.25}$$

10.3.1.1 The collision complex

Consider the special case in which the reactants form a long-lived collision complex, whose duration is sufficient so that the system has no memory of the initial directions of the vectors. Then the breakup of the complex is independent of its formation so that the dihedral angle between the $(\mathbf{v}, \mathbf{J})$ and $(\mathbf{v}', \mathbf{J})$ planes will be randomly distributed. Application of the spherical harmonic addition theorem yields

$$\langle P_n(\mathbf{v} \cdot \mathbf{v}') \rangle = \langle P_n(\mathbf{v} \cdot \mathbf{J}) P_n(\mathbf{v}' \cdot \mathbf{J}) \rangle \tag{10.26}$$

Let us examine some special cases that give us much insight. The first is the situation in which both the reagents and the products have no internal angular momentum. Then the total angular momentum of the collision system, $\mathbf{J}$, must coincide with the orbital angular momentum of the approaching reagents $\mathbf{L}$, which by conservation of angular momentum must also coincide with the orbital angular momentum $\mathbf{L}'$ of the products: $\mathbf{J} = \mathbf{L} = \mathbf{L}'$. The magnitudes of the vectors are therefore given by the product of the reduced mass, the impact parameter, and the relative velocity magnitude (speed) of the collision partners. It also follows that the direction of $\mathbf{J}$ is perpendicular to both $\mathbf{v}$ and $\mathbf{v}'$. We choose $\mathbf{J}$ to be along the z axis. Then

$$a_l = [P_l(0)]^2 \tag{10.27}$$

The Legendre polynomials have the property that $P_l(0)$ vanishes when l is odd. Hence, for this condition, the differential cross-section, Eq. (10.20)

$$I(\theta) = \frac{1}{4\pi} \sum_l (2l+1)[P_l(0)]^2 P_l(\cos\theta) \tag{10.28}$$

is symmetric about $\pi/2$, that is, shows equal forward–backward peaking. Problem D of Chapter 4 shows that $I(\theta)$ is proportional to $1/\sin\theta$, that is, it peaks at $\theta = 0$ and π (which are the poles), and reaches a minimum for $\theta = \pi/2$ (which is the equator). In the more general case the reactants and the products each have rotational angular momentum **j** and **j**′. Ordinarily, the direction of **j** is random (unpolarized) and the direction of **j**′ is unobserved. The spherical harmonic addition theorem may be used again to perform the appropriate averages over the random distribution of **j** and the unobserved distribution of **j**′. Because the random orientation of **j** ensures that the dihedral angle between the (**v**, **L**) plane and the (**L**, **J**) plane is randomly distributed, we have for the entrance channel

$$\langle P_n(\hat{\mathbf{v}} \cdot \hat{\mathbf{J}}) \rangle = P_n(\hat{\mathbf{v}} \cdot \hat{\mathbf{L}}) \langle P_n(\hat{\mathbf{L}} \cdot \hat{\mathbf{J}}) \rangle$$
$$= P_n(0) \langle P_n(\hat{\mathbf{L}} \cdot \hat{\mathbf{J}}) \rangle \tag{10.29}$$

For the exit channel an analogous formula holds involving **L**′. Thus,

$$a_n = (2n + 1)[P_n(0)]^2 \langle P_n(\hat{\mathbf{L}} \cdot \hat{\mathbf{J}}) \rangle \langle P_n(\hat{\mathbf{L}}' \cdot \hat{\mathbf{J}}) \rangle \tag{10.30}$$

It follows once again that the breakup of a long-lived complex must show forward–backward scattering symmetry.

10.3.1.2 *Photodissociation*

Another commonly encountered vector correlation is the angular distribution of photofragments. This correlation involves the correlation between the recoil velocity **v** of the photofragment with respect to the electric vector **E** of a beam of light that causes the photolysis. Already in Chapter 7 we used the angular distribution of the general form

$$I(\theta) = \frac{\sigma}{4\pi} [1 + \beta P_2(\cos\theta)] \tag{10.31}$$

where θ is the angle between **v** and **E**, P_2 is the second Legendre polynomial, σ is the photodissociation cross-section, and β is what we called the *asymmetry parameter*. Another related correlation is the direction of the internal angular momentum of the photofragment with respect to its recoil velocity. It might be expected that linearly polarized light could only cause alignment. This expectation is correct for either a parallel transition or a perpendicular transition. But if both a parallel transition and a perpendicular transition lead to the same final states of the separated photofragments, then interference is possible and orientation of the photofragments can occur.[15] A particularly striking example is displayed in the photodissociation of ICl by a beam of linearly polarized light, as discussed in Section 7.1.2.3.

Another example of the power of vector correlations to reveal dynamic behavior is the photolysis of glyoxal.[16] Three photolysis channels appear to participate, and in order of decreasing importance these are

$$CHOCHO \xrightarrow{h\nu} \begin{cases} H_2CO + CO \\ CO + CO + H_2 \\ HCOH + CO \end{cases}$$

By carefully measuring the Doppler line widths of the CO and H_2 photofragments, it is found that the CO departs with a recoil velocity **v** that is predominantly perpendicular to its rotational angular momentum $\mathbf{j}$. This result implies that the formation of CO is primarily from the dissociation of a planar geometry. The production of H_2 might at first seem quite surprising. The ground-state geometry of glyoxal places the two H atoms on opposite sides of the central C—C bond (*trans* or *E* configuration). How then does H_2 form? Measurement of the H_2 Doppler profile shows that its recoil velocity $\mathbf{v}$ is parallel to its rotational angular momentum vector $\mathbf{j}$. It seems that in the excited state glyoxal isomerizes to form the *cis* or *Z* configuration by rotation about the C—C bond. If the two CO fragments push off the H_2 fragment in the planar configuration (asymmetric breakup of the planar excited state), the resulting torque would spin the H_2 so that $\mathbf{j}$ is predominantly perpendicular to $\mathbf{v}$, contrary to what is observed. On the other hand, out-of-plane motion in the transition state, such as torsion about the C—C bond, would cause $\mathbf{j}$ to be predominantly parallel to $\mathbf{v}$, as is found experimentally. Thus, for this process in which glyoxal falls apart into three fragments in a concerted manner, we are still able to learn about how this dissociation occurs in detail.

It is anticipated that future experiments in photodissociation dynamics as well as inelastic and reactive collisions will involve *coincidence measurements*[17] in which vector quantities of more than one fragment are simultaneously measured. Such measurement would provide a wealth of detailed information on the nature of the transition state region. Many such experiments were unrealized because of the poor signal-to-noise considerations. With the advent of laser light sources with high repetition rates, however, coincidence measurements are becoming quite feasible.

Problems

A. Transformation of coordinates from reactants to products for an atom–diatom, $A + BC \rightarrow AB + C$ reaction. (a) By drawing the ABC triangle and locating the position of the center of mass of AB and BC show that the transformation is

$$\begin{pmatrix} \mathbf{R}' \\ \mathbf{r}' \end{pmatrix} = \begin{pmatrix} m_A/m_{AB} & (m_A m_C/m_{BC} m_{AB}) - 1 \\ 1 & m_C/m_{BC} \end{pmatrix} \begin{pmatrix} \mathbf{R} \\ \mathbf{r} \end{pmatrix}$$

Here $\mathbf{R}$ is the position vector from A to BC while $\mathbf{r}$ is the internal coordinate for BC. The primes denote products. $m_{AB} = m_A + m_B$ [F. T. Smith, *J. Chem. Phys.* **51**, 1352, (1959)]. (b) Hence derive the transformation Eq. (10.4) between the mass-skewed coordinates $\mathbf{Q}$. (c) Using either transformation verify that it conserves the kinetic energy. (d) Using either transformation derive Eq. (10.3) for the transformation of the angular momenta in the collision [I. R. Elsun and R. G. Gordon, *J. Chem. Phys.* **76**, 3009 (1982)]. (e) How is the initial kinetic energy of the approach motion of the reactants partitioned in the products? (f) If you like handling matrices show that the transformation is just a rotation of the coordinates by our ubiquitous angle β. From a mathematical point of view all the conservation laws follow from the transformation being an orthogonal one.

B. Transformation of coordinates from reactants to products for a diatom–diatom, $AB + CD \rightarrow AC + BD$ reaction. (a) As in Problem A but for the four-atom case:

$$
\begin{pmatrix} \mathbf{r}_{AC} \\ \mathbf{r}_{BD} \\ \mathbf{R}' \end{pmatrix} = \begin{pmatrix} \dfrac{m_B}{m_{AB}} & \dfrac{m_D}{m_{CD}} & 1 \\[2mm] \dfrac{m_A}{m_{AB}} & \dfrac{m_C}{m_{CD}} & -1 \\[2mm] \dfrac{m_A m_B M}{m_{AB} m_{AC} m_{BD}} & -\dfrac{m_C m_D M}{m_{CD} m_{AC} m_{BD}} & \dfrac{m_A m_D - m_B m_C}{m_{AC} m_{BD}} \end{pmatrix} \begin{pmatrix} \mathbf{r}_{AB} \\ \mathbf{r}_{CD} \\ \mathbf{R} \end{pmatrix}
$$

(b) Show that in the kinematic limit, for an $A_2 + B_2 \rightarrow AB + AB$ reaction, the kinetic energy of the approach motion is not converted into a kinetic energy for the separation of products. Only vibrational energy of the reactants can help the products to separate from one another. Does that ring a bell? (c) Where does the initial kinetic energy of the relative motion of the reactants go to? (d) Derive the transformation for the $A + BCD \rightarrow AB + CD$ reaction [check against T. Raz and R. D. Levine, *Chem. Phys. Lett.* **246**, 405 (1995)].

C. It is suggested that the cross-section for the $H + LiCl \rightarrow HCl + Li$ reaction will increase, roughly linearly, with the initial rotational energy of LiCl. What is the physical reason?

D. With laser detection of products one can examine the polarization of products in specified internal levels. Say we examine the $H + LiCl$ reaction where almost all the available energy is in the internal excitation of HCl and little energy is available for the relative translation. Will the angular momentum of HCl be preferentially polarized, and if so, how?

E. The $K + HBr \rightarrow KBr + H$ reaction is exoergic by about $18 \, \text{kJ mol}^{-1}$. In an early experiment, at a collision velocity of $1500 \, \text{m s}^{-1}$, the mean rotational energy of the KBr product was measured to be about $5 \, \text{kJ mol}^{-1}$. (a) Why did D. A. Case and D. R. Herschbach, *Mol. Phys.* **30**, 1537 (1976) expect that the angular momentum vector of KBr would be aligned perpendicular to the initial velocity vector? (b) Estimate the reaction cross-section. The rotational constant of KBr is $0.08 \, \text{cm}^{-1}$.

F. Brute force orientation requires cold molecules. Consider a dilute gas of diatomic molecules with a permanent dipole moment μ in an electrical field E. The interaction energy of an individual molecule with the field is $-\mu E \cos \theta$ where θ is the angle between the dipole and the field. In the ground state the gas is fully oriented but the thermal motion allows the occupation of higher-energy states where the dipole does not quite point out in the direction of the field. (a) Using the dimensionless variables $x \equiv \cos \theta$, $a \equiv \mu E / k_B T$ show that

$$
\langle \cos \theta \rangle \equiv \int_{-1}^{1} x \exp(ax) \, dx \bigg/ \int_{-1}^{1} \exp(ax) \, dx = \coth(a) - 1/a \rightarrow \begin{cases} a/3, & a \ll 1 \\ 1, & a \gg 1 \end{cases}
$$

(b) Take an extreme case, a dipole due to a charge of an electron separated by an atomic unit. What is the electrical field $(V\ cm^{-1})$ required to achieve a significant orientation at room temperature? At 10 K?

F. The DIPR model. The model improves on the simple kinematic limit for direct reactions by adding repulsion between the products. See P. Kuntz, *Trans. Faraday Soc.* **66**, 2980 (1970) for a detailed analytical treatment and Truhlar and Muckerman (1979) for a review. Here we just consider the instantaneous product repulsion. (a) Show that you can add an impulse along both the products' separation coordinates by modifying the equation in Problem A to

$$\begin{pmatrix} \dot{\mathbf{R}}' \\ \dot{\mathbf{r}}' \end{pmatrix} = \begin{pmatrix} m_A/m_{AB} & (m_A m_C/m_{BC} m_{AB}) - 1 \\ 1 & m_C/m_{BC} \end{pmatrix} \begin{pmatrix} \dot{\mathbf{R}} \\ \dot{\mathbf{r}} \end{pmatrix} + \begin{pmatrix} \mathbf{q}_{R'} \\ \mathbf{q}_{r'} \end{pmatrix}$$

where the dots indicate a time derivative. (b) Derive Eq. (10.6) and the vibrational energy of the products. (c) How is the angular momentum disposal, Eq. (10.4), modified? (d) Discuss the $\mathbf{v} - \mathbf{j}$ and $\mathbf{v} - \mathbf{j}'$ vector correlations, Section 10.2, in the kinematic limit.

Notes

1 The cone of acceptance plays a key role in describing steric effects in collisions, and many further extensions and modifications have appeared in the literature, see for example Levine (1990), Orr-Ewing (1996), Alexander *et al.* (1998), Simons (1999). For a general overview of earlier dynamical stereochemistry, see Bernstein (1988).

2 Using this approach, Brooks and Marcelin successfully carried out such an experiment, Figure 10.1, with many subsequent studies exploring and exploiting this effect (Bernstein, 1982; Stolte, 1982, 1988; Harren *et al.*, 1991), including for scattering of oriented molecules off surfaces, Kuipers *et al.* (1988).

3 On pendular states, see B. Friedrich and D. R. Herschbach, Nature **353**, 412 (1991); A. Durand, J. C. Loison, and J. Vigué, *J. Chem. Phys.* **102**, 7046 (1994). See Problem E.

4 For brute force experiments using cold molecules, see H. J. Loesch and A. Remscheid, *J. Chem. Phys.* **93**, 4779 (1990); and J. J. van Leuken *et al.*, *J. Phys. Chem.* **99**, 4360 (1995).

5 H. J. Loesch and J. Möller, *J. Chem. Phys.* **97**, 9016 (1992); H. J. Loesch and J. Möller, *J. Phys. Chem.* **97**, 2158 (1993).

6 Early beam-gas experiments seemed to find no unambiguous effect, which may be a consequence of thermal averaging washing the degree of alignment that may be prepared in the collision frame [J. A. Kettleborough and K. G. McKendrick, *J. Phys. Chem.* **95**, 8255 (1991)]. Under crossed-beam conditions, however, Loesch and co-workers observed small but unmistakable alignment effects in the reactions of $Sr + HF(v = 1, j)$ [H. J. Loesch and F. Stienkemeier, *J. Chem. Phys.* **100**, 740, 4308 (1994)] and $K + HF(v = 1, j)$ (Loesch, 1995). Optical preparation of aligned reagents can also be achieved through the interaction of the molecule's polarization anisotropy with intense electric fields of pulsed lasers, both on resonance and off resonance. For example, using stimulated Raman pumping, Zacharias and co-workers [R. Dopheide and H. Zacharias, *J. Chem. Phys.* **99**, 4864 (1993); J. B. Halpern, R. Dopheide, and H. Zacharias, *J. Phys. Chem.* **99**, 13611 (1995)] studied the inelastic collisions of, aligned acetylene prepared by stimulated Raman pumping, and Sitz

and Farrow [G. O. Sitz and R. L. Farrow, *J. Chem. Phys.* **101**, 4682 (1994)] prepared aligned N_2 molecules for surface scattering. Optical preparation of aligned reagents may also be achieved by photofragmentation, either by photodissociation into neutral fragments or by photoionization into charged fragments. For example, J. N. Greeley, J. S. Martin, J. R. Morris, and D. C. Jacobs, *J. Chem. Phys.* **102**, 4996 (1995) studied collisions of fast NO^+ with the (111) face of single-crystal silver, in which the NO^+ is prepared by resonance-enhanced multiphoton ionization. The polarization of the radiation is used to select whether the NO^+ ion preferentially collides with its plane of rotation face on or edge on with respect to the surface.

7 For the beginning of orbital control, see C. T. Rettner and R. N. Zare, *J. Chem. Phys.* **77**, 2416 (1982). Since then, many related studies have been carried out, almost exclusively with orbitally aligned atoms of the alkaline earths and mercury. See Simons (1987) for a review. A. G. Suits *et al.*, *J. Chem. Phys.* **95**, 8178 (1991) and A. G. Suits *et al.*, *J. Chem. Phys.* **96**, 2777 (1992) have prepared orbitally aligned Ba (^{1}P) atoms that collided, in a crossed-beam configuration, with O_3, NO_2, and Br_2. In the reaction with NO_2, the formation of Ba^+ was found to vary by more than a factor of two as the alignment of the p orbital was altered.

8 On Λ doublets, see Zare (1988, pp. 308–310).

9 Ch. Ottinger, R. Velasco, and R. N. Zare, *J. Chem. Phys.* **52**, 1636 (1970) showed that collisions with A' and A'' Λ-doublet levels of a $^1\Pi$ state behave quite differently. T. L. D. Collins *et al.*, *J. Chem. Phys.* **102**, 4419 (1995) have studied in detail the inelastic collisions of Li_2 $A^1\Pi$ with Xe.

10 A fuller treatment of hyperfine depolarization may be found in Zare (1988), pp. 239–241.

11 I. R. Elsum and R. G. Gordon, *J. Chem. Phys.* **76**, 3009 (1982).

12 A detailed account of the DIPR model P. Kuntz, *Trans. Faraday Soc.* **66**, 2980 (1970).

13 On rotational energy transfer, see Parmenter *et al.* (1997), McCaffery and Marsh (2001).

14 This behavior is the same type of rainbow as for structureless particles [see W. Schepper, U. Ross, and D. Beck, *Z. Phys. A* **290**, 131 (1979), H. J. Korsch and R. Schinke, *J. Chem. Phys.* **75**, 3850 (1981) and Murrell and Bosanac (1989)]. Rainbows are also observed in the scattering of diatomic molecules from flat surfaces [A. W. Kleyn, A. C. Luntz, and D. J. Auerbach, *Phys. Rev. Lett.* **47**, 1169 (1981)].

15 A. J. Alexander and R. N. Zare, *Acc. Chem. Res.* **33**, 199 (2000) review this type of behavior, which is found to occur in many dissociation processes.

16 L. M. Dobeck *et al.*, *J. Phys. Chem.* **103**, 10312 (1999); I. Burak *et al.*, *J. Chem. Phys.* **86**, 1258 (1987).

17 Coincidence experiments, see Continetti (2001), Hayden (2002), are likely to become more common.

Chapter 11
Dynamics in the condensed phase

The presence of a solvent interacting with a system throughout its evolution from reactants to products brings about qualitative changes from the corresponding gas-phase reaction. There are changes in both the reaction rate and the dynamics. The energetic effects due to the solvent reflect the electronic reorganization that takes place as the system transverses the reaction path.* The S_N2 ion–molecule reaction, shown in Figure 11.1 for the generic $X^- + CH_3X$ exchange reaction, provides an example in which the charge delocalization in the transition-state region causes qualitative changes in the energy profile along the reaction coordinate when the reaction is in the presence of a polar solvent. As becomes clear in this chapter, even the reaction coordinate itself is not exactly the same in solution as it is in the gas phase, because the solvent adaptation to the changing system must also be considered.

A solute molecule at room temperature undergoes of the order of 10^{13} collisions per second with solvent molecules. The solvent can therefore hinder the large-amplitude motions that often accompany chemical transformations (e.g., as in a twist isomerization, Figure 11.14). The cage effect, where the solvent hinders the separation of the products, Figure 1.8, or the approach of the reactants, was one of the first examples of the role of the solvent. The cage effect remains a major difference between gas-phase and solution dynamics. There can also be dynamical effects. They occur first and foremost because, not as in the gas phase, the solvent must provide the energy for a reaction to take place and must drain the energy from the nascent products. These couplings can be strong enough to make the solvent motion part of the definition of the reaction coordinate.**

---

 * Organic chemists call this concept the Hughes–Ingold proposition: polar solvents assist reactions that generate charge or increase charge separation. The opposite is true for reactions in which charge is annihilated or charge separation is reduced.

** We sometimes describe this by saying that the solvent motions cause the barrier to fluctuate. When we discuss charge transfer in Section 11.1.2, the solvent motion is the reaction coordinate.

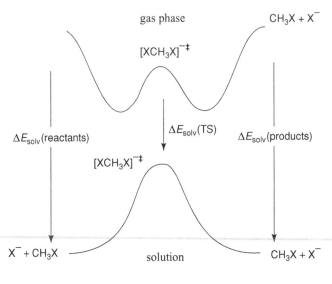

gas phase

$CH_3X + X^-$

$[XCH_3X]^{-\ddagger}$

$\Delta E_{solv}(\text{reactants})$

$\Delta E_{solv}(\text{TS})$

$\Delta E_{solv}(\text{products})$

$[XCH_3X]^{-\ddagger}$

$X^- + CH_3X$

solution

$CH_3X + X^-$

reaction coordinate $\longrightarrow$

Figure 11.1 The energy profile along the reaction path for a symmetric S_N2 reaction in the gas phase is contrasted with that in a solvent having a high dielectric constant such as water. The qualitative argument for the change seen in the figure is clear: in the gas phase a fairly deep well occurs in the reactants' or products' valley arising from the strong ion–molecule attractive force. There is a barrier to atom exchange, owing to the planarity of the CH_3 group in the transition state, see Figure 5.21, but owing to the strong forces in the entrance valley this barrier is lowered with respect to the energy of the reactants. In solution the localized charge on the reactant or product ion leads to considerable stabilization. In the transition state the charge is delocalized and so its solvation energy is not as high as that of the more compact ionic reactant or product. In Section 11.1.1 we discuss a simple model that quantitatively relates the stabilization energy to the size of the solvated ion. The rate of such reactions can be studied by isotopic labeling,[1] $^{35}Cl^- + CH_3{}^{37}Cl \rightarrow {}^{37}Cl^- + CH_3{}^{35}Cl$. For asymmetric S_N2 reactions, say $Br^- + CH_3Cl$, the picture is more complicated because the reactant and product ions are not identical. When the CH_3 group is substituted, the transition state is bulkier and the model of Section 11.1.1 suggests that it will be even less stabilized by solvation. The barrier to reaction in solution will therefore be higher, quite apart from any steric requirements due to the bulkier central group. For substitution reactions at metal ions, where the charge is localized on the metal, our considerations suggest that the energy profile will look rather the same as in the gas phase and in solution.

11.0.1 Many facets of the solvent

Why do we need different ways to look at a solvent? Can we present a unified picture? Because a solvent plays an interesting role when not weakly coupled to the reactants, it is common to regard reactants and their solvation shells together as one system. Trajectory simulations can easily keep track of the motion of all atoms, including those of the solvent, during the reaction, see Figure 5.23. So if we consider solvent and solute together, dynamics in the condensed phase does not differ from any other many-atom system.[2] But we seek to understand and interpret this dynamics in simple terms. In particular, to discuss the effect of, say, changing the solvent, we need to think in terms of reactants and separately of their immediate surroundings. This will not always be possible and we must be prepared to think differently about different aspects of the dynamics. The primary consideration is that of the time scale of the process under discussion. On very short time scales the solvent is effectively frozen; that is, it does not move appreciably. It takes time for a solvent to respond. For example, consider an atomic solvent in thermal equilibrium and then kick one atom. It takes time for the immediate neighbors to even recognize that the equilibrium has been disturbed. After the first collision between the disturbed atom and its neighbors, the excess energy needs to be spread further. How the energy spreads depends on the range and strength of the forces; is it a weakly bound liquid, an associated liquid, a glass, or a solid? Similarly it can take the reactants some time to probe what kind of solvated environment is present around them. On our own human time scale we have no trouble telling if a macroscopic system is rigid as a solid or floppy as a fluid, but for reactants descending from a barrier only a very limited time is available.*

We can therefore envisage that barrier-crossing dynamics spans a wide range of behaviors – from a rather diffusive motion along the reaction path[3] to a fully "ballistic" motion where the solvent cannot respond fast enough and is effectively frozen while the system transverses the barrier. Of course, the manner in which reactants acquire the energy necessary to overcome the barrier differs for different regimes. In a diffusive regime, the reactive system gains and loses small amounts of energy, undergoing a kind of random walk in its energy content until it manages to scale the barrier. To imagine the opposite limit, consider the products rapidly descending from the barrier, separating from one another at an increasing pace, repelled by their potential. During this fast descent the solvent has hardly the time to move. At the foothills of the barrier the products must run into the nearly stationary solvent molecules. The hot nascent products will therefore be rapidly decelerated there. Now consider the same dynamics but reversed in time. To scale the barrier the reactants acquire the needed energy in one or a few impulses delivered by the solvent at the foothills of the barrier.[4]

The time scale is a factor in another way. A polyatomic solvent has a range of frequencies with which it can respond because the solvent's intramolecular vibrations can be coupled, even if indirectly, to the solute. The translations and

---

* For special materials or if we can manage a fast probe, we can have trouble deciding between a solid and a liquid. Toy stores sell plastic balls that are as malleable as plasticine and can be shaped by a gentle slow pressure but that rebound as a new tennis ball when thrown hard at the floor. The same situation is familiar to a diver who jumps from a board into a pool: too high a jump and the water responds rather rigidly. The limit of a fast perturbation corresponds to the sudden regime of Chapter 9.

rotations are certainly coupled, because they are almost always the first way the solvent adjusts to changes in the solute. So the very same solvent can exhibit different dynamics for different solutes or even for different excitations of the same solute.

11.1 Solvation

We are considering both the equilibrium and the dynamics of solvation. Equilibrium here means constrained equilibrium as we hold the reactants at a given point along the reaction coordinate and allow all other degrees of freedom, both of the reactants and of the solvent, to adjust thermally. Thereby we can map a free energy profile along the reaction coordinate and compare it to the profile in the gas phase, as in Figure 11.1. To apply transition state theory to compare gas and condensed phase reactivity we need the energy difference between the solvated and the isolated transition states and the corresponding difference between the solvated and isolated reactants. It is the difference of these two differences that determines to what extent solvation causes the activation energy and hence the rate constant to differ from the gas-phase value.

We begin below with the energetic cost of solvation. We should not forget, however, the entropic term. The same thermodynamic cycle that we discuss for the energetics also applies to entropies. For example, entropic considerations tell us that unimolecular reactions have comparable reaction rates in the gas phase and in solution, assuming that other things are equal and no large-amplitude motions occur in the transition state. This conclusion is valid because a comparable entropy change (equal to the entropy of evaporation) occurs when a mole of reactant or a mole of the transition state is transferred from solution to the gas phase.* For bimolecular reactions, however, a real difference exists. Entropy favors such reactions in solution. The reason for this behavior is the loss of translational freedom of the gas-phase reactants when they form a transition state, as discussed in Section 6.1.4.1. In solution, the freedom of motion of the reactants is already much reduced compared with the separated reactants in the gas phase. So the entropy barrier to the formation of the transition state is lowered compared with that in the gas phase.** Similar considerations arise in identifying the factors contributing to *catalysis* and catalysis by enzymes in particular. A catalyst presumably lowers the free energy of activation of the rate-determining step. But

---

* Exceptions are possible. For example, photoisomerization around a double bond, Figure 8.6, involves a large-amplitude twist so that the assumption above fails.

** Other things being equal, when we compare the evaporation of two moles of reactants and one mole of transition states, the reaction rate in solution should be faster by $\exp(\Delta S_{vap}/R)$, which is of the order of 10^2. Another way to think about the difference is to compare the concentration of a reactant in the gas and the condensed phase. This equilibrium ratio equals $\exp(-\Delta G_{vap}/RT) = \exp(-\Delta H_{vap}/RT)\exp(\Delta S_{vap}/R)$ and is often larger than 10 at room temperature. For more on this point see Section 11.1.3.1 on the cage effect.

the relative importance of energetic and entropic factors is still under very active discussion.[5]

11.1.1 Electrostatic models for solvation

In the simplest model of solvation, the solvent is treated as a structureless and continuous medium of dielectric constant* ε. In 1920, Born developed the earliest polarizable continuum model. He treated the ion as a point charge q located in the center of a hollow sphere with radius R. The hollow charged sphere is embedded in a classical dielectric continuum having a relative dielectric constant ε_r. The electrostatic contribution to the free energy, evaluated in Section 11.1.1.1, is given by

$$\Delta G_{\text{solvation}} = -\frac{q^2}{4\pi \varepsilon_0 R}\left(\frac{\varepsilon_r - 1}{2\varepsilon_r}\right) \qquad \text{polarizable continuum model} \qquad (11.1)$$

This expression predicts that the free energy of solvation in any given solvent will increase directly proportionally to the square of the charge of the ion and

---

* The dielectric constant or permittivity describes the attenuation of the coulomb force $\mathbf{F}$ between two interacting charges q_1 and q_2 separated by a distance R. The magnitude of the force is given by

$$F = \frac{q_1 q_2}{4\pi \varepsilon R^2}$$

We are using SI units where the force is measured in $N = \text{kg m s}^{-2}$ and the charge is in coulombs. When the medium is a vacuum, it is conventional to use ε_0 for the permittivity. In SI units, $1/(4\pi \varepsilon_0) = 9.10^9$. The ratio of the force in a vacuum to the force in any other medium between the same pair of point charges separated by the same distance is called the relative permittivity ε_r of that medium. It follows that $\varepsilon_r = \varepsilon/\varepsilon_0$. Of course the relative permittivity (relative dielectric constant) is a dimensionless quantity. We can rewrite the magnitude of the force as

$$F = \frac{q_1 q_2}{4\pi \varepsilon_r \varepsilon_0 R^2}$$

in terms of the relative permittivity. This force can be considered to arise from the negative of the gradient of the electrical potential ϕ. At a distance R from a point charge q the electrical potential is

$$\phi = \frac{q}{4\pi \varepsilon_r \varepsilon_0 R}$$

The more polar the solvent, the higher is the dielectric constant and the smaller is the force that one charge exerts on the other in solution. The following table illustrates this trend:

Material	Relative dielectric constant, ε_r
Vaccum	1.0
Carbon tetrachloride	2.2
Acetone	20.7
Methanol	32.6
Water	78.5

decrease inversely proportionally to the ion's size.* The success of the Born model depends on the choice of the ionic radius R. With carefully chosen cavity radii, see Section 11.1.1.1, one can accurately reproduce the vacuum-to-water hydration-free energies of both cations and anions.

Why are ions so pervasive in solution? An analysis of NaCl provides a simple answer. The very stable NaCl molecule has a bond energy of $4.26\,eV = 411\,kJ\,mol^{-1}$ for dissociation to neutral Na and Cl atoms. To compute the dissociation energy needed to form Na^+ and Cl^- we must add the energy required for removing an electron from the Na atom and subtract the energy gained by giving the electron to the Cl atom. This electron transfer process costs us the ionization potential of the Na atom ($5.139\,eV$) minus the electron affinity of the Cl atom ($3.6\,eV$), for a net cost of $1.5\,eV = 145\,kJ\,mol^{-1}$. The energy required to separate NaCl into Na^+ and Cl^- in the gas phase is the sum of the dissociation to neutrals and electron transfer energies, that is, $556\,kJ\,mol^{-1}$. We conclude that the dissociation of NaCl into ions is a highly endoergic process in the gas phase. Suppose however that we let the Na^+ and Cl^- ions become hydrated. We gain the very considerable heats of hydration of the two ions, where, from the table ($99.3 + 86.1$) $= 185.4\,kcal\,mol^{-1} = 775.7\,kJ\,mol^{-1}$. This energy more than offsets the energy ($556\,kJ\,mol^{-1}$) needed to cleave the NaCl bond and make the Na atom a cation and the Cl atom an anion and the solvation energy of neutral NaCl. It is no wonder then that so many molecules ionize in water (Ando and Hynes, 1999).

We can improve upon the Born model for solvation by recognizing the discrete nature of the solvent and the specific interactions with the solute, but we have the added costs of more computation and more complexity. When the solvent is water, it is especially important to take into account the large dipole moment of the water molecule as well as the propensity of water molecules to form hydrogen-bonded networks. Consequently, the insertion of an ion into water causes water molecules in the immediate vicinity of the ion to orient in such a way that the part of the water molecule having the opposite charge to that of the ion faces the ion. This reorientation of the water molecules surrounding the ion costs entropy and energy. It leads to a structuring of the water molecules close to the ion, and this structuring destroys our simple model of a dielectric continuum. At the water–vapor interface, Section 11.3, the destruction of the hydrogen-bonded network at the interface is equally important.

An advanced treatment (Honig and Nicholls, 1995) shows that the water molecules are in different sheaths about the ion, a first hydration shell of rotationally hindered water molecules, an intermediate layer that is a transition between

---

* Note that our considerations refer to a neat solvent, which means in practice a solution of negligible ionic strength μ. If μ is not negligible, the field ϕ is reduced by a term proportional to $\sqrt{\mu}$ arising from the shielding by the other ions present. The effect of this term is familiar to us from the dependence of the activation energy for reactions between ions on the ionic strength (see Problem A).

it and the bulk, and finally a bulk sheath. The existence of rotationally hindered water molecules explains why some adjustment needs to be made in the radius of the hollow sphere for the ion to fit experiment better. Ultimately, we must take into account the molecular sizes and charge distributions of the solute and solvent. Various computer codes have been written to simulate ion solvation energies and solvation dynamics.

*11.1.1.1 The Born solvation model

It is instructive to review Born's derivation of the expression for $\Delta G_{\text{solvation}}$. Not surprisingly, he considered a four-step cycle as shown in Figure 11.2. Consider an ion in a vacuum, without solvent. First, the ion is discharged, so that it is electrically neutral. The work done to discharge the ion is denoted $W_{\text{discharging}}$. Second, the neutral species is introduced into the solvent. The work required to make this transfer is denoted W_{transfer}. Third, the neutral solute sphere is charged while it is in the solvent. The work done to charge it is denoted W_{charging}. Last, the ion is brought from the solvent to the vacuum. The work $W_{\text{solvation}}$ in going from the solvated ion to the bare ion is equal to the negative of the free energy change $\Delta G_{\text{solvation}}$. It is this quantity that we wish to calculate. Because these four steps constitute a closed cycle, we write:

$$W_{\text{discharging}} + W_{\text{transfer}} + W_{\text{charging}} - \Delta G_{\text{solvation}} = 0 \qquad (11.2)$$

Neglecting W_{transfer} because it is small in comparison with the electrostatic terms $W_{\text{discharging}}$ and W_{charging}, we have

$$\Delta G_{\text{solvation}} = W_{\text{discharging}} + W_{\text{charging}} \qquad (11.3)$$

The energy of charging (in solution) and discharging (which is just the reverse process of charging, but is here in the gas phase) requires us to evaluate the following integrals, where the relative dielectric constant ε_r is that of the solvent for charging and $\varepsilon_r = 1$, that of the vacuum, for discharging:

$$W_{\text{charging}} = \int_0^q \varphi \, dq' = \int_0^q \frac{q'}{4\pi \varepsilon_r \varepsilon_0 R} dq' = \frac{q^2}{8\pi \varepsilon_r \varepsilon_0 R}$$

$$W_{\text{discharging}} = \int_q^0 \varphi \, dq' = \int_q^0 \frac{q'}{4\pi \varepsilon_0 R} dq' = -\frac{q^2}{8\pi \varepsilon_0 R} \qquad (11.4)$$

Hence,

$$\Delta G_{\text{solvation}} = -\frac{q^2}{8\pi \varepsilon_0 R}\left(1 - \frac{1}{\varepsilon_r}\right) = -\frac{q^2}{4\pi \varepsilon_0 R}\left(\frac{\varepsilon_r - 1}{2\varepsilon_r}\right) \qquad (11.5)$$

which agrees with Eq. (11.1).

This expression is for a single ion. To find the expression for a mole of ions, multiply by Avogadro's number. If radius R is measured in units of nanometers

Figure 11.2 Born cycle
for ion solvation in a
dielectric medium.

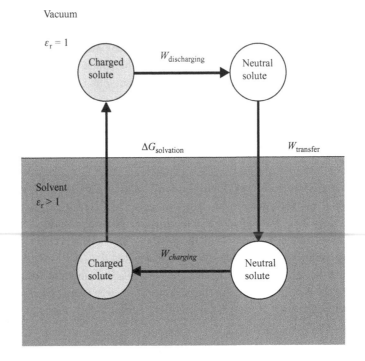

and charge q is measured in units of electron charge, then we can express the free energy change $(kJ\,mol^{-1})$ as

$$\Delta G_{\text{solvation}} = -69.0\frac{q^2}{R}\left(\frac{\varepsilon_{\text{r}} - 1}{\varepsilon_{\text{r}}}\right) \tag{11.6}$$

For a sodium ion Na^+ solvated in water, we take $R = 0.168$ nm and $q = 1$:

$$\Delta G_{\text{solvation}} = -\left(\frac{69.0}{0.168}\right)\left(\frac{77.5}{78.5}\right) = -405\,kJ\,mol^{-1} = -96.8\,kcal\,mol^{-1} \tag{11.7}$$

This simple calculation forcefully demonstrates how large free energies of solvation are and why solvation should be expected to play such an important role in reactions of ions in solution.

How well does the Born model for solvation compare with experiment? We are not able to measure $\Delta G_{\text{solvation}}$ directly, but we can use the relation $\Delta G_{\text{solvation}} = \Delta H_{\text{solvation}} - T\Delta S_{\text{solvation}}$ to make a judgment about the validity of this model. The term $\Delta H_{\text{solvation}}$ is the heat of the reaction, that is, the heat of transferring an ion from the vapor to the solvent (where we assume that the vapor is essentially equivalent to a vacuum). These data are available. In the expression for $\Delta G_{\text{solvation}}$ the only term that varies with temperature is the dielectric constant. Hence we can write

$$-\Delta S_{\text{solvation}} = \frac{\partial \Delta G}{\partial T} = N_{\text{a}}\frac{q^2}{8\pi\varepsilon_{\text{o}}R}\frac{1}{\varepsilon_{\text{r}}^2}\frac{\partial\varepsilon_{\text{r}}}{\partial T} \tag{11.8}$$

We solve for the heat of reaction and write

$$\Delta H_{\text{solvation}} = -N_a \frac{q^2}{8\pi\varepsilon_0 R} \left(1 - \frac{1}{\varepsilon_r} - \frac{T}{\varepsilon_r} \frac{\partial \varepsilon_r}{\partial T}\right) \qquad (11.9)$$

Thus, if we know the radius of the ion, its charge, the change of its dielectric constant with temperature, and the temperature, we can calculate $\Delta H_{\text{solvation}}$. The following table compares the solvation enthalpies calculated from the Born model with those enthalpies derived from experiment. The entry "corrected radius" is the radius of the cavity surrounding the ion, adjusted so as to reproduce solution thermochemistry:*

Ion	Corrected radius (nm)	$\Delta H_{\text{solvation}}$ (Born model) (kcal mol^{-1})	$\Delta H_{\text{solvation}}$ (Experiment) (kcal mol^{-1})
Li$^+$	0.1316	−126.7	−125.8
Na$^+$	0.1680	−99.3	−98.5
K$^+$	0.2172	−76.8	−78.4
F$^-$	0.1423	−117.2	−119.3
Cl$^-$	0.1937	−86.1	−85.3
Br$^-$	0.2087	−79.9	−78.8

11.1.2 Electron transfer reactions

Oxidation and reduction reactions abound in chemistry. Photosynthesis in plants, energy transduction in our body, corrosion of metals, and the operation of solid-state devices are just some of the processes that require electron transfer.[6] We have already discussed the harpoon mechanism as a long-range electron transfer in the gas phase. In Section 11.3 we discuss electrochemistry and other examples of electron transfer at an interface. Here we want to discuss electron transfer in solution and show how solvation plays a key role. But note that when a molecule gains or loses a charge, its own structure also changes. Consequently, it is not only the solvent that needs to reorganize when charge transfer takes place.

Why do we speak of solvent reorganization? The issue comes into focus when, using isotopes, we study thermoneutral reactions in solution. For example, consider electron transfer between two different aquated iron complexes

$$\text{Fe(H}_2\text{O)}_6^{+2} + \text{*Fe(H}_2\text{O)}_6^{+3} \rightarrow \text{Fe(H}_2\text{O)}_6^{+3} + \text{*Fe(H}_2\text{O)}_6^{+2}$$

in which *Fe represents an isotope of iron. This type of reaction was observed to occur at a finite rate and to require an activation energy.[7] Why cannot the electron easily hop from one center to the other? As written we can think that

---

* Only a semi-quantitative agreement would have been obtained if the ionic radius from the crystal data had been used for the cavity radius.

the reaction is a resonance transfer with a zero energy gap. The point is that the transfer is not really a resonance process, because the traditional way of writing the chemical change, as above, is misleading! Let us assume that before the transfer, the reactants are indeed in thermal equilibrium. Then immediately after the transfer, the products are not in equilibrium. The reason is that if the reactants are in equilibrium, the nascent products are an Fe^{+3} ion solvated as if it is still an Fe^{+2} ion and an $*Fe^{+2}$ ion as if it is still an $*Fe^{+3}$ ion. The transfer is energetically expensive because the solvent cannot adjust as quickly as an electron can jump. The solvent needs to adjust because the radius of the Fe^{+2} ion is larger than that of the Fe^{+3} ion so the water molecules that surrounded the Fe^{+2} ion reactant need to move closer in. For the Fe^{+3} ion reactant the water molecules need to move out. As part of this reorganization the dipoles of the water molecules need to rotate and optimize. In Section 11.1.3 we discuss experiments that probe the relaxation of the solvent motion to reach new solvation equilibrium. Here we characterize the sluggishness of the solvent response in the frequency domain. We distinguish between the dielectric constant under static conditions and the dielectric constant when the electric field varies so rapidly that the nuclei of the solvent cannot follow. We use the Born model, Section 11.1.1.1, and note that the electrons of the solvent do respond as quickly as it takes the valence electron to hop. So the electronic response of the solvent is fast. Hence the cost in solvation energy caused by the solvent failing to remain in equilibrium, known as the *reorganization energy*, is*

$$\lambda_0 = \frac{e^2}{4\pi\varepsilon_o}\left(\frac{1}{2R_A} + \frac{1}{2R_B} - \frac{1}{R_{AB}}\right)\left(\frac{1}{\varepsilon_{r\infty}} - \frac{1}{\varepsilon_{r0}}\right) \tag{11.10}$$

Here R_A and R_B are the cavity radii for solvation of the isolated donor and acceptor ions, where $A + B^+ \rightarrow A^+ + B$; R_{AB} is the A–B distance at contact; and ε_{r0} is the familiar (relative) dielectric constant ε_r, except that we add the subscript 0 to emphasize that we mean the static limit. The limiting value of the dielectric constant at high frequencies is $\varepsilon_{r\infty}$. This is the dielectric constant of the solvent when only its electrons can respond. Because $\varepsilon_{r\infty}$ is but a small fraction of ε_r, the reorganization energy is of the order of the energy of solvation of a charge and as such is high.

---

* Marcus (1964). This cost is the difference in solvation energy change throughout the process of charge transfer between the solvent in equilibrium and the solvent not in equilibrium. Assume first that the solvent can fully adjust. Then the process is to discharge one electron from reactant A, charge product B, then bring reactant and product to a distance R_{AB} from each other. Because the solvent is allowed to adjust throughout, the dielectric constant is at a low ($\rightarrow 0$) frequency of the electric field. As the frequency of oscillation of the electrical field increases, the dielectric constant decreases whenever a molecular motion is no longer able to follow the changing field. For electron transfer reactions only the electrons of the solvent can adjust to the rapid change in the field. We therefore need to use the limiting value of the dielectric constant at visible or higher frequencies, $\varepsilon_{r\infty}$. Otherwise the computation of solvation energy change is the same. The difference between equilibrium and non-equilibrium is the reorganization energy, Eq. (11.10), the additional energetic cost arising from non-equilibrium solvation.

So there is a high cost in energy, identified as the reorganization energy, λ_0, to transfer the charge if the reactants are in equilibrium with the solvent and the products are fully out of equilibrium, Figure 11.3. Marcus has shown that the reaction can proceed by means of a lower barrier.* This path requires a fluctuation that takes the reactants away from equilibrium and part of the way toward the equilibrium solvation configuration of the products. One still forms products not in equilibrium, but the extent of solvation disequilibrium in the nascent products is reduced.

The computation of the minimal barrier height is instructive. We use Figure 11.3 as an aid. For the purpose of our discussion the interaction with the solvent is described by a (single) displacement coordinate r for small deviations of the solvent from its equilibrium configuration.[8] The reorganization energy is defined in terms of the solvation energy cost for charge transfer when the reactants are at their equilibrium with the solvent, so that $r = R_A$. The equilibrium value for the solvation coordinate of the products is R_B. Assuming a harmonic approximation** for small displacements of the solvent about the ion (see Figure 11.3), the solvation energy of B^+ is larger than its equilibrium value by

$$\lambda_0 = \tfrac{1}{2}k(R_B - R_A)^2 \tag{11.11}$$

The reorganization energy λ_0 is the basic parameter that enters the theory. The aim of the development is to compare rates of reaction of given λ_0 and different values of the thermodynamic driving force ΔG^0. To do so we note, using Figure 11.3, that the lowest bottleneck for charge transfer occurs at that value of the solvent displacement $r^\ddagger$ given as the point of intersection of the two parabolas

$$\tfrac{1}{2}k(r^\ddagger - R_A)^2 = \Delta G^0 + \tfrac{1}{2}k(r^\ddagger - R_B)^2 \tag{11.12}$$

where ΔG^0 is the free energy difference between the products and reactants (at equilibrium) and equals zero for the self-exchange shown in Figure 11.3. Equation (11.12) determines the height of this minimal barrier to be

$$\Delta G^\ddagger = \tfrac{1}{2}k(r^\ddagger - R_A)^2 \tag{11.13}$$

Comparison with Eq. (11.11) shows that by crossing at the intermediate value, $r = r^\ddagger$ where neither reactants nor products are in equilibrium with the solvent, the minimal barrier height, $\Delta G^\ddagger$, is lower than the reorganization energy. Solving Eq. (11.12) for $r^\ddagger$ (see Problem B) and substituting the value in Eq. (11.13) leads to the Marcus equation for the (free-energy) barrier height or, equivalently, the

---

* See Marcus (1964) and Marcus (1993) for the Nobel Prize lecture giving more background to this early work.

** For large displacements the potential must be anharmonic, because there is exchange of solvent molecules between the solvation shell and the bulk. Isotopic labeling experiments show that even water molecules in the first solvation shell of polyvalent ions do exchange, albeit slowly, because the activation energy is high. For weaker solvation the exchange can be quite fast.

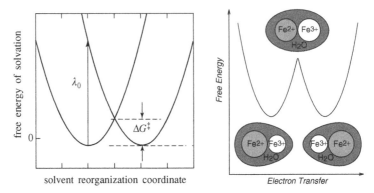

Figure 11.3 Left: Potential energy for displacing the solvent, along the solvent reorganization coordinate r, about its equilibrium configuration for the reactants A–A$^+$ and a second such plot, for the products A$^+$–A. For small displacement of the solvent from equilibrium (about either the reactants or the products) the energy rises quadratically, as shown. For this symmetric exchange the two parabolas have the same force constant but their equilibrium positions are not the same. The reorganization energy λ_0 is indicated as the energy necessary to form the products when the reactants are at equilibrium with the solvent. Reactants with the solvent displaced from equilibrium can cross over to products at a greatly reduced barrier, indicated as $\Delta G^\ddagger$. The letter G is used for the barrier height because, strictly speaking*, we need to plot the free energy rather than just the potential energy. The reaction exoergicity is ΔG^0 and is here zero. See Figure 11.4 for plots with three different signs of ΔG^0. Right: The need for solvent reorganization is shown in the cartoon of water molecules surrounding the doubly charged ion Fe^{2+} and the triply charged ion Fe^{3+}. Note the smaller cavity around the triple ion. At the location where the free energy for electron transfer is minimal the solvation is symmetric, as shown. An environment such as a protein has a lower reorganization energy than water and so the rate of electron transfer is higher. (Adapted from H. B. Gray.)

rate constant for charge transfer:

$$\Delta G^\ddagger = \frac{1}{4\lambda_0}(\lambda_0 + \Delta G^0)^2 \qquad \text{or} \qquad k(T) = A\exp(-\Delta G^\ddagger/RT) \qquad (11.14)$$

For a self-exchange reaction, $\Delta G^0 = 0$ and the entire barrier comes from the solvent reorganization energy $k(T) = A\exp(-\lambda_0/4RT)$. In this case the location of the barrier is midway between the reactants and products, and the non-equilibrium displacement about each contributes equally to the lowering of the barrier from the value λ_0, for reactants at equilibrium, to $\Delta G^\ddagger = \lambda_0/4$.

---

* Note that this plot is in the non-adiabatic limit, meaning that there is a separate electronic state of the reactant and product, each with its own nuclear motion. The same kind of plot was used in another context in Section 5.1.4. In the adiabatic limit the initial and final states are two different wells on the same ground-state adiabatic potential. This limit is appropriate here when the excited electronic state is separated by a gap that exceeds the available thermal energy. See also Problem H.

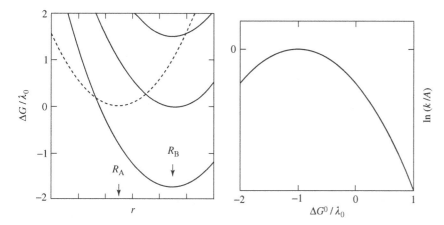

Figure 11.4 Left: determining the activation barriers for an endoergic, symmetric, and highly exothergic electron transfer process. The reactants' parabola is dashed, the barrier height is determined as the energy at which the reactants' and products' parabolas cross. The free energy is plotted in units of the reorganization energy. The *inverted regime* is when $-\Delta G^0/\lambda_0 > 1$ and, as the plot suggests, in this regime the barrier **increases** upon further increase in the exoergicity. For the thermoneutral case the barrier is $\lambda_0/4$. In the endoergic regime the barrier is higher than ΔG^0. How much higher depends on the value of λ_0. The plot of the rate constant vs. the thermodynamic driving force, Eq. (11.14), is shown on the right.

Generally it is customary to distinguish among different regimes as shown in Figure 11.4. In the endoergic regime ΔG^0 is positive and the barrier height increases with increasing ΔG^0. In the exoergic direction ΔG^0 is negative. As it becomes more negative, the barrier height decreases until the point at which the exoergicity can fully compensate for the solvent reorganization, $\lambda_0 = -\Delta G^0$, and the barrier height is zero, $\Delta G^\ddagger = 0$. The reaction rate is now very fast. It is so fast that the reaction itself need not be the rate-determining step for the observed change in the bulk concentrations. The approach of two reactants that are initially far apart can be slower than the reactive event. This condition is the limit of diffusion control as discussed further in Section 11.1.3.1.1 and in Problem D. The unexpected prediction of the Marcus equation is that for very exoergic reactions, those where $\lambda_0 < -\Delta G^0$, the barrier increases again while ΔG^0 becomes more negative. This *inverted regime* occurs because of the need to release a large exoergicity into the solvation modes. The Franck–Condon overlap between the motions in the two parabolas is then not optimal.[9]

The loss or gain of an electron by a molecule is typically accompanied by structural changes, which in turn also create a barrier to electron transfer. We are familiar with this effect in the gas phase when the transferred electron enters an antibonding orbital, Section 5.1.3.1. In solution, the barrier arising from intramolecular reorganization is known as an inner-sphere effect because it occurs within the solvation sphere. The barrier arising from solvent reorganization is then

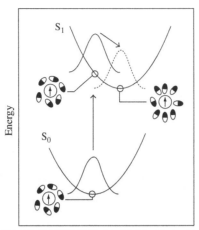

Energy

Solvation coordinate

Figure 11.5 Probing the dynamics of solvation. Shown is the potential for solvent displacement about a chromophore in its ground state, S_0, and the electronically excited state, S_1. When the dipole of the excited chromophore is larger than its value in the ground state, at equilibrium the dipolar solvent is much more organized around the electronically excited solute. An ultrafast excitation creates an excited molecule solvated as if it were in its ground state. As the solvent relaxes to its equilibrium structure, the energy decreases. This energy change is conveniently monitored by a shift to the red in the fast fluorescence of the solute, as shown in Figure 11.6 [adapted from Stratt and Maroncelli (1996)]. Below we discuss the large spectral shift, a manifestation of the interaction with the solvent, in terms of the dielectric constant. However, large spectral shifts are also observed in non-polar solvents such as benzene, see Reynolds *et al.* (1996). Understanding such shifts requires a more detailed examination of the long-range dispersion forces between solvent and solute. Quantum chemistry has yet to do a more detailed analysis of these forces; see Cramer and Truhlar (1996).

referred to as an outer-sphere effect. Section 11.4 seeks a unified approach to both aspects.

11.1.3   Dynamics of solvation

In Section 11.1.2 we argued that the solvation shell is not static. To probe the dynamics of the motion under the solvent–solute potential we need to displace the shell from equilibrium in a sudden manner. Ultrafast optical excitation of a solute from the ground state to an electronically excited state, particularly when the excited state has a very different dipole moment from the ground state, creates suitable initial conditions (see Figure 11.5).

As shown schematically in Figure 11.5, the failure of the solvent to readjust during the fast electronic transition prepares the solute in the excited state S_1 with a solvent configuration appropriate to equilibrium with the ground state S_0 of the solute. On a time scale short compared with any radiationless decay, the

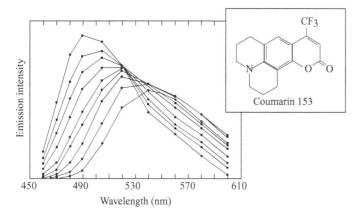

Figure 11.6 Time-resolved emission of a dye molecule (Coumarin 153, shown as an insert) in formaldehyde following an ultrashort electronic excitation. The spectra shown were taken with a delay of 0.0, 0.05, 0.1, 0.2, 0.5, 1, 2, 5, and 50 ps with respect to the excitation. These times are short compared with any radiationless decay of the excited state.* In reference to Figure 11.5, the results show that for a fast optical excitation the solvent has no time to adjust to the new electronic state of the solute. As time progresses, the spectra monotonically and continuously shift to the red [adapted from M. L. Horng *et al.*, *J. Phys. Chem.* **99**, 17311(1995)].

solute will fluoresce from its unequilibrated state. Monitoring this fluorescence as a function of time, Figure 11.6, shows the relaxation toward equilibrium.

For the purpose of a quantitative analysis we normalize data like that shown in Figure 11.5 using the relation

$$S_{\bar{\nu}(t)}(t) \equiv (\bar{\nu}(t) - \bar{\nu}(\infty))/(\bar{\nu}(0) - \bar{\nu}(\infty)) \tag{11.15}$$

where $\bar{\nu}(t)$ is the frequency of the peak position (or some other spectral feature) at time t. The results span a range of variations between different solvents, Figure 11.7, but do follow what one would expect from the dielectric continuum model that we have discussed: the characteristic time for the decay of $S_{\bar{\nu}(t)}(t)$ is shorter in the more polar solvents.

Quantitatively, we expect the temporal response of a solvent to be governed by the dynamics of the translation and reorientation of its molecules. This response changes the interaction of the anisotropic charge distribution of the solute, as characterized by the multipole moments (dipole, quadrupole, etc.) of its charge distribution, with the multipole moments of the solvent molecules. In a polar solvent we can seek to relate the time scale of the solvent's reorientation dynamics to the frequency dependence of the dielectric constant in the spectral region corresponding to nuclear motions.[10]

---

* The duration of the pump pulse is ultrashort but it needs to be long enough that it does not create vibrational coherence, cf. Chapter 7, in the solute itself. This is possible because the dephasing time of the solute is usually (but not always) shorter than the relaxation time of the solvation shell.

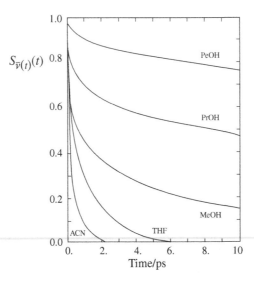

Figure 11.7 The normalized fluorescence emission, $S_{\bar{v}(t)}(t)$, vs. time for solvents of decreasing polarity [ACN (acetonitrile, where the decay has a time constant of about 100 fs), tetrahydrofuran (THF), methanol, 1-propanol, and 1-pentanol]. Not shown is 1-decanol, for which the time is already about 200 ps. Longer times are measured for even more viscous solvents such as glycerol, conforming to the rule of thumb that solvent viscosity is a useful measure for the strength of the solvent–solute coupling [adapted from Horng *et al.* (1995) loc.cit.].

The relaxation of the solvent configuration toward equilibrium as characterized by $S_{\bar{v}t}(t)$ is not simply exponential. At least two time scales are discerned in Figure 11.7. Aided by dynamical simulations a picture emerges to rationalize the separation of time scales into two regimes. The fast relaxation is essentially inertial, which means that there is not sufficient time for solvent molecules to be strongly coupled to one another. Each solvent molecule rotates independently so as to reduce the (dispersion force) multipole–multipole coupling between solvent and solute. Quite small rotations are often sufficient to bring about a significant reduction in the solvation energy and so the inertial relaxation can account for much of the reorganization. Simulations of solvation in water show times as short as 20 fs for this component. The longer relaxation is collective and dissipative in nature. Of course, any specific interaction between solute and solvent (such as strong hydrogen bonding, or charge flow to solvent) modifies this simple picture.

11.1.3.1 Cage effect
The term "cage effect" was coined to describe the low quantum yield for simple photodissociations in solution, such as

$$(I_2)_{\text{solvated}} \xrightarrow{h\nu} (I)_{\text{solvated}} + (I)_{\text{solvated}}$$

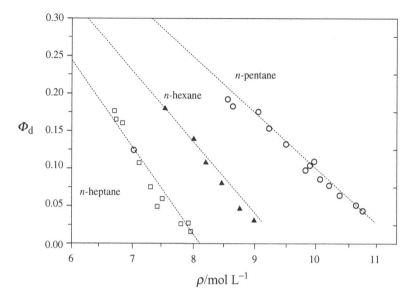

Figure 11.8 Cage effect in the photodissociation of I_2 in hydrocarbon solvents. Shown is the quantum yield vs. the density of the inert solvent [adapted from J. Schroeder, and D. Schwarzer, *Z. Phys. Chem.* **215**, 183 (2000)]. When we go to supercritical solvents, the yield decreases. Simple models for the density dependence are discussed in Schroeder and Troe (1993, 1987). The lines shown are for static caging.

There being no computers in those early days (Rabinowitch and Wood, 1936), two red billiard balls were placed next to each other on a big tray, surrounded by many white billiard balls, as in Figure 1.8. Vigorous shaking of the tray did not move the two red balls from each other. The conclusion was reached that the solvent was caging the products. The nascent I atoms, instead of separating, recombined* to reform I_2. The observed yield of I atoms is therefore far less than two per photon. Recent data are shown in Figure 11.8. Further experiments[11] and simulations show that caging can be multistaged. There is the primary cage and beyond it a secondary cage where a solvent atom gets between the two I atoms but these are still kept together by the surrounding solvent.

11.1.3.1.1 Diffusion control

An immediate implication of the possibility of caging is that an A + B barrier-crossing reaction in solution is not quite like the sequence of independent binary collisions seen in the gas phase. Rather, the A + B attempted barrier crossings are bunched together into encounters as shown schematically in Figure 11.9. This means that once A and B get together at the foothills of their chemical

---

* This recombination is known as geminate because the two sibling radicals meet each other. Time-resolved experiments (Harris *et al.*, 1988) have verified this mechanism with additional refinements. First, there are many electronic states of I_2 that dissociate to two open-shell I atoms in their ground states. So the geminate recombination is not necessarily back to the ground electronic state. Even when it is, the nascent I_2 molecule that is formed in this way is vibrationally energy-rich. Its cooling is by energy transfer to the solvent (Harris *et al.*, 1990), which is an interesting example of the exponential gap principle at play, because the vibrational spacings of an anharmonic molecule depend on its energy content.

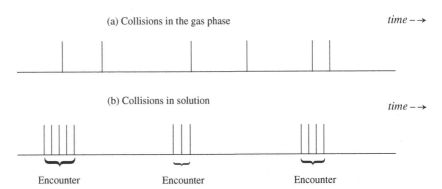

(a) Collisions in the gas phase *time* $-\!\!\rightarrow$

(b) Collisions in solution *time* $-\!\!\rightarrow$

Encounter   Encounter   Encounter

Figure 11.9 Contrasting the time sequence of attempted crossings of an activation barrier for reaction in the gas phase and in solution [adapted from Schwartz *et al.* (1994)]. (a) In the gas phase the collisions of A and B are randomly spaced in time. (b) In solution once A and B get together they are caged by the solvent and therefore can try several times to cross over to products. We refer to such a bunch of attempted crossings as an encounter.

interaction, they are caged by the solvent and can have more than one opportunity to attempt to cross the barrier to products. This behavior enhances the bimolecular chemical reaction rate, because every A + B meeting allows for several attempted crossings.

The caging of the reactants can be phenomenologically represented by the kinetic scheme

$$A + B \underset{k_{\text{dissociation}}}{\overset{k_{\text{diffusion}}}{\rightleftharpoons}} (AB)_{\text{encounter}} \overset{k}{\longrightarrow} \text{products}$$

The concentration of the encounter pair is low and so we make a steady-state approximation for it, $d[(AB)_{\text{encounter}}]/dt = 0$, and this yields for the rate of formation of products

$$\frac{d\,[\text{products}]}{dt} \equiv k\,\big[(AB)_{\text{encounter}}\big] = \frac{k\,k_{\text{diffusion}}}{k_{\text{dissociation}} + k}\,[A]\,[B] \qquad (11.16)$$

When the barrier crossing is very fast (so that $k > k_{\text{dissociation}}$), the reaction is diffusion-controlled with a rate constant given by $k = k_{\text{diffusion}}$. Thus, the getting together of A and B is the rate-determining step. Otherwise, if the barrier to reaction is high or the barrier crossing is slow for other reasons, most encounters are not fruitful and result in the separation of A and B, so that $k_{\text{dissociation}} > k$. Then the formation of the encounter pair is itself in equilibrium, with an equilibrium constant given by $K = k_{\text{dissociation}}/k_{\text{diffusion}}$ and the observed reaction rate constant is Kk, where k is the rate constant for barrier crossing.

This phenomenological treatment shows that there are really several reasons why reaction rates in solution differ from those in the gas phase. As seen before, the solvent can have an intrinsic effect, influencing the process of barrier crossing both because it modifies the potential and because of dynamical effects. Beyond that, the solvent also influences the transport of the reactants (and products) into the cage. Equation (11.16) is discussed further in Problem D, in terms of the rate of diffusion in solution with the result that $k_{\text{diffusion}} = 4\pi(D_A + D_B)\beta$, where the Ds are the diffusion coefficients, and β is the reaction radius for an A + B reaction. If there is no long-range interaction between A and B, β is the radius of the cage. For ion–molecule reactions and other examples where there is a long-range attraction, there is usually a precursor to the primary cage, where the reactants are separated by just one solvent molecule. This situation is referred to as a solvent-separated (ion) pair.

11.1.3.2 Caging dynamics

Ultrafast excitation experiments and corresponding dynamical simulations reveal that on the molecular time scale caging is sensitive both to the detailed molecular structure of the solvent surrounding the molecule and to the dynamics of the dissociation process. As an example let us take excitation of I_2 to an electronically bound state (B) that dissociates because of a non-adiabatic transition to a repulsive state, Figure 11.10. The transition is symmetry forbidden in the isolated homonuclear molecule but allowed in solution, where coupling to the solvent breaks the symmetry of the isolated molecule.*

11.1.3.3 Caging dynamics in clusters

Clustering of a solute by a small number of solvent molecules allows one more variable for testing our ideas about how properties scale with the number of molecules in the solvation shell.[12] For example, fast ionization of a neat cluster generates an ion in a non-equilibrium environment and is a way to explore the dynamics of ion solvation. Here we consider another aspect of the caging dynamics: in a series of experiments a dihalogen ion, solvated by n CO_2 molecules, is photoexcited above its dissociation limit. The quantum yield of atomic and molecular ions is determined as a function of n

$$XY^-(CO_2)_n \xrightarrow{h\nu} \begin{cases} X^-(CO_2)_m + Y + (n-m)CO_2, & \text{uncaged} \\ XY^-(CO_2)_m + (n-m)CO_2, & \text{caged} \end{cases}$$

For all dihalogens examined, there is a strong increase of caging efficiency with increasing cluster size n. Complete caging is observed once a single solvation shell is completed. Results for $I_2^-(CO_2)_n$ are shown in Figure 11.11 and compared with dynamical simulations.

That caging is effective even when the solvation shell is not completed, as seen in Figure 11.11, is observed in other clusters and solutes. Simulations suggest that more than one effect is at play. There is the purely mechanical effect that a solvent molecule is in the way and that, particularly when it is sufficiently heavy,

---

* This example is yet another instance of how the solvent induces qualitative changes in the electronic structure of the isolated system.

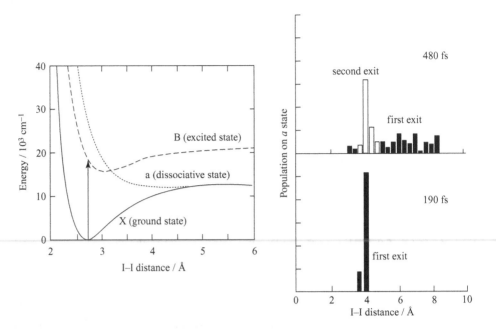

Figure 11.10 Potential energy curves of the isolated molecule, left panel, and snapshots of simulations of the dissociation dynamics of I_2 in a rare-gas solvent [adapted from M. Ben-Nun, R. D. Levine, and G. R. Fleming, *J. Chem. Phys.* **105**, 3035 (1996)]. The Franck–Condon region on the bound B state for excitation from the ground state is on the left from the crossing point to the dissociative a state, and the motion on the shallow dissociation asymptote is very slow. There are therefore distinct time windows during which the two I atoms can separate, just as in the gas-phase NaI experiment shown in Figure 9.12. The first and second such exits from the bound state are indicated. The essential difference is that in solution the solvation shell stops the atoms from escaping and it is the first collision of the fragment with the solvent cage that stops it. With time, the initially excited wave-packet on the B state dephases both because of its own very anharmonic motion and because of the interaction with the solvent. Only then is the dissociation possible at every time interval and we can describe the dissociation by first-order kinetics.

it can turn back the atom trying to escape. This effect is observed in studies of the photodissociation of I_3^- in alcohols,[13] ROH, where the quantum yield goes down as the size of R increases. In addition there can be an electronic effect arising from the solvation.* An asymmetric solvent configuration around the molecule polarizes the molecular charge distribution. When the bond extends, these forces are comparable to the chemical force between the two atoms. The potential energy is then no longer simply a function of the interatomic distance.

---

* Photodissociation of a chemical bond creates open-shell species. It is to be expected that such reactive intermediates will interact chemically with the solvent – particularly so for nascent species that are translationally hot and/or when the solvent itself has unpaired electrons.

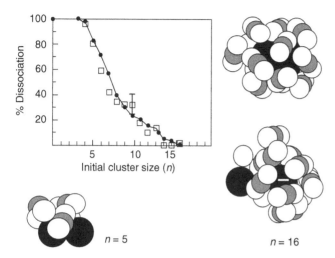

$n = 5$ $n = 16$

Figure 11.11 The quantum yield for photodissociation of $I_2^-(CO_2)_n$ vs. cluster size n. [Experimental results at 790 nm, dots, from Vorsa *et al.* (1997); simulations, squares, from J. Faeder *et al.*, *Chem. Phys.* **239**, 525 (1998).] The simulations include the effect of the longer-range ion–molecule polarization potential and this favors caging. The simulations also provide typical configurations; these are shown at equilibrium for $n = 5$ and $n = 16$. Also shown is a solvent-separated ion pair for $n = 16$. Note that already at $n = 5$, when the molecule is still completely unsolvated on one side, there is an onset of caging. Dissociation is fully suppressed by $n = 16$.

Instead the description of the dynamics is more complicated, meaning that there is time for the solvent to interfere with the exit motion.

Studies of I_2 ultrafast direct dissociation and recombination in Ar clusters[14] have shown that, as in solution, see Figure 11.10, the geminate recombination happens in two stages. The first stage is fast in which the solvent has no time to move appreciably. Its structure presents a rigid cage from which the wave-packet for I—I motion is reflected but with much retention of its coherence so that the atoms can recombine as the wave-packet vibrates within the solvent shell. The second stage is slower geminate recombination in which the solvent has time to move and then a solvent species can get in between. The relative importance of the two stages depends on the solvent density and temperature.

11.1.4 Vibrational relaxation

In this section we carry over the ideas examined for the gas phase in Chapter 9 to the condensed phase and look for the new features.

There are many examples of energy transfer to or from the solvent. Caging dissipates the translational energy after bond breaking and conversely, the solvent must provide the energy required to cross an activation barrier to a chemical reaction. Photodissociation can be used, such as the HgI product in the photo-dissociation of HgI_2 (Figure 8.3). Using fast laser pulses in the IR, we can pre-pare lower vibrational states that can be monitored, most typically by Raman spectroscopy (Laubereau and Kaiser, 1978; Seilmeier and Kaiser, 1988). Internal conversion (Section 7.0.3) provides one convenient way to put lots of vibrational energy in a solute. Vibrational relaxation dynamics in solution are therefore well studied.[15]

The condensed phase brings definite new features but, as in the gas phase, we continue to emphasize the exponential role of the energy gap. As an example, a vibrationally excited diatomic, such as O_2, survives intact in a cold atomic solvent for a very long time because of the large frequency mismatch. Further-more, it is observed that for such rather rigid molecules the relaxation rate scales linearly with solvent density, which suggests that a close-in collision with a particular solvent atom is required to affect the relaxation. This "independent binary collisions" approximation is not surprising, because one must dump a great deal of energy into translation. How then can the solvent play a special role? We shall see that the answer is that the solvent offers new ways for bridging the gap.

In a polar solvent it is natural to study ions. In a polar solvent, ions are strongly solvated, providing new modes into which the solute vibrational energy can flow. Indeed, ions in water undergo facile vibrational relaxation.[16] Furthermore, ionic species couple well to the intramolecular modes of the polar water molecules. Once we recognize this fact, we can ask a more sophisticated question: can the relaxation compete with, or even be faster than, the vibrational motion of the diatomic molecule? If it can, then we have a regime not familiar from the gas phase: it is no longer sufficient to just ask how fast the energy drains out, because during relaxation the molecule can retain a coherence that is built into the initial state by an optical excitation. Figure 11.12 shows several scenarios. They range over two extremes. In one case, called the slow relaxation limit, the initially prepared wave-packet oscillates many times and so dephasing is faster than vibrational energy relaxation. In the other extreme, called the ultrafast relaxation limit, the wave-packet simply sinks to the bottom of the potential before it has time to complete even one vibration.

The relatively low frequencies of solvent motions, the very reason why non-polar diatomic molecules relax slowly, is just what is needed to make polyatomic molecules relax more effectively. The reason is the possibility of solvent-aided intramolecular vibrational energy transfer. Recall that polyatomic molecules have ladders of vibrational states (e.g., Figure 7.18). At higher energies these ladders overlap and form a quasi-continuum. Consequently, the solvent-induced relax-ation does not need to remove much energy. It can effectively compete with the intramolecular vibrational relaxation. When the molecule is large enough or energy rich enough, or both, for the molecule to act as its own heat bath, the solvent provides an even bigger bath but qualitatively there is no essential difference: a

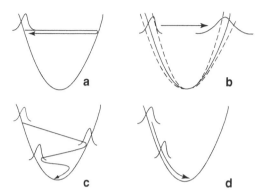

Figure 11.12 Limiting cases of vibrational relaxation of a wave-packet prepared at time zero with much vibrational energy. An example is the HgI diatomic fragment from an ultrafast dissociation of HgI_2, but carried out in a polar solvent to allow for faster relaxation. (a) Slow vibrational relaxation. The packet oscillates in the well, dephasing slowly owing to the anharmonicity of its potential, just as it would in the gas phase. Solvation may make the potential even more anharmonic because of the fluctuating solvent dipole moments as the ionic molecule vibrates. No other solvation effect is expected. (b) Slow vibrational relaxation. The packet oscillates in the well and at the same time rapidly dephases owing to the modulation of the potential by the solvent but without much loss of vibrational energy. (c) Fast vibrational relaxation, comparable to the vibrational period. Coherence is retained, at least in part, as the molecule is drained out of its vibrational energy. (d) Ultrafast vibrational relaxation. The motion is like that of an overdamped oscillator. The signature is that the motion in the relaxed state retains its coherence, that is, it remains localized [adapted from Voth and Hochstrasser (1996); see also Voth (2001)].

localized vibrational excitation will rapidly relax. It is at lower energies that the solvent plays a key role. This is the regime where there are many states but these are not quite so near in energy. Here the solvent can take up (or provide) the small energy mismatches. The solvent bridges the exponential gap and we discuss this intermediate regime further below. Finally, there must be a slow stage where the molecule returns to the ground state by dumping a fair amount of energy into the solvent. This stage is just the same as for collisional relaxation of polyatomics in the gas phase, where the lowest vibrational frequency determines how much energy needs to be lost by a V–T process.

The new feature in the vibrational relaxation of polyatomics in solution is the intermediate energy stage where the solvent significantly aids the intramolecular redistribution of vibrational energy, Section 7.2.4. Figure 11.13 shows experimental results for the relaxation of a C–H chromophore in acetonitrile (CH_3CN). The C–H vibration is a high-energy ($\sim 3000\,cm^{-1}$) mode that can be excited to $\nu = 1$ by an ultrafast IR pulse. In the neat liquid it relaxes very rapidly, a few picoseconds, as opposed to relaxation of CH $\nu = 1$ states in isolated small organic molecules that is observed in the gas phase to take a few nanoseconds. The

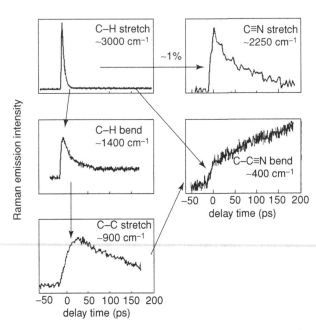

Figure 11.13 Time-resolved pump–probe vibrational relaxation of a C–H stretch vibration in liquid acetonitrile at room temperature detected by (Raman) emission from the excited state. In the isolated molecule such a $v = 1$ stretch mode will relax on the nanosecond time scale. The experiment maps out the energy transfer pathways in the liquid. The essential point is that the relaxation is by solvent-aided intramolecular transfer, as discussed in the text and indicated by arrows. Energy dumping into the liquid is far slower, requiring about 250 ps [adapted from Deak *et al.* (1998); see also Iwaki and Dlott (2001), Wang *et al.* (2002)].

essential point of the experiment is to identify where the energy goes. Monitoring the other vibrations by Raman spectroscopy, the decay of 3 ps is correlated (see Figure 11.13) with a 3 ps rise time of the C–H bend ($\sim$1400 cm^{-1}) and C–C$\equiv$N bend ($\sim$400 cm^{-1}), C–H stretch ($v = 1$) $\rightarrow$ C–H bend ($v = 1$) + C–C$\equiv$N bend (overtone).

A small energy gap requires excitation of the $v = 3, 4$ overtone vibrations of the C–C$\equiv$N bend. A wide-gap transfer from the C–H stretch to the C$\equiv$N stretch ($\sim$2250 cm^{-1}) is much less probable. There is also no small-gap process to drain energy out of the C$\equiv$N stretch so its lifetime is long, $\sim$80 ps. As seen in Figure 11.13, the C–H bend ($\sim$1400 cm^{-1}) decays by a small-gap process to the lower-energy ($\sim$900 cm^{-1}) C–C stretch and C–C$\equiv$N bend. The final dumping of vibrational energy into the solvent bath requires about 250 ps, as can be seen in the long-time heating of the C–C$\equiv$N bend and can also be detected through a temperature rise of the liquid. This behavior illustrates the hierarchy of different relaxation times.

11.2 Barrier-crossing dynamics

Understanding the participation of the solvent in crossing the barrier to reaction has challenged theorists ever since the birth of dynamics. We want to control not only the rate of reaction but also the nature of the products in the same way that synthetic chemists do. The more ambitious take their cue from enzymatic reactions, where the (protein) environment is able to affect changes in the rates by many orders of magnitude. That goal has proved challenging. The problem in part is that we are seeking a goal that contains an inherent contradiction. On the one hand, we think in terms of reactants and their surroundings and want to discuss how varying the surroundings can alter what the reactants do. On the other hand, as shown by the discussion of solvation, the reactants and their surroundings can be strongly coupled. The forces between solvent and solute, not necessarily weak, cannot be considered a small perturbation. A solvent that exercises a qualitative effect cannot be represented simply as the reactants' response to it; there are mutual influences. So it is useful to speak of a combined system. By treating the solute and solvent as one system we can often achieve a quantitative description. But when we want to change the solvent, such an approach requires that we start from the beginning. The use of transition state theory allows us, to a certain extent, to factor the dynamical behavior into two terms, one having to do with the reactants and another representing the solvation of the transition state. The problem is to go beyond this factorization. In principle, tools such as the Langevin equation (see also Section 5.2.2.1) allow us to represent the solute dynamics under the influence of the solvent in terms of friction and memory effects, but there is still a gap between the principle and the practice. Incisive experiments have yet to be done that critically test well-developed theoretical tools.

We begin with the potential along the reaction coordinate known as the *potential of mean force*. This allows us to compute the reaction rate using transition state theory. Then we discuss two dynamical regimes where respectively the solvent aids and hinders the crossing of the chemical barrier. Kramers[17] (1940) clearly identified these two complementary roles of the solvent and showed that, other things being equal,* they correspond to weak and strong solvent–solute coupling. The weak coupling regime is intuitively clear and is analogous** to the Lindemann regime, with which we are familiar: the molecule acquires energy by collisions with the fluid around it. For reaction in solution, the thermal reactants at the foothills of the chemical barrier need to be activated so that they can

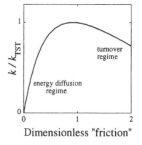

Dimensionless "friction"

---

* The problem for an experimental verification is that when we change the solvent it can be the case that we also change the barrier height and other reaction attributes.

** The Lindemann scheme is applied in the gas phase where the collisions are binary and isolated. Kramers (1940) couples the molecule to a medium with which it exchanges energy in a random fashion. This weaker coupling regime is known as the energy diffusion regime because the energy content of the motion along the reaction coordinate often varies up and down by small amounts.

cross the barrier. The weak coupling regime is where the energy flow from the solvent to the reaction coordinate is rate-determining and so the reaction rate is lower than what we would compute from transition state theory. Already noted in Section 11.0.1 is another scenario, namely, that the thermal reactants at the foothills of the barrier acquire the required energy in one rare but large impulse rather than in a diffusive manner. Coupling to the solvent is important also in deactivating the products in an isomerization reaction, as shown in Figure 11.14. The reason is that just after crossing the barrier from the reactants' well to the products' well there is energy released along the reaction coordinate during the descent from the barrier. Unless this energy is rapidly channeled into other degrees of freedom, the motion along the reaction coordinate can be reflected by the outer repulsive potential of the products' well, see Figure 11.14, causing a recrossing of the barrier back to the reactants.

As the friction increases, the role of the solvent changes and, as a function of the strength of coupling to the solvent, the reaction rate reaches a maximum and then decreases in what is known as the *turnover regime*. At very high friction it is the very motion along the barrier top that becomes diffusive (we shall interpret this as due to the participation of the solvent motion in the crossing of the barrier).

11.2.1   Potential of mean force

The potential of mean force, the technical term for the free energy along the reaction coordinate, is an essential ingredient in a correct application of transition state theory to processes in the condensed phase.[18] We need the potential of mean force for two purposes. First, to know the location of the bottleneck to reaction, the transition state. Except when it is dictated by symmetry, as in symmetric exchange, the transition state in the presence of a solvent is not necessarily the same as in the gas phase. Second, we need to know the height of the barrier.

To determine the potential of mean force we constrain the solute to be at a particular position along the reaction coordinate. Then the solvent–solute system is brought to thermal equilibrium and the energy is determined. So all degrees of freedom are allowed to vary and approach equilibrium, the solvent, the solute, and their relative motion. Only the position along the reaction coordinate is held fixed. We determine the free energy and then choose a new position along the reaction coordinate, equilibrate, determine the energy, and so on. This procedure is what we do in the gas phase, even when there is no solvent. The only practical difference between the gas phase and a solution is in how many degrees of freedom there are to deal with. In either case, we treat the whole system.[19] The one fixed coordinate is the minimum energy path. All other degrees of freedom, those of the solute, the solute–solvent configuration, and the solvent itself constitute the thermal bath. The potential of mean force, $W(q)$, is the work required to bring the solute to a particular location q along the reaction coordinate, averaged over the equilibrium distribution of all other degrees of freedom. We now turn to show that this potential indeed determines the mean (averaged over all other modes) force for the motion along the reaction coordinate.

*11.2.1.1 Evaluating the mean force

The potential of mean force, $W(q)$, at position q along the reaction coordinate enters into our quantitative considerations when we come to compute the partition function for the transition state. This information is an input into the expression for the barrier-crossing rate constant, Eq. (6.11). The partition function $Q^{\ddagger}$ is like the partition function for the whole system except that we do not allow motion along the reaction coordinate and we constrain the position along the reaction coordinate to be at the top of the barrier. Next comes an observation about partition functions that is generally valid in the limit when classical mechanics is applicable. One way of evaluating partition functions is by summing over the quantum states of the system, Eq. (A.6.4). In the classical limit we can also evaluate this sum as an integral of the Boltzmann factor over all n positions and n momenta

$$Q = \int \exp(-H(\mathbf{x}, \mathbf{p})/k_{\mathrm{B}}T)\,d\mathbf{x}\,d\mathbf{p}/h^{n} \tag{11.17}$$

As with transition state theory in the gas phase,[20] it is useful if we can[21] write the Hamiltonian as a sum of the kinetic energy, $T(\mathbf{p})$, a function only of the momenta and the potential energy, and $V(\mathbf{x})$, a function only of the positions. The $2n$-dimensional integral in Eq. (11.17) then factors into a product of an n-dimensional kinetic energy integral and an n-dimensional potential energy integral, known as the configuration integral Z. The configuration integral at the transition state is obtained by constraining the value of reaction coordinate q. So if n is the number of degrees of freedom of the system, we need to evaluate

$$Z^{\ddagger} \equiv \int \exp\left(-V(\mathbf{x})/k_{\mathrm{B}}T\right)\delta(q - q^{\ddagger})\,d\mathbf{x} \tag{11.18}$$

The delta function constrains the value of the reaction coordinate q to $q^{\ddagger}$, the location of the transition state along q. Remember that in solution q can involve solvent modes and need not be the same coordinate as in the gas phase.

How is this configuration integral related to the mean force? Let us compute the mean force along the reaction coordinate at $q = q^{\ddagger}$:

$$-\left\langle \frac{\partial V(\mathbf{x})}{\partial q} \right\rangle_{q=q^{\ddagger}} = -\frac{\int (\partial V(\mathbf{x})/\partial q)\exp\left(-V(\mathbf{x})/k_{\mathrm{B}}T\right)\delta(q - q^{\ddagger})\,d\mathbf{x}}{\int \exp\left(-V(\mathbf{x})/k_{\mathrm{B}}T\right)\delta(q - q^{\ddagger})\,d\mathbf{x}}$$

$$= k_{\mathrm{B}}T\left(\frac{\partial}{\partial q^{\ddagger}}Z^{\ddagger}\right)\bigg/ Z^{\ddagger} = k_{\mathrm{B}}T\frac{\partial \ln Z^{\ddagger}}{\partial q^{\ddagger}} \equiv \left(\frac{\partial W(q)}{\partial q}\right)_{q=q^{\ddagger}} \tag{11.19}$$

We have shown* why ($k_B T$ times) the logarithm of the configuration integral $Z^\ddagger$ acts as a potential from which the mean force can be determined. We are already familiar with a simpler version of this connection, that $k_B T$ times the logarithm of the partition function is the free energy, which is the (reversible) work available.

11.2.1.2 From gas phase to solution

When the potential of mean force is known, we can compute the transition state theory rate of barrier crossing in solution. It is customary to define $\Delta W(q)$ as the difference in the potential of mean force between the gas phase and solution and to write symbolically

$$k_B T \ln \frac{k_{\text{solution}}}{k_{\text{gas phase}}} = -(\Delta W(q^\ddagger) - \Delta W(q \text{ at reactants})) \qquad (11.20)$$

This shorthand summary is not without pitfalls. It suggests that the environment merely changes the energetics of the transition state relative to that of the reactants. But in solution the very configuration of the transition state and the displacement that is the reaction coordinate may differ in different solvents. On the other hand, computing the potential of mean force and determining the transition state for it often accounts for much of the solvent effect on the rate of reaction, an effect that would otherwise be attributed to the solvent playing a dynamical (rather than an equilibrium) role in crossing the barrier to reaction.

Having shown the application of the potential of mean force, we should remember that the purpose of transition state theory is not to describe the dynamics of the reaction. It is a theory for computing the rate of a single barrier crossing, starting with reactants in equilibrium and assuming that the passage over the barrier is the rate-determining step. If there is more than one barrier en route to products – say if the reactants are caged at the foothills of the chemical barrier – then the theory needs suitable modification. Similarly, if the reactants are not in equilibrium with the solvent, we cannot apply the theory. Finally, there may be true dynamical effects such as barrier recrossings and, in particular, barrier recrossings induced by the solvent.

11.2.2 Isomerization

The dynamics of the photoinduced conformational change shown in Figure 11.14 allows us to explore the role of the solvent in barrier crossing. The measured rate for isomerization of 1,1′ binaphtyl as a function of solvent viscosity is shown

---

* Note that

$$\partial \left(\int f(y)\delta(y-a)\,dy \right) \Big/ \partial a = \int f(y)(\partial\delta(y-a)/\partial a)\,dy$$

$$= -\int f(y)(\partial\delta(y-a)/\partial y)\,dy = \int \delta(y-a)(\partial f(y)/\partial y)\,dy$$

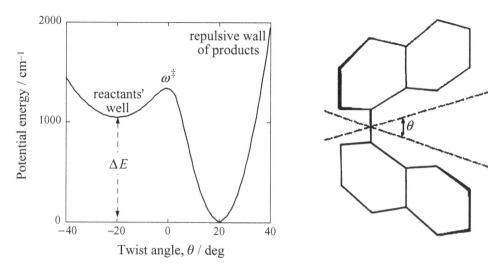

Figure 11.14 Isomerization in 1,1' binaphtyl. The conformational change is the twist of the angle θ, right panel. The potential for the twist motion in the first excited state, as fitted to the measured isomerization rate, is shown vs. the angle. The twist angle at the barrier is taken as the zero [adapted from D. P. Millar and K. B. Eisenthal, *J. Chem. Phys.* **83**, 5076 (1985)].

in Figure 11.15, under the premise that the viscosity of the neat solvent is a reasonable measure for the friction that is applied onto the solute. Note that the optical excitation prepares a non-stationary state localized over the left well. So the experiment directly probes the role of the solvent in hindering the isomerization and is not concerned with the activation stage because the energy is provided by the optical excitation.

Assuming that the potential at the bottom of the reactants' well is parabolic with a frequency ω and that the barrier can be fitted as an inverse parabola with a frequency $\omega^\ddagger$, see Figure 11.14 and Eq. (11.23) below, the decline of the rate constant due to coupling to the solvent can be written in the form that Kramers deduced:

$$k = \kappa \, k_{TST}$$

$$\kappa = [1 + (\gamma/2m\omega^\ddagger)^2]^{1/2} - (\gamma/2m\omega^\ddagger) \xrightarrow{\text{high } \gamma} (m\omega^\ddagger/\gamma) \qquad (11.21)$$

The transition state rate k_{TST} is for a one-dimensional solute reaction coordinate, a result that depends only on the barrier height and the reactants' well frequency ω. The mass for the motion along the reaction coordinate is m.* The transmission coefficient κ corrects the rate so as to include the effect of solvent friction γ. This treatment assumes one-dimensional motion along the reaction coordinate and a

---

* Because friction·velocity has the dimension of force γ/m has the dimension of frequency (1/time).

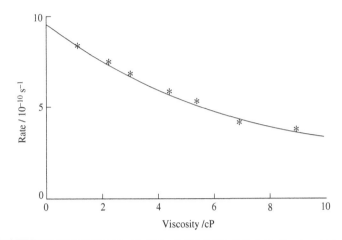

Figure 11.15 The barrier-crossing rate for the (fast) isomerization in 1,1' binaphtyl in its electronically excited state vs. the solvent shear viscosity (cP). The twist angle θ, see Figure 11.14, is about 40° different in the equilibrium configurations on the ground and first singlet excited state. A short pulse prepares the initial state that has the same configuration of the nuclei as the ground state. The subsequent change of conformation can be probed by fast absorption [adapted from D. P. Millar and K. B. Eisenthal, *J. Chem. Phys.* **83**, 5076 (1985)]. The change in geometry requires that the solvent be pushed out of the way, and the rate decreases with increasing viscosity of the solvent. The solid line is the fit by Millar and Eisenthal to Eq. (11.21). (At the high friction end, the rate does decrease as $1/\gamma$.) From this fit, the potential as shown in Figure 11.14 is deduced. For activation by the solvent in the weak-coupling regime where the rate of isomerization increases with the viscosity, see Campbell *et al.* (1992).

static friction, Problem F. The time delay in the response of the solvent can also be incorporated* and leads to the implicit equation

$$\lambda^\ddagger = \frac{(\omega^\ddagger)^2}{[\lambda^\ddagger + \gamma(\lambda^\ddagger)/m]} \qquad (11.22)$$

that needs to be solved for the reactive frequency $\lambda^\ddagger$ and $\kappa = \lambda^\ddagger/\omega^\ddagger$. The reactive frequency is the (imaginary) barrier-crossing frequency that takes into account the friction caused by the solvent. To determine $\lambda^\ddagger$ we need to determine the frequency-dependent friction evaluated at the frequency of the barrier crossing, denoted by $\gamma(\lambda^\ddagger)$ in Eq. (11.22). In Section 11.2.2.1 we discuss a toy model for barrier crossing that yields Eq. (11.22) as the exact result within transition state theory **provided** that we recognize that the reaction coordinate in the presence of the solvent should be modified from that of the pure solute. For details see Problems E and F.

* Known as the Grote–Hynes correction (Hynes, 1985a, 1985b). See Problem F.

*11.2.2.1 Toy model for barrier crossing

The model offers insights on several key points regarding the role of the solvent.[22] But in the final analysis it cannot substitute for realistic dynamical simulations. In particular, the model overlooks any role played by the internal modes of the solvent molecules or the internal modes of the solute. What the model provides is insight into the motion along the solvent-modified reaction coordinate. Specifically, the diffusion forward and backward across the barrier, that in the Kramers model occurs along the reaction coordinate of the isolated solute, is shown in the model to be equivalent to a single crossing of the transition state provided that we use a collective reaction coordinate, one that involves both solute and solvent motion.

The model assumes one-dimensional motion along a reaction coordinate q with a parabolic barrier:*

$$H_{\text{solute}} = \frac{p^2}{2m} + V(q)$$

$$V(q \text{ near } q^{\ddagger}) = E_{\text{barrier}} - \tfrac{1}{2} m (\omega^{\ddagger})^2 (q - q^{\ddagger})^2 \qquad (11.23)$$

The solvent is represented as a set of harmonic modes.[23] These solvent modes are coupled, by a term linear in their displacement, x, from equilibrium, to the solute

$$H = H_{\text{solute}} + \sum_{\text{solvent modes } j} \left[\frac{p_j^2}{2m_j} + \frac{m_j}{2} \left(\omega_j x_j + \frac{C_j}{m_j \omega_j} (q - q^{\ddagger}) \right)^2 \right] \qquad (11.24)$$

The solvent modes, labeled j, are not coupled to one another, and the form of the coupling to the solute is written to ensure that the barrier height is independent of the strength, C, of the coupling. The masses m_j and frequencies ω_j of the individual solvent modes enter into the frictional force (where each solvent mode contributes $C_j^2 / m\, m_j\, \omega_j^2$ to the static friction). The reduction in the rate due to the coupling to the solvent appears in the final result[24] for the transition state theory rate constant through the value of the reactive frequency $\lambda^{\ddagger}$. Specifically, the Hamiltonian given in Eq. (11.24) is, in the vicinity of the barrier, $q \approx q^{\ddagger}$, a sum of bilinear terms. Therefore, its normal modes can be analytically determined. The mode that is unstable, meaning that its frequency is purely imaginary, is the barrier-crossing mode. The reactive frequency is found to be the positive root of the equation

$$(\lambda^{\ddagger} / \omega^{\ddagger})^2 = \frac{1}{1 + \sum_j \left((C_j^2 / m\, m_j\, \omega_j^2) \, / \omega_j^2 \, (1 + (\lambda^{\ddagger} / \omega_j)^2) \right)} \qquad (11.25)$$

In summary, we have two ways to understand the reduction in the reaction rate at higher friction. One is to use a Langevin equation for the diffusive motion along

---

* The frequency $\omega^{\ddagger}$ at the barrier corresponds to an unstable motion. The harmonic frequency is defined as $\sqrt{k/m}$, where k is the force constant. At the barrier top k is negative so the harmonic frequency is purely imaginary and equals $i\omega^{\ddagger}$.

the reaction coordinate of the solute, as was done by Kramers. In general it is not practical to evaluate the friction terms exactly, but for the quadratic Hamiltonian of Eq. (11.24) this computation can be explicitly done and the Langevin equation can be written down with the details given in Problem E. The alternative is to locate the bottleneck for the crossing from reactants to products in the presence of the solvent and to do so along a reaction coordinate that allows for the coupling to the solvent. This is what variational transition state theory, Section 6.1.4.2, tells us to do. For the toy model this procedure yields results equivalent to the Langevin equation, as shown further in Problem F.

The toy model allows us also to examine the dynamics.[25] For this purpose we can rewrite the Hamiltonian such that the reaction coordinate is coupled to only one harmonic solvent mode. This solvent mode is in turn coupled to the other solvent modes, but we will neglect the coupling that is purely within the solvent. The model reduces to motion involving an anharmonic mode, the reaction coordinate, a mode whose frequency, $\omega(q)$, depends on the position along the coordinate, coupled to a harmonic (solvent) mode of frequency ω. We can now ask where, along the reaction coordinate, is the role of the solvent expressed? We know that two modes will effectively couple when the two frequencies are comparable. This condition is measured by the frequency ratio of the reactive vs. the solvent modes

$$\rho^2(q) \equiv \omega^2(q)/\omega^2 = \left(m^{-1}\partial^2 V(q)/\partial q^2\right)/\omega^2 \qquad (11.26)$$

When the energy variation along the reaction coordinate arises from chemical bonds being broken and reformed in concert, the spatial spread of the barrier is confined because chemical forces are short-range. But the height of the barrier can be significant. This means that chemical potential varies rapidly over a short range so that, say, for atom exchange reactions, $|\omega^2(q)|$ will be large. (An absolute value is needed because near the barrier $\omega^2(q)$ is negative.) The frequencies of unassociated solvents are generally lower than chemical-sized frequencies. Under such circumstances, the role of the solvent is largely confined to the foothills of the chemical barrier, where $\omega^2(q)$ is much reduced in magnitude compared with its value at the barrier top where $|\omega(q^\ddagger)| \equiv \omega^\ddagger$.

The potential energy in the toy model, Eq. (11.24), is such that at the barrier, where $q = q^\ddagger$, the solute mode is uncoupled to the solvent. At the very top of the barrier the crossing is a motion along the q coordinate alone. However, away from the barrier the reaction coordinate is slanted, meaning that it involves a displacement of both the solvent and the solute. The potential energy contours for the case of high friction are shown in Figure 11.16. Superimposed on the potential is a typical trajectory. It shows that the toy model recovers the failure of the solvent to move substantially during crossing the barrier. As a result, at large friction the barrier along the q coordinate is crossed many times and, as can be seen from the trajectory, this caging involves energy exchange between solvent

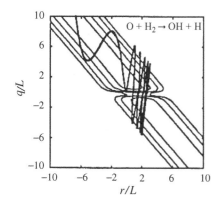

Figure 11.16 Potential energy contours for a Hamiltonian consisting of a reaction coordinate q and a single solvent mode r when the two are linearly and strongly coupled, as in Eq. (11.24). For weak coupling the contours run almost parallel to the q axis. The stronger is the coupling, the more is the solvent displacement involved in the dynamics (L is a range parameter). Superimposed on the potential contours is a trajectory showing caging on either side of the barrier. The motion over the barrier at high coupling is diffusive [adapted from Ben-Nun and Levine (1994)]. The potentials and masses are chosen to mimic the $O + H_2 \rightarrow OH + H$ reaction in supercritical Ar. The variational transition state theory discussed above seeks to rotate the coordinates so as to reduce the number of recrossings.

and solute. Figure 11.17 shows the same effects in a full molecular dynamics simulation.

11.3 Interfaces

Much that is central to our life and our technology occurs at interfaces, the interface between a solid or a liquid and a gas, the interface between two liquids (as in phase transfer catalysis), and the interface between a solid and a liquid. The solid–gas interface has received much attention because of the relation to catalysis, and it is the subject of Chapter 12. In this section we will specifically examine interfaces with a liquid (Benjamin, 1997; Kondow and Mafuné, 2000). We begin with the gas–liquid interface (Nathanson *et al.*, 1996), continue with liquid interfaces, and finally discuss fuel cells as an example where the dynamics at the electrode–electrolyte interface is rate-determining. Our purpose is to emphasize what is unique about such systems and how the inhomogeneity that is an essential characteristic is reflected in the dynamics.

11.3.1 The gas–liquid interface

Experiments probing gas-phase molecules hitting the surface of the liquid have shown both aspects unique to the interface, such as uptake of molecules into

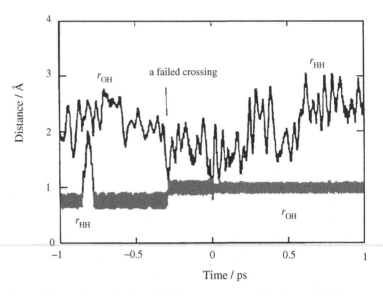

Figure 11.17   Molecular dynamics simulation of the $O + H_2 \rightarrow OH + H$ exchange reaction in supercritical Ar where the atomic structure of the solvent is fully retained in the simulations. Shown are the O–H and H–H bond distances* vs. time. The dense solvent cages the reactants (as well as the products). There are several tries to cross the barrier, but only the attempt at time zero is successful [adapted from Ben-Nun and Levine (1994)]. To perform such computations it is advantageous to initiate the trajectory at $t = 0$ at the barrier top and to integrate both forward and backward in time (Bennett, 1977).

the bulk** or the preferential orientation and the aggregation of molecules on the surface, as well as providing interfacial analogs of bulk solvation, hydrogen bonding, proton exchange and in general acid–base reactions, and oxidation–reduction.

As a concrete example, Figure 11.18 reports the different outcomes when a fast DCl molecule is incident on the hydrogen-bonding solvent glycerol. The first question that also arises in the gas–solid interface is whether the impinging HCl molecule recoils directly from the surface in one or a few bounces, or dissipates its incident kinetic energy and binds momentarily to the surface. If it is trapped, does it remain intact or does it fall apart at the surface of glycerol or wait until it is deeper in the bulk, and the mechanisms for interfacial dissociation and recombination differ from those in bulk solution. It is this ability to discuss both interfacial and bulk processes that is special to the liquid surface.

---

* These distances oscillate at a high frequency when bound and at a much lower frequency when the atoms are unbound but the reactants or products are caged.

** This is of much concern for the uptake of pollutants by water drops and in connection with the ozone cycle in the atmosphere.

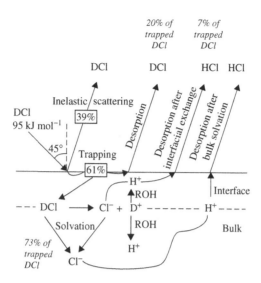

Figure 11.18 Observed channels for collisions of DCl with liquid glycerol (ROH) at a collision energy of 95 kJ mol⁻¹: 61% of the incident DCl molecules are trapped. DCl dissociation and recombination to form HCl may occur in the interfacial and bulk regions of the liquid [adapted from Ringeisen *et al.* (2002)].

Of the DCl molecules striking the surface of glycerol, 73% undergo D/H exchange and desorb as HCl. These DCl molecules first dissociate into D^+ and Cl^- and then the D^+ ion exchanges with glycerol and the resulting H^+ ion diffuses for long times (seconds) before eventually recombining with Cl^- and evaporating as HCl from the liquid. An additional 20% of the DCl molecules desorb immediately back into the gas phase as DCl, despite the measured -67 kJ mol⁻¹ exothermicity for dissociation. The remaining 7% of the DCl molecules striking the glycerol film desorb as HCl *within one microsecond*, most likely reacting within the top two monolayers. These DCl molecules choose to dissociate, undergo D/H exchange, recombine, and then leave without ever penetrating deeply below the surface. The interfacial $DCl(g) \rightarrow Cl^- + D^+ \rightarrow Cl^- + H^+ \rightarrow HCl(g)$ mechanism appears to be quite different from recombination in bulk glycerol, where D^+ and Cl^- ions originating from two different DCl molecules diffuse toward each other and combine. The experiments provide strong evidence for the following pathways: a DCl molecule first lands on the surface and dissociates within a near-surface cage of glycerol molecules. Nine out of ten times, the Cl^- ion and D^+ or an exchanged H^+ ion hop out of the cage, and the ions diffuse deeply into the liquid for several seconds. One out of ten times, however, a D/H exchanged H^+ ion within the surface cage reattaches to the solvated Cl^- and the nascent HCl desorbs into the vacuum. This interfacial exchange and desorption occurs within 10^{-6} s, or one million times faster than bulk dissolution and eventual HCl evaporation.

This study demonstrated that the interfacial region of a liquid can be regarded as a distinct reaction medium even when dissolution in the bulk is very favorable. The gas–liquid interface is not only the gateway for a gas molecule into the bulk, but a region where distinct reactions occur.

11.3.2   The liquid–liquid interface

At the interface of two immiscible liquids there are marked changes in the density, viscosity, and polarity. It is a reasonable zeroth approximation to take these properties to change abruptly at a sharp boundary between the two liquids, but simulations* suggest that on the molecular scale the dividing surface is rather rough so that theories developed for understanding the bulk must be applied with care. A prime example is the dielectric constant. Over what distance scale does it change across an interface between two different solvents? The results of an experiment that uses "molecular rulers" to measure this change is shown in Figure 11.19.

The experiments shown in Figure 11.19 suggest that there is a reordering of the organic solvent at the interface.[26] In Chapter 12 we discuss much additional evidence for reordering at the solid–gas interface and note that there can be a further, adsorbate-induced, reordering.

Another active research area where emphasis is shifting from static structures to dynamics is interfaces in biological systems, in particular at the membrane of the cell where so much critical chemistry is taking place.

11.3.3   Fuel cells

A *fuel cell* directly converts chemical energy into electrical energy.** The most familiar example is the cell using hydrogen and oxygen, Figure 11.20, with the cell reactions

$$\left. \begin{array}{l} \text{anode: } 2H_2O + H_2 \rightarrow 2H_3O^+ + 2e \\ \text{cathode: } \frac{1}{2}O_2 + 2H_3O^+ + 2e \rightarrow 3H_2O \end{array} \right\} \quad \text{net: } H_2 + \frac{1}{2}O_2 \rightarrow H_2O$$

The rate-determining step is that of electron transfer to the oxygen molecule at the cathode. The Marcus theory of Section 11.1.2 can be applied to this situation as follows (Marcus, 1964). We define the *overpotential* η as the excess free energy required to drive a current through the cell, $e\eta = \Delta G - \Delta G^0$, where the standard free energy refers to the absence of any net current passing through and e is the charge of the electron. We use Eq. (11.14) to compute the rate of electron transfer from the electrode to the molecule so that λ_0 is the reorganization energy about the molecule. To convert the rate constant to a current density we need to multiply it by the concentration $[c]$ and by the Faraday constant F. The difference in free energy between the oxidized and reduced species needs to take into account that there is a potential V across the cell. We can write for either oxidation or reduction

$$j^{\pm} = \pm F[c] A \exp(-(\lambda_0 \pm (\Delta G^0 - eV))^2 / 4\lambda_0 R T) \tag{11.27}$$

---

* Dynamics at interfaces is extensively discussed in Benjamin (1997a, 1997b, 2002).

** Unlike the familiar car battery, a fuel cell gets the fuel from an external source. There is therefore no need for charging it and the fuel cell will not go flat as long as the fuel is supplied.

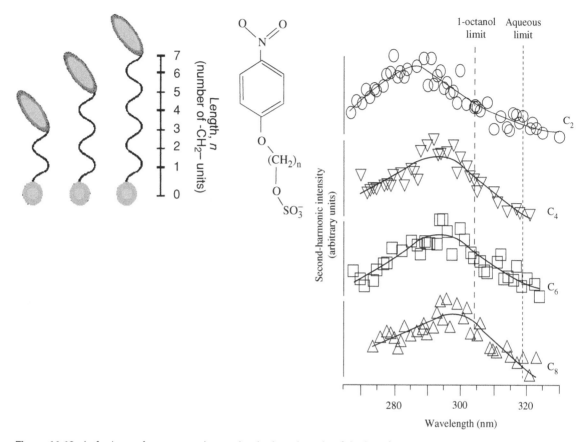

Figure 11.19 Left: the surfactants used to probe the length scale of the interface between water and an immiscible organic solvent. The anionic sulphate group is assumed to remain solvated in water. This group is attached by an alkyl spacer to *p*-nitroanisole whose absorption monotonically shifts as the solvent polarity changes. See the spectrum on the right that indicates the absorption maxima for water ($\varepsilon_r = 78.9$) and for 1-octanol ($\varepsilon_r = 10.3$). Longer spacers allow the aromatic group to penetrate deeper into the organic solvent. The second-harmonic generation spectrum used as a probe (Eisenthal, 1996) is surface-specific, meaning that only molecules at the anisotropic interface between the two solvents will give rise to a signal. The density of molecules at the surface is kept low so that they do not form a monolayer and hence act as a probe of the interface. For the non-associating water–cyclohexane interface the polarity is found to change smoothly from that of water to that of cyclohexane ($\varepsilon_r = 2$) in less than a nanometer. But for the hydrogen bonding but immiscible water–1-octanol interface, the spectrum, shown on the right, is quite different. All probes have shifts that show a polarity that is far less than for either solvent. As the ruler length increases, the local dielectric environment around the probe tends to that of the organic layer [adapted from W. H. Steel and R. A. Walker, *Nature* **424**, 296 (2003)].

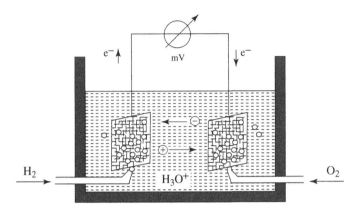

Figure 11.20 Schematic of a hydrogen/oxygen fuel cell. In a commercial alkaline cell of this type the electrolyte is phosphoric acid and it is operated at above room temperature. At even higher temperature the electrolyte is a molten carbonate and the species that migrates is CO_3^{2-}. At sufficiently high temperature a solid oxide can be used and the mobile ion is O_2^{-}.

and verify that these rates are equal when there is no overpotential, $\eta = eV - \Delta G^0 = 0$. The exponential rise of the net current, $j^+ + j^-$, with the overpotential is familiar as the Butler–Volmer law of electrochemistry.

11.4 Understanding chemical reactivity in solution

Let us bring together several key ideas that we have discussed. We seek a unified approach where both the role of the solvent and the rearrangement of the reactants to form products are taken into consideration. We further want the approach to center attention on the correlation of reactivity with structure, a theme that we started in Chapter 6.[27] In essence, we generalize the one-coordinate discussion of solvation in Section 11.1 to a two-dimensional world that consists of a solvation coordinate and a reaction coordinate.* Starting from the gas phase, what we do is generalize the one-coordinate Evans–Polanyi model to include the role of the solvent.

11.4.1 The reaction series

The systematics of the energy profile along the reaction coordinate are obtained from the Evans–Polanyi model. Recall, Sections 5.1.4 and 7.0.2, that this model regards the adiabatic potential energy profile leading from reactants to products

---

* As a dynamical approach, it is a toy model not unlike that of Section 11.2.2.1. But we do not mean to do dynamics. Rather we want to obtain an order-of-magnitude estimate for the rate constant. This is a more forgiving application because we are concerned with terms that are in the exponent rather than with the prefactor.

as determined by the mixing of two diabatic potentials,[28] one representing the electronic state of the reactants and the other of the products, as in Figure 5.9. The adiabatic energy is always lower than the diabatic one. Therefore we obtain an approximation to the location of the transition state as the crossing of the two diabatic potentials, Figure 11.21. To find this point and the height of the potential, we take the diabatic potential curves to be parabolas. This assumption may be unrealistic in the asymptotic reactants-and-products region in the gas phase. Because we intend to apply the results in solution, you can think of the bottoms of the parabolas as the reactants and the products inside the solvent cage.* Let us first examine the transition-state region for which the model of intersecting parabolas captures the essence. Next, and still following Evans and Polanyi, let us make a change in the nature of the products. A nicely studied reaction series[29] is that of substituted benzyl anions with CH_3Br

and we vary the substituent X. Figure 11.21 shows how the diabatic curve of the products shifts down as the reaction exoergicity is changing. The key to getting a simple answer is shifting the products' parabola up or down without changing its shape.

The approximation of the energies of different reactions in the series as congruent parabolas allows us to determine the free energy height of the barrier, in analogy to the derivation of the Marcus Eq. (11.13), as

$$\Delta G^{\ddagger} = (\lambda + \Delta G^0)^2/4\lambda \tag{11.28}$$

It is sometimes useful to rewrite this equation as

$$\Delta G^{\ddagger} = \underbrace{\Delta G_0^{\ddagger}}_{\text{intrinsic barrier}} + (\Delta G^0/2) + \left((\Delta G^0)^2/16\Delta G_0^{\ddagger}\right) \tag{11.29}$$

where $\Delta G_0^{\ddagger} = \lambda/4$ is the intrinsic barrier, independent of the energy requirements or release, and is regarded as a kinetic contribution. The thermodynamic contribution consists of two parts, one that depends on the sign of the energy release and is the thermodynamic driving term and another that is a *mixing term*. The so-called kinetic term is really a property of the energy profile along the reaction coordinate and as the derivation of Eq. (11.13) showed, see also Problem G, the intrinsic barrier is determined by the force constant of the parabolas of Figure 11.21. In structural terms λ is the energy necessary to distort the products (without allowing a chemical change) until they assume the configuration of the reactants, see Figure 5.9.

---

* But you also must consider the free-energy changes that accompany the separated solvated reactants coming together into the cage. We will introduce this correction.

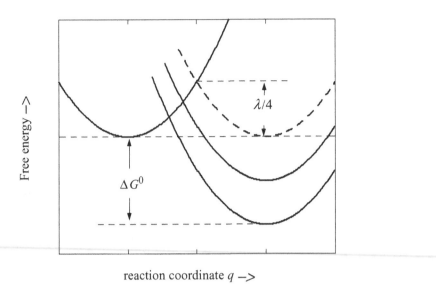

reaction coordinate $q \rightarrow$

Figure 11.21 The essence of the Evans–Polanyi model for systematic variations in the transition state location and height in a reaction series. For a given reactant's parabola three sets of products are shown. Dashed parabola: a symmetrical reaction* for which $\Delta G^0 = 0$. The barrier height originates from only the intrinsic electronic rearrangement necessary to go from reactants to products. By analogy to the Marcus theory of electron transfer, the barrier height is assigned the value $\lambda/4$, where λ is the free energy of the products at the configuration of the reactants, see Figure 11.3. The two solid parabolas on the products' side represent two reactions of increasing free-energy change. Note how the Hammond postulate arises inherently from this construction: the more exoergic reaction has the earlier transition state. Problem G derives the barrier height, Eq. (11.28), and shows that the Hammond postulate is a necessary quantitative intermediate result in the derivation.

11.4.2 The unified approach

So far, the treatment is one-dimensional, along reaction coordinate q. We have not discussed the role of the solvent except possibly as a motivation for why the reactants or products are located at the bottoms of parabolas and do not escape to infinity. If we only treat the solvent, we have the solvation coordinate r of Section 11.2. Here we go to a two-dimensional model where both the reaction coordinate q and the solvent coordinate r can vary. We have two (diabatic) parabolas, each two-dimensional, located over the reactants' and products' configuration, Figure 11.22.

---

* Not every reaction series has a physically clear example of such a symmetric reaction. So this reference case can be hypothetical, but it need not be. Well-studied reaction series for which there is an obvious example of a symmetric case are proton transfers between two bases, $AH + B^+ \rightarrow A^+ + HB$. Many-group transfer reactions and electron transfer reactions also offer examples. We

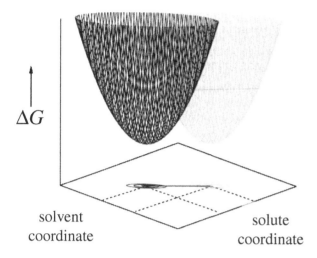

ΔG

solvent
coordinate

solute
coordinate

Figure 11.22 A simple unified model for free energy change of a chemical reaction in solution. Each parabola is a function of the solvent displacement coordinate r and the solute reaction coordinate q. The minimal energy path followed by the system, as shown, involves both coordinates, just as in the toy model of Section 11.2.2.1. Here we do not couple the two motions, although we could. To apply transition state theory we need to know where is the lowest energy barrier and what is its height.

As shown in Figure 11.22 there is a reaction coordinate that involves both solvent and solute motion. Proceeding in familiar steps, we recover again a parabolic expression for the minimal barrier height

$$\Delta G^{\ddagger} = (\lambda_u + \Delta G^0)^2 / 4\lambda_u \tag{11.30}$$

where in the unified model the intrinsic barrier is a sum of three terms, one from the solvent alone, one from the kinetic contribution alone, and a cross term. This cross term represents the shift of the location of the transition-state configuration, from where it would be in the gas phase to where it is in solution, or, equivalently, a shift in the solvation energy caused by the chemical change along the reaction coordinate.

We need one more step. The observed free energy change is not the change from the reactants at the foothills of the barrier to the products at the other side. Rather, it is from the reactants far apart in solution to the well-separated products, Figure 11.23. The same is true for the barrier height. From Figure 11.23 we see that

$$\Delta G^0_{obs} = \Delta G^0 + W_R - W_P = \Delta G^0 + \Delta W$$
$$\Delta G^{\ddagger}_{obs} = \Delta G^{\ddagger} + W_R \tag{11.31}$$

reiterate that the barrier we are discussing is for the isolated system, known as the inner-sphere contribution in the terminology of electron transfer reactions.

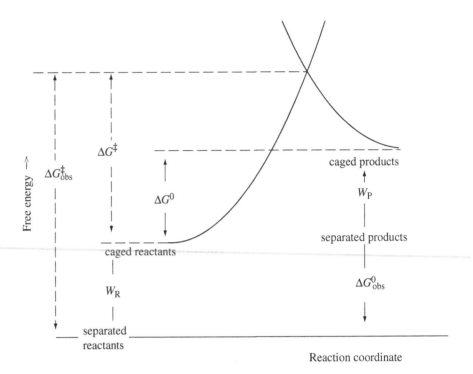

Figure 11.23 Free energy balance in going from separated reactants to separated products in solution. The observed quantities refer to the separated species. Without the subscript "obs" the free energies refer to Figure 11.20, where caged reactants cross the barrier to caged products. W is the free energy of caging and ΔW is the difference between products and reactants and is expected to be small in a given reaction series. In a careful treatment, e.g., Marcus (1997), it is useful to distinguish between the caged species and the caged species with the (old or new) bond stretched to the location corresponding to the minimum of the parabola.

The observed free energy height of the barrier, the one measured with respect to the separated reactants, is from Eqs. (11.30) and (11.31)

$$\Delta G^{\ddagger}_{obs} = W_R + \left(\lambda_u + \left(\Delta G^0_{obs} - \Delta W\right)\right)^2/4\lambda_u$$
$$\xrightarrow{\Delta W \text{small}} W_R + \left(\lambda_u + \Delta G^0_{obs}\right)^2/4\lambda_u \qquad (11.32)$$

11.4.3 Recapitulations

The unified approach is based on what we have learned about typical reaction processes in solution. The first assumption is a separation of time scales between caging and barrier crossing. The reactants reach the foothills of the barrier and remain caged there until solvent reorganization enables them to cross to the products over a low barrier. The barrier crossing is fast compared with the lifetime of the cage. The same behavior is observed in the products. Molecular dynamics simulations, starting with the system at the top of the barrier and watching it roll down in either direction, have provided support for this picture, but direct experimental evidence is still needed. We have kinetic evidence[30] that the three-step mechanism – association, reaction, and dissociation – can rationalize the measured reaction rates in a reaction series. But kinetics alone cannot prove a mechanism. Next is the assumption of a parabolic barrier. The precise dependence is not critical except that the simplicity of the quadratic relation between barrier height and net free energy change depends on it.* The congruence assumption is essential for the concept of a reaction series where all members share the same intrinsic barrier, $\lambda/4$. Finally there is the two-dimensional dynamics requiring a synchronous motion of solvent and system along the reaction coordinate. The simple result that we obtained above was by identifying a one-dimensional cut along which motion can be treated as one-dimensional. In Chapter 8 we cautioned that the one-dimensional picture, while making life simple, does sometimes over-look essential aspects of chemical reality. Consequently, care must be exercised here as well.

The future is surely for direct dynamical examination of the individual ele-mentary steps in such reactions. For proton transfer from excited electronic states this is already a reality,[31] and we look forward to additional work along the same lines of inquiry for other archetypal reactions.

Problems

A. Reaction rate and ionic strength, also known as the primary salt effect. For reactions between ions in solution, the presence of other ions influences the reaction rate. The Debye–Hückel theory provides an explanation for both the direction and the magnitude of the effect. It stems from the stabilization of an ion by a cloud of oppositely charged ions. The stabilization of an ion of charge z scales as $z^2\sqrt{\mu}$ where μ is the ionic strength, $\mu \equiv (1/2)\sum_i c_i z_i^2$ due to the other ions where the cs are concentrations. Physically, when like ions form an encounter complex it is more charged than the separate reactants and so it is more stabilized. The opposite is the case for unlike ions. Derive this conclusion quantitatively. This problem is treated in many textbooks. In fact, the agreement with experiment is better than what one has a right to expect since the Debye–Hückel theory is valid only in the limit of high dilution.

B. Compute $r^{\ddagger}$, the location of the barrier along the reaction coordinate in the Marcus theory of electron transfer. Show how, for a given solvent, the location shifts with the exoergicity of the transfer. Compare your conclusions here to what you did in Problem G of Chapter 5. Should one not be able to unify the two

---

* The parabolic assumption is critical in the prediction of an inverted region for very exoergic processes, where it shows that the rate slows down. In awarding Marcus the Nobel Prize for his work on electron transfer, the Nobel committee specifically cited this prediction.

aspects, the reorganization of the reactants and the reorganization of the solvent? We make a start in Section 11.4. In the meantime, separate the barrier height in the Marcus theory into an intrinsic part and a part that depends on the exoergicity. Note that the second term does depend on the sign of the exoergicity.

C. Show that the Marcus result for the reaction rate constant of electron transfer satisfies detailed balance.

D. Diffusion-controlled reactions. When the barrier crossing is fast the reaction rate in solution is determined by the diffusion of the reactants to form the encounter complex, Eq. (11.16). In this problem you are asked to compute the reaction rate in this limiting case, for an A + B reaction without any long-range potential acting between A and B. Then, get some help from the literature and derive the general result that the effective reaction rate constant is given by

$$k_{\text{effective}} = \frac{4\pi(D_A + D_B)\beta\, k}{k + 4\pi(D_A + D_B)\beta\, \exp(V(R)/k_B T)}$$

Here the Ds are the diffusion coefficients, R is the radius of the cage, and $V(R)$ is the A–B potential. If this potential vanishes beyond R, β, the reaction radius, equals R. (a) Show that the general result above is consistent with Eq. (11.16) of the text. (b) Derive the limiting result $k_{\text{effective}} \cong k_{\text{diffusion}} = 4\pi(D_A + D_B)R$ when the barrier crossing rate constant k is much larger than $k_{\text{diffusion}}$ and there is no long-range potential between A and B. To do this we proceed as follows. Draw a sphere of radius R around A. Let $[B](r)$ be the concentration of B at a distance r from A. The flux of B molecules is given by Fick's law of diffusion as $-(D_A + D_B)\, d[B](r)/dr$. The rate of diffusion through the sphere of radius R is $4\pi R^2$ times the flux. Since there are $[A]$ molecules per unit volume, the rate of diffusion through all spheres is that much higher. At a steady state, the rate of reaction, $k[A][B]$, is balanced by the rate of diffusion, $k[B] = 4\pi R^2(D_A + D_B)\, d[B](r)/dr$. $[B]$ is the bulk concentration of B. Now integrate to obtain $[B](r)$. You will need boundary conditions. Here is where the key approximation comes in: we assume that reaction is so fast that no B molecule lingers and so at the sphere $[B](R) = 0$. This should enable you to get the reaction rate. (c) Now allow the diffusion to take place in a potential gradient. This will require defining a reaction radius β. (d) Finally do not require that $[B](R) = 0$.

E. The Langevin equation. The purpose is to derive a Langevin equation for the motion along the reaction coordinate q in the presence of a solvent. The coupling to the solvent will be introduced implicitly but exactly, through a friction term. The Hamiltonian is given in Eq. (11.24) and the derivation follows Lindenberg and West (1990), Section 4.3.1. The first step is to solve for the motion of the normal modes of the solvent. (a) Write down the equation of motion and hence show that the solution is $x_j(t) = x_j^0(t) - C_j \int_0^t d\tau (\sin \omega_j(t - \tau)/m_j \omega_j) q(\tau)$ where $x_j^0(t)$ is the normal mode vibration in the absence of the solute. (b) Substitute the solution from (a) in the equation of motion for $q(t)$. Integrate the equation by parts and you get the Langevin equation

$$m\frac{d^2q(t)}{dt^2} = -\frac{\partial V(q)}{\partial q} - \int_0^t \eta(t')\dot{q}(t-t')\,dt' + F(t)$$

The first contribution to the force is just from the motion along the reaction coordinate in the absence of the solvent. The second part is the friction term. Note how it depends on the past motion of the system. Hence the friction part is known as the *memory kernel*. (c) Show that the time-dependent friction is given by $\eta(t) = \sum_j \left(C_j/m_j\omega_j^2\right)\cos(\omega_j t)$. Hence conclude that it is a symmetric function of time. (d) $F(t)$ is the so-called *random force* and it represents the role of the fluctuation of the solvent. You can solve for it if only to show that for a solvent in thermal equilibrium, $\langle F(t)\rangle = 0$. (e) The Langevin equation is exact. Hence, for a finite number of solvent modes it shows no damping. You can verify that, with or without coupling to the solvent, the equation propagates into the past just as into the future. On the other hand, and as we discussed in connection with the picket fence model in Chapter 7, for the finite times that are of interest to us, the equation does exhibit damping.

F. Transition state theory for reactions in solution. We have two ways to correct for the role of the solvent. One is to write a Langevin equation for the motion along the reaction coordinate as in Problem E. This gives rise to the Grote–Hynes [R. F. Grote and J. T. Hynes, *J. Chem. Phys.* **73**, 2715 (1980)] correction, Eq. (11.22). The role of the solvent is to introduce a transmission factor κ. This corrects for the recrossings of the barrier. The other option is to work out the transition state theory result when the solvent is explicitly taken into account in the definition of the reaction coordinate and the structure of the transition state. This is what we ask you to do in this problem, for the Hamiltonian given by Eq. (11.24). For this case the two ways give the same answer for the reaction rate constant. This need not be so in general but it reinforces the point that one must be careful before interpreting the role of the solvent as a dynamical effect. The derivation follows E. Pollak, *J. Chem. Phys.* **85**, 865 (1986) and we assume that initially the system is in thermal equilibrium. (a) No frictional coupling to the solvent. Let ω_0 be the frequency of the solute in the well (of the cage) before the reaction, Hamiltonian given by Eq. (11.23). Show that the TST reaction rate constant for a one-dimensional reaction coordinate will be $(\omega_0/2\pi)\exp(-E_{\text{barrier}}/k_B T)$. (b) Show that in the presence of coupling to the solvent this is modified to $(\omega_0/2\pi)(\Pi_i\omega_i/\Pi_i\lambda_i)\exp(-E_{\text{barrier}}/k_B T)$ where, as in the Hamiltonian Eq. (11.24), the ω_is are the frequencies of the solvent. The λ_i's are the frequencies of the bound motions at the transition state. These motions involve both the solvent and the solute and the key to the final answer is to evaluate the required product. (c) Let $\omega^\ddagger$ be the (imaginary) frequency of the pure solute at the transition state as defined in Eq. (11.23). $\lambda^\ddagger$ is the (imaginary) frequency across the barrier but in the presence of the coupling to the solvent. It appears first in the text in Eq. (11.22). Either before the reaction or at the barrier we have a Hamiltonian of coupled harmonic oscillators. It can be exactly diagonalized. Show that $(\Pi_i \omega_i / \Pi_i \lambda_i) = \lambda^{\ddagger}/\omega^{\ddagger}$. This recovers Eq. (11.22) for κ. (d) Finally solve for $\lambda^{\ddagger}$. You will really have to diagonalize the Hamiltonian Eq. (11.24) at $q = q^{\ddagger}$ and to identify the one imaginary frequency. The answer recovers Eq. (11.25).

G. The unified model for barrier crossing in solution. (a) Construct the two intersecting two-dimensional parabolas, each being a function of the solvation coordinate r and the solute coordinate q. (b) Determine the reaction coordinate and the lowest barrier. (c) Determine the intrinsic barrier, when the reaction is symmetrical. (d) Derive Eq. (11.29) for the free energy at the barrier.

H. Electron transfer with strong coupling to the solvent. As in Problem F we need to choose the reaction coordinate to incorporate the role of the solvent [Warshel, 1991; I. Benjamin and E. Pollak, *J. Chem. Phys.* **105**, 9093 (1996); Barzykin *et al.*, 2002]. The purpose is to show that the solvent can modify the dynamics so that the diabatic picture used in Section 11.1.2, Figure 11.3 in particular, needs to be replaced by a diabatic picture [L. D. Zusman, *Chem. Phys.* **49**, 250 (1980)].

Notes

1 For Steric vs. Solvation, see Regan *et al.* (2002).

2 In certain experiments we do not displace the combined solute–solvent system very far from equilibrium. The theoretical tool for handling such situations is the correlation function, which is the equilibrium expectation value of the time evolution of the dynamical variable of interest. This function is usually referred to as a linear response approach (Gordon *et al.*, 1968; Mukamel, 1990). For example, vibrational relaxation of a diatomic molecule in an atomic solvent occurs because of the force F that the solvent applies along the bond axis. The relevant time correlation function is $\langle F(t)F(0) \rangle$, which represents how fast the force varies that determines the efficiency of the energy transfer. Explicitly, this variation needs to be comparable to the vibrational frequency ω. How can we test this condition? By taking the Fourier transform of the time correlation function. The Fourier component at the frequency ω cannot be small. There are however many situations where the coupling to the solvent is strong enough that we cannot use linear response.

3 Diffusive motion occurs when the frictional force, caused by coupling to the solvent, is the dominant driving term. The force of friction is proportional to the velocity, and when it dominates, the motion reaches a steady velocity. The equation of motion is then $ma = dp/dt = 0$, where a is the acceleration and ma is the sum of the force derived from the potential and the force arising from friction. This regime is known as the Smoluchowski limit (Gardiner, 1983). In this limit the reaction rate slows down with increasing coupling to the solvent. If we regard the viscosity as a measure of the friction that the solvent can apply, the reaction rate will slow down with increasing viscosity.

4 For a more detailed discussion of this mechanism, see K. R. Wilson and R. D. Levine, *Chem. Phys. Lett.* **152**, 435 (1988).

5 See, for example, the series of mini-reviews in Vol. 273 of *JBC* and Villa *et al.* (2000), Warshel and Parson (2001).

6 Understanding electron transfer in solution is an active research topic. Critical reviews include Bagchi and Gayathri (1999), Balzani (2001), Walker and Beratan (2001), Newton (1999), Bixon and Jortner (1999), Formosinho *et al.* (1998), Iversen *et al.* (1999), Barbara *et al.* (1996), Hynes (1994), Marcus (1993), Mikkelsen and Ratner (1987), Newton and Sutin (1984), Tributsch and Pohlmann (1998), Ulstrup (1979), Wasielewski (1992), Yoshihara (1999). Even so, ever shorter probes continue to reveal new aspects; see, e.g., Barthel *et al.* (2001).

7 See Taube (1970) for a review of the early work on electron transfer.

8 To obtain a molecular understanding of the nature of coordinate r, one can probe the charge transfer to the solvent itself (Barthel *et al.*, 2001).

9 One can ask why the electron transfer needs to be **over** the barrier. Can an electron not tunnel through it? It can, but just as the case of over the barrier, the solvent needs to adjust to the new parabolic potential. The propensity for doing so is maximal when the vibrational turning points in the two wells coincide. This occurs at the crossing distance $r^{\ddagger}$. There is however an exponentially small but finite transfer probability also by tunneling under the barrier. For a discussion of solvent-mediated coupling see Zimmt and Waldeck (2003).

10 For more on solvation dynamics, see P. G. Wolynes, *J. Chem. Phys.* **86**, 5133 (1987); I. Rips, J. Klafter, and J. Jortner, *J. Chem. Phys.* **89**, 4288 (1988); Bagchi (1989, 1999) Nandi *et al.* (2000).

11 Schwartz *et al.* (1994), Barthel *et al.* (2001).

12 The study of the size effects (Jortner, 1992) applies not only to static properties such as the solvation energy (Castleman and Keesee, 1986; Castleman and Bowen, 1996) but also to dynamical properties (Castleman and Wei, 1994; Zhong and Castleman, 2000).

13 E. Gershgorn, U. Banin and S. Ruhman, *J. Phys. Chem.* **A 102**, 9.

14 J. K. Wang, Q. Liu and A. H. Zewail, *J. Phys. Chem.* **99**, 11309 (1995).

15 Vibrational energy relaxation in liquids is reviewed by Chesnoy and Gale (1984), Elsaesser and Kaiser (1991), Harris *et al.* (1990), Iwaki and Dlott (2001), Owrutsky *et al.* (1994), Oxtoby (1981), Stratt and Maroncelli (1996).

16 Klein (2001), Kropman and Bakker (2001). P. Hamm *et al.*, *J. Chem. Phys.* **107**, 10523 (1997); P. K. Walhout *et al.*, *J. Phys. Chem.* **100**, 5188 (1996).

17 Activated barrier crossing is reviewed by Fleming and Hanggi (1993), Hanggi *et al.* (1990), Hynes (1985), Nitzan (1988), Schroeder and Troe (1987, 1993), Talkner and Hanggi (1995), Pollak (1996).

18 On the transition state approximation in solution D. Chandler, *J. Chem. Phys.* **68**, 2959 (1978).

19 There is the practical question of how to compute the free energy when the solvent extends to large distances. In a practical computation we describe the solvent using so-called periodic boundary conditions that allow for mimicking the effect of many solvent molecules. Alternatively we can treat the solvent beyond the first layer as a continuum (Allen and Tildesley, 1987).

20 See the elegant applications of the factorization in Johnston (1966). For application in the condensed phase see Ben-Nun and Levine (1995).

21 A caveat is that while this is possible for the reactants it is not quite so easy to implement at the transition state.

22 Pollak (1996).

23 For the normal modes of the solvent and their application, see Stratt (1995), Stratt and Maroncelli (1996).

24 E. Pollak, *J. Chem. Phys.* **85**, 865 (1986).

25 For dynamics in the toy model, see Ben-Nun and Levine (1994).

26 For simulations, see also J. Viceli, and I. Benjamin, *J. Phys. Chem. B* **106**, 7898 (2002); S. R. DaRocha, and P. J. Rossky, *J. Phys. Chem. B* **106**, 13250 (2002); S. Senapati, and M. L. Berkowitz, *Phys. Rev. Lett.* **87**, 176101 (2001).

27 A very detailed account of such a unified approach with emphasis on structure–reactivity correlations can be found in the last two chapters of Caldin (2001). Other key reviews include Marcus (1964) and Albery (1980). Marcus (1997) is a careful delineation of all the relevant considerations.

28 Can we obtain a model with three diabatic states: reactants, intermediate, products? See Pross (1995), but more work is needed.

29 For a discussion of a reaction series, see J. A. Dodd and J. I. Brauman, *JACS* **106**, 5356 (1984). See also Chabinyc *et al.* (1998). For many other examples, see Caldin (2001).

30 As summarized by Caldin (2001) and references therein. See also Albery (1980).

31 Fast, Photo-induced proton transfer is an active area, see, for example, Agmon (2003), Arnaut and Formosinho (1993), Gutman and Nachliel (1997); M. Rini *et al.*, *Chem. Phys. Lett.* **281**, E. Pines and E. J. J. Nibbering, *Science*, **301**, 349 (2003); Stolow (2003); J. T. Hynes, *Nature*, **397**, 565 (1999), Tolbert and Solntsev (2002a).

Chapter 12
Dynamics of gas–surface interactions and reactions

Heterogeneous catalysis is responsible for a significant fraction of the output of the chemical industry. The kinetics of surface chemical reactions have therefore been studied extensively.[1] One must not however think that surface processes are synonymous with catalysis. The study of lubrication, known under the modern name of tribology, is very active. For example, what determines the rate of information storage and retrieval is the speed with which the reading head can move over the hard disk at a very low elevation. Other familiar examples of surface processes, such as corrosion, come to mind. Microelectronics and nanostructures on surfaces are also benefiting from and contributing to progress in surface science. An important development is that imaging techniques initially introduced for the probing of surface structure on the atomic scale are revealing details about reaction dynamics. Scanning tunneling microscopy (STM) has been particularly useful.[2]

Our intent in this chapter, as in the rest of this volume, is to examine the molecular-level description. We shall thus make no attempt to review the extensive literature on the macrolevel description but proceed immediately to the microlevel, considering first elastic and energy transfer collisions and then "pre-reactive" and reactive collisions. Many of the experimental and theoretical techniques are closely related to those used to study collisions in the gas phase, but both the experiments and the collision dynamics per se are more complicated because the surface is never really passive and may even be one of the reactants, as in the etching of silicon for microelectronics. Nevertheless, a great deal has been learned, and we have witnessed a molecular "takeover" of the chemistry component of the field of surface science.[3] Much of the heterogeneous chemistry of real-world interest is taking place on amorphous structures or on supported catalysts, etc. Yet much of the fundamental science that we discuss below is for well-characterized surface structures. This is necessary both for the interpretation of the results and the strict reproducibility of the experiment. As in the case of liquids, Section 11.3, the understanding of chemistry on irregular interfaces is just emerging. What is also in progress is the bridging of the gap between the understanding of elementary events and real-world catalytic processes.[4] We begin this process by highlighting a few truths.

12.0.1 A clean surface?

The number of surface atoms per unit area is of the order of $10^{15}\,\text{cm}^{-2}$. But as soon as we generate a pristine surface, say by cleaving a solid, a surface tension is generated, therefore molecules from the gas phase get absorbed on the bare surface and, as we discuss below, are reluctant to leave. It is worthwhile to note how quickly a clean surface gets covered. The flux, I, of molecules coming from the gas phase and striking the surface is $I = n\langle v\rangle/4$ where n is the number density and $\langle v\rangle$ is the mean velocity of molecules at thermal equilibrium. Problem B shows where the factor of $1/4$ comes from. In practical units and relating number density to pressure,

$$I\,(\text{atoms}\,\text{cm}^{-2}\text{s}^{-1}) = 3.5{\cdot}10^{22}\,P\,(\text{torr})/\sqrt{M\,(\text{g}\,\text{mol}^{-1})T(\text{K})}$$

For N_2 at room temperature and at the quite low pressure of 10^{-6} torr, the surface will be covered in a few seconds, assuming that every molecule sticks. For this reason, ultrahigh vacuum ($\approx 10^{-10}$ torr) conditions are necessary in experiments that study reactions on clean surfaces. Even then the surface will be contaminated in about an hour. Otherwise, a surface is always covered and under the higher-pressure conditions (required in industrial catalytic processes to achieve a high throughput of material) there can be several absorbed layers, one on top of the other.*

12.0.2 The reconstructed surface

It is natural to assume that when we cleave a solid so as to expose a fresh surface, the atomic arrangement is that which it was in the bulk. As a rough rule of thumb, this is not so. The newly exposed atoms will rearrange their positions, perhaps only slightly, often more. This reconstruction should be expected. The equilibrium arrangement of the bulk solid is a result of an optimization process where each atom seeks to maximize its attraction to its neighbors while reducing the short-range repulsion. When atoms suddenly find themselves on the surface, the number of their near neighbors is reduced. The surface atoms will therefore re-optimize by seeking a possible configuration of lower energy. Also the second (and even deeper) layer can reconstruct.** This means that when molecules are absorbed on the surface, and particularly so when it is a chemisorption process so that the adsorbate–surface forces are strong, there can be an adsorbate-induced (second) reconstruction of the surface. In short, the surface is flexible and can

---

* Often, only the first layer is strongly bound to the surface and the additional layers can easily be evaporated. Water is an example to the contrary because of the hydrogen bonding (Henderson, 2002). See Problem C for an example of water-saturated air in equilibrium with Ru.

** Clusters of atoms serve as a useful model for this and other surface processes. Small clusters exhibit non-monotonic variations in reactivity with size reflecting this optimization process that leads to different packings for different numbers of atoms.

accommodate to the adsorbate (Somorjai and Rupprechter, 1998). Sometimes one says that "rough surfaces do the chemistry" because it is irregularity of packing of surface atoms that makes for surface sites that are not optimally bonded.

12.0.3 The electronically active surface

Metallic solids have mobile electrons that can be excited at low energies. These are the electrons of the partly full conduction band and the excited electron leaves a hole at the lower energy.* Electronic excitation of the surface can therefore take place at lower collision energies than between molecules in the gas phase. Similarly, charge transfer between an incident projectile and the surface is more common than in collisions between molecules in the gas phase.[5] This alters the bonding in the molecule and facilitates bond stretching and dissociation. Non-adiabatic processes are more common in surface-induced chemistry.

The simplistic discussion of long-range forces in Section 2.1.9 suggests that the low excitation energy of surface charges leads to a physical attraction between a molecule and the surface that is stronger. The interaction also has a longer range,** typically of the form

$$V(z) = -C/z^3$$

where z is the distance from the surface. At shorter range where the (Pauli) repulsion sets in, the potential reflects the atomic arrangement on the surface, that is, the surface is corrugated and the potential depends on the precise point of impact. The net result is a physisorption well whose depth can be significant, particularly on metal surfaces (where the low excitation energies of the conduction electrons make the polarizability and hence the coefficient C large). The well depth can exceed $k_B T$ at room temperature even for rare gases. For molecules, the physisorption potential can be quite anisotropic.†

12.1 Surface scattering

Just as for gas-phase molecular collisions, gas–surface encounters can be elastic, inelastic, or reactive in nature. A wide range of scattering behavior is observed depending upon the gas molecules, the composition, structure, and temperature

---

* In a semiconductor the valence band is full and the conduction band is empty. The band gap need not be large and furthermore it contains impurity levels. So the creation of electron–hole pairs is also possible at low-energy collisions with semiconductors.

** This is because the incoming molecule samples not just the interaction with the site directly where it will land but with the entire surface layer. The attraction to sites further away is weaker but there are more such distant sites.

† As for a molecule solvated in a liquid, the physical potential between the molecule and its environment can be modeled by considering the anisotropic charge distribution on the molecule. See, for example, D. B. Whitehouse *et al.*, *J. Phys. Chem.* **95**, 8175 (1991).

of the surface, the translational and internal energies of the molecules, and even on the orientation of the colliding molecules with respect to the array of surface atoms.

The processes that need to be considered include two kinds of elastic scattering. In either there is no energy loss, and in both cases the component of the projectile momentum in the direction perpendicular to the surface reverses its sign but without any change in magnitude. For *specular scattering*, there is also no change in the component of momentum parallel to the surface. The angle of scattering is thus equal to the incident angle. *Diffraction scattering* is a quantal interference phenomenon and occurs when the de Broglie wavelength of the incident projectile is comparable to the spacing of atoms at the surface. Diffraction scattering is thus seen mostly for light projectiles, typically He atoms, at low velocities. In Problem D you are asked to show that it is characterized by constructive interference of scattering from two adjacent surface sites, namely a "Bragg-like" condition familiar from X-ray structure determination. The important practical difference is that X-rays do penetrate into the bulk solid while low-energy atoms scatter from the surface. Hence diffraction scattering probes the periodicity of the surface structure.

12.1.1 Inelastic scattering

Next we recognize three kinds of inelastic scattering involving a change in the component of the momentum of the projectile in the direction normal to the surface. Two are of the direct mode and are discussed in this section. The indirect kind is the subject of Section 12.1.1.1. First we deal with atomic, "structureless," projectiles where the only possible direct inelastic process is energy exchange between the incident translational motion and the solid degrees of freedom (phonons but possibly also the electronic excitation). A simple picture that provides a realistic prediction of the trends in energy transfer is the *hard-cube* model: the incident atom is assumed to undergo a binary elastic collision with a hard cube that is viewed as a surface atom but with an effective mass M that reflects the interaction of that surface atom with its neighbors. The hard cube has a thermal velocity distribution in the direction normal to the surface. The velocity of the incident atom, v, is changed only in the normal direction and the magnitude of the change is determined, Problem E, from conservation of momentum in the collision between the projectile and the hard cube. The outgoing velocity is given by, Problem E,

$$v'_\perp = ((\mu - 1)/(\mu + 1))\, v_\perp + (2/(\mu + 1))\, u$$

where u is the thermal velocity of the hard cube and μ is the mass ratio m/M, where m is the incident mass. Typically $\mu < 1$ and the smaller is μ the more rigid is the surface. The effect of the second term, due to the thermal motion of the hard cube, becomes more important the lower is the incident energy compared with

$k_B T$, meaning that the thermal motion of the cube is comparable to the duration of the collision. At higher temperatures the thermal component can overwhelm the first term and cause the incident atom to emerge from the collision with a higher translational energy.

Molecular projectiles offer the possibility of additional direct inelastic channels, namely, the excitation or de-excitation of the molecular internal modes, much as for gas-phase molecular inelastic scattering. Unlike a gas-phase collision of a molecule with a structureless projectile, here the energy balance of the internal modes of the molecule need not be met entirely by the translation. The participation of the surface degrees of freedom is possible and the low-energy modes of the surface, not only phonons but also electron–hole pairs, are particularly important in bridging the gap (remember the exponential gap principle) and thereby making such inelastic collisions quite efficient.

12.1.1.1 Trapping at the surface

Ubiquitous to interfaces is the possibility of "sticking" collisions where the incident projectile is temporarily bound to the surface by the strong physical attraction before departing. This is similar to the longer-living complexes in gas-phase collisions. Here the projectile is initially trapped in the attractive well of the physical potential to the surface, a well that is often deep on the scale of thermal energies.* The energy released by this binding is taken up by the surface and so eventually trapping can be followed by desorption where the energy is returned to the translation along the z direction that is normal to the surface. But owing to the long residence time of the projectile on the surface, considerable energy exchange can take place and the exit velocity of the projectile can be much lower than the incident velocity. This is shown both schematically and for experimental data in Figure 12.1.

The signature of the trapping can be seen not only in the translational energy of the outgoing particles, Figure 12.1, but also in the angular distribution of the scattered particles, Figure 12.2. The reason is that projectiles that have long been trapped at the surface can be taken to have forgotten the initial direction of incidence. So their angular distribution will not be specular. Rather, trapped species desorb preferentially in the direction of the normal to the surface, as shown by the following argument. Consider a gas at thermal equilibrium inside a cubic container and apply **detailed** balance. At equilibrium, the flux of projectiles impinging on the surface while making an angle θ with the normal is equal to the flux of trapped projectiles desorbing into that direction. The incident flux is determined by the velocity of projectiles in the direction normal to the surface. At equilibrium the incident flux is proportional to $\cos \theta$. Hence the *law of cosines*: the flux of desorbing atoms varies as $\cos \theta$ where θ is the angle measured with respect

---

* The analogous situation in the gas phase is ion–molecule collisions, Section 6.2.3.1, where there is a deep well in both the entrance and exit valleys.

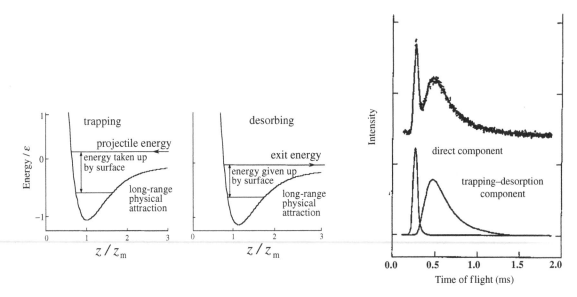

Figure 12.1 Left: the mechanism of trapping–desorption. Right: experimental results showing that even atoms can be trapped at the surface. Shown are the time-of-flight spectra of Ar atoms at an incident energy of $12.5\,kJ\,mol^{-1}$, scattered from the 111 surface of Pt at 110 K. The incident angle is $60°$ while the outgoing atoms are detected at $40°$ of the normal. The distribution is due to two processes, direct inelastic scattering that leads to faster outgoing atoms and trapping followed by desorption that results in higher energy uptake by the surface. The curve through the data is a sum of two separate components [adapted from M. Head-Gordon *et al.*, *J. Chem. Phys.* **94**, 1516 (1991)]. A model for the trapping process by adding an attractive potential to the interaction of the hard cube and the incident Ar atom is discussed in Problem E.

to the normal direction. Experimentally, Figure 12.2, one typically observes both a direct, that is, specular, component in the direction complementary to incidence and a diffuse, law of cosines, component such that the observed scattering is the sum of the two.

Projectiles can react with surface atoms or with adsorbed molecules, but more complex processes involving interaction via diffusion of adsorbed species are also possible. Diffusion of absorbed species on the surface is often fast on the time scale of how long a molecule remains trapped on the surface. This allows us to assume that the molecule has had time to sample the available binding sites and configurations and can therefore be regarded as having reached thermal equilibrium with the surface. One often says that the molecule has *accommodated* to the surface. When these molecules desorb they are found to do so with unexpectedly fast rates and this can be traced not to a low activation energy but rather to a high Arrhenius A factor. Can we use transition state theory to rationalize this observation? To apply the theory we need to compare the partition function of the

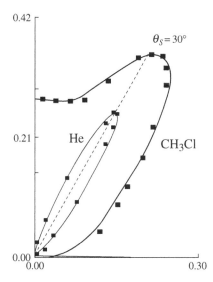

0.42

$\theta_S = 30°$

0.21

He

CH_3Cl

0.00
0.00   0.30

Figure 12.2 The two components in the angular distribution of the polar molecule CH_3Cl scattered off graphite, compared with the scattering of He from the same surface. The incident angle, θ_s, for both is $30°$. The curve shown through the scattering intensity of CH_3Cl is a sum of a direct specular component, like that for He, and a cosine law component for the trapped–desorbed molecules [adapted from T. J. Curtiss, R. S. Mackay, and R. B. Bernstein, *J. Chem. Phys.* **93**, 7387 (1990)].

reactant, namely a molecule bound to the surface, and that of the transition state. Let us consider a diatomic molecule with a high vibrational frequency so that the vibrational partition function is comparable to unity in either state. The reaction coordinate is presumably the normal distance to the surface. So that leaves the two translational modes of the center of mass in the two directions along the surface and the two rotations. The large anisotropy of the potential means that the adsorbed molecule is rather restricted in its ability to rotate. Experiments with oriented molecules indeed show a strong steric effect and spectroscopic probing of adsorbed molecules confirms that a particular end of the molecule is typically preferentially bound to the surface, e.g., the N end of NO is more strongly bound to metals. On the other hand, in the transition state the molecule is further away from the surface and its rotation is far less restricted. The rotational partition function will therefore be significantly larger in the transition state. The same is true for the motion of the center of mass. For the molecule bound to the surface this motion is frustrated but it is almost free in the transition state. The two effects together mean that the pre-exponential factor can be larger* by three to four orders of magnitude than $k_B T / h$ ($\cong 5 \cdot 10^{12}$ s^{-1} at room temperature, Problem F).

---

* In Section 6.1.4.1 we have seen how transition state theory can account for the pre-exponential factor being smaller than $k_B T / h$. There it was due to the loss of entropy in forming the transition state due to the steric requirements. Here the steric requirements work in the opposite direction. It is

12.1.2   Collision-induced surface processes

Projectiles incident from the gas phase can induce both inelastic and reactive processes in adsorbed molecules.[6] Such processes are typical of incident molecules from the high-energy tail of the thermal energy distribution in bulk systems. Desorption or dissociation of adsorbed molecules, CID, can be induced by projectiles and it is only to be expected that collisions can also cause adsorbed molecules to migrate along the surface, CIM, a process that usually has a lower activation energy than outright desorption. The fast projectiles can also be generated photochemically on the surface, see Section 12.2.4. The chemistry that is induced by collisions with species from the gas phase is known as an Eley–Rideal mechanism, Section 12.2.2.

12.2   Dynamics on surfaces

12.2.1   Dissociative adsorption

A key surface elementary process is dissociative adsorption wherein the incident molecule undergoes bond rupture (Darling and Holloway, 1995). The energy required to break a bond in the incident molecule (typically 1–5 eV) is provided by the formation of new bonds with the (unsaturated) surface atoms. The dissociative adsorption of H_2 on clean copper can serve as an example:

$$
\begin{array}{ccccc}
H_2(g) & & H\!-\!H & & H \quad\; H \\
+ & \rightarrow & |\;\; | & \rightarrow & |\quad\; | \\
Cu\cdots Cu\cdots & & CuCuCu\cdots Cu\cdot & & CuCu\cdots CuCu\cdots
\end{array}
$$

Such processes are confirmed by observations of facile isotopic exchange, for example, detection of scattered HD when the incident beam consists of a mixture of H_2 and D_2.

The mechanism of dissociative adsorption necessarily involves the crossing of (at least) two potential energy hypersurfaces (Section 3.2.4.3) since both the undissociated and dissociated molecule can be at the (physical) surface. A simplified potential curve-crossing diagram is shown in Figure 3.6 for a diatomic molecule approaching the (physical) surface. Given sufficient energy, the molecule can pass from the weaker (often van der Waals) well, binding the undissociated molecule to the surface and over the barrier E_b, if any, to the region of a deep chemical well corresponding to the formation of two new bonds to the surface (at the expense of breaking the molecular bond). The energy barrier, E_b, for dissociative (or "activated") adsorption can be measured as a translational energy threshold for the dependence of the sticking probability S upon the component of kinetic energy normal to the surface, i.e., $E_\perp = E_T \cos^2 \theta_i$. As in the

the reactant that is more constrained. For both situations transition state theory correctly rationalizes the observable trends.

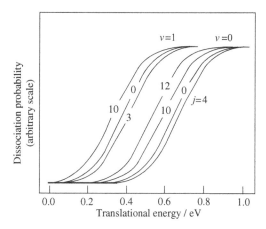

Figure 12.3 Dissociative sticking probability S_d vs. normal component of the translational energy of the incident D_2 molecules in a given initial vibrational and rotational state incident on a Cu(111) surface [adapted from H. A. Michelsen *et al.*, *J. Chem. Phys.* **98**, 8294 (1993); see also Rettner *et al.* (1996), Darling *et al.* (2001)]. The initial vibrational excitation of D_2 serves to lower the barrier to dissociation. The effect of rotation is in two opposing directions. Dissociation occurs preferentially when the molecule is parallel to the surface, so rotation should hinder dissociation. On the other hand, the higher the initial rotational state, the more energy is brought in by the molecule. The latter effect increases quadratically with the quantum number and so wins out at the higher j's.

gas phase, the barrier height may well be different for different energies (e.g., translational vs. vibrational) of the incident molecule and will also depend on the orientation of the molecule with respect to the surface, as seen in Figure 12.3. So far, everything is much the same as in the gas phase. A new twist is that the barrier can also differ for different faces of the crystal (which differ in the detailed atomic configuration at the surface).

Another indicator of the presence of the barrier is the energy disposal in the desorbed molecules. From our considerations so far we expect the barrier energy to appear as translational energy of the emitted molecules. The velocity distributions of H_2 (and D_2) molecules desorbing from Cu surfaces are indeed found to be peaked at high speeds. Time-of-flight distributions (Figure 12.4) are narrower than for a Maxwellian velocity distribution at the surface temperature T_s and have a mean energy that is about four times $k_B T_s$. In addition, the angular distribution is also narrow and is peaked at the normal. For the internal energy of the desorbing H_2 (and D_2) molecules it is found that the rotational distribution is only mildly deviant from a thermal one at T_s but the vibrational excitation is substantially higher than the thermal.

To better explain the observations above, the simple curve-crossing picture of Figure 3.6 is clearly inadequate. To begin with, we must consider at least a potential energy surface, showing the dependence not only on the distance

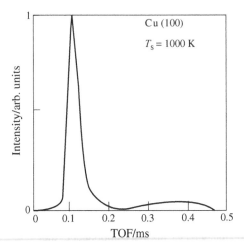

Figure 12.4 Time-of-flight spectrum of D_2 molecules desorbing in the normal direction from a Cu(100) surface at $T_s = 1000$ K. The deuterium was supplied to the surface "from the rear" by permeation through a thin Cu foil (as D atoms), recombining at the surface and desorbing into the forward direction. From an analysis of the time-of-flight data the mean translational energy of the desorbing D_2 molecules is about four times larger than T_s [adapted from G. Comsa and R. David, *Surf. Sci.* **117**, 77 (1982)].

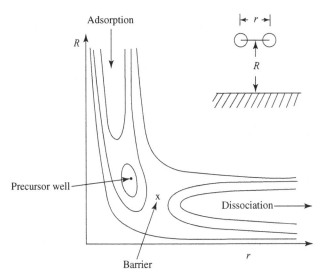

Figure 12.5 Schematic potential energy surface for dissociative adsorption–desorption [adapted from Ertl (1982)]. In the drawing, the barrier to dissociation-recombination occurs when the molecule is already at the surface and is along the bond distance. This will lead to vibrationally excited desorbed molecules. Note also the precursor well along the approach to the (physical) surface. This well will slow down the approaching reactant but may not be deep enough to insure that it fully accommodates to the surface.

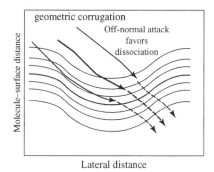

Figure 12.6 Surface corrugation. Potential energy contours, schematic, as a function of the distance of the molecule from the surface and the lateral position along the surface. The thicker line is the location of the barrier to dissociation. Its height and its location along the surface normal are seen to depend on the lateral position. Such a picture helps in understanding why different surface sites may be important for inelastic collisions and for dissociation and why also momentum along the surface can help overcome the barrier to dissociation.

from the surface but also on the bond distance of the diatomic molecule as drawn in Figure 12.5. Then one needs to recognize that the interaction energy can be strongly dependent on the orientation of the molecule with respect to the surface, so that just two variables are not sufficient. Then, unique to gas–surface interaction, there is the effect of surface corrugation: the interaction energy will depend on the precise lateral location on the surface, Figure 12.6. Finally, we must also remember that the surface itself is neither static nor two-dimensional.

The many factors that come into play make it difficult to draw quantitative, microlevel conclusions from desorption data alone and simulations play an important role in unraveling the dynamics. This is even more so when it comes to the understanding of reactive processes on the surface.

12.2.2 Heterogeneous chemical reactivity

In a gas–surface reaction the surface atoms can be reagents and therefore can be consumed in the reaction or the surface can serve as a catalyst. In the latter case, one often distinguishes between the Eley–Rideal mechanism, where the molecules from the gas react with surface chemisorbed reagents, or the Langmuir–Hinshelwood mechanism, where both reagents are chemisorbed prior to reaction. Of course, in view of the often multistaged nature of the adsorption process, there can be intermediate situations as well, where the absorbed molecule is in a precursor state, see Figure 12.5, so that it is on the surface but it is not fully accommodated.[7]

Reactions of gaseous molecules with surface atoms are often of practical importance. They include the oxidation of graphite, $O_2 + C(gr) \rightarrow CO + O$, and other, so-called, corrosion reactions that lead to desorbing products at the temperature of the experiment, e.g., $Cl_2 + Ni(s) \rightarrow NiCl(g) + Cl$. The pre-exponential factors for such reactions can be estimated (to within an order of magnitude) from transition state theory adapted to the two-dimensional world of surface chemistry. In so doing, one must identify correctly the rate-determining step (whether reagent chemisorption or surface diffusion or reaction or product desorption).

Simple surface-catalysed reactions that have been well studied include isotopic exchange reactions,

$$H_2 + D_2 \xrightarrow{\text{on a Pt surface}} 2HD \quad \text{or} \quad {}^{14}N^{14}N + {}^{15}N^{15}N \xrightarrow{\text{on a W surface}} 2\,{}^{14}N^{15}N$$

bimolecular reactions,

$$CO + O_2 \xrightarrow{\text{surface}} CO_2 + O \quad \text{or} \quad C_2H_4 + H_2 \xrightarrow{\text{surface}} C_2H_5 + H$$

and dissociation,

$$N_2O \xrightarrow{\text{surface}} N_2 + O \quad \text{or} \quad HCOOH \xrightarrow{\text{surface}} H_2 + CO_2$$

The exchange reactions are perhaps the most instructive since they require only dissociative adsorption followed by surface diffusion prior to recombination. Thus, when a beam containing a mixture of H_2 and D_2 is incident on a Pt surface, the integrated scattered yield of HD can be as high as 35% depending on the angle of incidence *and* the detailed atomic arrangement at the surface. Figure 12.7 compares the HD yield from a smooth (111) or stepped (332) Pt surface. The data imply that the exchange probability per unit surface is some sevenfold greater for "step" sites than for "terrace" sites, and that this difference is due to a higher probability for dissociative adsorption on the stepped surface.[8] Subsequent processes are probably the same for both types of surface plane.

For N_2 on W(100), the yield of ${}^{14}N^{15}N$ is essentially identical to the directly measured sticking probability of the molecules on the surface. This is to be expected because the reaction rate at low energies is dominated by the formation of a precursor.[9] The role of dissociative adsorption is also demonstrated in the absence of any detectable CO_2 formation when O_2 is incident on a Pt surface that is saturated with CO. The converse is not the case. However, since the O atom saturation coverage is low, the question of whether CO is adsorbed prior to reaction needs to be considered. By modulating the CO beam it is found that the mean residence time on the surface is very long. The reaction occurs between two adsorbed species (a Langmuir–Hinshelwood mechanism) to produce adsorbed CO_2. That the adsorbed species need not be in its gas-phase equilibrium configuration is revealed by measuring the energy disposal and angular distribution, as discussed in Section 12.2.3. The many intertwined steps during a surface reaction mean that the kinetics can be made to exhibit oscillations and pattern formation as discussed in Section 12.3.

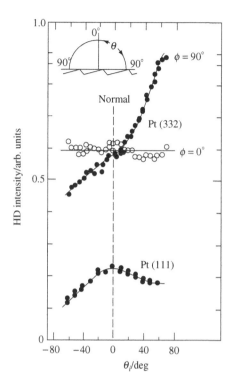

Figure 12.7 HD formation as a function of the angle of incidence θ_i, measured from the normal to the surface, when a mixed H_2/D_2 effusive beam (300 K) is scattered from clean surfaces of Pt at a surface temperature $T_s = 1100$ K. The bottom curve is for a smooth Pt(111) surface, for which the HD yield is relatively small and there is no dependence on azimuthal angle* ϕ. The two upper data sets are for a stepped Pt(332) surface at the same T_s. At $\phi = 0$ the HD yield is independent of θ_i. However, for $\phi = 90°$ the incident beam strikes the step edges; here a strong dependence of the scattered HD intensity upon θ_i is found, with greatly enhanced yields at large positive θ_i and decreased yields at negative θ_i [adapted from Somorjai (1981)].

Many reactions on the surface occur between reagents that have come to equilibrium with the surface and this is particularly so when diffusion along the surface is facile. As a practical definition one can regard the Eley–Rideal mechanism as the situation when this is not the case. The primary signature of such a "direct" surface reaction (Rettner and Auerbach, 1994) is, as in direct reactions in the gas phase, the distribution of the products, both in space and in energy, and the dependence of these distributions on the initial state of the gas-phase reactant. Even reactions that otherwise will proceed by a

---

* Consider a beam incident on the surface at a given angle θ. As we vary the azimuthal angle ϕ the beam will span a cone. If the surface is uniform it makes no difference what is the value of ϕ. But if the surface is rough, it very much does matter. The experiment shown was an early direct demonstration of what we now take for granted, namely that surface irregularities are active sites.

Langmuir–Hinshelwood mechanism can be direct if the surface is saturated with one reactant. The near specular exit of HD from H (or D) atoms incident at a given angle on D/Cu(111) (or H/Cu(111)) is a well-studied example. Further applications of reaction dynamics to elucidate the mechanism are discussed next.

12.2.3 Dynamics of gas–surface reactions

The many relevant variables in gas–surface reactions make it almost imperative to probe the dynamics before conclusions on the molecular mechanism can be reached. If anywhere, it is here that laser probing to determine the velocity and angular distribution for each individual final quantum state (Section 7.1.3) will prove decisive.

The beam-scattering studies of the oxidation of CO on Pt in the presence of adsorbed O_2 have not only detected a long residence time prior to CO_2 desorption but have also shown that the emerging CO_2 molecules are translationally fairly hot and vibrationally excited (as revealed by infrared emission). The angular distribution of the desorbing CO_2 molecules is sharper than cosine, and this also is indicative of repulsive forces operating in the exit valley. In contrast, the angular distribution of NO from the Ni(100)-catalysed dissociation of N_2O is cosine, suggestive of accommodation of NO prior to desorption.

A simpler reaction that is more easily interpreted is the dissociation of fast molecules impacting a hard surface. Figure 12.8 shows the dissociation probability of I_2 on MgO(100) measured as a function of the initial kinetic energy. The fraction of molecules dissociated rises rapidly from the thermochemical threshold ($\cong$ the dissociation energy of I_2, 1.54 eV) and tends to level off at higher energies. Although a rigid-surface model can account for this trend, time-of-flight data on the undissociated I_2 show a large average energy transfer to the solid requiring a non-rigid description of the solid. Such a model, which fits the inelasticity data, also recovers the energy dependence of the dissociation probability as shown in Figure 12.8. Note that the dissociation yield saturates at well below 100%. This is an example of the steric effect that we already alluded to in connection with the role of reagent rotation: in order to dissociate, the molecule should preferentially hit the surface sideways.

The theoretical approach that so far has been most effective in describing the dynamics of adsorption–desorption and of reactive gas–surface collisions is based on the method of classical trajectories. The essence of the problem is to provide a tractable yet realistic approach to the coupling of the molecular and surface (and bulk) degrees of freedom. In principle, one can introduce a (often, semi-empirical) potential energy, which is a function of the positions of all atoms, both those of the molecule and those of the surface. The classical equations of motion can then be solved. Since each atom of the solid is interacting with its neighbors, the number of coupled differential equations that need to be solved in

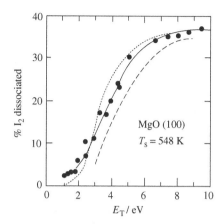

Figure 12.8 Dissociation probability (percentage) for I_2 molecules striking a MgO(100) surface at $T_s = 548$ K as a function of their incident translational energy. Experimental data: points connected by a solid curve. Theoretical calculations: dotted curve, rigid surface model; dashed curve, non-rigid [adapted from E. Kolodney *et al.*, *Chem. Phys. Lett.* **111**, 366 (1984)].

order to keep track is enormous. Moreover, the solution needs to be repeated many times with different initial conditions, so that, for example, thermal averaging from the solid's degrees of freedom can be carried out. This is computationally unrealistic and is also unreasonable on physical grounds: surely the outcome of the surface reaction cannot depend on the precise details of the motion of an atom several layers below the surface? This situation is already familiar to us from dynamics in the liquid and the same way out is available: we average over the solid's degrees of freedom first and then integrate for the time evolution of the molecular degrees of freedom. This can be done in a manner that is formally exact and leads to the result that the presence of the surface introduces two additional terms in the equation of motion for the molecular degrees of freedom (known as the generalized Langevin equation). The two extra terms are a dissipative or frictional force leading to energy flow into (out of) the surface and a random force reflecting the thermal fluctuations in the positions and momenta of the surface atoms.

As in the liquid, evaluating the two additional terms exactly is as difficult as the original problem. However, because of their physical interpretation it is possible to provide simple yet realistic approximations for them. In practice, one sometimes solves the equations of motion not only for the molecular degrees of freedom but also for those surface atoms to which they are directly coupled. Only the rest of the solid is averaged over. This structureless, "pillow-like" description of the environment has enabled the method of classical trajectories to be applied not only to reactions at the surface but also in solution. Unique to the surface is the need to allow for electron–hole pair excitations.[10]

12.2.4 Laser-induced processes

Finally, we consider the interaction of photons with adsorbed molecules and, specifically, laser-induced desorption and laser-induced photofragmentation processes. The potential practical applications of the former have resulted in considerable activity, but definitive generalizations on the dynamics are still not at hand. Photochemistry at surfaces is being actively explored (Dai and Ho, 1995), and the ability of the surface to align the absorbed molecules (Polanyi, 2001) allows reactions to occur with restricted geometries of approach as shown in Figure 12.9.

Laser desorption using CO_2 lasers for heating the substrates has been widely used to produce evaporated species ranging from metal atoms to complex organic polyatomic molecules. By sweeping these in a beam of cold gas such as He one can study large molecules in the gas phase. This method has also been used to produce clusters of atoms or molecules for spectroscopic or kinetic study. Visible and ultraviolet lasers providing larger photon energies are sometimes preferred when attempting vaporization of refractory metals – some of the rare earths, transition metals, etc. C_{60} was first detected in this way.

An ultrafast laser can excite the surface electrons and these can induce chemistry as well as desorption prior to their being rapidly ($\sim$1 ps) thermalized with the phonons (Bonn *et al.*, 1999). The electronic mechanism for laser-induced desorption (Gomer, 1983; Gadzuk, 1988; Avouris and Walkup, 1989) is through the temporary formation of the negative ion of the adsorbate. The ion is pulled strongly toward the surface while its equilibrium distance tends to increase. Shortly thereafter the charge is returned to the surface and a vibrationally excited neutral is ejected.

An early example of photodissociation is a study of the CH_3Br molecule chemisorbed on a LiF(100) surface. Using an ultraviolet laser tuned to 222 nm, the time-of-flight velocity of the CH_3 radical from the photodissociation of the adsorbed CH_3Br was determined. The key observation is that the translational energy distribution of the CH_3 normal to the surface extends to the theoretical upper limit, assuming all the excess energy (i.e., the photon minus the H_3C-Br bond dissociation energy) goes into CH_3 recoil, implying that the CH_3 is recoiling from a more massive "particle" than Br, namely, the "near-infinite" mass of the surface. The inference is that the H_3C-Br bond of the chemisorbed CH_3Br is perpendicular to the surface.[*]

An elegant example of a surface-aligned reaction is the oxidation of CO with O atoms produced by photodissociation of O_2 when both reactants are localized parallel to step sites on a Pt surface. CO molecules absorbed on terraces of the same surface are less than half as reactive. The principle of the experiment and the results of probing the CO reactant before and after irradiation are shown in Figure 12.9.

---

[*] The CH_3Br bond is perpendicular to the surface but whether it is the Br atom or the CH_3 group that is near the surface depends on the surface coverage.

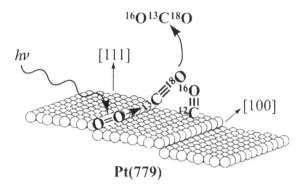

Pt(779)

Figure 12.9   Schematic of the surface-aligned photo-oxidation experiment on a highly corrugated Pt surface and spectroscopic probing of the depletion of CO, showing the preferential reactivity of CO molecules that are along a step edge. The first stage in the experiment is to cover about 50% of the step sites with O_2. Next, isotopic $^{13}C^{18}O$ is introduced to saturate the step sites. Then isotopic $^{12}C^{16}O$ is absorbed on the terraces. The width of the terrace is about 17 Å between two edges. Spectroscopic monitoring confirmed that at low temperatures (88 K) there is no exchange of CO molecules between the two types of site. Photodissociation of O_2 produces O atoms preferentially along the step edge and these react with the coadsorbed and coaligned $^{13}C^{18}O$ molecules, which then desorb as $^{16}O^{13}C^{18}O$ molecules. There is some reaction of the $^{12}C^{16}O$ molecules adsorbed on the terrace, but their depletion is significantly less. The concentration of the different isotopomers is measured by their different absorptions and this can also distinguish between the absorption of the same isotopomer at the two different sites. Using this (small) shift one notes that as reaction preferentially depletes $^{13}C^{18}O$ molecules from the edge sites, $^{12}C^{16}O$ molecules move into these vacated sites [adapted from E. M. Tripa and J. T. Yates Jr., *J. Chem. Phys.* **112**, 2463 (2000)].

12.3   Chaos and pattern formation: spatiotemporal aspects of surface reactivity

Chaotic behavior requires a nonlinearity in the equations of motion. For conservative mechanical systems, of which computing classical trajectories is, for us, the prime example, Section 5.2.2.1, the nonlinearity is due to the anharmonicity of the potential. In chemical kinetics[11] there are two sources of nonlinearity. One is when the concentrations are not uniform throughout the system so that diffusion must be taken into account. The other is if there is a feedback so that, for example, formation of products influences the reaction rate, see Problem H. As we shall see, this type of nonlinearity occurs naturally in many surface reactions and this is why we chose catalytic processes as an example. In both mechanical and chemical kinetics systems there is one more way to add nonlinear terms and this is by an external perturbation. For surface reactions this additional control can be implemented, for example, by modulating the gas-phase pressures of reactants and/or products.[12]

The full manifestation of chaotic behavior requires that at least three equations of motion are coupled. This is not a strong requirement for a mechanical system. Each degree of freedom gives rise to two (Hamilton) equations of motion (or one, but second-order, Newton equation). So two coupled (anharmonic) oscillators can already exhibit chaotic behavior. Solving the trajectory for an atom colliding with a diatom requires six equations. If there are only two variables, one can get oscillatory solutions but not chaos.

Whether the behavior is fully chaotic or not depends on both the initial conditions and the magnitude of the parameters that appear in the equations of motion. In chemical kinetics these parameters are rate constants and their value can be varied, most simply by changing the temperature. En route to chaos the system goes through eye-catching patterns and these are our topic here. For surface reactions such patterns have been extensively observed, particularly in oxidation reactions using O_2 or NO as the oxidants on transition metal catalysts.[13]

12.3.1   The $CO + O_2$ reaction on Pt(110)

The reason for picking a particular Pt surface is that, unlike the hexagonal close-packed Pt(111) surface, the more open, so-called missing row, Pt(110) surface reconstructs, as shown schematically in Figure 12.10. As further emphasized in the figure, the sticking of O_2 is different for the two phases, being higher for the unreconstructed surface. The reconstruction of the bare surface can be driven back by increasing the concentration of CO and so O_2 sticks preferentially on the bare metal areas when the coverage by CO is higher. The fraction of reconstructed surface provides the essential third variable besides the coverage of the surface by CO and by O atoms.

The kinetic equations describing the reaction need to take into account the following observations (Ertl, 1982). A surface saturated by CO will not lead to CO_2 formation when O_2 is introduced. So the mechanism cannot be an Eley–Rideal with O_2 coming from the gas phase. The O_2 molecules need to first dissociate on the surface. A surface saturated with O_2 will, however, lead to CO_2 formation when it is exposed to CO. To rule out an Eley–Rideal mechanism with CO coming from the gas phase to react with adsorbed O atoms it was necessary to show that CO can still be absorbed on an O-saturated surface. The explanation is that the O atom coverage is rather open with an O—O distance of about 5.5 Å. Using S to denote a surface site, the mechanism, without taking surface reconstruction into account, and neglecting any desorption of O_2, is

$$CO + S \underset{k_d}{\overset{\kappa_{CO}}{\rightleftharpoons}} CO_{ad}$$

$$O_2 + 2S \xrightarrow{\kappa_O} 2O_{ad}$$

$$O_{ad} + CO_{ad} \xrightarrow{k_r} CO_2 + 2S$$

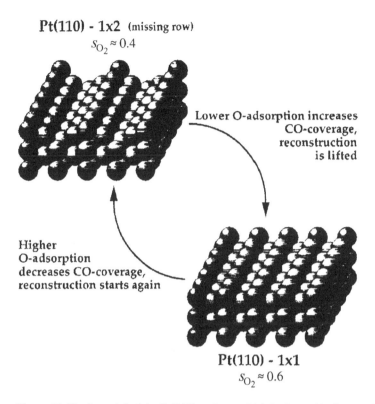

Pt(110) - 1x2 (missing row)
$$s_{O_2} \approx 0.4$$

Lower O-adsorption increases
CO-coverage,
reconstruction
is lifted

Higher
O-adsorption
decreases CO-coverage,
reconstruction starts again

Pt(110) - 1x1
$$s_{O_2} \approx 0.6$$

Figure 12.10 A model of the Pt(110) surface, which is the stable form at high CO coverage, and the reconstructed surface. Note the missing row in the 110 surface and the steps on the reconstructed surface; s_{O_2} is the sticking coefficient of molecular oxygen and it is higher for the unreconstructed (i.e., CO covered) surface [adapted from Imbihl and Ertl (1995)].

where the subscript ad means an adsorbed species.* This reaction scheme can exhibit oscillatory kinetics because at a low partial pressure of CO the reaction rate will rise with increasing concentration of adsorbed CO while at a higher pressure the surface will become saturated with CO and reaction will cease. Using θ to denote coverage, with a subscript sat for the coverage at saturation, coupled rate equations for the assumed mechanism are

$$\frac{d\theta_{CO}}{dt} = \kappa_{CO} P_{CO} \left(1 - (\theta_{CO}/\theta_{COsat})\right) - k_d \theta_{CO} - k_r \theta_{CO} \theta_O$$

$$\frac{d\theta_O}{dt} = s_{O_2} \kappa_O P_{O_2} \left(1 - (\theta_{CO}/\theta_{COsat}) - (\theta_O/\theta_{Osat})\right)^2 - k_r \theta_{CO} \theta_O$$

---

* The observed rate of adsorption of CO does not quite have the expected Langmuir form where the rate of adsorption is proportional to the fraction of empty sites. This is because CO is initially trapped in a mobile precursor site. Quantitative agreement with the experimental oscillations requires using $(\theta_{CO}/\theta_{COsat})$ with a power higher than unity, say between 3 and 4.

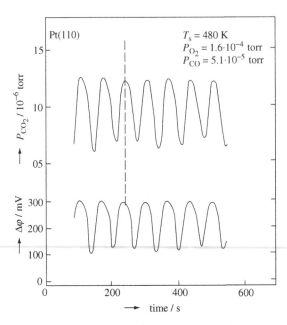

Figure 12.11 Experimental oscillations in the catalytic oxidation of CO by absorbed O atoms on Pt(110) [adapted from M. Eiswirth *et al.*, *J. Chem. Phys.* **90**, 510 (1989)]. Shown is the partial pressure of CO_2 which, because of the high pumping speed, is proportional to the reaction rate and the variation of the surface work function. The correlation between these two variables is expected for the proposed mechanism.

To really complete the model we cannot use a sticking coefficient for O_2 that is the mean value for the two surface forms but must allow for the fraction of Pt(110) surface to be a dynamical variable, coupled to the concentration of CO. The qualitative role of surface reconstruction can be discussed with reference to Figure 12.10. Starting, say, with a CO-covered and hence unreconstructed surface, the adsorption of oxygen and therefore the catalytic oxidation rate will be high. This consumes adsorbed CO molecules and if these cannot be replenished quickly by adsorption, their coverage will decrease below a critical value (CO coverage of about 0.2) and the surface will reconstruct. The sticking probability of O_2 will then go down so that the concentration of adsorbed CO can rise and we return to the starting conditions. It is possible to monitor both the partial pressure of the CO_2 product and the concentration of adsorbed O atoms because the surface work function is high in proportion to the O atom coverage. The oscillations in both are shown in Figure 12.11.

12.3.1.1 Imaging surface processes

The experimental evidence for the coupling between surface structure and catalytic reaction rate uses low-energy electron diffraction, LEED, to monitor the surface structure and the reaction rate is followed through changes in the work

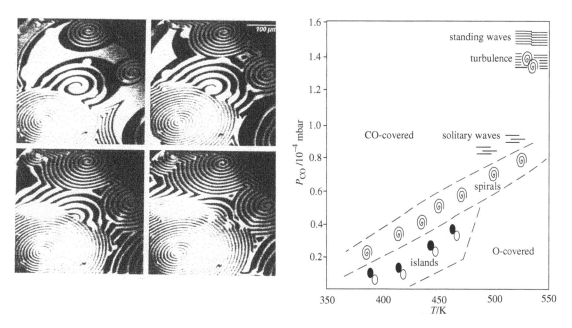

Figure 12.12   Left panel: spiral patterns, a few micrometers in size, measured at different times, for oxidation of CO on Pt (110). Shown are the local variations in the work function, low at CO-covered surface, bright areas, and high for O atom coverage, dark areas. The time interval between the frames is 30 s and as can be seen different spirals rotate at different rates. Right panel: the different spatial patterns that have been analyzed as a function of temperature and partial pressure of CO, for a low ($4 \cdot 10^{-4}$ mbar) O_2 partial pressure. [Both adapted from Rotermund (1997)].

function. Examining patterns also in space and not only in time requires probing with high lateral resolution. Photoemission electron microscopy, PEEM, allows submicrometer resolution (Rotermund, 1997). Anisotropic patterns can arise owing to the anisotropy of the surface diffusion of CO, which is fast along the troughs in the surface, cf. Figure 12.10, but slow in the perpendicular direction. Defects on the surface can serve as the nuclei for spiraling waves, as seen in Figure 12.12. As a reminder that not only the partial pressure of the reactants but also the temperature (that enters through the rate constants) can affect the patterns, we also show a bifurcation diagram, exhibiting the nature of the different patterns.

Problems

A. The Langmuir adsorption isotherm and kinetics of reactions on surfaces. Consider a molecule A coming from the gas phase and absorbed on an empty surface site S that then becomes occupied, $A + S \rightarrow A - S$. The fraction of surface sites that are covered by A molecules is denoted θ. (a) At equilibrium the rate of adsorption, $k_a[A](1 - \theta)$, should equal the rate of desorption, $k_d\theta$ (see Problem E for computing the rate constants). Derive the Langmuir expression for the equilibrium coverage, $\theta = K[A]/(1 + K[A])$. What is K? Plot θ against the pressure of A assuming an ideal gas law. (b) Deviations from the simple adsorption isotherm are not uncommon. One possible reason is that the molecule dissociates on the surface so that each adsorbed molecule occupies two surface sites $A_2 + S + S \rightarrow A - S + A - S$. Show that the equilibrium surface coverage is $\theta = \sqrt{K[A_2]}/(1 + \sqrt{K[A_2]})$. (c) Say that adsorbed A molecules undergo a unimolecular reaction. Then the reaction rate is $k\theta$. Suppose however that A undergoes a bimolecular, Langmuir–Hinshelwood type, reaction with another adsorbed molecule B. Solve first for the equilibrium coverage of A and B and hence show that the bimolecular reaction rate is $kK_AK_B[A][B]/(1 + K_A[A] + K_B[B])^2$ and define all the symbols. (d) Plot the reaction rate as a function of the pressure of B when the pressure of A remains constant. In Section 12.3 we saw a similar behavior when the pressure of one reactant is varied. (e) But suppose that the mechanism is Eley–Rideal, where an absorbed molecule A reacts with a molecule B coming from the gas phase. What then is the expression for the reaction rate?

B. The flux of gas molecules striking a surface. Consider a gas in thermal equilibrium in a container shaped like a cube. Focus attention on a wall of the container that is perpendicular to the z direction. Compute the mean velocity of molecules striking this surface and hence show that the flux is $I = n\langle v \rangle/4$. The factor $1/4$ comes from two different considerations, each contributing a factor of $1/2$.

C. How many water molecules are absorbed on Ru (desorption energy 10 kcal mol^{-1}) at atmospheric pressure (partial pressure of water 20 torr) and room temperature.

D. Diffraction scattering. Say that the flux of He atoms reaching a detector at an angle θ to the surface normal comes from atoms scattered from two different surface sites that are a distance d apart. Compute the path lengths that each atom travels from the surface to the detector. (a) Since the detector is far away from the surface show that the path difference is $d \sin \theta$. (b) Argue that if λ is the de Broglie wavelength of the He atoms, the two events interfere constructively or destructively at different angles θ. At what angle is the main constructive interference? (c) What is the velocity of the He atoms to observe an effect for a realistic value of d? Can we use Xe atoms?

E. The hard-cube model. Adapted from E. K. Grimmelman, J. C. Tully, and M. J. Cardilo, *J. Chem. Phys.* **72**, 1039 (1980). See also Harris (1987). An incident atom of mass m undergoes a binary elastic collision with a hard cube that is viewed as a surface atom with an effective mass M. The velocity of the incident atom, v, is changed only in the direction normal to the surface. (a) Using conservation of momentum show that the outgoing velocity of the atom in the direction normal to the surface is given by $v'_\perp = ((\mu - 1)/(\mu + 1))v_\perp + (2/(\mu + 1))u$ where

u is the velocity of the hard cube and μ is the mass ratio m/M. The velocity in the direction parallel to the surface is assumed not to change. (b) Say that the velocity of the hard cube has a (one-dimensional) thermal distribution. Compute the average kinetic energy of the outgoing atom. (c) We have emphasized the importance of the attraction to the surface. Modify the hard sphere repulsion by adding a square-well attraction as in Problem D of Chapter 5. If the well depth is D and the incident atom comes in with energy E_T, show that the energy loss to the surface is $4\mu(E_T + D)/(1 + \mu)^2$. To get this simple result, and also in (d), neglect the thermal motion of the hard cube. (d) Compute the (low) translational energy below which the incident atom is trapped at the surface. *(e) Include both the thermal motion of the hard cube and the attraction in the model.

F. Transition state theory for adsorption and desorption. (a) Adsorption with an early transition state. The reactants are A and S as in Problem A and they retain their identity in the transition state. There is no barrier. The reaction coordinate is the translation perpendicular to the surface. Show that the transition state theory rate constant is $k_a = (k_B T/2\pi m)^{1/2}$. (b) The result looks familiar. What is it? (Remember the factor of 4 from Problem B. Show that it implies every molecule that hits an empty site, sticks.) (c) When a surface is already partially covered, an incoming atom can be attracted (or repelled) by atoms that are already absorbed. Sketch how you would include this effect in computing the rate constant and conclude that the activation energy can be coverage-dependent. (d) An atom desorbing from a surface. First assume that the problem is one-dimensional and derive the rate constant $k_d = \nu \exp(-D/k_B T)$, where ν is the vibrational frequency of the atom bound to the surface (assume a harmonic binding potential) and D is the well depth. Of course, the harmonic assumption can be made only if $D \gg k_B T$. (e) An atom bound to the surface has two other degrees of freedom that we did not consider in (d). What are they? If we include them in the theory will the desorption rate constant be larger or smaller? (f) A diatomic molecule desorbing from a surface. Modify the discussion in (d) to conclude that if the surface-bound molecule cannot freely rotate, the pre-exponential factor will be significantly higher than ν. There are two factors that are relevant. One is the same as in (e). Identify the other one.

G. Potential energy surface. H_2 molecules desorbing from a Cu surface are observed to be translationally hot, Figure 12.4. Modify the sketch of the potential energy surface shown in Figure 12.5 so that it can rationalize this observation.

H. Nonlinear kinetics: an autocatalytic process. Contrast the catalytic reaction R $\xrightarrow{X}$ P and the autocatalytic reaction R $\xrightarrow{X}$ P + X in terms of the rate of production of the product P as a function of the concentration of the catalyst X for an excess of reagent R. It can be done by the methods of chemical kinetics or we can quickly get to the point by noting that Avogadro's number is about 2^{79} and that the autocatalytic reaction doubles the concentration of X every time it takes place.

Notes

1 Textbooks of this technologically important topic include Kolasinski (2002), Masel (1996), Somorjai (1994), van Santen (1991), Grunze and Kreuzer (1987), Boudart and Djéga-Mariadassou (1984), King and Woodruff (1983). Modern texts of chemical kinetics, e.g., Houston (2001), Steinfeld *et al.* (1999), also pay attention to this key topic.

2 For more on STM in surface processes, see, for example, Ho (1998) for a review and J. R. Hern and W. Ho, *Phys. Rev. Let.* **87** (2001) for a study of the oxidation of CO.

3 See the general overviews Somorjai (1981), Ho (1996), Ertl (1993, 1982).

4 Examples include Goodman (1996), Ho (1996), Hodgson (2000), Gellman (2000), van Santen (1991).

5 Reviews of surface non-adiabatic processes include Tully (2000), Greeley *et al.* (2002), Kroes *et al.* (2002), Kasemo (1996), Avouris and Walkup (1989).

6 For projectile induced processes, see Ceyer (1990), Asscher and Zeiri (2003).

7 For more about precursors, see Harris and Kasemo (1981).

8 The ultimate proof is to block the steps using adsorbed metal atoms and measure the orders of magnitude change in reactivity, S. Dahl *et al.*, *Phys. Rev. Lett.* **83**, 1814 (1999).

9 Rettner *et al.* (1990).

10 For the important special case of dynamics at metal surfaces, see Tully (2000), Greeley *et al.* (2002), Kroes *et al.* (2002).

11 On chaos and non-linear chemical kinetics, see Epstein and Pojman (1998), Gray and Scott (1994); Kapral and Fraser (2001), Slinko and Jaeger (1994), Kapral and Showalter (1995), Scott, (1994), Scott (1987), Noyes (1989).

12 For more on the control of surface reactions, see Kim *et al.* (2001).

13 On pattern formation in oxidation reactions on surfaces, see Imbihl and Ertl (1995), Rotermund (1997).

14 G. A. Somorjai and A. L. Marsh, 2004 symposium.

Epilogue

C—H bond activation,[14] a key stage in the catalytic transformation of organic molecules by transition metals is one example that takes us from the present to the future. Dissociative adsorption is accompanied by restructuring of the metal surface and in the presence of hydrogen the subsequent course of the reaction can be altered. Surface mobility is necessary to free up catalytically active surface sites and this diffusion can be inhibited by CO adsorption. Similar understanding, and with it the options for control, is likely to become increasingly available. Other technologically important areas will also be transformed by the application of the tools of reaction dynamics.

The tools, experimental, computational, and conceptual, are being both honed and developed with new and more complex questions in mind. Imaging techniques are providing ever-increasing resolution and we are moving towards the single molecule limit. The remarkable progress of applied quantum chemistry will allow us to complement the experimental innovations by viewing elementary reactions with unprecedented detail. We will also pay increasing attention to processes where it is the motion of the electrons that determines the time scale for the chemical change and also for physical processes including nanoelectronics. At the other end, the time scale for both experimental and computational examination will increase all the way through the μ sec range thereby bringing biological processes, material response, and molecular machines fully within the scope of dynamics. This will require a better understanding of the dynamical implications of the topography of potential energy landscapes of many-atom systems that is replete with many shallow minima and regio-selective and hindered regions. Both for the inherent interest and for technological need to know reaction dynamics under extreme conditions, very cold collisions and also very hot ones and chemistry at higher densities will receive even more consideration.

This introduction to molecular reaction dynamics is over but the music of time flows on and the molecules continue their dance. From here on you are invited to give directions.

Bibliography

Abraham, M. H. (1974). "Solvent effects on transition states and reaction rates." *Prog. Phys. Org. Chem.* **11**, 1.

Adelman, S. A. (1980). "Generalized Langevin equations and many-body problems in chemical dynamics." *Adv. Chem. Phys.* **44**, 143.

Adelman, S. A. and J. D. Doll (1977). "Brownian motion and chemical dynamics on solid surfaces." *Acc. Chem. Res.* **10**, 378.

Agmon, N. (2003). "Elementary steps in excited-state proton transfer." *J. Phys. Chem. A* (submitted).

Agranovich, V. M. and M. D. Galanin (1982). *Electronic Excitation Energy Transfer in Condensed Matter*. Amsterdam, Elsevier/North Holland.

Aharoni, A., B. Hou, *et al.* (2001). "Non-isomerizable artificial pigments: implications for the primary light-induced events in bacteriorhodopsin." *Biochemistry* **66**, 1499.

Aker, P. M. and J. J. Valentini (1993). "Experimental characterization and computational simulation of chemical reaction dynamics." *Int. Rev. Phys. Chem.* **12**, 363.

Albery, W. J. (1980). "The application of the Marcus relation to reactions in solution." *Ann. Rev. Phys. Chem.* **31**, 227.

Alexander, A. J., M. Brouard, *et al.* (1998). "Chemistry with a sense of direction – the stereodynamics of bimolecular reactions." *Chem. Soc. Rev.* **27**, 405.

Alexander, A. J. and R. N. Zare (1998). "Anatomy of elementary chemical reactions." *J. Chem. Educ.* **75**, 1105.

Allen, M. P. and D. J. Tildesley (1987). *Computer Simulation of Liquids*. Oxford, Clarendon Press.

Althorpe, S. C. and D. C. Clary (2003). "Quantum scattering calculations on chemical reactions." *Ann. Rev. Phys. Chem.* **54**, 493.

Althorpe, S. C., F. Fernandez-Alonso, *et al.* (2002). "Observation and interpretation of a time-delayed mechanism in the hydrogen exchange reaction." *Nature* **416**, 67.

Amirav, A., U. Even, *et al.* (1983). "Rotational cooling of aniline in axisymmetric and planar pulsed expansions." *Canad. J. Phys.* **61**, 278.

Ammal, S. C., H. Yamataka, *et al.* (2003). "Dynamics-driven reaction pathway in an intramolecular rearrangement." *Science* **299**, 1555.

Anderson, J. G. (1987). "Free radicals in the Earth's atmosphere: their measurement and interpretation." *Ann. Rev. Phys. Chem.* **38**, 489.

Anderson, S. M., R. G. Sadygov, *et al.* (2001). Quantum dynamics and spectroscopy. In *Encyclopedia of Chemical Physics and Physical Chemistry*, J. H. Moore and N. D. Spencer (eds.). Bristol, IOP Publishing, Vol. 2, p. 2027.

Ando, K. and J. T. Hynes (1999). "Acid–base proton transfer and ion pair formation in solution." *Adv. Chem. Phys.* **110**, 381.

Andrews, D. L. (1990). *Lasers in Chemistry*. Berlin, Springer-Verlag.

Anfinrud, P. A. (2002). "Ultrafast infrared studies of biological systems." *Ann. Rev. Phys. Chem.* **53**.

Aquilanti, V., A. Giardini-Guidoni, *et al.* (eds.) (1994). *Photon-Induced Molecular Dynamics*. Chem. Phys. Amsterdam, Elsevier.

Armentrout, P. B. (1990). "Electronic state-specific transition metal ion chemistry." *Ann. Rev. Phys. Chem.* **41**, 313.

(2001). "Reactions and thermochemistry of small transition metal cluster ions." *Ann. Rev. Phys. Chem.* **52**, 423.

Armentrout, P. B. and T. Baer (1996). "Gas-phase ion dynamics and chemistry." *J. Phys. Chem.* **100**, 12 866.

Armstrong, M. R., *et al.* (2003). *Proc. Natl. Acad. Sci.* **100**, 4990.

Arnaut, L. G. and S. J. Formosinho (1993). "Excited-state proton transfer reactions I. Fundamentals and intermolecular reactions." *J. Photochem. Photobiol. A: Chem.* **75**, 1.

Arumainayagam, C. R. and R. J. Madix (1991). "Molecular beam studies of gas–surface collision dynamics." *Prog. Surf. Sci.* **38**, 1.

Ashfold, M. N. R. and J. E. Baggott (eds.) (1989). *Bimolecular Collisions*. Advances in Gas-Phase Photochemistry and Kinetics. London, Royal Society of Chemistry.

(1997). *Molecular Photodissociation Dynamics*. London, Royal Society of Chemistry.

Ashfold, M. N. R., K. Obi, *et al.* (eds.) (1998). *New Aspects of Photochemistry and Reaction Dynamics*. Chem. Phys. Amsterdam, Elsevier.

Ashkenazi, G., U. Banin, *et al.* (1997). "Quantum description of the impulsive photodissociation dynamics of I_3^- in solution." *Adv. Chem. Phys.* **100**, 229.

Asscher, M. and Y. Zeiri (2003). "Surface processes induced by collisions with adsorbates." *J. Phys. Chem. B* **107**, 6903.

Ausloos, P. (ed.) (1979). *Kinetics of Ion–Molecule Reactions*. New York, Plenum Press.

Avouris, P., W. M. Gelbart, *et al.* (1977). "Nonradiative electronic relaxation under collision-free conditions." *Chem. Rev.* **27**, 793.

Avouris, P. and R. E. Walkup (1989). "Fundamental mechanisms of desorption and fragmentation induced by electronic transitions at surfaces." *Ann. Rev. Phys. Chem.* **40**, 173.

Bacic, Z. and R. E. Miller (1996). "Molecular clusters: structure and dynamics of weakly bound systems." *J. Phys. Chem.* **100**, 12 945.

Baede, A. P. M. (1975). "Charge transfer between neutrals at hyperthermal energies." *Adv. Chem. Phys.* **30**, 463.

Baer, M. (ed.) (1985). *The Theory of Chemical Reaction Dynamics*. Boca Raton, FL, CRC Press.

(2002). "Introduction to the theory of electronic non-adiabatic coupling terms in molecular systems." *Phys. Rep.* **358**, 75.

Baer, T. (1986). "The dissociation dynamics of energy-selected ions." *Adv. Chem. Phys.* **64**, 111.

Baer, T. and W. L. Hase (1996). *Unimolecular Reaction Dynamics. Theory and Experiments*. New York, Oxford University Press.

Bagchi, B. (1989). "Dynamics of solvation and charge transfer reactions in dipolar fluids." *Ann. Rev. Phys. Chem.* **40**, 115.

Bagchi, B. and R. Biswas (1999). "Polar and nonpolar solvation dynamics, ion diffusion, and vibration relaxation: role of biphasic solvent response in chemical dynamics." *Adv. Chem. Phys.* **109**, 207.

Bagchi, B. and A. Chandra (1991). "Collective orientational relaxation in dense dipolar liquids." *Adv. Chem. Phys.* **80**, 1.

Bagchi, B. and G. R. Fleming (1990). "Dynamics of activationless reactions in solution." *J. Phys. Chem.* **94**, 9.

Bagchi, B. and N. Gayathri (1999). "Interplay between ultrafast polar solvation and vibrational dynamics in electron transfer reactions: role of high-frequency vibrational modes." *Adv. Chem. Phys.* **107**, 1.

Balucani, N., O. Asvany, *et al.* (2000). "Laboratory investigation on the formation of unsaturated nitriles in Titan's atmosphere." *Planet. Space Sci.* **48**, 447.

Balzani, V. (ed.) (2001). *Electron Transfer in Chemistry*. Chichester, John Wiley & Sons.

Balzani, V., A. Credi, *et al.* (2000). "Artificial molecular machines." *Angew. Chem.-Int. Ed. Engl.* **39**, 3348.

Balzani, V., M. Venturi, *et al.* (2003). *Molecular Devices and Machines: A Journey into the Nanoworld*. Weinheim, Wiley-VCH.

Bandrauk, A. D. (1999). "Intense field ionization of molecules by ultra-short laser pulses – charge resonance enhanced ionization and coulomb explosion." *Comm. Mod. Phys.* **1**(3), 97.

Banin, U., A. Bartana, *et al.* (1994). "Impulsive excitation of coherent vibrational motion ground surface dynamics induced by intense short pulses." *J. Chem. Phys.* **107**, 8461.

Bar, I. and S. Rosenwaks (2001). "Controlling bond cleavage and probing intramolecular dynamics via photodissociation of rovibrationally excited molecules." *Int. Rev. Phys. Chem.* **20**, 711.

Barbara, P. F., T. J. Meyer, *et al.* (1996). "Contemporary issues in electron transfer research." *J. Phys. Chem.* **100**, 13 148.

Bardeen, C. J. (2001). "Sometimes you can go home again." *Science* **293**, 444.

Bardeen, C. J., J. Che, *et al.* (1997). "Quantum control of NaI photodissociation reaction product states by ultrafast tailored light pulses." *J. Phys. Chem. A* **101**, 3815.

Bardeen, C. J., Q. Wang, *et al.* (1998). "Femtosecond chirped pulse excitation of vibrational wave packets in bacteriorhodopsin." *J. Phys. Chem. A* **102**, 2759.

Barker, J. A. and D. J. Auerbach (1985). "Gas–surface interactions and dynamics: thermal energy atomic and molecular beam studies." *Surf. Sci. Rep.* **4**, 1.

Barker, J. R. (1984). "Direct measurements of energy transfer involving large molecules in the electronic ground state." *J. Phys. Chem.* **88**, 11.

Barker, J. R., J. D. Brenner, *et al.* (1995). "The vibrational deactivation of large molecules by collisions and by spontaneous infrared emission." *Adv. Chem. Kin. Dyn.* **2B**, 393.

Barron, L. D. and A. D. Buckingham (2001). "Time reversal and molecular properties." *Acc. Chem. Res.* **34**, 781.

Barthel, E. R., I. B. Martini, *et al.* (2001). "How does the solvent control electron transfer? Experimental and theoretical studies of the simplest charge transfer reaction." *J. Phys. Chem. B*, 12 230.

Barzykin, A. V., P. A. Frantsuzov, *et al.* (2002). "Solvent effects in nonadiabatic electron-transfer reactions: theoretical aspects." *Adv. Chem. Phys.* **123**, 511.

Basov, N. G., A. S. Bashkin, *et al.* (1990). *Chemical Lasers*. Berlin, Springer-Verlag.

Bauer, S. H. (1978). "How energy accumulation and disposal affect the rates of reactions." *Chem. Rev.* **78**, 147.

(1979). "Four center metathesis reactions." *Ann. Rev. Phys. Chem.* **30**, 271.

Baumert, T., J. Helbing, *et al.* (1997). "Coherent control with femtosecond laser pulses." *Adv. Chem. Phys.* **101**, 47.

Becker, O. M. and M. Karplus (1997). "The topology of multidimensional potential energy surfaces: theory and application to peptide structure and kinetics." *J. Chem. Phys.* **106**, 1495.

Ben-Amotz, D., A. Gift, *et al.* (2003). "Updated principle of corresponding states." *J. Chem. Educ.* **81**, 142.

Benkovic, S. J. and S. Hammes-Schiffer (2003). "A perspective on enzyme catalysis." *Science* **301**, 1196.

Ben-Nun, M. and R. D. Levine (1994). "Liquid state control of chemical reactions – toward a molecular description." *Acc. Chem. Res.* **27**, 166.

(1995). "Kinetics and dynamics of reactions in liquids." *Int. Rev. Phys. Chem.* **14**, 215.

Ben-Nun, M. and T. J. Martinez (2002). "Ab initio quantum molecular dynamics." *Adv. Chem. Phys.* **121**, 439.

Ben-Nun, M., J. Quenneville, *et al.* (2000). "Ab initio multiple spawning: photochemistry from first principles quantum molecular dynamics." *J. Phys. Chem. A* **104**, 5161.

Ben-Reuven, A. (1975). "Spectral line shapes in gases in the binary-collision approximation." *Adv. Chem. Phys.* **33**, 235.

Ben-Shaul, A., Y. Haas, *et al.* (1981). *Lasers and Chemical Change*. New York, Springer-Verlag.

Benjamin, I. (1997a). "Molecular structure and dynamics at liquid–liquid interfaces." *Ann. Rev. Phys. Chem.* **48**, 407.

(1997b). "Molecular dynamics simulations in interfacial electrochemistry." *Mod. Asp. Electrochem.* **31**, 115.

(1998). "Solvent effects on electronic spectra at liquid interfaces. A continuum electrostatic model." *J. Phys. Chem. A* **102**, 9500.

(2002). "Chemical reaction dynamics at liquid interfaces." *Prog. React. Kin. Mech.* **27**, 87.

Bennett, C. H. (1977). *Molecular Dynamics and Transition State Theory: The Simulation of Infrequent Events*. ACS Symposium, Algorithms for Chemical Computation. Washington, D.C., American Chemical Society.

Bergman, R. G. (1984). "Activation of alkanes with organotransition metal complexes." *Science* **223**, 902.

Bergmann, K., H. Theuer, *et al.* (1998). "Coherent population transfer among quantum states of atoms and molecules." *Rev. Mod. Phys.* **70**, 1003.

Bernardi, F., M. Olivucci, *et al.* (1996). "Potential energy surface crossing in organic photochemistry." *Chem. Soc. Rev.* **25**, 321.

Bernstein, E. R. (1995). "Dynamics and photochemistry of neutral van der Waals clusters." *Ann. Rev. Phys. Chem.* **46**, 197.

Bernstein, R. B. (1966). "Quantum effects in elastic molecular scattering." *Adv. Chem. Phys.* **10**, 75.

(ed.) (1979). *Atom-Molecular Collision Theory: A Guide for the Experimentalist*. New York, Plenum Press.

(1982). *Chemical Dynamics via Molecular Beam and Laser Techniques*. New York, Oxford University Press.

(1988a). *Molecular Beams in Chemistry: A Subjective Account*. The Robert A. Welch Foundation Conference on Chemical Research, Houston, TX.

(1988b). Selectivity in elementary chemical reactions. In *Selectivity in Chemical Reactions*, J. C. Whitehead (ed.). Dordrecht, Kluwer.

Bernstein, R. B., D. R. Herschbach, *et al.* (1987). "Dynamical aspects of stereochemistry." *J. Phys. Chem.* **91**, 5365.

Bernstein, R. B. and A. H. Zewail (1988). "Real-time laser femtochemistry: viewing the transition from reagents to products." *Chem. Eng. News* **66**(45), 24.

Berry, R. S. (1993). "Potential surfaces and dynamics: what clusters tell us." *Chem. Rev.* **93**, 2379.

Bersohn, R. (1976). Reactions of electronically excited atoms: superalkalis and superhalogens. In *Molecular Energy Transfer*, R. D. Levine and J. Jortner (eds.). New York, John Wiley & Sons.

(2003). "Some pleasures in chemical physics." *Ann. Rev. Phys. Chem.* **54**, 1.

Bersohn, R. and S. H. Lin (1969). "Orientation of targets by beam excitation." *Adv. Chem. Phys.* **16**, 67.

Beswick, J. A. and J. Jortner (1981). "Intramolecular dynamics of van der Waals molecules." *Adv. Chem. Phys.* **47**, 363.

Beynon, J. H. and J. R. Gilbert (1984). *Application of Transition State Theory to Unimolecular Reactions*. New York, John Wiley & Sons.

Billing, G. D. and K. V. Mikkelsen (1996). *Introduction to Molecular Dynamics and Chemical Kinetics*. New York, John Wiley & Sons.

(1997). *Advanced Molecular Dynamics and Chemical Kinetics*. New York, John Wiley & Sons.

Birge, R. R. (1990). "Photophysics and molecular electronic applications of the rhodopsins." *Ann. Rev. Phys. Chem.* **41**, 683.

Birnbaum, G. (1976). "Microwave pressure broadening and its application to intermolecular forces." *Adv. Chem. Phys.* **12**, 487.

(ed.) (1985). *Phenomena Induced by Intermolecular Interactions*. New York, Plenum Press.

Bixon, M. and J. Jortner (eds.) (1999). Electron transfer – from isolated molecules to biomolecules. *Adv. Chem. Phys.* **106**.

Blanchet, V., M. Z. Zgierski, *et al.* (1999). "Discerning vibronic molecular dynamics using time-resolved photoelectron spectroscopy." *Nature* **401**, 52.

Bolhuis, P. G., D. Chandler, *et al.* (2002). "Transition path sampling: throwing ropes over rough mountain passes, in the dark." *Ann. Rev. Phys. Chem.* **53**, 291.

Bonacic-Koutecky, V., J. Koutecky, *et al.* (1987). "Neutral and charged biradicals, zwitterions, funnels in S_1, and proton translocation: their role in photochemistry, photophysics, and vision." *Angew. Chem.-Int. Ed. Engl.* **26**, 170.

Bonn, M., S. Funk, *et al.* (1999). "Phonon- versus electron-mediated desorption and oxidation of CO on Ru(0001)." *Science* **285**, 1042.

Bonnet, L. and J. C. Rayez (1999). "Some key factors of energy distributions in the products of complex-forming elementary reactions." *Phys. Chem. Chem. Phys.* **1**, 2383.

Borden, W. T., R. J. Loncharich, *et al.* (1988). "Synchronicity in multibond reactions." *Ann. Rev. Phys. Chem.* **39**, 213.

Born, M. (1920). "Volumen und Hydratationswaerme der Ionen." *Z. Phys.* **1**, 45.

Boudart, M. and G. Djéga-Mariadassou (1984). *Kinetics of Heterogeneous Catalytic Reactions*. Princeton, Princeton University Press.

Bowers, M. T. (ed.) (1979–84). *Gas Phase Ion Chemistry*. New York, Academic Press.

Bowman, J. M. (ed.) (1983). *Molecular Collision Dynamics. Topics in Current Physics*. Berlin, Springer-Verlag.

(1998). "Resonances: bridge between spectroscopy and dynamics." *J. Phys. Chem. A* **102**, 3006.

(2002). "Overview of reduced dimensionality quantum approaches to reactive scattering." *Theor. Chem. Acc.* **108**, 125.

Bowman, J. M. and G. C. Schatz (1995). "Theoretical studies of polyatomic bimolecular reaction dynamics." *Ann. Rev. Phys. Chem.* **46**, 169.

Boyer, P. D. (1997). "The ATP synthase – a splendid molecular machine." *Ann. Rev. Biochem.* **66**, 717.

(1999). "What makes ATP synthase spin?" *Nature* **402**, 247.

Breckenridge, W. H. (1989). "Chemical reactions and energy transfer processes of electronically excited group IIB metal atoms: full collisions and half collisions." *Acc. Chem. Res.* **22**, 21.

Breckenridge, W. H. and H. Umemoto (1982). "Collisional quenching of electronically excited metal atoms." *Adv. Chem. Phys.* **50**, 325.

Brixner, T., N. H. Damrauer, *et al.* (2001). "Femtosecond quantum control." *Adv. Mol. Opt. Phys.* **46**, 1.

Brixner, T. and G. Gerber (2003). "Quantum control of gas-phase and liquid-phase reactivity." *Chem. Phys. Chem.* **4**, 419.

Broeckhove, J. and L. Lathouwers (eds.) (1992). *Time-Dependent Molecular Dynamics*. New York, Plenum Press.

Bromberg, S. E., H. Yang, *et al.* (1997). "The mechanism of a C─H bond activation reaction in room-temperature alkane solution." *Science* **278**, 260.

Brooks, P. R. (1988). "Spectroscopy of transition region species." *Chem. Rev.* **88**, 407.

Brouard, M., P. O'Keeffe, *et al.* (2002). "The state resolved dynamics of elementary reactions." *J. Phys. Chem.* **106**, 3629.

Brouard, M. and J. P. Simmons (1995). *Chemical Dynamics and Kinetics of Small Radicals*. Singapore, World Scientific.

Brown, N. J. and J. A. Miller (1984). "Collisional energy transfer in the low-pressure-limit unimolecular dissociation of HO_2." *J. Chem. Phys.* **80**, 5568.

Brumer, P. (1981). "Intramolecular energy transfer: theories for the onset of statistical behavior." *Adv. Chem. Phys.* **47**, 201.

Brumer, P. and M. Shapiro (1988). "Chaos and reaction dynamics." *Adv. Chem. Phys.* **70**, 365.

(1992). "Laser control of molecular processes." *Ann. Rev. Phys. Chem.* **43**, 257.

(1995). "Laser control of chemical reactions." *Sci. Am.* **272**, 34.

(1997). "Coherent control of bimolecular scattering." *Adv. Chem. Phys.* **101**, 295.

Brunner, T. A. and D. Pritchard (1982). "Fitting laws for rotationally inelastic collisions." *Adv. Chem. Phys.* **50**, 589.

Buchachenko, A., N. Halberstadt, *et al.* (2003). "Ar-12: a model system for complex dynamics." *Int. Rev. Phys. Chem.* **22**, 153.

Buckingham, A. D. (1967). "Permanent and induced molecular moments and long-range intermolecular forces." *Adv. Chem. Phys.* **12**, 107.

Bunker, D. L. (1966). *Theory of Elementary Gas Reaction Rates*. New York, Pergamon Press.

Butkovskaya, N. I. and D. W. Setser (2003). "Infrared chemiluminescence from water-forming reactions: characterization of dynamics and mechanisms." *Int. Rev. Phys. Chem.* **22**, 1.

Butler, L. J. (1998). "Chemical reaction dynamics beyond the Born–Oppenheimer approximation." *Ann. Rev. Phys. Chem.* **49**, 125.

Butler, L. J. and D. M. Neumark (1996). "Photodissociation dynamics." *J. Phys. Chem.* **100**, 12 801.

Caldin, E. F. (2001). *The Mechanism of Fast Reactions in Solution*. Amsterdam, IOS Press.

Callear, A. B. (1978). "An overview of molecular energy transfer in gases." *Spec. Per. Rep.* **3**, 82.

Campargue, R. (ed.) (2001). *Atomic and Molecular Beams: The State of the Art 2000*. Berlin, Springer-Verlag.

Campbell, E. E. B. and R. D. Levine (2000). "Delayed ionization and fragmentation en route to thermionic emission: statistics and dynamics." *Ann. Rev. Phys. Chem.* **51**, 65.

Campbell, E. E. B., H. Schmidt, *et al.* (1988). "Symmetry and angular momentum in collisions with laser excited polarised atoms." *Adv. Chem. Phys.* **72**, 37.

Canning, N. D. S. and R. J. Madix (1984). "Toward an organometallic chemistry of surfaces." *J. Phys. Chem.* **88**, 2437.

Cannon, W. R., S. F. Singleton, *et al.* (1996). "A perspective on biological catalysis." *Nature Struct. Biol.* **3**, 821.

Carpenter, B. K. (1998). "Dynamic behavior of organic reactive intermediates." *Angew. Chem.-Int. Ed. Engl.* **37**, 3340.

Carrette, L., K. A. Friedrich, *et al.* (2000). "Fuel cells: principles, types, fuels and applications." *Chem. Phys. Chem.* **1**, 162.

Carrington, T. (1974). "The geometry of intersecting potential surfaces." *Acc. Chem. Res.* **7**, 20.

Casavecchia, P. (2000). "Chemical reaction dynamics with molecular beams." *Rep. Prog. Phys.* **63**, 355.

Casavecchia, P., N. Balucani, *et al.* (1998). "Reactive scattering of oxygen and nitrogen atoms." *Acc. Chem. Res.* **32**, 503.

(1999). "Crossed-beam studies of reaction dynamics." *Ann. Rev. Phys. Chem.* **50**, 347.

Castillo, J. F. (2002). "The dynamics of the H + H$_2$O reaction." *Chem. Phys. Chem.* **3**, 320.

Castleman, A. W. (2001). An ultrafast look at cluster dynamics. In *The Physics and Chemistry of Clusters*, E. E. B. Campbell and M. Larsson (eds.). Singapore, World Scientific.

Castleman, A. W. and K. H. Bowen (1996). "Clusters: structure, energetics and dynamics of intermediate states of matter." *J. Phys. Chem.* **100**, 12 911.

Castleman, A. W. and R. G. Keesee (1986). "Clusters – properties and formation." *Ann. Rev. Phys. Chem.* **37**, 525.

Castleman, A. W. and S. Wei (1994). "Cluster reactions." *Ann. Rev. Phys. Chem.* **45**, 685.

Ceyer, S. T. (1988). "Dissociative chemisorption: dynamics and mechanisms." *Ann. Rev. Phys. Chem.* **39**, 479.

(1990). "New mechanisms for chemistry at surfaces." *Science* **249**, 133.

(2001). "The unique chemistry of hydrogen beneath the surface: catalytic hydrogenation of hydrocarbons." *Acc. Chem. Res.* **34**, 737.

Chabinyc, M. L., S. L. Craig, *et al.* (1998). "Gas-phase ionic reactions: dynamics and mechanism of nucleophilic displacements." *Science* **279**, 1882.

Chandler, D. W. and P. L. Houston (1987). "Two-dimensional imaging of state-selected photodissociation products by multiphoton ionization." *J. Chem. Phys.* **87**, 1445.

Chandler, D. W. and D. H. Parker (1999). "Velocity mapping studies of highly excited molecules." *Adv. Photochem.* **25**, 59.

Chao, S. D. and R. T. Skodje (2002). "Signatures of reactive resonances: three case studies." *Theor. Chem. Acc.* **108**, 273.

Cheatham III, T. E. and P. A. Kollman (2000). "Molecular dynamics simulation of nucleic acids." *Ann. Rev. Phys. Chem.* **51**, 435.

Chelkowski, S. and A. D. Bandrauk (1995). "Two-step coulomb explosions of diatoms in intense laser fields." *J. Phys. B: At. Mol. Opt. Phys.* **28**, L723.

Chen, P. and T. J. Meyer (1998). "Medium effects on charge transfer in metal complexes." *Chem. Rev.* **98**, 1439.

Chergui, M. (ed.) (1996). *Femtochemistry*. Singapore, World Scientific.

Chesnoy, J. and G. M. Gale (1984). "Vibrational energy relaxation in liquids." *Ann. Phys., Paris* **9**, 893.

Child, M. S. (1974). *Molecular Collision Theory*. London, Academic Press.

(1991). *Semiclassical Mechanics with Molecular Applications*. Oxford, Clarendon Press.

Child, M. S. and L. O. Halonen (1984). "Overtone frequencies and intensities in the local mode picture." *Adv. Chem. Phys.* **57**, 1.

Cho, M. and G. R. Fleming (1999). Electron transfer and solvent dynamics in two- and three-state systems. In *Electron Transfer: From Isolated Molecules to Biomolecules*, J. Jortner and M. Bixon (eds.). John Wiley & Sons, Part II, p. 311.

Christov, S. G. (ed.) (1980). *Collision Theory and Statistical Theory of Chemical Reactions*. Lecture Notes in Chemistry. Berlin, Springer-Verlag.

Clark, A. P., A. S. Dickinson, *et al.* (1977). "The correspondence principle in heavy-particle collisions." *Adv. Chem. Phys.* **36**, 63.

Clary, D. C. (ed.) (1986). *The Theory of Chemical Reaction Dynamics*. Boston, Reidel.

(1990). "Fast chemical reactions: theory challenges experiment." *Ann. Rev. Phys. Chem.* **41**, 61.

(1998). "Quantum theory of chemical reaction dynamics." *Science* **279**, 1879.

Clary, D. C., R. G. Gilbert, *et al.* (1995). "Mechanisms for supercollisions." *Faraday Disc.* **102**, 423.

Closs, G. L. and J. R. Miller (1988). "Intramolecular long distance electron transfer in organic molecules." *Science* **240**, 440.

Cohen, R. C. and R. J. Saykally (1991). "Multidimensional intermolecular potential surfaces from vibration–rotation tunneling (VRT) spectra of van der Waals complexes." *Ann. Rev. Phys. Chem.* **42**, 369.

Collins, M. (1996). "The interface between electronic structure theory and reaction dynamics by reaction path methods." *Adv. Chem. Phys.* **93**, 389.

Comsa, G. and R. David (1985). "Dynamical parameters of desorption." *Surf. Sci.* **5**, 145.

Cong, P. and J. D. Simon (eds.) (1994). *Introduction to Ultrafast Laser Spectroscopic Techniques Used in the Investigation of Condensed Phase Chemical Reactivity.* Ultrafast Dynamics of Chemical Systems. Dordrecht, Kluwer.

Conroy, D., V. Aristov, *et al.* (2001). "Competitive pathways via nonadiabatic transitions in photodissociation." *Acc. Chem. Res.* **34**, 625.

Constant, E., H. Stapelfeldt, *et al.* (1999). "Using time resolved coulomb explosion imaging to measure dissociative molecular wave packets." *Comm. At. Mol. Phys.* **D1**, 85.

Continetti, R. E. (1998). "Photoelectron–photofragment coincidence studies of dissociation dynamics." *Int. Rev. Phys. Chem.* **17**, 227.

(2001). "Coincidence spectroscopy." *Ann. Rev. Phys. Chem.* **52**, 165.

Corey, E. J. (1991). "The logic of chemical synthesis – multistep synthesis of complex carbogenic molecules." *Angew. Chem.-Int. Ed. Engl.* **30**, 455.

Corkum, P. B., M. Y. Ivanov, *et al.* (1997). "Subfemtosecond processes in strong laser fields." *Ann. Rev. Phys. Chem.* **48**, 387.

Cramer, C. J. (2002). *Essentials of Computational Chemistry.* Chichester, John Wiley & Sons.

Cramer, C. J. and D. G. Truhlar (1996). *Continuum Solvation Models.* Dordrecht, Kluwer.

Crane, J. C., H. Nam, *et al.* (1998). "Stimulated emission pumping spectra and intramolecular vibrational dynamics of DFCO(S_0) from 9000 to 20 000 cm^{-1}." *J. Phys. Chem. A* **102**, 9433.

Crim, F. F. (1984). "Selective excitation studies of unimolecular reaction dynamics." *Ann. Rev. Phys. Chem.* **35**, 657.

(1993). "Vibrationally mediated photodissociation: exploring excited-state surfaces and controlling decomposition pathways." *Ann. Rev. Phys. Chem.* **44**, 397.

(1996). "Bond-selected chemistry: vibrational state control of photodissociaation and bimolecular reaction." *J. Phys. Chem.* **100**, 12 725.

(1999). "Vibrational state control of bimolecular reactions: discovering and directing the chemistry." *Acc. Chem. Res.* **32**, 877.

(2001). "Close encounters." *Science* **293**, 2014.

Dagdigian, P. J. (1997). "State-resolved collision-induced electronic transitions." *Ann. Rev. Phys. Chem.* **48**, 95.

(2001). Reactive scattering. In *Encyclopedia of Chemical Physics and Physical Chemistry*, J. H. Moore and N. D. Spencer (eds.). Bristol, IOP Publishing, Vol. 2, p. 1819.

Dai, D., C. C. Wang, *et al.* (2003). "Interference of quantized transition-state pathways in the H + D_2 → D + HD chemical reaction." *Science* **300**, 1730.

Dai, E. H. L. and W. Ho (1995). *Laser Spectroscopy and Photochemistry at Metal Surfaces.* Singapore, World Scientific.

Dantus, M. (2001). "Coherent nonlinear spectroscopy: from femtosecond dynamics to control." *Ann. Rev. Phys. Chem.* **52**, 639.

Darling, G. R. and S. Holloway (1995). "The dissociation of diatomic molecules." *Rep. Prog. Phys.* **58**, 1595.

Darling, G. R., S. Holloway, *et al.* (2001). Molecular reaction dynamics: surfaces. In *Encyclopedia of Chemical Physics and Physical Chemistry*, J. H. Moore and N. D. Spencer (eds.). Bristol, IOP Publishing, Vol. 1, p. 775.

Dauben, W. G., L. Salem, *et al.* (1975). "A classification of photochemical reactions." *Acc. Chem. Res.* **8**, 41.

Davidson, E. R. (1993). "Molecular mechanics and modeling: overview." *Chem. Rev.* **93**, 2337.

Davies, J. W. and M. J. Pilling (1989). Association reactions. In *Bimolecular Collisions*, M. N. R. Ashfold and J. E. Baggott (eds.). London, Royal Society of Chemistry, p. 105.

Davis, M. J., H. Koizumi, *et al.* (1994). "Experimental and theoretical study of the O + HCl transition state region by photodetachment of OHCl⁻." *J. Chem. Phys.* **101**, 4708.

DeBoeij, W. P., M. S. Pshenichnikov, *et al.* (1998). "Ultrafast solvation dynamics explored by femtosecond photon echo spectroscopies." *Ann. Rev. Phys. Chem.* **49**, 99.

DeFeyter, S., E. W.-G. Diau, *et al.* (2000). "Femtosecond dynamics of Norrish type-II reactions: nonconcerted hydrogen-transfer and diradical intermediacy." *Angew. Chem.* **39**, 260.

Dellago, C., P. G. Bolhuis, *et al.* (2002). "Transition path sampling." *Adv. Chem. Phys.* **123**, 1.

Delone, N. B. and V. P. Krainov (1985). *Atoms in Strong Light Fields*. Berlin, Springer-Verlag.

Demidov, A. A. and D. L. Andrews (2001). Electronic energy transfer in condensed phases. In *Encyclopedia of Chemical Physics and Physical Chemistry*, J. H. Moore and N. D. Spencer (eds.). Bristol, IOP Publishing, Vol. 3, p. 2701.

Demtroder, W. (1981). *Laser Spectroscopy – Basic Concepts and Instrumentation*. Berlin, Springer-Verlag.

DePristo, A. E. and A. Kara (1990). "Molecule–surface scattering and reaction dynamics." *Adv. Chem. Phys.* **77**, 163.

DePristo, A. E. and H. Rabitz (1980). "Vibrational and rotational collision processes." *Adv. Chem. Phys.* **42**, 271.

DeSchryver, F. C., S. DeFeyter, *et al.* (2001). *Femtochemistry*. Weinheim, Wiley-VCH.

Desouter-Lecomte, M., D. Dehareng, *et al.* (1985). "Nonadiabatic unimolecular reactions of polyatomic molecules." *J. Phys. Chem.* **89**, 214.

deVivie-Riedle, R., H. Rabitz, *et al.* (eds.) (2001). *Laser Control of Quantum Dynamics*. Chem. Phys. Amsterdam, North-Holland.

DeWit, A. (1999). "Spatial patterns and spatiotemporal dynamics in chemical systems." *Adv. Chem. Phys.* **109**, 435.

Diau, E. W.-G., J. L. Herek, *et al.* (1998). "Femtosecond activation of reactions and the concept of nonergodic molecules." *Science* **279**, 847.

Diestler, D. (1980). "Theoretical studies of vibrational relaxation of small molecules in dense media." *Adv. Chem. Phys.* **42**, 305.

Dietrich, P. and P. B. Corkum (1992). "Ionization and dissociation of diatomic molecules in intense infrared laser fields." *J. Chem. Phys.* **97**, 3187.

Dill, K. A. and H. S. Chan (1997). "From Levinthal to pathways to funnels." *Nat. Struct. Biol.* **4**, 10.

Dixon, R. N. (1994). "The dynamics of photodissociation." *Chem. Soc. Rev.* **23**, 375.

Dixon, R. N., D. H. Hwang, *et al.* (1999). "Chemical 'double slits': dynamical interference of photodissociation pathways in water." *Science* **285**, 1249.

Dlott, D. D. (1999). "Ultrafast spectroscopy of shock waves in molecular materials." *Ann. Rev. Phys. Chem.* **50**, 251.

Dobson, C. M., A. Sali, *et al.* (1998). "Protein folding: a perspective from theory and experiment." *Angew. Chem.* **37**, 869.

Dolbier Jr, W. R., H. Koroniak, *et al.* (1996). "Electronic control of stereoselectivities of electrocyclic reactions of cyclobutenes: a triumph of theory in the prediction of organic reactions." *Acc. Chem. Res.* **29**, 471.

Doll, J. D. and A. F. Voter (1987). "Recent developments in the theory of surface diffusion." *Ann. Rev. Phys. Chem.* **38**, 413.

Domcke, W., P. Hanggi, *et al.* (eds.) (1997). *Dynamics of Driven Quantum Systems*. Chem. Phys. Amsterdam, Elsevier.

Domcke, W. and G. Stock (1997). "Theory of ultrafast nonadiabatic excited-state processes and their spectroscopic detection in real time." *Adv. Chem. Phys.* **100**, 1.

Donahue, N. M. (2001). "Revisiting the Hammond postulate: the role of reactant and product ionic states in regulating barrier heights, locations, and transition state frequencies." *J. Phys. Chem. A* **105**, 1489.

Doren, D. J. (1996a). "Kinetics and dynamics of hydrogen adsorption and desorption on silicon surfaces." *Adv. Chem. Phys.* **95**, 1.

  (1996b). "Kinetics and dynamics of hydrogen adsorption on silicon surfaces." *Adv. Chem. Phys.* **95**, 1.

Douhal, A. and J. Santamaria (eds.) (1999). *Femtochemistry in Molecular Science*. Toledo, World Scientific.

Dressler, R. A. (ed.) (2001). *Chemical Dynamics in Extreme Environments*. Singapore, World Scientific.

Duley, W. W. and D. A. Williams (1984). *Interstellar Chemistry*. London, Harcourt Brace Jovanovich.

Duncan, M. A. (2003). "Infrared spectroscopy to probe structure and dynamics in metal ion–molecule complexes." *Int. Rev. Phys. Chem.* **22**, 407.

Dunning, F. B. and R. F. Stebbings (1982). "Collisions of Rydberg atoms with molecules." *Ann. Rev. Phys. Chem.* **33**, 173.

Duren, R., E. Hasselbrink, *et al.* (1982). "On the interaction of excited alkali atoms with rare gas targets in scattering processes." *Z. Phys. A* **307**, 1.

Dykstra, C. E. (2003). "Intermolecular interaction: from properties to potentials and back." *Adv. Chem. Phys.* **126**, 1.

Eisenthal, K. B. (1996). "Photochemistry and photophysics of liquid interfaces by second harmonic spectroscopy." *J. Phys. Chem.* **100**, 12 997.

El-Sayed, M. A., I. Tanaka, *et al.* (eds.) (1995). *Ultrafast Processes in Chemistry and Photobiology*. Oxford, Blackwell.

Elber, R., A. Cardenas, *et al.* (2003). "Bridging the gap between long time trajectories and reaction pathways." *Adv. Chem. Phys.* **126**, 93.

Elber, R. and M. Karplus (1987). "Multiple conformational states of proteins: a molecular dynamics analysis of myoglobin." *Science* **235**, 318.

Elsaesser, T. and W. Kaiser (1991). "Vibrational and vibronic relaxation of large polyatomic molecules in liquids." *Ann. Rev. Phys. Chem.* **42**, 83.

Eppink, A. T. J. B. and D. H. Parker (1997). "Velocity map imaging of ions and electrons using electrostatic lenses; application in photoelectron and photofragment ion imaging of molecular oxygen." *Rev. Sci. Instrum.* **68**, 3477.

Epstein, I. R. (1989). "The role of flow systems in far-from-equilibrium dynamics." *J. Chem. Educ.* **66**, 191.

Epstein, I. R. and J. A. Pojman (1998). *An Introduction to Nonlinear Chemical Dynamics: Oscillations, Waves, Patterns, and Chaos*. Oxford, Oxford University Press.

Epstein, I. R. and K. Showalter (1996). "Nonlinear chemical dynamics: oscillations, patterns, and chaos." *J. Phys. Chem.* **100**, 13 132.

Ertl, G. (1982). "Chemical dynamics in surface reactions." *Ber. Bunsenges. Phys. Chem.* **86**, 425.

(1991). "Oscillatory kinetics and spatio-temporal self-organization in reactions at solid surfaces." *Science* **254**, 1750.

(1993). "Reactions at well-defined surfaces." *Surf. Sci.* **299/300**, 742.

Eu, B. C. (1984). *Semiclassical Theories of Molecular Scattering*. Berlin, Springer-Verlag.

Even, U., J. Jortner, *et al.* (2000). "Cooling of large molecules below 1 K and He clusters." *J. Chem. Phys.* **112**, 8068.

Eyring, H., J. Walter, *et al.* (1944). *Quantum Chemistry*. New York, John Wiley & Sons.

Farrar, J. M. (1995). "Ion reaction dynamics." *Ann. Rev. Phys. Chem.* **46**, 525.

Faubel, M. (1983). "Vibrational and rotational excitation in molecular collisions." *Adv. At. Mol. Phys.* **19**, 345.

Faubel, M. and J. P. Toennies (1977). "Scattering studies of rotational and vibrational excitation of molecules." *Adv. At. Mol. Phys.* **13**, 229.

Fayer, M. D. (2001). "Fast protein dynamics probed with infrared vibrational echo experiments." *Ann. Rev. Phys. Chem.* **52**, 315.

Felker, P. M. and A. H. Zewail (1988). "Picosecond time-resolved dynamics of vibrational-energy redistribution and coherence in beam-isolated molecules." *Adv. Chem. Phys.* **70**, 265.

Fenn, J. B., M. Mann, *et al.* (1989). "Electrospray ionization for mass spectrometry of large biomolecules." *Science* **246**, 64.

Ferguson, E. E. (1986). "Vibrational quenching of small molecular ions in neutral collisions." *J. Phys. Chem.* **90**, 731.

Fernandez-Alonso, F. and R. N. Zare (2002). "Scattering resonances in the simplest chemical reaction." *Ann. Rev. Phys. Chem.* **53**, 67.

Feynman, R. P., R. B. Leighton, *et al.* (1966). *The Feynman Lectures on Physics*. Reading, Addison-Wesley.

Field, R. W., J. P. O'Brien, *et al.* (1997). "Intramolecular dynamics in the frequency domain." *Adv. Chem. Phys.* **101**, 463.

Finlayson-Pitts, B. J. and J. J. N. Pitts (1986). *Atmospheric Chemistry: Fundamentals and Experimental Techniques*. New York, Wiley-Interscience.

Fischer, I. and P. Chen (2002). "Allyl – a model system for the chemical dynamics of radicals." *J. Phys. Chem. A* **106**, 711.

Fisk, G. A. and F. F. Crim (1977). "Single collision studies of vibrational energy transfer mechanisms." *Acc. Chem. Res.* **10**, 73.

Fleming, G. and P. Hanggi (eds.) (1993). *Activated Barrier Crossing*. New Jersey, World Scientific.

Fleming, G. R. (1986). *Chemical Applications of Ultrafast Spectroscopy*. Oxford, Oxford University Press.

Fleming, G. R. and M. Cho (1996). "Chromophore–solvent dynamics." *Ann. Rev. Phys. Chem.* **47**, 109.

Fleming, G. R., T. Joo, *et al.* (1997). "Femtosecond chemical dynamics in condensed phases." *Adv. Chem. Phys.* **101**, 141.

Fleming, G. R. and P. G. Wolynes (1990). "Chemical dynamics in solution." *Physics Today*, 36.

Flygare, W. H. (1978). *Molecular Structure and Dynamics*. Englewood Cliffs, NJ, Prentice-Hall.

Flygare, W. H. and T. G. Schmalz (1976). "Transient experiments and relaxation processes involving rotational states." *Acc. Chem. Res.* **9**, 385.

Flynn, G. W. (1981). "Collision-induced energy flow between vibrational modes of small polyatomic molecules." *Acc. Chem. Res.* **14**, 334.

(2001). Energy transfer in gases. In *Encyclopedia of Chemical Physics and Physical Chemistry*, J. H. Moore and N. D. Spencer (eds.). Bristol, IOP Publishing, Vol. 3, p. 2681.

Flynn, G. W., C. S. Parmenter, *et al.* (1996). "Vibrational energy transfer." *J. Phys. Chem.* **100**, 12 817.

Flynn, G. W. and J. R. E. Weston (1986). "Hot atoms revisited: laser photolysis and product detection." *Ann. Rev. Phys. Chem.* **37**, 551.

Fontijn, A. (ed.) (1985). *Gas-Phase Chemiluminescence and Chemi-Ionization*. Amsterdam, North-Holland.

Foote, C. S., J. S. Valentine, *et al.* (1995). *Active Oxygen in Chemistry*. Glasgow, Chapman & Hall.

Ford, K. W. and J. A. Wheeler (1959). *Ann. Phys.* **7**, 259.

Formosinho, S. J., L. G. Arnaut, *et al.* (1998). "A critical assessment of classical and semi-classical models for electron transfer reactions in solution." *Prog. React. Kinet.* **23**, 1.

Forst, W. (1973). *Theory of Unimolecular Reactions*. New York, Academic Press.

Fourkas, J. T. (2002). "Higher-order optical correlation spectroscopy in liquids." *Ann. Rev. Phys. Chem.* **53**, 17.

Frauenfelder, H., S. G. Sliger, *et al.* (1991). "The energy landscapes and motions of proteins." *Science* **254**, 1598.

Freed, K. F. (1978). "Radiationless transitions in molecules." *Acc. Chem. Res.* **11**, 74.

(1980). "Collisional effects on electronic relaxation processes." *Adv. Chem. Phys.* **42**, 207.

(1981). "Collision induced intersystem crossing." *Adv. Chem. Phys.* **47**(2), 291.

Freeman, D. L. and J. D. Doll (1987). "The quantum mechanics of clusters." *Adv. Chem. Phys.* **70**, 139.

(1996). "Computational studies of clusters: methods and results." *Ann. Rev. Phys. Chem.* **47**, 43.

Frenkel, D. and B. Smit (2002). *Understanding Molecular Simulation*. San Diego, Academic Press.

Frey, E. (2002). "On the physics of biopolymers and molecular motors." *Chem. Phys. Chem.* **3**, 270.

Friedrich, B. and D. Herschbach (1996). "Statistical mechanics of pendular molecules." *Int. Rev. Phys. Chem.* **15**, 325.

Frommhold, L. (1981). "Collision-induced scattering of light and the diatom polarizabilities." *Adv. Chem. Phys.* **46**, 1.

Fueno, T. (1999). *The Transition State: A Theoretical Approach*. Tokyo, Gordon and Breach.

Fuke, K., K. Hashimoto, *et al.* (1999). "Structures, spectroscopies, and reactions of atomic ions with water clusters." *Adv. Chem. Phys.* **110**, 431.

Fukui, K. (1971). "Recognition of stereochemical paths by orbital interaction." *Acc. Chem. Res.* **4**, 57.

(1982). "The role of frontier orbitals in chemical reactions." *Angew. Chem.-Int. Ed. Engl.* **21**, 801.

Gadzuk, J. W. (1988). "The semiclassical way to molecular dynamics at surfaces." *Ann. Rev. Phys. Chem.* **39**, 395.

Gai, F., K. C. Hasson, *et al.* (1998). "Chemical dynamics in proteins: the photoisomerization of retinal in bacteriorhodopsin." *Science* **279**, 1886.

Gao, J. and D. G. Truhlar (2002). "Quantum mechanical methods for enzyme kinetics." *Ann. Rev. Phys. Chem.* **53**, 467.

Gardiner, C. W. (1983). *Handbook of Stochastic Methods for Physics, Chemistry, and the Natural Sciences*. Berlin, Springer-Verlag.

Gardiner, W. C. (ed.) (1984). *Combustion Chemistry*. New York, Springer-Verlag.

Garrison, B. J. and D. Srivastava (1995). "Potential energy surfaces for chemical reactions at solid surfaces." *Ann. Rev. Phys. Chem.* **46**, 373.

Gaspard, P. and I. Burghardt (eds.) (1997). "Chemical Reactions and their Control on the Femtosecond Time Scale." *Adv. Chem. Phys.* **101**.

Gellman, A. J. (2000). "Transition states for surface-catalyzed chemistry." *Acc. Chem. Res.* **33**, 19.

George, T. F., I. H. Zimmermann, *et al.* (1977). "A new concept in laser-assisted chemistry: electronic-field representation." *Acc. Chem. Res.* **10**, 449.

Gerber, R. B., A. B. McCoy, *et al.* (1994). "Photochemical reactions in weakly bound clusters." *Ann. Rev. Phys. Chem.* **45**, 275.

Gerber, R. B. and M. A. Ratner (1988). "Self-consistent-field methods for vibrational excitations." *Adv. Chem. Phys.* **70**, 97.

Gianturco, F. A. (1979). *The Transfer of Molecular Energies by Collision: Recent Quantum Treatments*. Berlin, Springer-Verlag.

(ed.) (1982). *Atomic and Molecular Collision Theory*. New York, Plenum Press.

Gilbert, R. G. and S. C. Smith (1990). *Theory of Unimolecular and Recombination Reactions*. Oxford, Blackwell Scientific.

Glasstone, S., K. J. Laidler, *et al.* (1941). *Theory of Rate Processes*. New York, McGraw-Hill.

Gobeli, D. A., J. J. Yang, *et al.* (1985). "Laser multiphoton ionization–dissociation spectrometry." *Chem. Rev.* **85**, 529.

Goldstein, H. (1950). *Classical Mechanics*. New York, Addison-Wesley.

Gomer, R. (1983). Mechanisms of electron-stimulated desorption. In *Desorption Induced by Electronic Transitions DIET I*, N. H. Tolk, M. M. Traum, *et al.* (eds.). Berlin, Springer-Verlag, Vol. 24, p. 40.

Goodman, D. W. (1996). "Correlations between surface science models and 'real world' catalysis." *J. Phys. Chem.* **100**, 13 090.

Gordon, R. G., W. A. Klemperer, *et al.* (1968). "Vibrational and rotational relaxation." *Ann. Rev. Phys. Chem.* **19**, 215.

Gordon, R. J. and G. E. Hall (1996). "Applications of Doppler spectroscopy in photofragmentation." *Adv. Chem. Phys.* **96**, 1.

Gordon, R. J. and S. A. Rice (1997). "Active control of the dynamics of atoms and molecules." *Ann. Rev. Phys. Chem.* **48**, 601.

Gordon, R. J., L. Zhu, *et al.* (1999). "Coherent control of chemical reactions." *Acc. Chem. Res.* **32**, 1007.

(2001). "Using the phase of light as a photochemical tool." *J. Phys Chem. A* **105**, 4387.

Gray, P. and S. K. Scott (1994). *Chemical Oscillations and Instabilities*. Oxford, Oxford University Press.

Greeley, J., J. K. Nirskov, *et al.* (2002). "Electronic structure and catalysis on metal surfaces." *Ann. Rev. Phys. Chem.* **53**, 319.

Green, R. J. and S. L. Anderson (2001). "Complex formation and decay in ion–molecule reactions: mode-selective scattering as a dynamical probe." *Int. Rev. Phys. Chem.* **20**, 165.

Green, W. H., C. B. Moore, *et al.* (1992). "Transition states and rate constants for unimolecular reactions." *Ann. Rev. Phys. Chem.* **43**, 591.

Greene, C. H. and R. N. Zare (1982). "Photofragment alignment and orientation." *Ann. Rev. Phys. Chem.* **33**, 119.

Grice, R. (1975). "Reactive scattering." *Adv. Chem. Phys.* **30**, 247.

Grice, R. and A. H. Zewail (eds.) (1996). *Laser and Molecular Beam Studies of Chemical Reaction Dynamics*. Chem. Phys. Amsterdam, Elsevier.

Gross, A. (1998). "Reactions at surfaces studied by ab initio dynamics calculations." *Surf. Sci. Rep.* **32**, 291.

Gross P. and Dantus M. (1997). *J. Chem. Phys.* **106**, 801.

Grosser, J., O. Hoffmann, *et al.* (1999). "Direct observation of collisions by laser excitation of the collision pair." *Comm. Mod. Phys.* **1**, 117.

Gruebele, M. (1999). "The fast protein folding problem." *Ann. Rev. Phys. Chem.* **50**, 485. (2003). "Mechanism and control of molecular energy flow." *Theor. Chem. Acc.* **109**, 53.

Gruebele, M. and R. Bigwood (1998). "Molecular vibrational energy flow: beyond the Golden Rule." *Int. Rev. Phys. Chem.* **17**, 91.

Grunze, M. and H. J. Kreuzer (1987). *Kinetics of Interface Reactions*. New York, Springer-Verlag.

Guo, X.-C. and R. J. Madix (2003). "Real-time observation of surface reactivity and mobility with scanning tunneling microscopy." *Acc. Chem. Res.* **36**, 471.

Gust, D. A., T. A. Moore, *et al.* (1993). "Molecular mimicry of photosynthetic energy and electron transfer." *Acc. Chem. Res.* **26**, 198.

Gutman, M. and E. Nachliel (1997). "Time-resolved dynamics of proton transfer in proteinous systems." *Ann. Rev. Phys. Chem.* **48**, 329.

Haas, Y., M. Klessinger, *et al.* (eds.) (2000). *Conical Intersections in Photochemistry, Spectroscopy, and Chemical Dynamics*. Chem Phys. Amsterdam, Elsevier.

Hall, G. E. and P. L. Houston (1989). "Vector correlations in photodissociation dynamics." *Ann. Rev. Phys. Chem.* **40**, 375.

Halonen, L. O. (1998). "Local mode vibrations in polyatomic molecules." *Adv. Chem. Phys.* **104**, 41.

Hamaguchi, H. and T. L. Gustafson (1994). "Ultrafast time-resolved spontaneous and coherent Raman spectroscopy: the structure and dynamics of photogenerated transient species." *Ann. Rev. Phys. Chem.* **45**, 593.

Hamilton, C. E., J. L. Kinsey, *et al.* (1986). "Stimulated emission pumping: new methods in spectroscopy and molecular dynamics." *Ann. Rev. Phys. Chem.* **37**, 493.

Hammes-Schiffer, S. (2001). "Theoretical perspectives on proton-coupled electron transfer reactions." *Acc. Chem. Res.* **34**, 273. (2002). "Comparison of hydride, hydrogen atom, and proton-coupled electron transfer reactions." *Chem. Phys. Chem.* **3**, 33.

Hanggi, P., P. Talkner, *et al.* (1990). "Reaction-rate theory: fifty years after Kramers." *Rev. Mod. Phys.* **62**, 251.

Harren, F., D. H. Parker, *et al.* (1991). "Hexapole orientation and scattering of oriented gas-phase molecules." *Comm. At. Mol. Phys.* **26**, 109.

Harris, A. L., J. K. Brown, *et al.* (1988). "The nature of simple photodissociation reactions in liquids on ultrafast time scales." *Ann. Rev. Phys. Chem.* **39**, 341.

Harris, C. B., D. E. Smith, *et al.* (1990). "Vibrational relaxation of diatomic molecules in liquids." *Chem. Rev.* **90**, 481.

Harris, J. (1987). "Notes on the theory of atom–surface scattering." *Phys. Scr.* **36**, 156.

Harris, J. and B. Kasemo (1981). "On precursor mechanisms for surface reactions." *Surf. Sci.* **105**, L281.

Hase, W. L. (1976). Dynamics of unimolecular reactions. In *Modern Theoretical Chemistry, Dynamics of Molecular Collisions*, W. H. Miller (ed.). New York, Plenum Press, Part B.
(1998). "Some recent advances and remaining questions regarding unimolecular rate theory." *Acc. Chem. Res.* **31**, 659.
(2001). Statistical mechanical description of chemical kinetics: RRKM. In *Encyclopedia of Chemical Physics and Physical Chemistry*, J. H. Moore and N. D. Spencer (eds.). Bristol, IOP Publishing, Vol. 1, p. 865.

Hase, W. L. and D. M. Wardlaw (1989). Transition state theory rate constants for association reactions without potential energy barriers. In *Bimolecular Collisions*, M. N. R. Ashfold and J. E. Baggott (eds.). London, Royal Society of Chemistry, p. 171.

Hay, P. J., W. R. Wadt, *et al.* (1979). "Theoretical studies of molecular electronic transition lasers." *Ann. Rev. Phys. Chem.* **30**, 311.

Hayden, C. (2002). "Ultrafast photoelectron/photoion coincidence." *Ann. Rev. Phys. Chem.* **53**.

Hayes, S. C., P. M. Wallace, *et al.* (2002). "Investigating the phase-dependent photochemical reaction dynamics of chlorine dioxide using resonance Raman spectroscopy." *Int. Rev. Phys. Chem.* **21**, 405.

Hayward, S. and N. Go (1995). "Collective variable description of native protein dynamics." *Ann. Rev. Phys. Chem.* **46**, 223.

Head-Gordon, M. (1996). "Quantum chemistry and molecular processes." *J. Phys. Chem.* **100**, 13 213.

Heck, A. J. R. and D. W. Chandler (1995). "Imaging techniques for the study of chemical reaction dynamics." *Ann. Rev. Phys. Chem.* **46**, 335.

Heilweil, E. J., M. P. Casassa, *et al.* (1989). "Picosecond vibrational energy transfer studies of surface adsorbates." *Ann. Rev. Phys. Chem.* **40**, 143.
(eds.) (1993). *Vibrational Energy Dynamics*. Chem Phys. Amsterdam, Elsevier.

Heller, E. J. (1981). "The semiclassical way to molecular spectroscopy." *Acc. Chem. Res.* **14**, 368.

Hemminger, J. C. (1999). "Heterogeneous chemistry in the troposphere: a modern surface chemistry approach to the study of fundamental processes." *Int. Rev. Phys. Chem.* **18**, 387.

Henderson, D. (ed.) (1981). *Theoretical Chemistry: Theory of Scattering*. New York, Academic Press.

Henderson, M. A. (2002). "The interaction of water with solid surfaces: fundamental aspects revisited." *Surf. Sci. Rep.* **46**, 1.

Henriksen, N. E. and V. Engel (2001). "Femtosecond pump–probe spectroscopy: a theoretical analysis of transient signals and their relation to nuclear wave-packet motion." *Int. Rev. Phys. Chem.* **20**, 93.

Henry, B. R. (1977). "Use of local modes in the description of highly vibrationally excited molecules." *Acc. Chem. Res.* **10**, 207.

Henry, B. R. and M. Kasha (1968). "Radiationless molecular electronic transitions." *Ann. Rev. Phys. Chem.* **19**, 161.

Herbst, E. (1995). "Chemistry in the interstellar medium." *Ann. Rev. Phys. Chem.* **46**, 27.

Herek, J., W. Wohlleben, *et al.* (2002). "Quantum control of energy flow in light harvesting." *Nature* **417**, 533.

Herman, M., J. Lievin, *et al.* (eds.) (1999). "Global and Accurate Vibration Hamiltonians from High-Resolution Molecular Spectroscopy." *Adv. Chem. Phys.* **108**.

Herman, M. F. (1994). "Dynamics by semiclassical methods." *Ann. Rev. Phys. Chem.* **45**, 83.

Herschbach, D. (2000). "Fifty years in physical chemistry: homage to mentors, methods, and molecules." *Ann. Rev. Phys. Chem.* **51**, 1.

Herschbach, D. R. (1966). "Reactive scattering in molecular beams." *Adv. Chem. Phys.* **10**.
(1973). "Reactive scattering." *Faraday Disc.* **55**, 233.
(1987). "Molecular dynamics of elementary chemical reactions." *Angew. Chem.-Int. Ed. Engl.* **26**, 1221.

Hertel, I. V. (1982). "Progress in electronic-to-vibrational energy transfer." *Adv. Chem. Phys.* **50**, 475.

Hertel, I. V., H. Schmidt, *et al.* (1985). "Angular momentum transfer and charge cloud alignment in atomic collisions: intuitive concepts, experimental observations and semiclassical models." *Rep. Prog. Phys.* **48**, 375.

Herzberg, G. (1950). *Molecular Spectra and Molecular Structure I. Spectra of Diatomic Molecules*. Princeton, Van Nostrand.
(1966). *Molecular Spectra and Molecular Structure III. Electronic Spectra and Electronic Structure of Polyatomic Molecules*. Princeton, Van Nostrand.

Hill, T. L. (1989). *Free Energy Transduction and Biochemical Cycle Kinetics*. New York, Springer-Verlag.

Hines, M. A. (2003). "Understanding the highly defect-selective chemistry of anisotropic etching." *Ann. Rev. Phys. Chem.* **54**, 29.

Hinze, J. (ed.) (1983). *Energy Storage and Redistribution in Molecules*. New York, Plenum Press.

Hippler, H. and J. Troe (1989). Recent direct studies of collisional energy transfer in vibrationally highly excited molecules in the ground electronic state. In *Bimolecular Collisions*, M. N. R. Ashfold and J. E. Baggott (eds.). London, Royal Society of Chemistry, p. 209.

Hirschfelder, J. O. (ed.) (1967). *Intermolecular Forces*. New York, John Wiley & Sons.

Hirschfelder, J. O., C. F. Curtiss, *et al.* (1954). *Molecular Theory of Gases and Liquids*. New York, John Wiley & Sons.

Hirst, D. M. (1985). *Potential Energy Surfaces–Molecular Structure and Reaction Dynamics*. London, Taylor and Francis.

Hla, S. W. and K.-H. Rieder (2003). "STM control of chemical reactions: single-molecule synthesis." *Ann. Rev. Phys. Chem.* **54**, 307.

Ho, W. (1996). "Reactions at metal surfaces induced by femtosecond lasers, tunneling electrons, and heating." *J. Phys. Chem.* **100**, 13 050.

(1998). "Inducing and viewing bond selected chemistry with tunneling electrons." *Acc. Chem. Res.* **31**, 567.

Hochstrasser, R. M. (2002). "Two-dimensional infrared spectroscopy." *Ann. Rev. Phys. Chem.* **53**.

Hodgson, A. (2000). "State resolved desorption measurements as a probe of surface reactions." *Prog. Surf. Sci.* **63**, 1.

Hoffmann, R. (1971). "Interaction of orbitals through space and through bonds." *Acc. Chem. Res.* **4**, 1.

Hoffmann, R. A. (1999). "A really moving story." *Am. Sci.* **87**, 21.

Holbrook, K. A., M. J. Pilling, *et al.* (1996). *Unimolecular Reactions*. Chichester, John Wiley & Sons.

Hollebeek, T., T.-S. Ho, *et al.* (1999). "Constructing multi-dimensional molecular potential energy surfaces from ab initio data." *Ann. Rev. Phys. Chem.* **50**, 537.

Holloway, S. (1993). "Dynamics of gas–surface interactions." *Surf. Sci.* **299/300**, 656.

Honig, B. and A. Nicholls (1995). "Classical electrostatics in biology and chemistry." *Science* **268**, 1144.

Hou, H., S. J. Gulding, *et al.* (1997). "The stereodynamics of a gas–surface reaction." *Science* **277**, 80.

Houston, P. L. (1987). "Vector correlations in photodissociation dynamics." *J. Phys. Chem.* **91**, 5388.

(1989). "Correlated photochemistry: the legacy of Johann Christian Doppler." *Acc. Chem. Res.* **22**, 309.

(1995). "Snapshots of chemistry: product imaging of molecular reactions." *Acc. Chem. Res.* **28**, 453.

(1996). "New laser-based and imaging methods for studying the dynamics of molecular collisions." *J. Phys. Chem.* **100**, 12 757.

(2001). *Chemical Kinetics and Reaction Dynamics*. New York, McGraw-Hill.

Howard, J. (2001). *Mechanics of Motor Proteins and the Cytoskeleton*. Sunderland, MA, Sinauer Associates.

Hunt, K. L. C., P. M. Hunt, *et al.* (1990). "Nonlinear dynamics and thermodynamics of chemical reactions far from equilibrium." *Ann. Rev. Phys. Chem.* **41**, 409.

Hurley, S. M. and A. W. Castleman (2001). "Keeping reactions under quantum control." *Science* **292**, 648.

Husain, D. and G. Roberts (1989). Bimolecular collisions involving electronically excited alkaline earth atoms. In *Bimolecular Collisions*, M. N. R. Ashfold and J. E. Baggott (eds.). London, Royal Society of Chemistry, p. 263.

Hutson, J. M. (1990). "Intermolecular forces and the spectroscopy of van der Waals molecules." *Ann. Rev. Phys. Chem.* **41**, 123.

(2001). Van der Waals molecules. In *Encyclopedia of Chemical Physics and Physical Chemistry*, J. H. Moore and N. D. Spencer (eds.). Bristol, IOP Publishing, Vol. 3.

Hynes, J. T. (1985a). The theory of reactions in solution. In *Theory of Chemical Reaction Dynamics*, M. Baer (ed.). Boca Raton, FL, CRC Press, Vol. IV, p. 171.

(1985b). "Chemical reaction dynamics in solution." *Ann. Rev. Phys. Chem.* **36**, 573.

(ed.) (1994). "Charge Transfer Reactions and Solvation Dynamics." *Ultrafast Dynamics of Chemical Systems*. Dordrecht, Kluwer.

Imbihl, R. and G. Ertl (1995). "Oscillatory kinetics in heterogeneous catalysis." *Chem. Rev.* **95**, 697.

Imre, D., J. L. Kinsey, *et al.* (1984). "Chemical dynamics studied by emission spectroscopy of dissociating molecules." *J. Phys. Chem.* **88**, 3956.

Ishikawa, H., R. W. Field, *et al.* (1999). "HCP ↔ CPH isomerization: caught in the act." *Ann. Rev. Phys. Chem.* **50**, 443.

Iversen, G., Y. I. Kharkats, *et al.* (1999). "Fluctuations and coherence in long-range and multicenter electron transfer." *Adv. Chem. Phys.* **106**, 453.

Iwaki, L. K. and D. D. Dlott (2001). Vibrational energy transfer in condensed phases. In *Encyclopedia of Chemical Physics and Physical Chemistry*, J. H. Moore and N. D. Spencer (eds.). Bristol, IOP Publishing, Vol. 3, p. 2717.

Jackson, B. (1995). "Time-dependent wave packet approach to Quantum Reactive Scattering." *Ann. Rev. Phys. Chem.* **46**, 251.

Jacobs, D. C. (1995). "The role of internal energy and approach geometry in molecule–surface reactive scattering." *J. Phys.: Condens. Matter* **7**, 1023.

(2002). "Reactive collisions of hyperthermal energy molecular ions with solid surfaces." *Ann. Rev. Phys. Chem.* **53**, 379.

Jakubetz, W. (ed.) (2001). *Methods in Reaction Dynamics*. Berlin, Springer-Verlag.

Janda, K. C. (1984). "Predissociation of polyatomic van der Waals molecules." *Adv. Chem. Phys.* **60**, 201.

Janssen, M. H. M., J. W. G. Mastenbroek, *et al.* (1997). "Imaging of oriented molecules." *J. Phys. Chem.* **101**, 7605.

Jencks, W. P. (1986). *Catalysis in Chemistry and Enzymology*. New York, Dover.

Johnson, M. R. and G. J. Kearley (2000). "Quantitative atom–atom potentials from rotational tunneling: their extraction and their use." *Ann. Rev. Phys. Chem.* **51**, 297.

Johnson, P. M. (1980). "Molecular multiphoton ionization spectroscopy." *Acc. Chem. Res.* **13**, 20.

Johnson, R. E. (1982). *Introduction to Atomic and Molecular Collisions*. New York, Plenum Press.

Johnston, H. S. (1966). *Gas Phase Reaction Rate Theory*. New York, Ronald.

Jonas, D. M. (2003). "Two-dimensional femtosecond spectroscopy." *Ann. Rev. Phys. Chem.* **54**, 425.

Jortner, J. (1982). "Is the Marcus equation valid only in the classical limit?" *Faraday Disc. Chem. Soc.* **74**, 306.

(1992). "Cluster size effects." *Z. Phys. D* **24**, 247.

Jortner, J. and R. D. Levine (1981). "Photoselective chemistry." *Adv. Chem. Phys.* **47**, 1.

(1990). "Selective chemistry." *Isr. J. Chem.* **30**, 207.

(eds.) (1981). Photoselective Chemistry. *Adv. Chem. Phys.* **47**.

Jortner, J., R. D. Levine, *et al.* (eds.) (1991). *Mode Selective Chemistry*. Dordrecht, Kluwer.

(eds.) (1993). *Reaction Dynamics in Clusters and Condensed Phases*. Dordrecht, Kluwer.

(1988). "Level structure and dynamics from diatomics to clusters." *Adv. Chem. Phys.* **70**, 1.

Jortner, J. and S. Mukamel (1975). "Radiationless transitions." *Int. Rev. Sci. Phys. Chem.* **1**, 329.

Jortner, J. and B. Pullman (eds.) (1982). *Intramolecular Dynamics*. Dordrecht, Reidel.

Jortner, J., S. A. Rice, *et al.* (1969). "Radiationless transitions in photochemistry." *Adv. Photochem.* **7**, 149.

Jouvet, C., M. Boivineau, *et al.* (1987). "Photochemistry in excited states of van der Waals complexes." *J. Phys. Chem.* **91**, 5416.

Kaiser, R. I. (2002). "An experimental investigation on the formation of carbon-bearing molecules in the interstellar medium via neutral–neutral reactions." *Chem. Rev.* **102**, 1309.

(2003). *The Chemical Evolution of the Interstellar Medium: From Astrochemistry to Astrobiology.* Cambridge, Cambridge University Press.

Kaiser, R. I. and N. Balucani (2001). "The formation of nitriles in hydrocarbon-rich atmospheres of planets and their satellites: laboratory investigations by the crossed molecular beam technique." *Acc. Chem. Res.* **34**, 699.

Kaiser, R. I., C. Ochsenfeld, *et al.* (1998). "Combined crossed molecular beams and ab initio investigation of the formation of carbon-bearing molecules in the interstellar medium via neutral–neutral reactions." *Faraday Disc.* **109**, 183.

Kaiser, W. (ed.) (1988). *Ultrashort Laser Pulses and Applications.* Berlin, Springer-Verlag.

Kajimoto, O. (1999). "Solvation in supercritical fluids: its effects on energy transfer and chemical reactions." *Chem. Rev.* **99**, 355.

Kaledin, A. L. (2001). "Recent developments in the theory of surface diffusion." *J. Phys. Chem. A* **105**, 2731.

Kapral, R. (1981). "Kinetic theory of chemical reactions in liquids." *Adv. Chem. Phys.* **48**, 71.

Kapral, R. and S. J. Fraser (2001). Chaos and complexity in chemical systems. In *Encyclopedia of Chemical Physics and Physical Chemistry*, J. H. Moore and N. D. Spencer (eds.). Bristol, IOP Publishing, Vol. 3, p. 2737.

Kapral, R. and K. Showalter (eds.) (1995). *Chemical Waves and Patterns.* Dordrecht, Kluwer.

Karplus, M. (2002). "Molecular dynamics simulations of biomolecules." *Acc. Chem. Res.* **35**, 321.

Kasemo, B. (1996). "Charge transfer, electronic quantum processes, and dissociation dynamics in molecule–surface collisions." *Surf. Sci.* **363**, 22.

Kawashima, H., M. M. Wefers, *et al.* (1995). "Femtosecond pulse shaping, multiple-pulse spectroscopy, and optical control." *Ann. Rev. Phys. Chem.* **46**, 627.

Keske, J., D. A. McWhorter, *et al.* (2000). "Molecular rotation in the presence of intramolecular vibrational energy redistribution." *Int. Rev. Phys. Chem.* **19**, 363.

Keske, J. C. and B. H. Pate (2000). "Decoding the dynamical information embedded in highly mixed quantum states." *Ann. Rev. Phys. Chem.* **51**, 323.

Khundkar, L. R. and A. H. Zewail (1990). "Ultrafast molecular reaction dynamics in real-time: progress over a decade." *Ann. Rev. Phys. Chem.* **41**, 15.

Kim, M., M. Bertram, *et al.* (2001). "Controlling chemical turbulence by global delayed feedback: pattern formation in catalytic CO oxidation on Pt(110)." *Science* **292**, 1357.

Kim, S. K., J. Guo, *et al.* (1996). "Femtosecond chemically activated reactions: concept of nonstatistical activation at high thermal energies." *J. Phys. Chem.* **100**, 9202.

King, D. A. and D. P. Woodruff (eds.) (1983). *The Chemical Physics of Solid Surfaces and Heterogeneous Catalysis: Adsorption at Solid Surfaces.* Amsterdam, Elsevier.

Kinsey, J. L. (1977). "Laser-induced fluorescence." *Ann. Rev. Phys. Chem.* **28**, 349.

Kinsey, J. L. and B. R. Johnson (1998). "Angular distributions of products in the photodissociation of diatomic molecules." *J. Phys. Chem. A* **102**, 9660.

Kinsey, J. L. and R. D. Levine (1979). Information theoretic approach: application to molecular collisions. In *Atom-Molecular Collision Theory*, R. B. Bernstein (ed.). New York, Plenum Press.

Kleiber, P. D., W. C. Stwalley, *et al.* (1993). "Scattering-state spectroscopy as a probe of molecular dynamics." *Ann. Rev. Phys. Chem.* **44**, 13.

Klein, M. L. (2001). "Water on the move." *Science* **291**, 2106.

Klemperer, W. H. (1995). "Some spectroscopic reminiscences." *Ann. Rev. Phys. Chem.* **46**, 1.

Klessinger, M. and J. Michl (1995). *Excited States and Photochemistry of Organic Molecules.* New York, VCH Publications.

Kleyn, A. W., J. Los, *et al.* (1982). "Vibronic coupling at intersections of covalent and ionic states." *Phys. Rep.* **90**, 1.

Kleyn, A. W. and A. M. C. Moutinho (2001). "Negative ion formation in alkali-atom–molecule collisions." *J. Phys. B: At. Mol. Opt. Phys.* **34**, R1.

Kleyn, A. W., A. M. C. Moutinho, *et al.* (eds.) (1994). *Vibronic Processes in Gas Phase and Surface Scattering.* Chem Phys. Amsterdam, Elsevier.

Kohguchi, H., T. Suzuki, *et al.* (2001). "Fully state-resolved differential cross sections for the inelastic scattering of the open-shell NO molecule by Ar." *Science* **294**, 832.

Kohler, B., J. L. Krause, *et al.* (1995). "Controlling the future of matter." *Acc. Chem. Res.* **28**, 133.

Kohn, W., A. D. Becke, *et al.* (1996). "Density functional theory of electronic structure." *J. Phys. Chem.* **100**, 12 974.

Kolasinski, K. W. (2002). *Surface Science: Foundations of Catalysis and Nanoscience.* New York, John Wiley & Sons.

Komatsuzaki, T. and R. S. Berry (2002). "Chemical reaction dynamics: many-body chaos and regularity." *Adv. Chem. Phys.* **123**, 79.

Kondow, T. and F. Mafuné (2000). "Structures and dynamics of molecules on liquid beam surfaces." *Ann. Rev. Phys. Chem.* **51**, 731.

Koppel, H., W. Domcke, *et al.* (1984). "Multimode molecular dynamics beyond the Born–Oppenheimer approximation." *Adv. Chem. Phys.* **57**, 59.

Kosloff, R. (1994). "Propagation methods for quantum molecular dynamics." *Ann. Rev. Phys. Chem.* **45**, 145.

Kouri, D. J. (1985). The general theory of reactive scattering: the integral equation approach. In *The Theory of Chemical Reaction Dynamics*, M. Baer (ed.). Boca Raton, FL, CRC Press.

Kroes, G. J. (1999). "Six-dimensional quantum dynamics of dissociative chemisorption of H-2 on metal surfaces." *Prog. Surf. Sci.* **60**, 1.

Kroes, G. J., A. Gross, *et al.* (2002). "Quantum theory of dissociative chemisorption on metal surfaces." *Acc. Chem. Res.* **35**, 193.

Kropman, M. F. and H. J. Bakker (2001). "Dynamics of water molecules in aqueous solvation shells." *Science* **291**, 2118.

Kuipers, E. W., M. G. Tenner, *et al.* (1988). "Observation of steric effects in gas–surface scattering." *Nature* **334**, 420.

Kuppermann, A. and E. F. Greene (1968). "Chemical reaction cross sections and rate constants." *J. Chem. Educ.* **45**, 361.

Laane, J. (1994). "Vibrational potential energy surfaces and conformations of molecules in ground and excited electronic states." *Ann. Rev. Phys. Chem.* **45**, 179.

(1999). "Spectroscopic determination of ground and excited state vibrational potential energy surfaces." *Int. Rev. Phys. Chem.* **18**, 301.

Lagana, A. and A. Riganelli (eds.) (1999). *Reaction and Molecular Dynamics*. Berlin, Springer-Verlag.

Laidler, K. J. (1969). *Theories of Chemical Reaction Rates*. New York, McGraw-Hill.
(1984). "The development of the Arrhenius equation." *J. Chem. Educ.* **61**, 494.

Lambert, J. D. (1977). *Vibrational and Rotational Relaxation in Gases*. Oxford, Clarendon Press.

Landau, L. D. and L. M. Lifschitz (1971). *Quantum Mechanics*. Reading, MA, Addison-Wesley.

Last, I. and J. Jortner (1998). "Multielectron ionization of large rare gas clusters." *J. Phys. Chem. A* **102**, 9655.

Lau, A. M. F. (1982). "The photon-as-catalyst effect in laser induced predissociation and autoionization." *Adv. Chem. Phys.* **50**, 191.

Laubereau, A. and W. Kaiser (1978). "Vibrational dynamics of liquids and solids investigated by picosecond light pulses." *Rev. Mod. Phys.* **50**, 607.

Lawless, M. K., P. J. Reid, *et al.* (eds.) (1994). *Analysis of Condensed Phase Photochemical Reaction Mechanisms with Resonance Raman Spectroscopy*. Ultrafast Dynamics of Chemical Systems. Dordrecht, Kluwer.

Lawley, K. P. (ed.) (1975). Molecular Beam Scattering. *Adv. Chem. Phys.* **30**.
(ed.) (1985). Photodissociation and Photoionization. *Adv. Chem. Phys.* **60**.

Lawrence, W. D., C. B. Moore, *et al.* (1985). "Understanding molecular dynamics quantum-state by quantum-state." *Science* **227**, 895.

Leach, A. R. (2001). *Molecular Modeling: Principles and Applications*. Englewood Cliffs, Prentice-Hall.

Lee, Y.-P. (2003). "State-resolved dynamics of photofragmentation." *Ann. Rev. Phys. Chem.* **54**, 215.

Lee, Y. T. (1987a). "Molecular beam studies of elementary chemical processes." *Science* **236**, 793.
(1987b). "Molecular beam studies of elementary chemical processes." *Science* **236**, 789.

Lee, Y. T. and Y. R. Shen (1980). "Studies with crossed lasers and molecular beams." *Phys. Today* **33**, 52.

Lehmann, K. H., G. Scoles, *et al.* (1994). "Intramolecular dynamics from eigenstate-resolved infrared spectra." *Ann. Rev. Phys. Chem.* **45**, 241.

Lehmann, K. K., M. Herman, *et al.* (eds.) (1995). *Overtone Spectroscopy and Dynamics*. Chem Phys. Amsterdam, Elsevier.

Lemont, S. and G. W. Flynn (1977). "Vibrational state analysis of electronic to vibrational energy transfer." *Ann. Rev. Phys. Chem.* **28**, 261.

Lennard-Jones, J. E. (1932). "Processes of adsorption and diffusion on solid surfaces." *Trans. Faraday Soc.* **28**, 333.

Leone, S. R. (1982). "Photofragmentation dynamics." *Ann. Rev. Phys. Chem.* **50**, 255.
(1984). "State-resolved molecular reaction dynamics." *Ann. Rev. Phys. Chem.* **35**, 109.
(1985). "Laser probing of chemical reaction dynamics." *Science* **227**, 889.

Lester, M. I. (1996). "Vibrational predissociation dynamics of van der Waals complexes: product rotational state distributions." *Adv. Chem. Phys.* **96**, 51.

Letokhov, V. S. (1983). *Nonlinear Laser Chemistry*. Berlin, Springer-Verlag.

(1989). *Laser Spectroscopy of Highly Vibrationally Excited Molecules.* Bristol, Adam Hilger.

Levine, I. N. (2000). *Quantum Chemistry.* New Jersey, Prentice-Hall.

Levine, R. and J. Jortner (eds.) (1976). *Molecular Energy Transfer.* New York, John Wiley & Sons.

Levine, R. D. (1969). *Quantum Mechanics of Molecular Rate Processes.* London, Oxford University Press.

(1988). "Fluctuations in spectral intensities and transition rates." *Adv. Chem. Phys.* **70**, 53.

(1990). "The chemical shape of molecules – an introduction to dynamical stereochemistry." *J. Phys. Chem.* **94**, 8872.

(2001). Chemical reaction dynamics looks to the understanding of complex systems. In *Chemistry for the 21st Century*, E. Keinan and I. Schechter (eds.). Weinheim, Wiley-VCH.

Levine, R. D. and R. B. Bernstein (1976). Thermodynamic approach to collision processes. In *Dynamics of Molecular Collisions*, W. H. Miller (ed.). New York, Plenum Press.

(1987). *Molecular Reaction Dynamics and Chemical Reactivity.* New York, Oxford University Press.

Levine, R. D. and G. A. Somorjai (1990). "Kinetic model for cooperative dissociative chemisorption and catalytic activity via surface restructuring." *Surf. Sci.* **232**, 407.

Levine, R. D. and M. Tribus (eds.) (1979). *Maximum Entropy Formalism.* Cambridge, MA, MIT Press.

Levis, R. J. (1994). "Laser desorption and ejection of biomolecules from the condensed phase into the gas phase." *Ann. Rev. Phys. Chem.* **45**, 483.

Levis, R. J. and M. J. DeWitt (1999). "Photoexcitation, ionization, and dissociation of molecules using intense near-infrared radiation of femtosecond duration." *J. Phys. Chem. A* **103**, 6493.

Levis, R. J., G. M. Menkir, *et al.* (2001). "Selective bond dissociation and rearrangement with optimally tailored, strong-field laser pulses." *Science* **292**, 709.

Levy, D. H. (1981). "Van der Waals molecules." *Adv. Chem. Phys.* **47**, 323.

(1984). "The spectroscopy of supercooled gases." *Sci. Am.* **245**, 68.

(1999). "Charge transfer in bichromophoric molecules in the gas phase." *Adv. Chem. Phys.* **106**, 203.

Li, Y., J. Oslonovitch, *et al.* (2001). "Turing-type patterns on electrode surfaces." *Science* **291**, 2395.

Lifshitz, C. (1989). "Recent developments in application of RRKM-QET." *Adv. Mass. Spec.* **11A**, 713.

Light, J. C. (1979). Complex mode chemical reactions: statistical theories of bimolecular reactions. *Atom-Molecular Collision Theory: A Guide for the Experimentalist*, R. B. Bernstein (ed.). New York, Plenum Press.

Lin, J. J., J. Zhou, *et al.* (2003). "State-specific correlation of coincident product pairs in the F + CD$_4$ reaction." *Science* **300**, 966.

Lin, K.-C. and R. Vetter (2002). "Alkali–hydrogen reactions." *Int. Rev. Phys. Chem.* **21**, 357.

Lin, M. C. and G. Ertl (1986). "Laser probing of molecules desorbing and scattering from solid surfaces." *Ann. Rev. Phys. Chem.* **37**, 587.

Lin, S. H., C. H. Chang, *et al.* (2002). "Ultrafast dynamics and spectroscopy of bacterial photosynthetic reaction centers." *Adv. Chem. Phys.* **121**, 1.

Lindenberg, K. and B. C. West (1990). *The Nonequilibrium Statistical Mechanics of Open and Closed Systems.* New York, VCH-Publications.

Liu, K. (2001a). "Crossed-beam studies of neutral reactions: state-specific differential cross sections." *Ann. Rev. Phys. Chem.* **52**, 139.

(2001b). "Excitation functions of elementary chemical reactions: a direct link from crossed-beam dynamics to thermal kinetics?" *Int. Rev. Phys. Chem.* **20**, 189.

Liu, K., R. T. Skodje, *et al.* (2002). "Resonances in bimolecular chemical reactions." *Phys. Chem. Commun.* **4**, 27.

Liu, R. S. H. (2001). "Photoisomerization by Hula-twist: a fundamental supramolecular photochemical reaction." *Acc. Chem. Res.* **34**, 555.

Llorente, J. M. G. and E. Pollak (1992). "Classical dynamics methods for high energy vibrational spectroscopy." *Ann. Rev. Phys. Chem.* **43**, 91.

Loesch, H. J. (1980). "Scattering of non-spherical molecules." *Adv. Chem. Phys.* **42**, 421.

(1995). "Orientation and alignment in reactive beam collisions: recent progress." *Ann. Rev. Phys. Chem.* **46**, 555.

Loesch, H. J., J. Bulthuis, *et al.* (1996). "Molecules oriented by brute force." *Europhys. News* **27**, 12.

Logan, S. R. (1982). "The origin and status of the Arrhenius equation." *J. Chem. Educ.* **59**, 279.

(1996). *Fundamentals of Chemical Kinetics.* Essex, Longman House.

Loomis, R. A. and M. I. Lester (1997). "OH—H_2 entrance channel complexes." *Ann. Rev. Phys. Chem.* **48**, 643.

Lorenz, K. T., D. W. Chandler, *et al.* (2001). "Direct measurement of the preferred sense of NO rotation after collision with argon." *Science* **293**, 2063.

Lorenz, K. T., M. Westley, *et al.* (2000). "Velocity map imaging of inelastic scattering processes." *Phys. Chem. Chem. Phys.* **2**, 481.

Lorquet, J. C. (1994). "Whither the statistical theory of mass spectra?" *Mass Spectr. Rev.* **13**, 233.

(1996). Non-adiabatic processes in ionic dissociation dynamics. In *The Structure, Energetics, and Dynamics of Organic Ions*, T. Baer, C. Y. Ng and I. Powis (eds.). New York, John Wiley & sons.

(2000). "Landmarks in the theory of mass spectra." *Int. J. Mass Spectrom.* **200**, 43.

Lovejoy, E. R., S. K. Kim, *et al.* (1992). "Observation of transition-state vibrational thresholds in the rate of dissociation of ketene." *Science* **256**, 1542.

Lupo, D. W. and M. Quack (1987). "IR-laser photochemistry." *Chem. Rev.* **87**, 181.

Machida, K. (1999). *Principles of Molecular Mechanics.* New York, John Wiley & Sons.

MacRitchie, F. (1990). *Chemistry at Interfaces.* San Diego, Academic Press.

Madix, R. J. and J. Benzinger (1978). "Kinetic processes on metal single-crystal surfaces." *Ann. Rev. Phys. Chem.* **29**, 285.

Mahan, B. H. (1973). "Recombination of gaseous ions." *Adv. Chem. Phys.* **23**, 1.

(1975). "Electronic-structure and chemical dynamics." *Acc. Chem. Res.* **8**, 55.

Mahapatra, S., N. Chakrabarti, *et al.* (1999). "Time correlation function and its unifying role in molecular structure and dynamics." *Int. Rev. Phys. Chem.* **18**, 235.

Maitland, G. C., M. Rigby, *et al.* (1987). *Intermolecular Forces: Their Origin and Determination.* Oxford, Clarendon Press.

Makri, N. (1999). "Time-dependent quantum methods for large systems." *Ann. Rev. Phys. Chem.* **50**, 167.

Malinovsky, A. L., I. Y. Petrova, *et al.* (1998). "Transition spectra in the vibrational quasicontinuum of polyatomic molecules: Raman spectra of highly excited SF_6." *J. Phys. Chem. A* **102**, 9353.

Manolopoulos, D. E. (2002). "A delayed reaction." *Nature* **419**, 266.

Manolopoulos, D. E., K. Stark, *et al.* (1993). "The transition state of the $F + H_2$ reaction." *Science* **262**, 1852.

Manz, J. and J. A. W. Castleman (eds.) (1993). Femtosecond Chemistry. *J. Phys. Chem. A* **97**.

Manz, J. and C. S. Parmenter (eds.) (1989). *Mode Selectivity in Unimolecular Reactions.* Chem Phys. Amsterdam, North-Holland.

Manz, J. and L. Woste (eds.) (1995). *Femtosecond Chemistry.* Weinheim, Verlag Chemie.

Marcus, R. A. (1964). "Chemical and electrochemical electron-transfer theory." *Ann. Rev. Phys. Chem.* **15**, 155.

(1992). "Skiing the reaction rate slopes." *Science* **256**, 1523.

(1993). "Electron-transfer reactions in chemistry – theory and experiment." *Rev. Mod. Phys.* **65**, 599.

(1997a). "Solvent dynamics and RRKM theory of clusters." *Adv. Chem. Phys.* **101**, 391.

(1997b). "Theory of rates of $S(N)_2$ reactions and relation to those of outer sphere bond rupture electron transfers." *J. Phys. Chem. A* **101**, 4072.

Marquardt, R. and M. Quack (2001). Energy redistribution in reacting systems. In *Encyclopedia of Chemical Physics and Physical Chemistry*, J. H. Moore and N. D. Spencer (eds.). Bristol, IOP Publishing, Vol. 1, p. 897.

Martini, I. B., E. R. Barthel, *et al.* (2001). "Optical control of electrons during electron transfer." *Science* **293**, 462.

Masel, R. I. (1996). *Principles of Adsorption and Reaction at Solid Surfaces.* New York, John Wiley & Sons.

(2001). *Chemical Kinetics and Catalysis.* New York, John Wiley & Sons.

Mathies, R. A., C. H. B. Cruz, *et al.* (1988). "Direct observation of the femtosecond excited-state cis–trans isomerization in bacteriorhodopsin." *Science* **240**, 777.

May, V. (2000). *Charge and Energy Transfer Dynamics in Molecular Systems: A Theoretical Introduction.* New York, John Wiley & Sons.

McCaffery, A. J., M. J. Proctor, *et al.* (1986). "Rotational energy transfer: polarization and scaling." *Ann. Rev. Phys. Chem.* **37**, 223.

McCaffery, A. J. (1981). "Reorientation by elastic and rotationally inelastic transitions." *Spec. Per. Rep.* **4**, 47.

McCaffery, A. J. and R. J. Marsh (2001). "Molecules as products; the good, the bad, and the eccentric." *Phys. Chem. Commun.* **24**, 1.

McCammon, J. A. and S. C. Harvey (1983). *Dynamics of Proteins and Nucleic Acids.* Cambridge, Cambridge University Press.

McCammon, J. A. and M. Karplus (1983). "The dynamic picture of protein structure." *Acc. Chem. Res.* **16**, 187.

McCoy, A. B. and R. Naaman (1999). "Reactions of oxygen atoms with van der Waals molecules in crossed molecular beams." *Int. Rev. Phys. Chem.* **18**, 459.

McGowan, J. W. (ed.) (1975, 1981). *The Excited State in Chemical Physics.* New York, John Wiley & Sons.

Mchale, J. L. (2001). "Subpicosecond solvent dynamics in charge-transfer transitions: challenges and opportunities in resonance Raman spectroscopy." *Acc. Chem. Res.* **34**, 265.

Meerts, W. L., J. Jortner, *et al.* (eds.) (1991). *Nonradiative Processes*. Chem Phys. Amsterdam, Elsevier.

Menzinger, M. and R. Wolfgang (1969). "The meaning and use of the Arrhenius activation energy." *Angew. Chem.-Int. Ed. Engl.* **8**, 438.

Mestdagh, J.-M., B. Soep, *et al.* (2003). "Transition state in metal atom reactions." *Int. Rev. Phys. Chem.* **22**, 285.

Metz, R. B., S. E. Bradforth, *et al.* (1992). "Transition state spectroscopy of bimolecular reactions using negative ion photodetachment." *Adv. Chem. Phys.* **81**, 1.

Meyer, S. and V. Engel (2000). "Femtosecond time-resolved CARS and DFWM spectroscopy on gas-phase I_2: a wave-packet description." *J. Raman Spectrosc.* **31**, 33.

Michalak, A. and T. Ziegler (2001). "First-principle molecular dynamic simulations along the intrinsic reaction paths." *J. Phys. Chem. A* **105**, 4333.

Michelsen, H. A. (2001). "The reaction of Cl with CH_4: a connection between kinetics and dynamics." *Acc. Chem. Res.* **34**, 331.

Michl, J. (1972). "Photochemical reactions of large molecules. I. A simple physical model of photochemical reactivity." *J. Mol. Photochem.* **4**, 243.

Michl, J. and V. Bonacic-Koutecky (1990). *Electronic Aspects of Organic Photochemistry*. New York, John Wiley & Sons.

Mikkelsen, K. V. and M. A. Ratner (1987). "Electron-tunneling in solid-state electron-transfer reactions." *Chem. Rev.* **87**, 113.

Miller, J. A. (2002). "Combustion chemistry: concluding remarks." *Faraday Disc.* **119**, 461.

Miller, J. A., R. J. Kee, *et al.* (1990). "Chemical kinetics and combustion modeling." *Ann. Rev. Phys. Chem.* **41**, 345.

Miller, R. E. (2001). "Comparative studies of cluster dynamics in the gas and condensed phases." *Faraday Disc.* **118**, 1.

Miller, R. J. D. and J. C. Tully (eds.) (1996). *Surface Reaction Dynamics*. Chem Phys. Amsterdam, Elsevier.

Miller, W. H. (1974). "Classical-limit quantum mechanics and the theory of molecular collisions." *Adv. Chem. Phys.* **25**, 69.

(ed.) (1976a). *Dynamics of Molecular Collisions*. New York, Plenum Press.

(1976b). "Importance of nonseparability in quantum mechanical transition state theory." *Acc. Chem. Res.* **9**, 306.

(1990). "Recent advances in quantum mechanical reactive scattering theory, including comparison of recent experiments with rigorous calculations of state-to-state cross sections for the $H/D + H_2 \rightarrow H_2HD + H$ reactions." *Ann. Rev. Phys. Chem.* **41**, 245.

(1998a). "'Direct' and 'correct' calculation of canonical and microcanonical rate constants for chemical reactions." *J. Phys. Chem. A* **102**, 793.

(1998b). "Quantum and semiclassical theory of chemical reaction rates." *Faraday Disc.* **110**, 1.

Mirny, L. and E. Shakhnovich (2001). "Protein folding theory: from lattice to all-atom models." *Ann. Rev. Biophys. Biomol. Struct.* **30**, 361.

Moore, C. B. (1973). "Vibration $\rightarrow$ vibration energy transfer." *Adv. Chem. Phys.* **23**, 41.

(ed.) (1974–1980). *Chemical and Biochemical Applications of Lasers*. New York, Academic Press.

Moore, C. B. and I. W. M. Smith (1996). "State-resolved studies of reactions in the gas phase." *J. Phys. Chem.* **100**, 12 848.

Moore, C. B. and P. F. Zittel (1973). "State selected kinetics from laser-excited fluorescence." *Science* **182**, 541.

Moore, J. W. and R. G. Pearson (1981). *Kinetics and Mechanism*. New York, John Wiley & Sons.

Morokuma, K. (1982). "Potential energy surface of the SN_2 reaction in hydrated clusters." *J. Am. Chem. Soc.* **104**, 3732.

Mott, N. F. and H. S. W. Massey (1965). *The Theory of Atomic Collisions*. Oxford, Clarendon Press.

Mukamel, S. (1990). "Femtosecond optical spectroscopy: a direct look at elementary chemical events." *Ann. Rev. Phys. Chem.* **41**, 647.

  (1995). *Principles of Non-linear Optical Spectroscopy*. New York, Oxford University Press.

  (2000). "Multidimensional femtosecond correlation spectroscopies of electronic and vibrational excitations." *Ann. Rev. Phys. Chem.* **51**, 691.

Mukamel, S. and R. M. Hochstrasser (eds.) (2001). *Multidimensional Spectroscopies*. Chem Phys. Amsterdam, Elsevier.

Mullin, A. S. and G. C. Schatz (eds.) (1997). *Highly Excited Molecules: Relaxation, Reaction and Structure*. Washington, D.C., ACS Books.

Murrell, J. N. and S. D. Bosanac (1989). *Introduction to the Theory of Atomic and Molecular Collisions*. New York, John Wiley & Sons.

Musaev, D. G. and K. Morokuma (1996). "Potential energy surfaces of transition-metal-catalyzed chemical reactions." *Adv. Chem. Phys.* **95**, 61.

Myers, A. B. (1998). "Molecular electronic spectral broadening in liquids and glasses." *Ann. Rev. Phys. Chem.* **49**, 267.

Nakamura, H. (1997). "Theoretical studies of chemical dynamics: overview of some fundamental mechanisms." *Ann. Rev. Phys. Chem.* **48**, 299.

  (2002). *Nonadiabatic Transition*. Singapore, World Scientific.

Nandi, N., K. Bhattacharya, *et al.* (2000). "Dielectric relaxation and solvation dynamics of water in complex chemical and biological systems." *Chem. Rev.* **100**, 2013.

Nathanson, G. M., P. Davidovits, *et al.* (1996). "Dynamics and kinetics at the gas–liquid interface." *J. Phys. Chem.* **100**, 13 007.

Nauenberg, M., C. Stroud, *et al.* (1994). "The classical limit of an atom." *Sci. Am.* **270**, 44.

Nelson, P. (2003). *Biological Physics*. New York, W. H. Freeman.

Nesbitt, D. J. (1994). "High-resolution, direct infrared-laser absorption spectroscopy in slit supersonic jets: intermolecular forces and unimolecular vibrational dynamics in clusters." *Ann. Rev. Phys. Chem.* **45**, 367.

Nesbitt, D. J. and R. W. Field (1996). "Vibrational energy flow in highly excited molecules: role of intramolecular vibrational redistribution." *J. Phys. Chem.* **100**, 12 735.

Neumark, D. M. (1992). "Transition state spectroscopy of bimolecular chemical reactions." *Ann. Rev. Phys. Chem.* **43**, 153.

  (1993). "Transition state spectroscopy via negative-ion photodetachment." *Acc. Chem. Res.* **26**, 33.

  (2001a). Molecular reaction dynamics in the gas phase. In *Encyclopedia of Chemical Physics and Physical Chemistry*, J. H. Moore and N. D. Spencer (eds.). Bristol, IOP Publishing, Vol. 1, p. 745.

(2001b). "Time-resolved photoelectron spectroscopy of molecules and clusters." *Ann. Rev. Phys. Chem.* **52**, 255.

Newton, M. D. (1991). "Quantum chemical probes of electron-transfer kinetics – the nature of donor–acceptor interactions." *Chem. Rev.* **91**, 767.

(1999). "Control of electron transfer kinetics: models for medium reorganization and donor–acceptor coupling." *Adv. Chem. Phys.* **106**, 303.

Newton, M. D. and N. Sutin (1984). "Electron transfer reactions in condensed phases." *Ann. Rev. Phys. Chem.* **35**, 437.

Newton, R. G. (1982). *Scattering Theory of Waves and Particles*. New York, Springer-Verlag.

Newton, T. W. (1968). "The kinetics of oxidation–reduction reactions – an alternative derivation of Marcus' cross relation." *J. Chem. Educ.* **45**, 571.

Ng, C. Y. (2002). "Vacuum ultraviolet spectroscopy and chemistry using photoionization and photoelectron methods." *Ann. Rev. Phys. Chem.* **54**, 101.

Ng, C. Y. and M. Baer (eds.) (1992). *State-Selected and State-to-State Ion-Molecular Reaction Dynamics*. New York, John Wiley & Sons.

Nicolis, G. (2002). "Nonlinear kinetics: at the crossroads of chemistry, physics, and life sciences." *Faraday Disc.* **120**, 1.

Nikitin, E. E. (1974). *Theory of Elementary Atomic and Molecular Processes in Gases*. Oxford, Clarendon Press.

(1975). "Theory of nonadiabatic collision processes including excited alkali atoms." *Adv. Chem. Phys.* **28**, 317.

(1999). "Nonadiabatic transitions: what we learned from old masters and how much we owe them." *Ann. Rev. Phys. Chem.* **50**, 1.

Nikitin, E. E. and S. Y. Umanskii (1984). *Theory of Slow Atomic Collisions*. Berlin, Springer-Verlag.

Nitzan, A. (1988). "Activated rate processes in condensed phases: the Kramers theory revised." *Adv. Chem. Phys.* **70**, 489.

Nobusada, K., O. I. Tolstikhin, *et al.* (1998). "Quantum reaction dynamics of heavy–light–heavy systems: reduction of the number of potential curves and transitions at avoided crossings." *J. Phys. Chem. A* **102**, 9445.

Nohmi, T. and J. B. Fenn (1992). "Electrospray mass spectrometry of poly(ethylene glycols) with molecular weights up to 5 million." *J. Am. Chem. Soc.* **114**, 3241.

Noid, D. W., M. L. Koszykowski, *et al.* (1981). "Quasi-periodic and stochastic behavior in molecules." *Ann. Rev. Phys. Chem.* **32**, 267.

Northrup, F. J. and T. J. Sears (1992). "Stimulated emission pumping: applications to highly vibrationally excited transient molecules." *Ann. Rev. Phys. Chem.* **43**, 127.

Noyes, R. M. (1989). "Some models of chemical oscillators." *J. Chem. Educ.* **66**, 190.

Nyman, G. and H. G. Yu (2000). "Quantum theory of bimolecular chemical reactions." *Rep. Prog. Phys.* **63**, 1001.

Ogilby, P. R. (1999). "Solvent effects on the radiative transitions of singlet oxygen." *Acc. Chem. Res.* **32**, 512.

Ohmine, I. and S. Saito (1999). "Water dynamics: fluctuation, relaxation, and chemical reactions in hydrogen bond network rearrangement." *Acc. Chem. Res.* **32**, 741.

Omberg, K. M., P. Chen, *et al.* (1999). "Spectroscopic determination of electron transfer barriers and rate constants." *Adv. Chem. Phys.* **106**, 553.

Onuchic, J. N., Z. Luthey-Schulten, *et al.* (1997). "Theory of protein folding: the energy landscape perspective." *Ann. Rev. Phys. Chem.* **48**, 545.

Oref, I. (1995). Supercollisions. In *Advances in Chemical Kinetics and Dynamics*, J. Barker (ed.). Greenwich, CT, JAI Press, Vol. 2B, p. 285.

(1998). "Selective chemistry redux." *Science* **279**, 820.

Oref, I. and B. S. Rabinovitch (1979). "Do highly excited reactive polyatomic molecules behave ergodically." *Acc. Chem. Res.* **12**, 166.

Oref, I. and D. C. Tardy (1990). "Energy transfer in highly excited large polyatomic molecules." *Chem. Rev.* **90**, 1407.

Orr-Ewing, A. J. (1996). "The dynamical stereochemistry of bimolecular reactions." *J. Chem. Soc. Faraday Trans.* **92**, 881.

Orr-Ewing, A. J. and R. N. Zare (1994). "Orientation and alignment of reaction products." *Ann. Rev. Phys. Chem.* **45**, 315.

Oudejans, L. and R. E. Miller (2001). "Photofragment translational spectroscopy of weakly bound complexes: probing the interfragmentary correlated final state distributions." *Ann. Rev. Phys. Chem.* **52**, 607.

Owrutsky, J. C., D. Raftery, *et al.* (1994). "Vibrational relaxation dynamics in solutions." *Ann. Rev. Phys. Chem.* **45**, 519.

Oxtoby, D. W. (1981). "Vibrational relaxation in liquids." *Ann. Rev. Phys. Chem.* **32**, 77.

Ozkan, I. and L. Goodman (1979). "Coupling of electronic and vibrational motions in molecules." *Chem. Rev.* **79**, 275.

Parker, D. H. (2000). "Laser photochemistry of molecular oxygen." *Acc. Chem. Res.* **33**, 563.

Parker, D. H. and R. B. Bernstein (1989). "Oriented molecule beams via the electrostatic hexapole: preparation, characterization, and reactive scattering." *Ann. Rev. Phys. Chem.* **40**, 561.

Parmenter, C. S. (1972). "Radiative and nonradiative processes in benzene." *Adv. Chem. Phys.* **22**, 365.

Parmenter, C. S., S. M. Clegg, *et al.* (1997). "Insightful experiment demonstrating that kinetic factors govern rotational transfer probabilities." *Proc. Natl. Acad. Sci. USA* **94**, 8387.

Parr, R. G. and W. Wang (1989). *Density Functional Theory of Atoms and Molecules*. Oxford, Oxford University Press.

Pauly, H. (1979). Elastic scattering cross sections. In *Atom-Molecular Collision Theory: A Guide for the Experimentalist*, R. B. Bernstein (ed.). New York, Plenum Press.

(2000). *Atom, Molecule, and Cluster Beams: Basic Theory, Production, and Detection of Thermal Energy Beams*. Berlin, Springer-Verlag.

Pausch, R., M. Heid, *et al.* (2000). "Quantum control by stimulated Raman scattering." *J. Raman Spectrosc.* **31**, 7.

Pechukas, P. (1976). Statistical approximation in collision theory. In *Dynamics of Molecular Collision*, W. H. Miller (ed.). New York, Plenum Press, Part B, p. 299.

(1981). "Transition-state theory." *Ann. Rev. Phys. Chem.* **32**, 159.

Pechukas, P. and U. Weiss (eds.) (2001). *Quantum Dynamics of Open Systems*. Chem Phys. Amsterdam, Elsevier.

Pedersen, S., J. L. Herek, *et al.* (1994). "The validity of the diradical hypothesis – direct femtosecond studies of the transition-state structures." *Science* **266**, 1359.

Peng, T., D. H. Zhang, *et al.* (2000). "Dynasol: a visual quantum dynamics package." *Comput. Phys. Commun.* **128**, 492.

Petek, H. and S. Ogawa (2002). "Surface femtochemistry: observation and quantum control of frustrated desorption of alkali atoms from noble metals." *Ann. Rev. Phys. Chem.* **53**, 507.

Pilling, M. J. (1996). "Radical–radical reactions." *Ann. Rev. Phys. Chem.* **47**, 81.

Pilling, M. J. and S. H. Robertson (2003). "Master equation models for chemical reactions of importance in combustion." *Ann. Rev. Phys. Chem.* **54**, 245.

Pilling, M. J., S. H. Robertson, *et al.* (1995). "Elementary radical reactions and autoignition." *J. Chem. Soc. Faraday Trans.* **91**, 4179.

Pilling, M. J. and I. W. M. Smith (eds.) (1987). *Modern Gas Kinetics: Theory, Experiment, and Application*. Oxford, Blackwell.

Pimentel, G. C. (ed.) (1985). *Opportunities in Chemistry*. Washington, D.C., National Academy Press.

Pogrebnya, S. K., *et al.* (2000). *PCCP* **2**, 693.

Polanyi, J. C. (1972). "Some concepts in reaction dynamics." *Acc. Chem. Res.* **5**, 161.

(1987). "Some concepts in reaction dynamics." *Angew. Chem.* **99**, 981.

(2001). Macro, micro, and nanobeams. In *Atomic and Molecular Beams: The State of the Art 2000*, R. Campargue (ed.). Berlin, Springer-Verlag.

Polanyi, J. C. and A. H. Zewail (1995). "Direct observation of the transition state." *Acc. Chem. Res.* **28**, 119.

Polanyi, M. (1932). *Atomic Reactions*. London, Williams and Norgate.

Pollak, E. (1985). Periodic orbits and the theory of reactive scattering. In *The Theory of Chemical Reaction Dynamics*, M. Baer (ed.). Boca Raton, FL, CRC Press, Vol. III.

(1996). Theory of activated rate processes. In *Dynamics of Molecules and Chemical Reactions*, J. Zhang and R. E. Wyatt (eds.). New York, Marcel Dekker, p. 617.

Pollard, W. T., A. K. Felts, *et al.* (1996). "The Redfield equation in condensed-phase quantum dynamics." *Adv. Chem. Phys.* **93**, 77.

Pollard, W. T. and R. A. Mathies (1992). "Analysis of femtosecond dynamic absorption spectra of nonstationary states." *Ann. Rev. Phys. Chem.* **43**, 497.

Porter, R. N. and L. M. Raff (1976). Classical trajectory methods in molecular collisions. In *Dynamics of Molecular Collisions*, W. H. Miller (ed.). New York, Plenum Press.

Pratt, D. W. (1998). "High resolution spectroscopy in the gas phase: even large molecules have well-defined shapes." *Ann. Rev. Phys. Chem.* **49**, 481.

Pritchard, H. O. (1984). *The Quantum Theory of Unimolecular Reactions*. New York, Cambridge University Press.

Pross, A. (1985). "A general approach to organic reactivity – the configuration mixing model." *Adv. Phys. Org. Chem.* **21**, 99.

(1995). *Theoretical and Physical Principles of Organic Reactivity*. New York, John Wiley & Sons.

Pross, A. and S. S. Shaik (1983). "A qualitative valence-bond approach to organic reactivity." *Acc. Chem. Res.* **16**, 363.

Pulay, P. (1995). Analytical derivative techniques and the calculation of vibrational spectra. In *Modern Electronic Structure Theory*, D. Yarkony (ed.). Singapore, World Scientific, p. 1191.

Pulay, P. and J. Baker (2001). Optimization and reaction path algorithms. In *Encyclopedia of Chemical Physics and Physical Chemistry*, J. H. Moore and N. D. Spencer (eds.). Bristol, IOP Publishing, Vol. 2, p. 2061.

Pullman, B. (ed.) (1978). *Intermolecular Interactions from Diatomics to Biopolymers*. New York, John Wiley & Sons.

Pullman, B., J. Jortner, *et al.* (eds.) (1984). *Dynamics on Surfaces*. Boston, Reidel.

Quack, M. (1982). "Reaction dynamics and statistical mechanics of the preparation of highly excited states by infrared radiation." *Adv. Chem. Phys.* **50**, 395.

      (1990). "Spectra and dynamics of coupled vibrations in polyatomic molecules." *Ann. Rev. Phys. Chem.* **41**, 839.

Quack, M. and W. Kutzelnigg (1995). "Molecular spectroscopy and molecular dynamics – theory and experiment." *Ber. Bunsenges. Phys. Chem. Chem. Phys.* **99**, 231.

Rabinovitch, B. S. and D. W. Sester (1964). "Unimolecular decomposition and some isotope effects of simple alkanes and alkyl radicals." *Adv. Photochem.* **3**, 1.

Rabitz, H. (1974). "Rotation and rotation–vibration pressure-broadened spectral lineshapes." *Ann. Rev. Phys. Chem.* **25**, 155.

Rabitz, H., R. DeVivie-Riedle, *et al.* (2000). "Whither the future of controlling quantum phenomena." *Science* **288**, 824.

Rabitz, H. and W. Zhu (2000). "Optimal control of molecular motion: design, implementation, and inversion." *Acc. Chem. Res.* **33**, 572.

Radeke, M. R. and E. A. Carter (1997). "Ab initio dynamics of surface chemistry." *Ann. Rev. Phys. Chem.* **48**, 243.

Raghavachari, K. and J. Hutson (2002). "Theory of molecular clusters." *Ann. Rev. Phys. Chem.* **53**.

Rahman, N. K. and C. Guidotti (eds.) (1984). *Photon-Assisted Collisions and Related Topics*. New York, Harwood.

Raineri, F. O. and H. L. Friedman (1999). "Solvent control of electron transfer reactions." *Adv. Chem. Phys.* **107**, 81.

Rakitzis, T. P., S. A. Kandel, *et al.* (1998). "Photofragment helicity caused by matter–wave interference from multiple dissociative states." *Science* **281**, 1346.

Rapaport, D. C. (1997). *The Art of Molecular Dynamics Simulation*. Cambridge, Cambridge University Press.

Rapp, D. and T. Kassal (1969). "The theory of vibrational energy transfer between simple molecules in nonreactive collisions." *Chem. Rev.* **69**, 61.

Rappe, A. K. and C. J. Casewit (1997). *Molecular Mechanics Across Chemistry*. Fort Collins, University of Colorado Science Publishers.

Rauk, A. (1994). *Orbital Interaction Theory of Organic Chemistry*. New York, John Wiley & Sons.

Ravishankara, A. R., G. Hancock, *et al.* (1998). "Photochemistry of ozone: surprises and recent lessons." *Science* **280**, 60.

Regan, C. K., S. L. Craig, *et al.* (2002). "Steric effects and solvent effects in ionic reactions." *Science* **295**, 2245.

Reid, K. L. (2003). "Photoelectron angular distributions." *Ann. Rev. Phys. Chem.* **54**, 397.

Reid, S. A. and H. Reisler (1996). "Experimental studies of resonances in unimolecular decomposition." *Ann. Rev. Phys. Chem.* **47**, 495.

Reisler, H. and C. Wittig (1986). "Photo-initiated unimolecular reactions." *Ann. Rev. Phys. Chem.* **37**, 307.

Rettner, C. T. and M. N. R. Ashfold (1991). *Dynamics of Gas–Surface Interactions*. London, Royal Society of Chemistry.

Rettner, C. T. and D. J. Auerbach (1994). "Distinguishing the direct and indirect products of a gas–surface reaction." *Science* **263**, 365.

Rettner, C. T., D. J. Auerbach, *et al.* (1996). "Chemical dynamics at the gas–surface interface." *J. Phys. Chem.* **100**, 13 201.

Rettner, C. T., E. K. Schweizer, *et al.* (1990). "Dynamics of chemisorption of N_2 on W(100): precursor-mediated and activated dissociation." *J. Chem. Phys.* **93**, 1442.

Reynolds, L., J. A. Gardecki, *et al.* (1996). "Dipole solvation in nondipolar solvents: experimental studies of reorganization energies and solvation dynamics." *J. Phys. Chem.* **100**, 10 337.

Rhodes, W. (1983). "Nonradiative relaxation and quantum beats in the radiative decay dynamics of large molecules." *J. Phys. Chem.* **87**, 30.

Rice, S. A. (1981a). "Collision induced intramolecular energy transfer in electronically excited polyatomic molecules." *Adv. Chem. Phys.* **47**(2), 237.

(1981b). An overview of the dynamics of intramolecular transfer of vibrational energy. In *Photoselective Chemistry*, J. Jortner, R. D. Levine, and S. A. Rice (eds.). New York, John Wiley & Sons, Vol. 1, p. 117.

(1997). "Perspectives on the control of quantum many-body dynamics: application to chemical reactions." *Adv. Chem. Phys.* **101**, 213.

Rice, S. A. and M. Zhao (2000). *Optical Control of Molecular Dynamics*. New York, John Wiley & Sons.

Ritz, T., A. Damjanovic, *et al.* (2002). "The quantum physics of photosynthesis." *Chem. Phys. Chem.* **3**, 243.

Robb, M. A., M. Garavelli, *et al.* (2000). *A Computational Strategy for Organic Photochemistry*. Reviews in Computational Chemistry, K. B. Lipkowitz and D. B. Boyd (eds.). New York, Wiley-VCH, Vol. 15, p. 87.

Rohrbacher, A., N. Halberstadt, *et al.* (2000). "The dynamics of noble gas – halogen molecules and clusters." *Ann. Rev. Phys. Chem.* **51**, 405.

Rosker, M. J., M. Dantus, *et al.* (1988). "Femtosecond clocking of the chemical bond." *Science* **241**, 1200.

Ross, J. (ed.) (1966). "Molecular Beams." *Adv. Chem. Phys.* **10**.

Rotermund, H. H. (1997). "Imaging of dynamic processes on surfaces by light." *Surf. Sci. Rep.* **29**, 265.

Rowland, F. S. (1991). "Stratospheric ozone depletion." *Ann. Rev. Phys. Chem.* **42**, 731.

Rubahn, H. G. and K. Bergmann (1990). "The effect of laser-induced vibrational bond stretching in atom–molecule collisions." *Ann. Rev. Phys. Chem.* **41**, 735.

Ruhman, S. and N. Scherer (eds.) (1998). *Coherence in Chemical Dynamics*. Chem Phys. Amsterdam, Elsevier.

Sachs, C., M. Hildebrand, *et al.* (2001). "Spatiotemporal self organization in a surface reaction: from the atomic to the mesoscopic scale." *Science* **293**, 1635.

Sagdeev, R. Z., D. A. Usikov, *et al.* (eds.) (1988). *Nonlinear Physics From the Pendum to Turbulence and Chaos*. Chur, Harwood.

Sandstroem, V. (ed.) (1997). *Femtochemistry and Femtobiology*. Singapore, World Scientific.

Sathyamurthy, N. (1983). "Effect of reagent rotation on elementary exchange reactions." *Chem. Rev.* **83**, 601.

(ed.) (1991). *Reaction Dynamics: Recent Advances*. Berlin, Springer-Verlag.

Sato, H. (2001). "Photodissociation of simple molecules in the gas phase." *Chem. Rev.* **101**, 2687.

Schatz, G. C. (1988). "Quantum effects in gas phase chemical reactions." *Ann. Rev. Phys. Chem.* **39**, 317.

(1989). "The analytical representation of electronic potential-energy surfaces." *Rev. Mod. Phys.* **61**, 669.

(1993). "Transition states of chemical reactions." *Science* **262**, 1828.

(1996). "Scattering theory and dynamics: time-dependent and time-independent methods." *J. Phys. Chem.* **100**, 12 839.

(2000a). "Detecting resonances." *Science* **288**, 1599.

(2000b). "Stretched water is more reactive." *Science* **290**, 950.

(2001). Quantum mechanics of interacting systems: scattering theory. In *Encyclopedia of Chemical Physics and Physical Chemistry*, J. H. Moore and N. D. Spencer (eds.) Bristol, IOP Publishing, Vol. 1, p. 827.

Schatz, G. C. and M. A. Ratner (1993). *Quantum Mechanics in Chemistry*. Englewood Cliffs, Prentice-Hall.

Schinke, R. (1988). "Rotational distributions in direct molecular photodissociation." *Ann. Rev. Phys. Chem.* **39**, 39.

(1993). *Photodissociation Dynamics*. Cambridge, Cambridge University Press.

Schinke, R. and J. M. Bowman (eds.) (1983). *Molecular Collision Dynamics*. Berlin, Springer-Verlag.

Schinke, R., H.-M. Keller, *et al.* (1997). "Resonances in unimolecular dissociation: from mode-specific to statistical behavior." *Adv. Chem. Phys.* **101**, 745.

Schlag, E. W. and R. D. Levine (1997). "ZEKE spectroscopy of ions, radicals, reactive intermediates, and clusters and the dynamics of high molecular Rydberg states." *Comm. At. Mol. Phys.* **33**, 159.

Schlag, E. W. and H.-J. Neusser (1983). "Multiphoton mass spectrometry." *Acc. Chem. Res.* **16**, 355.

Schlag, E. W., S. Schneide, *et al.* (1971). "Lifetimes in excited states." *Ann. Rev. Phys. Chem.* **22**, 465.

Schlag, E. W., R. Weinkauf, *et al.* (eds.) (1998). *Molecular Clusters*. Chem Phys. Amsterdam, Elsevier.

Schlegel, H. B. (1995). Geometry optimization on potential energy surfaces. In *Modern Electronic Structure Theory*, D. Yarkony (ed.). Singapore, World Scientific, p. 459.

Schleyer, P. v R. (ed.) (1998). *Encyclopedia of Computational Chemistry*. Chichester, John Wiley & Sons.

Schlick, T. (2002). *Molecular Modeling and Simulation*. Berlin, Springer-Verlag.

Schlier, C. (ed.) (1970). *Molecular Beams and Reaction Kinetics*. New York, Academic Press.

Schmidt, H. and I. V. Hertel (1988). "Symmetry and angular momentum in collisions with laser excited, polarised atoms." *Adv. Chem. Phys.* **72**, 37.

Schnieder, L., K. Seekamp-Rahn, *et al.* (1991). "Hydrogen exchange reaction $H + D_2$ in crossed beams." *Faraday Disc. Chem. Soc.* **91**, 259.

Schnieder, L., K. Seekamp-Rahn, *et al.* (1995). "Experimental studies and theoretical predictions for the $H + D_2 \rightarrow HD + D$ reaction." *Science* **269**, 207.

Scholes, G. D. (2003). "Long-range resonance energy transfer in molecular systems." *Ann. Rev. Phys. Chem.* **54**, 57.

Schowen, R. L. (2003). "How an enzyme surmounts the activation energy barrier." *Proc. Natl. Acad. Sci. USA* **100**, 11 931.

Schroeder, J. (2001). Chemical kinetics in condensed phases. In *Encyclopedia of Chemical Physics and Physical Chemistry*, J. H. Moore and N. D. Spencer (eds.). Bristol, IOP Publishing, Vol. 1, p. 711.

Schroeder, J. and J. Troe (1987). "Elementary reactions in the gas–liquid transition range." *Ann. Rev. Phys. Chem.* **38**, 163.

    (1993). Solvent effects in the dynamics of dissociation, recombination and isomerization reactions. In *Activated Barrier Crossing*, G. R. Fleming and P. Hanggi (eds.). Singapore, World Scientific, p. 206.

Schulz, P. A., A. S. Sudbo, *et al.* (1979). "Multi-photon dissociation of polyatomic-molecules." *Ann. Rev. Phys. Chem.* **30**, 379.

Schunemann, V. and H. Winkler (2000). "Structure and dynamics of biomolecules studied by Mossbauer spectroscopy." *Rep. Prog. Phys.* **63**, 263.

Schuster, H. G. (1988). *Deterministic Chaos: An Introduction*. Weinheim, VCH.

Schwartz, B. J., J. C. King, *et al.* (1994). The molecular basis of solvent caging. In *Ultrafast Dynamics of Chemical Systems*, J. D. Simon (ed.). Dordrecht, Kluwer.

Scoles, G. (ed.) (1988). *Atomic and Molecular Beam Methods*. New York, Oxford University Press.

Scott, S. K. (1987). "Oscillations in simple models of chemical systems." *Acc. Chem. Res.* **20**, 186.

    (1994). *Oscillations, Waves and Chaos in Chemical Kinetics*. Oxford, Oxford University Press.

    (2001). Nonlinear reactions, feedback and self-organizing reactions. In *Encyclopedia of Chemical Physics and Physical Chemistry*, J. H. Moore and N. D. Spencer (eds.). Bristol, IOP Publishing, Vol. 1, p. 937.

Secrest, D. (1973). "Theory of rotational and vibrational energy transfer in molecules." *Ann. Rev. Phys. Chem.* **24**, 379.

Seideman, T. (2002). "Time-resolved photoelectron angular distributions: concepts, applications, and directions." *Ann. Rev. Phys. Chem.* **53**, 41.

Seideman, T., M. Y. Ivanov, *et al.* (1995). "Role of electron localization in intense-field molecular ionization." *Phys. Rev. Lett.* **75**, 2819.

Seilmeier, A. and W. Kaiser (eds.) (1988). *Ultrashort Intramolecular and Intermolecular Vibrational Energy Transfer of Polyatomic Molecules in Liquids*. Topics in Applied Physics. Berlin, Springer-Verlag.

Shaik, S. S. (1985). "The collage of S_N2 reactivity patterns." *Prog. Phys. Org. Chem.* **15**, 197.

Shapiro, M. and P. Brumer (1989). "Coherent chemistry – controlling chemical reactions with lasers." *Acc. Chem. Res.* **22**, 407.

    (1994). "Coherent and incoherent laser control of photochemical reactions." *Int. Rev. Phys. Chem.* **13**, 187.

    (2003). *Principles of the Quantum Control of Molecular Processes*. New York, John Wiley & Sons.

Shatz, C. J. (1996). "Emergence of order in visual system development." *Proc. Natl. Acad. Sci. USA* **93**, 602.

Sheehy, B. and L. F. DiMauro (1996). "Atomic and molecular dynamics in intense optical fields." *Ann. Rev. Phys. Chem.* **47**, 463.

Shepard, R. (1987). "The multiconfiguration self-consistent field method." *Adv. Chem. Phys.* **69**, 63.

Shin, S. K., Y. Chen, *et al.* (1991). "Photoinitiated reactions in weakly bonded complexes." *Adv. Photochem.* **16**, 249.

Siebrand, W. (1976). Nonradiative processes in molecular systems. In *Dynamics of Molecular Collisions*, W. H. Miller (ed.). New York, Plenum Press.

Silva, M., R. Jongma, *et al.* (2001). "The dynamics of 'stretched molecules': experimental studies of highly vibrationally excited molecules with stimulated emission pumping." *Ann. Rev. Phys. Chem.* **52**, 811.

Simons, J. (2001). Quantum structural methods for atoms and molecules. In *Encyclopedia of Chemical Physics and Physical Chemistry*, J. H. Moore and N. D. Spencer (eds.). Bristol, IOP Publishing, Vol. 1, p. 827.

Simons, J. P. (1984). "Photodissociation, a critical survey." *J. Phys. Chem.* **88**, 1287.

(1987). "Dynamic stereochemistry and the polarization of reaction products." *J. Phys. Chem.* **91**, 5378.

(1997). "The ins and outs of collision complexes." *J. Chem. Soc. Faraday Trans.* **93**, 4095.

(1999). "Stereochemistry and control in molecular reaction dynamics." *Faraday Disc.* **113**, 1.

Sims, I. R. and I. W. M. Smith (1995). "Gas-phase reactions and energy transfer at very low temperatures." *Ann. Rev. Phys. Chem.* **46**, 109.

Sinha, A., M. C. Hsiao, *et al.* (1991). "Controlling bimolecular reactions – mode and bond selected reaction of water with hydrogen atoms." *J. Chem. Phys.* **94**, 4928.

Skodje, R. T. (1993). "The adiabatic theory of heavy–light–heavy chemical reactions." *Ann. Rev. Phys. Chem.* **44**, 145.

Skouteris, D., J. F. Castillo, *et al.* (2000). "ABC: a quantum reactive scattering program." *Comp. Phys. Com.* **133**, 128.

Skouteris, D., D. E. Manolopoulos, *et al.* (1999). "Van der Waals interactions in the Cl + HD reaction." *Science* **286**, 1713.

Skowronek, S. and A. G. Urena (1999). "Spectroscopy and dynamics of the laser induced intracluster $(Ba..FCH_3)^* \rightarrow BaF^* + CH_3$ and $Ba^* + FCH_3$ reaction." *Prog. React. Kin. Mech.* **24**, 101.

Slinko, M. M. and N. Jaeger (1994). *Oscillatory Heterogeneous Catalytic Systems*. Amsterdam, Elsevier.

Smith, I. W. M. (1976). "Relaxation in collisions of vibrationally excited molecules with potentially reactive atoms." *Acc. Chem. Res.* **9**, 161.

(1980a). *Kinetics and Dynamics of Elementary Gas Reactions*. London, Butterworth.

(ed.) (1980b). *Physical Chemistry of Fast Reactions: Reaction Dynamics*. New York, Plenum Press.

(1989). Vibrational excitation in bimolecular reactions. In *Bimolecular Collisions*, M. N. R. Ashfold and J. E. Baggott (eds.). London, Royal Society of Chemistry, p. 53.

(1990). "Vibrational adiabaticity in chemical reactions." *Acc. Chem. Res.* **23**, 101.

(2002). "Chemistry amongst the stars: reaction kinetics at a new frontier." *Chem. Soc. Rev.* **31**, 137.

Smith, I. W. M. and F. F. Crim (2002). "The chemical kinetics and dynamics of the prototypical reaction: $OH + H_2^- \rightarrow H_2O + H^-$." *Phys. Chem. Chem. Phys.* **4**, 3543.

Smith, I. W. M. and B. R. Rowe (2000). "Reaction kinetics at very low temperatures: laboratory studies and interstellar chemistry." *Acc. Chem. Res.* **33**, 261.

Somorjai, G. A. (1981). *Chemistry in Two Dimensions: Surfaces*. Ithaca, NY, Cornell University Press.

(1994a). "Surface reconstruction and catalysis." *Ann. Rev. Phys. Chem.* **45**, 721.

(1994b). *Introduction to Surface Chemistry and Catalysis*. New York, John Wiley & Sons.

(2000). "The development of molecular surface science and the surface science of catalysis: the Berkeley contribution." *J. Phys. Chem. B* **104**, 2969.

Somorjai, G. A. and G. Rupprechter (1998). "The flexible surface." *J. Chem. Educ.* **75**, 161.

Speiser, S. (1996). "Photophysics and mechanisms of intramolecular electronic energy transfer in bichromophoric molecular systems: solution and supersonic jet studies." *Chem. Rev.* **96**, 1953.

Spudich, J. A. (2001). "The myosin swinging cross-bridge model." *Nat. Cell Biol.* **2**, 387.

Stapelfeldt, H. and T. Seideman (2003). "Aligning molecules with strong laser pulses." *Rev. Mod. Phys.* **75**, 543.

Steinfeld, J. I. (1985). *Molecules and Radiation*. Cambridge, MA, MIT Press.

Steinfeld, J. I., J. S. Francisco, *et al.* (1999). *Chemical Kinetics and Dynamics*. Upper Saddle River, Prentice-Hall.

Stolow, A. (2003a). "Femtosecond time-resolved photoelectron spectroscopy of polyatomic molecules." *Ann. Rev. Phys. Chem.* **54**, 89.

(2003b). "Time-resolved photoelectron spectroscopy: non-adiabatic dynamics in polyatomic molecules." *Int. Rev. Phys. Chem.* **22**, 377.

Stolte, S. (1982). "Reactive scattering studies on oriented molecules." *Ber. Bunsenges. Phys. Chem.* **86**, 413.

(1988). Scattering experiments with state selectors. In *Atomic and Molecular Beam Methods*, G. Scoles (ed.). New York, Oxford University Press.

Stoneham, A. M. and J. H. Harding (1986). "Interatomic potentials in solid state chemistry." *Ann. Rev. Phys. Chem.* **37**, 53.

Straatsma, T. P. and J. A. McCammon (1992). "Computational alchemy." *Ann. Rev. Phys. Chem.* **43**, 407.

Stratt, R. M. (1995). "The instantaneous normal modes of liquids." *Acc. Chem. Res.* **28**, 201.

Stratt, R. M. and M. Maroncelli (1996). "Nonreactive dynamics in solution: the emerging molecular view of solvation dynamics and vibrational relaxation." *J. Phys. Chem.* **100**, 12 981.

Strazisar, B. R., C. Lin, *et al.* (2000). "Mode-specific energy disposal in the four-atom reaction $OH + D_2 \rightarrow HOD + D$." *Science* **290**, 958.

Suits, A. G. and R. E. Continetti (eds.) (2001). *Imaging in Chemical Dynamics*. ACS Symposium Series. Washington, D.C., American Chemical Society.

Sullivan, J. H. (1967). "Mechanism of the 'bimolecular' hydrogen–iodine reaction." *J. Chem. Phys.* **46**, 73.

Sumi, H. (1999). "Solvent-fluctuation control of solution reactions and its manifestation in protein functions." *Adv. Chem. Phys.* **107**, 601.

Sun, L., K. Song, *et al.* (2002). "A S_N2 reaction that avoids its deep potential energy minimum." *Science* **296**, 875.

Suzuki, T. and B. J. Whitaker (2001). "Non-adiabatic effects in chemistry revealed through charged particle imaging." *Int. Rev. Phys. Chem.* **20**, 313.

Szabo, A. and N. S. Ostlund (1996). *Modern Quantum Chemistry*. New York, Dover.

Taatjes, C. A. (1999). "Time-resolved infrared absorption measurements of product formation in Cl atom reactions with alkenes and alkynes." *Int. Rev. Phys. Chem.* **18**, 419.

Taatjes, C. A. and J. F. Hershberger (2001). "Recent progress in infrared absorption techniques for elementary gas-phase reaction kinetics." *Ann. Rev. Phys. Chem.* **52**, 41.

Taft, C. A., T. C. Guimaraes, *et al.* (1999). "Adsorption and dissociation of diatomic molecules on transition-metal surfaces." *Int. Rev. Phys. Chem.* **18**, 163.

Talkner, P. and P. Hanggi (eds.) (1995). *New Trends in Kramers' Reaction Rate Theory*. Dordrecht, Kluwer.

Tannor, D. J. (2003). *Introduction to Quantum Mechanics: A Time Dependent Perspective*. Mill Valley, CA, University Science Books.

Tannor, D. J. and S. Garashchuk (2000). "Semiclassical calculation of chemical reaction dynamics in wave-packet correlation functions." *Ann. Rev. Phys. Chem.* **51**, 553.

Tannor, D. J. and S. A. Rice (1988). "Coherent pulse sequence control of product formation in chemical reactions." *Adv. Chem. Phys.* **70**, 441.

Tardy, D. C. and B. S. Rabinovitch (1977). "Intermolecular vibrational energy transfer in thermal unimolecular systems." *Chem. Rev.* **77**, 369.

Taube, H. (1970). *Electron Transfer Reactions in Solution*. New York, Academic Press.

Thomas, J. M. and Thomas, W. J. (1997). *Principles and Practice of Heterogeneous Catalysis*, Wiley-VCH, Berlin.

Thompson, D. L. (ed.) (1998a). *Modern Methods for Multidimensional Dynamics Computations in Chemistry*. Singapore, World Scientific.

(1998b). "Practical methods for calculating rates of unimolecular reactions." *Int. Rev. Phys. Chem.* **17**, 547.

Toda, M. (2002). "Dynamics of chemical reactions and chaos." *Adv. Chem. Phys.* **123**, 153.

Toennies, J. P. (1976). "The calculation and measurement of cross sections for rotational and vibrational excitation." *Ann. Rev. Phys. Chem.* **27**, 225.

Toennies, J. P. and A. F. Vilesov (1998). "Spectroscopy of atoms and molecules in liquid helium." *Ann. Rev. Phys. Chem.* **49**, 1.

Tolbert, L. M. and K. M. Solntsev (2002). "Excited-state proton transfer: from constrained systems to 'super' photoacids to superfast proton transfer." *Acc. Chem. Res.* **35**, 19.

Tramer, A. and A. Nitzan (1981). "Collisional effects in electronic relaxation." *Adv. Chem. Phys.* **47**, 337.

Tributsch, H. and L. Pohlmann (1998). "Electron transfer: classical approaches and new frontiers." *Science* **279**, 1891.

Troe, J. (1987). "Elementary reactions in the gas–liquid transition range." *Ann. Rev. Phys. Chem.* **38**, 163.

(1992). "Statistical aspects of ion molecule reactions." *Adv. Chem. Phys.* **82**, 485.

(1997). "Recent advances in statistical adiabatic channel calculations of state-specific dissociation dynamics." *Adv. Chem. Phys.* **101**, 819.

Truhlar, D. G. (1978). "Interpretation of activation energy." *J. Chem. Educ.* **55**, 309.

(ed.) (1981). *Potential Energy Surfaces and Dynamics Calculations*. New York, Plenum Press.

(1998). "Chemical reaction theory." *Faraday Disc.* **110**, 521.

Truhlar, D. G. and D. A. Dixon (1979). Direct mode chemical reactions: classical theories. In *Atom-Molecular Collision Theory: A Guide for the Experimentalist*, R. B. Bernstein (ed.). New York, Plenum Press.

Truhlar, D. G., J. Gao, *et al.* (2002). "The incorporation of quantum effects in enzyme kinetics modeling." *Acc. Chem. Res.* **35**, 341.

Truhlar, D. G. and B. C. Garrett (1984). "Variational transition state theory." *Ann. Rev. Phys. Chem.* **35**, 159.

Truhlar, D. G., B. C. Garrett, *et al.* (1996). "Current status of transition-state theory." *J. Phys. Chem.* **100**, 12 771.

Truhlar, D. G., W. J. Howe, *et al.* (eds.) (1999). *Rational Drug Design.* New York, Springer-Verlag.

Truhlar, D. G. and J. T. Muckerman (1979). Reactive scattering cross sections III: quasiclassical and semiclassical methods. In *Atom-Molecular Collision Theory*, R. B. Bernstein (ed.). New York, Plenum Press, p. 505.

Truong, T. N. (1998). "Quantum modelling of reactions in solution: An overview of the dielectric continuum methodology." *Int. Rev. Phys. Chem.* **17**, 525.

Tse, J. S. (2002). "Ab initio molecular dynamics with density functional theory." *Ann. Rev. Phys. Chem.* **53**, 249.

Tuckerman, M. E., P. J. Ungar, *et al.* (1996). "Ab initio molecular dynamics simulations." *J. Phys. Chem.* **100**, 12 878.

Tully, J. C. (1980). "Semiempirical diatomics-in-molecules potential energy surfaces." *Adv. Chem. Phys.* **42**, 63.

  (1981). "Dynamics of chemical processes at surfaces." *Acc. Chem. Res.* **14**, 188.

  (1998). Mixed quantum-classical dynamics: mean-field and surface hopping. In *Classical and Quantum Dynamics in Condensed Phase Simulations*, B. J. Berne, G. Ciccotti and D. F. Coker (eds.). Singapore, World Scientific, p. 489.

  (2000). "Chemical dynamics at metal surfaces." *Ann. Rev. Phys. Chem.* **51**, 153.

Turro, N. J. (1991). *Modern Molecular Photochemistry.* Mill Valley, CA, University Science Books.

Ulstrup, J. (1979). *Charge Transfer Processes in Condensed Media.* New York, Springer-Verlag.

Urena, A. G. (1987). "Influence of translational energy upon reactive scattering cross section: neutral–neutral collisions." *Adv. Chem. Phys.* **66**, 337.

Urena, A. G. and R. Vetter (1996). "Dynamics of reactive collisions by optical methods." *Int. Rev. Phys. Chem.* **15**, 375.

Uzer, T. (1991). "Theories of intramolecular vibrational-energy transfer." *Phys. Rep.* **199**, 73.

Vaida, V., H. G. Kjaergaard, *et al.* (2003). "Hydrated complexes: relevance to atmospheric chemistry and climate." *Int. Rev. Phys. Chem.* **22**, 203.

Vale, R. D. (2003). "Myosin V motor proteins: marching stepwise towards a mechanism." *J. Cell Biol.* **163**, 445.

Valentini, J. J. (2001). "State-to-state chemical reaction dynamics in polyatomic systems: case studies." *Ann. Rev. Phys. Chem.* **52**, 15.

Valentini, J. J. and D. L. Phillips (1989). Experimental and theoretical studies of the dynamics of the hydrogen exchange reaction: the maturation of a prototype. In *Bimolecular Collisions*, M. N. R. Ashfold and J. E. Baggott (eds.). London, Royal Society of Chemistry, p. 1.

Van Hecke, G. R. and K. K. Karukstis (1998). *A Guide to Lasers in Chemistry.* Boston, Jones and Bartlett.

van Santen, R. A. (1991). *Theoretical Heterogeneous Catalysis.* Singapore, World Scientific.

Varandas, A. J. C. (2000). "Four-atom bimolecular reactions with relevance in environmental chemistry: theoretical work." *Int. Rev. Phys. Chem.* **19**, 199.

Vardi, A. and M. Shapiro (2001). "Theory of laser catalysis with pulses." *Comm. Mod. Phys. D* **2**, 233.

Verhoeven, J. W. (1999). "From close contact to long-range intramolecular electron transfer." *Adv. Chem. Phys.* **106**, 603.

Verlet, J. R. R. and H. H. Fielding (2001). "Manipulating electron wave packets." *Int. Rev. Phys. Chem.* **20**, 283.

Vetter, R. and J. Vigue (eds.) (1986). *Recent Advances in Molecular Reaction Dynamics.* Paris, Editions CNRS.

Viggiano, A. A. and T. M. Miller (2001). Ion chemistry. In *Encyclopedia of Chemical Physics and Physical Chemistry*, J. H. Moore and N. D. Spencer (eds.). Bristol, IOP Publishing, Vol. 1, p. 683.

Villa, J., M. Strajbl, *et al.* (2000). "How important are entropic contributions to enzyme catalysis?" *Proc. Natl. Acad. Sci. USA* **97**, 11 899.

Vitanov, N. V., T. Halfmann, *et al.* (2001). "Laser-induced population transfer by adiabatic passage techniques." *Ann. Rev. Phys. Chem.* **52**, 763.

Vivie-Riedle, D., H. Rabitz, *et al.* (eds.) (2001). *Laser Control of Quantum Dynamics.* Chem Phys. Amsterdam, Elsevier.

Vivie-Riedle, R. (2002). "Theory of laser control of molecular dynamics." *Ann. Rev. Phys. Chem.* **53**.

Voth, G. A. (2001). Molecular reaction dynamics in condensed phases. In *Encyclopedia of Chemical Physics and Physical Chemistry*, J. H. Moore and N. D. Spencer (eds.). Bristol, IOP Publishing, Vol. 1, p. 759.

Voth, G. A. and R. M. Hochstrasser (1996). "Transition state dynamics and relaxation processes in solutions: a frontier of physical chemistry." *J. Phys. Chem.* **100**, 13 034.

Wagner, A. F., I. R. Slagle, *et al.* (1990). "Experimental and theoretical studies of the $C_2H_5 + O_2$ reaction kinetics." *J. Phys. Chem.* **94**, 1853.

Wales, D. J. (2001). "A microscopic basis for the global appearance of energy landscapes." *Science* **293**, 2067.

Wales, D. J., M. A. Miller, *et al.* (1998). "Archetypal energy landscapes." *Nature* **394**, 758.

Walker, G. C. and D. N. Beratan (2001). Electron transfer reactions. In *Encyclopedia of Chemical Physics and Physical Chemistry*, J. H. Moore and N. D. Spencer (eds.). Bristol, IOP Publishing, Vol. 3, p. 2657.

Wall, M. E., S. C. Gallagher, *et al.* (2000). "Large-scale shape changes in proteins and macromolecular complexes." *Ann. Rev. Phys. Chem.* **51**, 355.

Walsh, A. D. (1953). "The electronic orbitals, shapes, and spectra of polyatomic molecules. Part II. Non-hydride AB_2 and BAC molecules." *J. Chem. Soc.*, 2266.

Wang, Q., R. W. Schoenlein, *et al.* (1994). "Vibrationally coherent photochemistry in the femtosecond primary event of vision." *Science* **266**, 422.

Wang, Z., A. Pakoulev, *et al.* (2002). "Watching vibrational energy transfer in liquids with atomic spatial resolution." *Science* **296**, 2201.

Wardlaw, D. M. and R. A. Marcus (1988). "On the statistical theory of unimolecular processes." *Adv. Chem. Phys.* **70**, 231.

Warman, J. M., M. P. DeHaas, *et al.* (1999). "Photoinduced electron transfer within donor–spacer–acceptor molecular assemblies studied by time-resolved microwave conductivity." *Adv. Chem. Phys.* **106**, 571.

Warneck, P. (1988). *Chemistry of the Natural Atmosphere*. New York, Academic Press.

Warren, W. S., H. Rabitz, *et al.* (1993). "Coherent control of quantum dynamics: the dream is alive." *Science* **259**, 1581.

Warshel, A. (1991). *Computer Modeling of Chemical Reactions in Enzymes and Solutions*. New York, John Wiley & Sons.

   (2002). "Molecular dynamics simulations of biological reactions." *Acc. Chem. Res.* **35**, 385.

Warshel, A. and W. W. Parson (1991). "Computer simulations of electron-transfer reactions in solution and in photosynthetic reaction centers." *Ann. Rev. Phys. Chem.* **42**, 279.

   (2001). "Dynamics of biochemical and biophysical reactions: insight from computer simulations." *Q. Rev. Biophys.* **34**, 563.

Waschewsky, G. C. G., P. W. Kash, *et al.* (1994). "What Woodward and Hoffmann didn't tell us – the failure of the Born–Oppenheimer approximation in competing reaction pathways." *J. Chem. Soc. Faraday Trans.* **90**, 1581.

Wasielewski, M. R. (1992). "Photoinduced electron transfer in supramolecular systems for artificial photosynthesis." *Chem. Rev.* **92**, 435.

Weaver, M. J. (1992). "Dynamic solvent effects on activated electron-transfer reactions – principles, pitfalls, and progress." *Chem. Rev.* **92**, 463.

Wegewijs, B. and J. W. Verhoeven (1999). "Long-range charge separation in solvent-free donor–bridge–acceptor systems." *Adv. Chem. Phys.* **106**, 221.

Weinberg, W. H. (1996). "Eley–Rideal surface chemistry: direct reactivity of gas phase atomic hydrogen with adsorbed species." *Acc. Chem. Res.* **29**, 479.

Weitz, E. and G. W. Flynn (1974). "Laser studies of vibrational and rotational relaxation in small molecules." *Ann. Rev. Phys. Chem.* **25**, 275.

   (1981). "Vibrational energy flow in the ground electronic states of polyatomic molecules." *Adv. Chem. Phys.* **47**, 185.

Wenthold, P. G. and W. C. Lineberger (1999). "Negative ion photoelectron spectroscopy studies of organic reactive intermediates." *Acc. Chem. Res.* **32**, 597.

Werner, H. J. (1987). "Matrix-formulated direct multiconfigurational self-consistent field and multireference configuration interaction methods." *Adv. Chem. Phys.* **69**, 1.

Weston Jr, R. E. and G. W. Flynn (1992). "Relaxation of molecules with chemically significant amounts of vibration energy: the dawn of the quantum state resolved era." *Ann. Rev. Phys. Chem.* **43**, 559.

Wheeler, M. D., D. T. Anderson, *et al.* (2000). "Probing reactive potential energy surfaces by vibrational activation of H_2—OH entrance channel complexes." *Int. Rev. Phys. Chem.* **19**, 501.

Whetten, R. L., G. S. Ezra, *et al.* (1985). "Molecular dynamics beyond the adiabatic approximation: new experiments and theory." *Ann. Rev. Phys. Chem.* **36**, 277.

Whitaker, B. J. (ed.) (1989). "Orientation and Polarization Effects in Reactive Collisions." *Faraday Trans.* **113**.

   (ed.) (2003). *Imaging in Molecular Dynamics: Technology and Application*. Cambridge, Cambridge University Press.

Whitehead, J. C. (1988). *Selectivity in Chemical Reactions*. Dordrecht, Kluwer Academic.

Whitten, J. L. and H. Yang (1996). "Theory of chemisorption and reactions on metal surfaces." *Surf. Sci. Rep.* **24**, 55.

Williams, D. H. and J. H. Beynon (1976). "The concept and role of charge localization in mass spectrometry." *Org. Mass Spectrom.* **11**, 103.

Wittig, C., S. Sharpe, *et al.* (1988). "Photoinitiated reactions in weakly bonded complexes." *Acc. Chem. Res.* **21**, 341.

Wolff, J., A. G. Papathanasiou, *et al.* (2001). "Spatiotemporal addressing of surface activity." *Science* **294**, 134.

Wolfrum, J. (2002). "Advanced laser spectroscopy in combustion chemistry: from elementary steps to practical devices." *Faraday Disc.* **119**, 1.

Wolfrum, J., H.-R. Volpp, *et al.* (eds.) (1996). *Gas Phase Chemical Reaction Systems*. Berlin, Springer-Verlag.

Wolynes, P. G. (1987). "Linearized microscopic theories of nonequilibrium solvation." *J. Chem. Phys.* **86**, 5133.

Wolynes, P. G., J. N. Onuchic, *et al.* (1995). "Navigating the folding routes." *Science* **267**, 1619.

Woodward, R. B. and R. Hoffmann (1968). "Conservation of orbital symmetry." *Acc. Chem. Res.* **1**, 17.

     (1970). *The Conservation of Orbital Symmetry*. Weinheim, Verlag Chemie.

Woolley, R. G. (ed.) (1980). *Quantum Dynamics of Molecules*. New York, Plenum Press.

Worth, G. A. and L. S. Cederbaum (2004). "Beyond Born–Oppenheimer: molecular dynamics through a conical intersection." *Ann. Rev. Phys. Chem.* **55**, 127.

Wyatt, R. E., C. Iung, *et al.* (1995). "Toward ab initio intramolecular dynamics." *Acc. Chem. Res.* **28**, 423.

Wyatt, R. E. and J. Z. H. Zhang (eds.) (1996). *Dynamics of Molecules and Chemical Reactions*. New York, Marcel Dekker.

Wynne, K. and R. M. Hochstrasser (1999). "Coherence and adiabaticity in ultrafast electron transfer." *Adv. Chem. Phys.* **107**, 263.

Yang, H., K. T. Kotz, *et al.* (1999). "Ultrafast infrared studies of bond activation in organometallic complexes." *Acc. Chem. Res.* **32**, 551.

Yardley, J. T. (1980). *Introduction to Molecular Energy Transfer*. New York, Academic Press.

Yarkony, D. R. (1996a). "Current issues in nonadiabatic chemistry." *J. Phys. Chem.* **100**, 18 612.

     (1996b). "Diabolical conical intersections." *Rev. Mod. Phys.* **68**, 985.

     (1998). "Conical intersections: diabolical and often misunderstood." *Acc. Chem. Res.* **31**, 511.

Yonekura, N., C. Gebauer, *et al.* (1999). "A crossed molecular beam apparatus using high-resolution ion imaging." *Rev. Sci. Instrum.* **70**, 3265.

Yoshihara, K. (1999). "Ultrafast intermolecular electron transfer in solution." *Adv. Chem. Phys.* **107**, 371.

Yurovsky, V. A. and A. Ben-Reuven (1998). "Channel interference in optical collisions of cold atom beams." *J. Phys. Chem. A* **102**, 9476.

Zaera, F. (2002). "Infrared and molecular beam studies of chemical reactions on solid surfaces." *Int. Rev. Phys. Chem.* **21**, 433.

Zare, R. N. (1972). "Photoejection dynamics." *Mol. Photochem.* **4**, 1.

(1982). "Optical preparation of aligned reagents." *Ber. Bunsenges. Phys. Chem.* **86**, 422.

(1988). *Angular Momentum*. New York, Wiley-Interscience.

(1998). "Laser control of chemical reactions." *Science* **279**, 1875.

Zare, R. N. and R. B. Bernstein (1980). "State-to-state reaction dynamics." *Phys. Today* **33**, 43.

Zare, R. N. and P. J. Dagdigian (1974). "Tunable laser fluorescence method for product state analysis." *Science* **185**, 739.

Zare, R. N. and D. R. Herschbach (1963). "Doppler line shape of atomic fluorescence excited by molecular photodissociation." *Proc. IEEE* **51**, 173.

Zewail, A. H. (1980). "Laser-selective chemistry – is it possible?" *Phys. Today* **33**, 27.

(1988). "Laser femtochemistry." *Science* **242**, 1645.

(1994). *Ultrafast Dynamics of the Chemical Bond*. Singapore, World Scientific.

(1996). "Femtochemistry: recent progress in studies of dynamics and control of reactions and their transition states." *J. Phys. Chem.* **100**, 12 701.

(2000). "Femtochemistry: atomic-scale dynamics of the chemical bond." *J. Phys. Chem. A* **104**, 5660.

(2001). "Chemistry at the uncertainty limit." *Angew. Chem.-Int. Ed. Engl.* **40**, 4371.

Zewail, A. H. and R. B. Bernstein (1988). "Special report – real-time laser femtochemistry – viewing the transition from reagents to products." *Chem. Eng. News* **66**, 24.

Zhang, D. H., M. A. Collins, *et al.* (2000). "First-principles theory for the H + H_2O, D_2O reactions." *Science* **290**, 961.

Zhang, D. H., M. Yang, *et al.* (2002). "Probing the transition state via photoelectron and photodetachment spectroscopy of H_3O^-." *Proc. Natl. Acad. Sci. USA* **99**, 11 579.

Zhang, J. and R. E. Wyatt (eds.) (1996). *Dynamics of Molecules and Chemical Reactions*. New York, Marcel Dekker.

Zhang, J. Z. H. (1999). *Theory and Application of Quantum Molecular Dynamics*. Singapore, World Scientific.

Zhong, Q. and J. A. W. Castleman (2000). "An ultrafast glimpse of cluster solvation effects on reaction dynamics." *Chem. Rev.* **100**, 4039.

Zhu, C., Y. Teranishi, *et al.* (2001). "Nonadiabatic transitions due to curve crossings: complete solutions of the Landau–Zener–Stueckelberg problems and their applications." *Adv. Chem. Phys.* **117**, 127.

Zhu, L., V. Kleiman, *et al.* (1995). "Coherent laser control of the product distribution obtained in the photoexcitation of HI." *Science* **270**, 77.

Zhu, L., J. T. Sage, *et al.* (1994). "Observation of coherent reaction dynamics in heme-proteins." *Science* **266**, 629.

Zimmerman, H. E. (1971). "Mobius–Huckel concept in organic chemistry – application to organic molecules and reactions." *Acc. Chem. Res.* **4**, 272.

Zimmt, M. B. and D. H. Waldeck (2003). "Exposing the solvent's role in electron transfer reactions: tunneling pathways and solvation." *J. Phys. Chem.* **107**, 3580.

Zwier, T. S. (1996). "The spectroscopy of solvation in hydrogen-bonded aromatic clusters." *Ann. Rev. Phys. Chem.* **47**, 205.

Index

Printed in the United States
By Bookmasters